Recent Advances in the Chemistry and Biochemistry of Plant Lipids

Annual Proceedings of the Phytochemical Society

1. **Biosynthetic Pathways in Higher Plants**
   Edited by J. B. Pridham and T. Swain

2. **Comparative Phytochemistry**
   Edited by T. Swain

3. **Terpenoids in Plants**
   Edited by J. B. Pridham

4. **Plant Cell Organelles**
   Edited by J. B. Pridham

5. **Perspectives in Phytochemistry**
   Edited by J. B. Harborne and T. Swain

6. **Phytochemical Phylogeny**
   Edited by J. B. Harborne

7. **Aspects of Terpenoid Chemistry and Biochemistry**
   Edited by T. W. Goodwin

8. **Phytochemical Ecology**
   Edited by J. B. Harborne

9. **Biosynthesis and its Control in Plants**
   Edited by B. V. Milborrow

10. **Plant Carbohydrate Biochemistry**
    Edited by J. B. Pridham

11. **The Chemistry and Biochemistry of Plant Proteins**
    Edited by J. B. Harborne and C. F. Van Sumere

PROCEEDINGS OF THE PHYTOCHEMICAL SOCIETY NUMBER 12

Recent Advances in the Chemistry and Biochemistry of Plant Lipids

PROCEEDINGS OF A SYMPOSIUM ARRANGED BY
THE PHYTOCHEMICAL SOCIETY AND THE LIPID GROUP OF THE
BIOCHEMICAL SOCIETY, UNIVERSITY OF EAST ANGLIA, NORWICH
APRIL, 1974

Edited by

T. GALLIARD

A.R.C. Food Research Institute, Norwich, England

and

E. I. MERCER

Department of Biochemistry and Agricultural Biochemistry,
University College of Wales, Aberystwyth, Wales

1975

ACADEMIC PRESS

LONDON NEW YORK SAN FRANCISCO

A Subsidiary of Harcourt Brace Jovanovich, Publishers

ACADEMIC PRESS INC. (LONDON) LTD.
24/28 Oval Road,
London NW1

United States Edition published by
ACADEMIC PRESS INC.
111 Fifth Avenue
New York, New York 10003

Library of Congress Catalog Card Number: 75 12654
ISBN: 0 12 274050 5

PRINTED IN GREAT BRITAIN BY
WILLMER BROTHERS LIMITED, BIRKENHEAD

Contributors

L-Å. APPELQVIST, *Department of Food Hygiene, The Royal Veterinary College, Stockholm 50, Sweden* (p. 247).

H. BEEVERS, *Divison of Natural Sciences, University of California, Santa Cruz, California 95064, U.S.A.* (p. 287).

C. DEMANDRE, *Université de Paris, Laboratoire de Physiologie cellulaire, 12 Rue Cuvier, 75005 Paris, France* (p. 301).

D. DOUADY, *Université de Paris, Laboratoire de Physiologie cellulaire, 12 Rue Cuvier, 75005 Paris, France* (p. 301).

T. GALLIARD, *Agricultural Research Council Food Research Institute, Colney Lane, Norwich, NR4 7UA England* (p. 319).

R. E. GARCIA, *Department of Biochemistry and Air Pollution Research Centre, University of California, Riverside, California 92502, U.S.A.* (p. 161).

F. D. GUNSTONE, *Department of Chemistry, University of St. Andrews, St. Andrews KY16 9ST, Scotland* (p. 21).

J. L. HARWOOD, *Department of Biochemistry, University College, Cardiff, Wales* (p. 43).

C. HITCHCOCK, *Biosynthesis Division, Unilever Research Laboratories, Colworth House, Sharnbrook, Bedford MK44 1LQ, England* (p. 1).

J. C. KADER, *Université de Paris, Laboratoire de Physiologie cellulaire, 12 Rue Cuvier 75005 Paris, France* (p. 301).

M. KATES, *Department of Biochemistry, University of Ottawa, Ottawa, Canada KIN 6N5* (p. 115).

P. E. KOLATTUKUDY, *Department of Agricultural Chemistry, Washington State University, Pullman, Washington 99163, U.S.A.* (p. 203).

P. J. C. KUIPER, *Laboratory of Plant Physiological Research, Agricultural University, Wageningen, The Netherlands* (p. 359).

M. O. MARSHALL, *Department of Biochemistry, University of Ottawa, Ottawa, Canada K1N 6N5* (p. 115).

P. MAZLIAK, *Université de Paris, Laboratoire de Physiologie cellulaire, 12 Rue Cuvier 75005 Paris, France* (p. 301).

J. B. MUDD, *Department of Biochemistry and Air Pollution Research Centre, University of California, Riverside, California 92502, U.S.A.* (p. 161).

P. K. STUMPF, *Department of Biochemistry and Biophysics, University of California, Davis, California 95616, U.S.A.* (p. 95).

Preface

This volume contains chapters contributed by each of the twelve invited speakers at an international symposium held in Norwich on 8–10 April 1974 and entitled "Recent Advances in the Chemistry and Biochemistry of Plant Lipids". The meeting was sponsored jointly by the Phytochemical Society and the Lipid Group of the Biochemical Society and organized by Drs T. Galliard, M. I. Gurr, E. I. Mercer and M. J. C. Rhodes.

The word "lipid" has several definitions, broadest of which includes any naturally occurring material soluble in fat solvents. Under this broad (and, from a biological viewpoint, unsatisfactory) heading would be included steroids, carotenoids and other isoprenoids which have been the subject of two previous symposia in this series. The organizers therefore decided that the present symposium would concentrate mainly on lipids containing fatty acids, i.e. acyl lipids.

A glance at the contents page will show that, even within these restricted terms of reference, the field of chemistry and biochemistry of lipids in plants is very wide and the organizers were delighted that so many internationally renowned experts in the various areas were able to contribute to the symposium. The truly international nature of the meeting is shown by the fact that eight of the twelve contributors, and many participants, were from overseas.

Those engaged in research on plant lipids have suffered in the past at biochemical meetings in which colleagues working with animal or microbial systems have dominated the proceedings. One of us has resorted to using only the term *Solanum tuberosum* when referring to the humble potato in the hope that the non-phytochemical majority might think this was the name of a micro-organism and thus be acceptable to the biochemical establishment. This symposium enabled those interested in plants and lipids to meet uninhibited by such constraints.

The standard text in the field is "Plant Lipid Biochemistry" by C. Hitchcock and B. W. Nichols (Academic Press, 1971) and the first chapter in the present volume, contributed by one of the above authors, C. Hitchcock, is intended to bring up to date the early chapters of that book by presenting recent information of the structure and distribution of lipids in plants.

Ten years ago, our knowledge of the composition, structure and biochemistry of lipids was meagre indeed. The recent major developments have been due mainly to the availability of new analytical and chromatographic techniques. The chapter by F. D. Gunstone reviews the powerful techniques which are now available for the study of the acyl moieties of lipids.

The "essential fatty acids" (linoleic and linolenic acids) in animal diets are of plant origin. Since these fatty acids are not produced by animal or bacterial systems, studies on their biosynthesis can only be made with plants. The unique biosynthetic mechanisms for plant fatty acids have challenged many laboratories. Although these systems are obviously complex and are not yet fully understood, major advances have been made in the last few years and J. L. Harwood's review of the present state of knowledge is timely. Many of the major advances in our understanding of fatty acid biosynthesis in plants have resulted from investigations in the laboratory of P. K. Stumpf, who contributes a chapter on the role of chloroplasts in the biosynthesis of acyl lipids.

The functions of plant lipids may be roughly divided into three categories: (a) as essential components of the many different membrane structures in cells; (b) as components of the surfaces of tissues and (c) as reserve sources of carbon and energy, particularly in seeds. The two major classes of membrane lipids are phospholipids and glycolipids. Chapters on the biosynthesis of each of these lipid classes come from laboratories at the forefront of research in this area. Although both laboratories have similar interests, for the purpose of this symposium, M. Kates and M. O. Marshall review phospholipid biosynthesis whilst J. B. Mudd and R. E. Garcia cover the various groups of glycolipids.

Until very recently, only superficial information was available on the composition of surface lipids of plants and virtually nothing was known of their biochemistry. The technical problems in handling the waxes and the polymeric cutin and suberin components were daunting. Modern analytical and separation techniques have, however, facilitated studies in this area and the rapid progress which has recently resulted is due, in no small part, to the prolific work of P. E. Kolattukudy whose chapter in this volume reviews the recent advances in both structure determination and elucidation of the biosynthesis of surface lipids.

Storage lipids as seed oils are of major commercial importance. Although the chemical nature of this group of lipids has occupied lipid chemists for some time, only recently have biochemical and physiological studies assumed equal importance and the two complementary contributions in this symposium review these areas: L-Å. Appelqvist deals with these problems in developing seeds and H. Beevers covers recent advances in our understanding of the processes of gluconeogenesis and development in germinating seeds, so extending his pioneering work in this area.

The interdependence of different subcellular organelles in lipid metabolism has recently become clear and the contribution by P. Mazliak and his colleagues describes these interactions in the formation of membrane lipids in plant tissues. Enzymic breakdown of plant lipids has received less attention than biosynthetic aspects. Apart from the importance of lipid catabolism in many physiological processes, degradation of lipids is also important commercially in the loss of quality and the development of undesirable flavours in edible plants. Both hydrolytic and oxidative degradation systems in plants are reviewed by T. Galliard. Too rarely does the lipid chemist or biochemist fully understand the physiological implications of his work. However the chapter by P. J. C. Kuiper on the role of lipids in physiological processes not only reviews this area but also gives the non-physiologist insight into this aspect of plant lipid research.

The organizers wish to acknowledge the support and assistance of all those who helped to provide a successful meeting. Professors T. W. Goodwin, A. T. James and B. E. Folkes, P. K. Stumpf and M. Kates kindly acted as Chairmen of sessions. Many members of staff of the ARC Food Research Institute and of the University of East Anglia gave invaluable help.

Generous financial support from the Royal Society, The Underwood Fund of the Agricultural Research Council, The Potato Marketing Board, Tate and Lyle Ltd. and Unilever (UK) Ltd. enabled the organizers to include so many overseas contributors.

The editors wish to thank the contributors for the time and effort which they have put into producing thorough and up-to-date reviews and to acknowledge the cooperation of the Director of the Food Research Institute and the assistance given to us by the staff of Academic Press.

April, 1975

T. GALLIARD
E. I. MERCER

EDITORIAL NOTE

The nomenclature and conventions used in this volume are based in general on the recommendations of the International Union of Pure and Applied Chemistry and the International Union of Biochemistry. Although full systematic nomenclature of lipids is impractical for general usage, those trivial names which are likely to cause confusion (e.g. "lecithin" and "cephalin") have been avoided. Conventional abbreviations and symbols used throughout this volume to denote fatty acid structures are explained on p. 6.

Contents

CHAPTER 3

Fatty Acid Biosynthesis

J. L. Harwood

CHAPTER 4

Biosynthesis of Fatty Acids in Spinach Chloroplasts

P. K. Stumpf

CHAPTER 5

Biosynthesis of Phosphoglycerides in Plants

M. Kates and M. O. Marshall

CHAPTER 6

Biosynthesis of Glycolipids

J. B. Mudd and R. E. Garcia

CHAPTER 7

Biochemistry of Cutin, Suberin and Waxes—the Lipid Barriers on Plants

P. E. Kolattukudy

CHAPTER 8

Biochemical and Structural Aspects of Storage and Membrane Lipids in Developing Oil-Seeds

L-Å. Appelqvist

CHAPTER 9

Organelles from Castor Bean Seedlings: Biochemical Roles in Gluconeogenesis and Phospholipid Biosynthesis

H. Beevers

CHAPTER 10

Exchange Processes between Organelles Involved in Membrane Lipid Biosynthesis

P. Mazliak, D. Douady, C. Demandre and J. C. Kader

CHAPTER 11

Degradation of Plant Lipids by Hydrolytic and Oxidative Enzymes

T. Galliard

CHAPTER 12

Role of Lipids in Water and Ion Transport

P. J. C. Kuiper

CHAPTER 1

Structure and Distribution of Plant Acyl Lipids

C. HITCHCOCK

Unilever Research, Sharnbrook, England

I. Introduction

Lipids are a heterogeneous class of compounds whose general solubility in organic solvents and insolubility in water distinguishes them from other cellular constituents such as proteins, carbohydrates and nucleic acids. The acyl lipids are all derivatives of long-chain fatty acids (R.COOH). The structure and distribution of plant lipids has been the subject of a recent review by Hitchcock and Nichols (1971) to which the reader is referred for greater detail than is presented here. The purpose of this chapter is two-fold: to bring the above review up to date with more recent information and to serve as a reference on the structures and occurrence of the lipids to be discussed in later chapters of this volume. A wide variety of fatty acids has been isolated from higher and lower plants and characterized by methods described in Chapter 2; in their natural state lipids occur as compounds also containing alcohols (especially glycerol), bases, phosphate esters, sugars, sterols or combinations of these.

II. Plant Acyl Lipids

Each lipid class contains a broad selection of fatty acids in proportions depending on the lipid and its source, so that each lipid is represented by a large number of possible molecular species. These may be analysed by techniques developed particularly over the past decade, often involving fractionation of the lipid classes by liquid chromatography, followed by separation of the component fatty acids of each class by gas chromatography.

A. Glycerides

1. *Neutral Glycerides*

Fatty acids often occur as esters of the trihydric alcohol, glycerol. The fully esterified triglycerides (**1**) are typical of plant seed-oils, the properties of which depend on the fatty acids present, their relative amounts and their positions in the molecule. If one of the acids is a hydroxyacid, further

$$\begin{array}{rl} & CH_2O\cdot CO\cdot R^1 \\ & | \\ R^2CO\cdot O\cdot & CH \\ & | \\ & CH_2O\cdot CO\cdot R^3 \end{array}$$

(**1**) Triglyceride

acylation at this hydroxyl group can give rise to a tetra-acid triglyceride, but this occurs only rarely (Sprecher *et al.*, 1965; Phillips and Smith, 1970; Kleiman *et al.*, 1972b). Though an important intermediate of glyceride metabolism (Chapters 5 and 6), 1,2-diglyceride does not accumulate in plants to any extent.

2. *Phosphoglycerides*

Lipids based on phosphatidic acid (**2**; X = H) are considerably more polar than those related to glycerol itself and tend to be associated with the membranes of plant cells. Phosphatidylcholine (**2**; X = $CH_2\cdot CH_2\ \overset{+}{N}(CH_3)_3$) was the first plant phospholipid to be characterized; it has gained the trivial name "lecithin". This, together with lesser quantities of phosphatidylethanolamine (**2**; X = $CH_2\cdot CH_2\cdot NH_2$), comprises the main phospholipid of non-photosynthetic plant tissue. Chloroplasts contain in addition a high proportion of phosphatidylglycerol (**2**; X = $CH_2\cdot CHOH\cdot CH_2OH$). Less abundant lipids, in decreasing order of general occurrence, include phosphatidylinositol (**2**; X = *myo*-inositol), phosphatidylserine (**2**; X = $CH_2\cdot CH(NH_2)\cdot CO_2H$), diphosphatidylglycerol (cardiolipin) and *N*-acylphosphatidylethanolamine (**2**; X = $CH_2\cdot CH_2\cdot NHCOR^3$). Monoacyl phos-

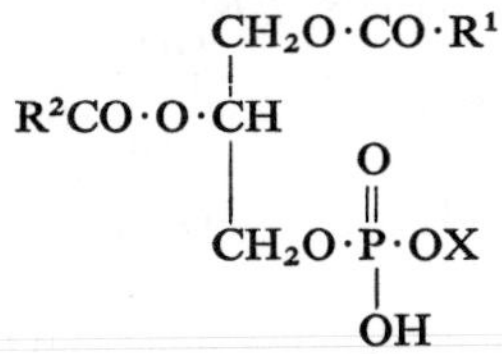

(2) Phosphoglyceride

pholipids (lysophosphatides) may be minor natural plant lipids, and have for instance been found in wheat starch granules (Wren and Merryfield, 1970). Phosphatidic acid itself (**2**; X = H) is an important metabolite (Chapter 5), but occurs only in trace amounts in plant tissues.

3. *Glycosyldiglycerides*

Plant glycolipids fall into two classes of which those based on glycerol are more important than the sphingolipids (q.v.). Monogalactosyldiglyceride (1,2-diacyl-[β-D galactopyranosyl-(1′ → 3)]-*sn*-glycerol; **3**) and digalacto-

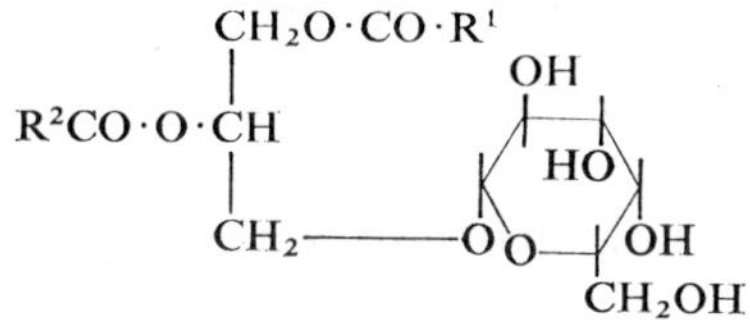

(3) Monogalactosyldiglyceride

syldiglyceride (1,2-diacyl-[α-D-galactopyranosyl-(1″ → 6′)-β-D-galactopyranosyl-(1′ → 3)]-*sn*-glycerol; **4**) occur in a wide variety of plant tissues. Small amounts of tri- and tetragalactosyldiglycerides have been occasionally

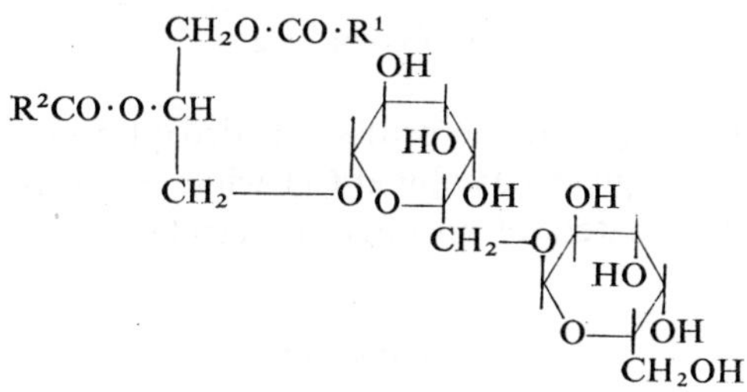

(4) Digalactosyldiglyceride

found. A unique sulpholipid present in algae and leaves has been characterized as 1,2-diacyl-[6-sulpho-α-D-quinovopyranosyl-(1′ → 3)]-*sn*-glycerol (sulphoquinovosyl diglyceride: **5**).

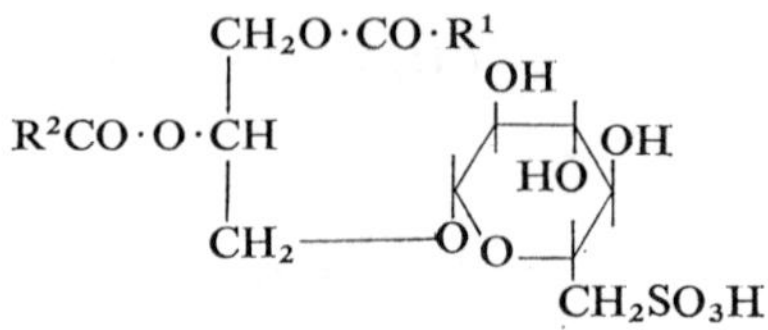

(5) Sulphoquinovosyldiglyceride

B. OTHER ACYL ESTERS

Plant waxes usually contain esters of fatty acids with monohydric *n*-alcohols of chain lengths of about 10–30 carbon atoms in addition to long-chain hydrocarbons (see Chapter 7). Esters of dihydric alcohols (ethandiol, propandiol, butandiol) occur with triglycerides as minor components (Bergelson, 1969). The occurrence of esters of hydroxynitriles has been reported (Mikolajczak and Smith, 1971). Most plants appear to contain small quantities of sterol esters and of sterol glycosides in which the sugar is esterified. Isoprenoid esters have been isolated from flowers.

C. SPHINGOLIPIDS

Several long-chain amino-alcohols such as 4-sphingenine (i.e. sphingosine or D-*erythro*-1,3-dihydroxy-2-amino-*trans*-4-octadecene) occur in plants to a minor degree, carrying fatty acid attached through an amide bond to the

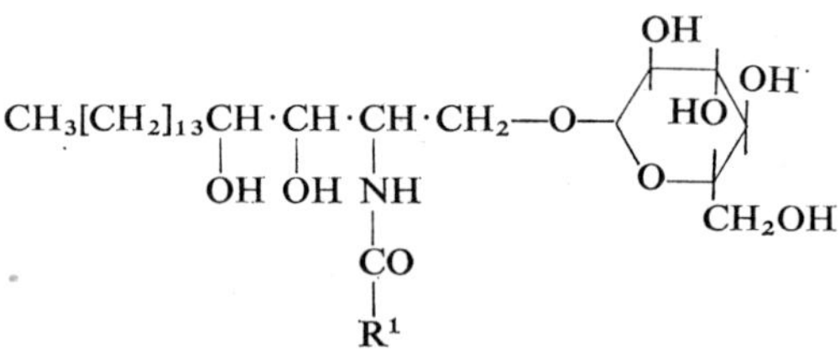

(6) Cerebroside

amino group, and a sugar at the primary hydroxyl group. Cerebrosides (**6**) contain glucose bound glycosidically at C-1 while more complex phytoglycolipids are sphingosine esters of an oligosaccharide phosphate (Carter *et al.*, 1969).

D. OTHER LIPIDS

Plants contain sterols and their glycosides as well as acylated derivatives of these (Chapter 6). Long-chain polyhydric alcohols are presumably biosynthesized via fatty acids, and occur in primitive plants as glycosides. Bryce *et al.* (1972) have isolated the monoglucosyl and monogalactosyl derivatives of 1,3,25-trihydroxyhexacosane and other members of this lipid class from nitrogen-fixing heterocystous algae; it is absent from the corresponding vegetative cells and also from other algae, higher plants and bacteria.

III. Plant Fatty Acids

Over 200 different fatty acids have been isolated from higher and lower plants, but the majority of these may be classed as "unusual" since they occur in only a few plant species. Of the widely distributed acids, a few are present in large quantities in most plant lipids, and qualify for recognition as "major" fatty acids; it is convenient to classify the remainder as "minor" acids, since they are fairly ubiquitous but usually present in smaller proportions, often being disregarded in analytical reports. While the distinction between "minor" and "unusual" acids is somewhat arbitrary, the "major" and "minor" acids are evidently biosynthesized by analogous pathways (Chapter 3), while the biogenesis of an "unusual" acid requires an additional enzyme system to provide the unusual feature.

A. MAJOR FATTY ACIDS

The major fatty acids are all saturated or unsaturated monocarboxylic acids with an unbranched even-numbered carbon chain. The saturated homologues lauric (dodecanoic), myristic (tetradecanoic), palmitic (hexadecanoic) and stearic (octadecanoic) acids all occur in plants, but even more abundant are the unsaturated analogues oleic (*cis*-9-octadecenoic), linoleic (*cis*-9-, *cis*-12-octadecadienoic) and α-linolenic (*cis*-9, *cis*-12, *cis*-15-octadecatrienoic) acids; structures are given in Table I. These seven acids are usually all found in the lipids from all parts of all plants, often with palmitate, oleate and linoleate predominating. Chloroplasts are particularly rich in α-linolenate.

TABLE I

Structures of the major fatty acids

Common name	Symbol[a]	Structure
lauric	12:0	$CH_3 \cdot [CH_2]_{10} \cdot COOH$
myristic	14:0	$CH_3 \cdot [CH_2]_{12} \cdot COOH$
palmitic	16:0	$CH_3 \cdot [CH_2]_{14} \cdot COOH$
stearic	18:0	$CH_3 \cdot [CH_2]_{16} \cdot COOH$
oleic	18:1(9*c*)	$CH_3 \cdot [CH_2]_7 \cdot CH \overset{c}{:} CH \cdot [CH_2]_7 \cdot COOH$
linoleic	18:2(9*c*,12*c*)	$CH_3 \cdot [CH_2]_3 \cdot [CH_2 \cdot CH \overset{c}{:} CH]_2 \cdot [CH_2]_7 \cdot COOH$
linolenic	18:3(9*c*,12*c*,15*c*)	$CH_3 \cdot [CH_2 \cdot CH \overset{c}{:} CH]_3 \cdot [CH_2]_7 \cdot COOH$

[a] For abbreviations, see p. 6.

The structure of many fatty acids is conveniently written as a symbol of two numbers separated by a colon: the first is the number of carbon atoms in the normal chain and the second the number of unsaturated centres. The position of unsaturation may be given by additional numbers in brackets after the main symbol; these are usually counted from the carboxyl group (C-1) and followed by *c*, *t*, *a* or *e* to represent *cis*-olefinic, *trans*-olefinic, acetylenic or ethylenic bonds. Occasionally it is expedient to number from the opposite end of the carbon chain, and the numbers so generated are distinguished by the prefix ω*. Substituted acids are denoted by a suitable prefix. In this and subsequent chapters, omission of a symbol for the configuration of unsaturation normally indicates a *cis*-olefinic bond. Thus, e.g. *cis*-9, *cis*-12-octadecadienoic acid will be represented by linoleic acid (18:2 9*c*,12*c*) or 18:2(9*c*,12*c*) or 18:2(9,12).

B. MINOR FATTY ACIDS

Lipids often contain small but significant proportions of structures similar to the major fatty acids, but differing in chain length, e.g. saturated acids and *cis*-9-monoenoates. Other unsaturated minor acids are classified as belonging to ω9, ω6 or ω3 families depending on the position (counted from the methyl end of the molecule) of the first characteristic double bond; these families are therefore related to oleate, linoleate and α-linolenate respectively (Table II). Occasionally, a higher plant is found in which a "minor" acid accumulates in abnormally large amounts, but lower plants often contain a wide variety in small quantities.

C. UNUSUAL FATTY ACIDS

When fatty acids with features unrelated to the usual structures exist naturally, they are often found only in the seeds of related plants, and may be restricted to a few individual species, a genus or a whole family; however, within this narrow distribution they may represent the principal acid present. Their structures and occurrence are surveyed in Tables III to VII, and have been reviewed by Smith (1970), Hitchcock and Nichols (1971), Hopkins (1972) and Galliard (1974).

1. *Non-conjugated Ethylenic Acids*

Some unsaturated plant acids are unusual only in the position or geometry of their double bonds, which are neither allenic nor conjugated. Over thirty

* The ω-convention to denote the position of a double bond relative to the methyl terminal group of a fatty acid chain is widely used by lipid biochemists. In the IUPAC–IUB recommendations (see *Biochem. J.* (1967), **105**, 897–902), the ω-system is replaced by (n$-$x) where n = the no. of carbon atoms in the fatty acid chain. The position of the double bond in, e.g. erucic acid (*cis*-13-docosenoic acid), would be indicated by ω9 or by (n-9) in the respective systems.

TABLE II

Some fatty acids present in lower plants[a]
After Hitchcock and Nichols (1971)

Saturated														
even carbon number		10:0		12:0		14:0		16:0		18:0		20:0	22:0	24:0
odd carbon number	9:0		11:0		13:0		15:0		17:0		19:0			
Δ9-Family														
Δ9-monoenoates						14:1		16:1		18:1				
					13:1		15:1		17:1					
Δ9-dienoates										18:2				
									17:2		19:2			
Δ-9-trienoates										18:3				
									17:3					
ω9-Family														
ω9-monoenoates						14:1		16:1		18:1		20:1	22:1	
ω9-dienoates										18:2		20:2		
ω6-Family														
ω6-dienoates								16:2		18:2		20:2	22:2	
ω6-trienoates								16:3		18:3		20:3	22:3	
ω6-tetraenoates												20:4	22:4	
ω6-pentaenoates													22:5	
ω3-Family														
ω-3-trienoates								16:3		18:3		20:3		
ω3-tetraenoates								16:4		18:4		20:4	22:4	
ω3-pentaenoates												20:5	22:5	
ω3-hexaenoates													22:6	

[a] For abbreviations, see p. 6.

structures have been characterized (Table III); more recent work has included the isolation of *cis*-5-enoic acids from seeds (Kleiman *et al.*, 1972a; Spencer and Earle, 1972), from conifer leaves (Jamieson and Reid, 1972) and of *trans*-5-enoic acids (Rankov *et al.*, 1971). New monoenoic acids have been found in *Thunbergia alata* seed oil, which is also a rich source of *cis*-6-hexadecenoate (Spencer *et al.*, 1971).

2. *Conjugated Ethylenic Acids*

Polyenoic acids with conjugated (1,3-spaced) double bonds are clearly distinguishable from the usual methylene-interrupted (1,4-spaced) isomers; they have been reviewed by Hopkins (1972), and are listed in Table IV.

They fall into four families recognizable by characteristic *cis*-9-, *cis*-12-, *trans*-9- or *trans*-12- double bonds, the structures within each family suggesting metabolic significance (Hitchcock and Nichols, 1971; Gardner *et al.*, 1973).

TABLE III

Structures and distribution of unusual non-conjugated ethylenic acids[a], [b]

Structure[d]	Occurrence
18:3(2*t*,9*c*,12*c*)	mixed pollen
16:1(3*t*)	photosynthetic tissue
	Helenium bigelowii (10%)
	Grindelia oxylepsis (14%)
18:1(3*t*)	*Grindelia oxylepsis* (2%)
18:3(3*t*,9*c*,12*c*)	*Calea urticaefolia* (31%)
	Aster alpinus (14%)
	Arctium minus (10%)
18:4(3*t*,9*c*,12*c*,15*c*)	*Tecoma stans* (19%)
10:1(4*c*)	*Lindera umbellata* (4%)
12:1(4*c*)	*Lindera umbellata* (47%)
14:1(4*c*)	*Lindera umbellata* (5%)
	Thunbergia alata[c] (0·2%)
14:1(5)	*Thunbergia alata*[c] (0·2%)
16:1(5*c*)	*Carlina acaulis* (2%)
	Dioscoreophyllum cumminsii[c] (1%)
	Chenopodiaceae[c] (4 spp.) (5 to 12%)
	bacteria
16:2(5*c*,9*c*)	Chenopodiaceae[c] (4 spp.) (1%)
	slime mould
17:2(5*c*,9*c*)	slime mould
18:1(5*c*)	*Carlina acaulis* (21%)
	Caltha palustris (18%)
	Dioscoreophyllum cumminsii[c] (84%)
	Chenopodiaceae[c] (4 spp.) (1%)
	bacteria
18:2(5*c*,9*c*)	conifer leaves[c]
	slime mould
18:2(5*c*,11*c*)	slime mould
18:3(5*c*,9*c*,12*c*)	Chenopodiaceae[c] (4 spp.) (1%)
	conifer leaves[c]
18:4(5*c*,9*c*,12*c*,15*c*)	conifer leaves[c]
20:1(5*c*)	*Limnanthes douglasii* (65%)
	Caltha palustris (3%)
20:3(5*c*,11*c*,14*c*)	*Caltha palustris* (23%)
	conifer leaves[c]
20:4(5*c*,11*c*,14*c*,17*c*)	*Caltha palustris* (1%)
	conifer leaves[c]
22:1(5*c*)	*Limnanthes douglasii* (7%)
22:2(5*c*,13*c*)	*Limnanthes douglasii* (10%)
16:1 (5*t*)	*Thalictrum* spp.[c] (1–4%)
	conifer leaves[c]
18:1(5*t*)	*Thalictrum* spp.[c] (6–23%)
18:1(5*t*,9*c*)	*Thalictrum* spp.[c] (1–6%)
18:3(5*t*,9*c*,12*c*)	*Thalictrum* spp.[c] (40–57%)
16:1(6*c*)	*Thunbergia alata*[c] (82%)
	Picramnia sellowii

TABLE III *continued*

Structure[2]	Occurrence
18:1(6*c*)	Umbelliferae spp.
	Araliaceae spp.
	Picramnia sellowii
20:1(6*c*)	*Alvarado amorphoides*[c] (1%)
16:1(6*t*)	*Alvarado amorphoides*[c]
18:1(6*t*)	*Alvarado amorphoides*[c]
20:1(6*t*)	*Alvarado amorphoides*[c]
16:1(7)	*Thunbergia alata*[c] (2%)
18:1(8*c*)	*Thunbergia alata*[c] (2%)
18:1(11)	*Dioscoreophyllum cumminsii*[c] (1%)
18:1(11*c*)	*Doxantha unguis-cati* (15%)
	Asclepias syriaca (15%)
20:1(13*c*)	*Brassica napus*
18:2(9*c*,12*t*)	*Dimorphotheca sinuata* (1%)
	Crepis rubra (3%)
18:2(9*t*,12*t*)	*Chilopsis linearis* (15%)

[a] After Hitchcock and Nichols (1971) where references are given; see Smith (1970).
[b] Where the source is a seed oil, the percentage weight of the unusual acid (relative to the total weight of acids present in a sample) is given if available.
[c] Recent references are given in the text.
[d] For abbreviations, see p. 6.

3 *Acetylenic Acids*

Several hundred natural acetylenes have been isolated from fungi and higher plants; their structure and distribution has been reviewed by Bohlmann *et al.* (1973). While fatty acids are often precursors, only a few acetylenic fatty acids accumulate. One distinct group, recognizably derived from stearolic (9-octadecynoic) acid by virtue of the remaining ω9-acetylenic bond, is found in the seed-oils of two plant families (Santalaceae and Olacaceae, both of the order Santales). The structures of the known acids suggest biogenesis from oleic acid by sequential desaturations, hydroxylations and/or α-oxidations (Fig. 1).

Another group of shorter chain acetylenes typified by dehydromatricaria ester (methyl *cis*-2-decen-4,6,8-triynoate), exhibit a characteristic ω6-acetylenic bond evidently originating in crepenynic (*cis*-9-octadecen-12-ynoic) acid (Fig. 2). The rare allenic and furanoid acids may be biochemically related to acetylenes. Tariric (octadec-6-ynoic) acid, hitherto thought to be unique in the position of its triple bond, has now been found to co-exist in *Alvarado amorphoides* with octadec-17-en-6-ynoic acid and traces of eicos-6-ynoic acid by Pearl *et al.* (1973).

4. *Substituted Acids*

Substitution of hydrogen by oxygen in the fatty acid chain leads to hydroxyacids, polyhydroxyacids, ketoacids, epoxyacids and dicarboxylic acids

TABLE IV

Structure and distribution of unusual conjugated ethylenic fatty acids[a, b]

Structural family (see text)	Structure[c]	Occurrence
cis-9	D-13-hydroxy-18:2(9*c*,11*t*)	*Coriaria nepalensis* (66–68 %)
		Dimorphotheca sinuata
		Xeranthemum annuum
	L-13-hydroxy-18:2(9*c*,11*t*)	*Monnina emarginata* (30 %)
	13-hydroxy-18:3(3*t*,9*c*,11*t*)	*Stenachaenium macrocephalum*
	18:3(9*c*,11*t*,13*t*)	*Aleurites* spp. (72–82 %)
		Kentranthus macrosiphon (50 %)
		Curcurbitaceae (3 spp.) (49–65 %)
	4-oxo-18:3(9*c*,11*t*,13*t*)	*Chrysobalanus icaco* (10 %)
	18:3(9*c*,11*t*,13*c*)	*Punica granatum* (71 %)
		Curcurbitaceae (3 spp.) (32–56 %)
	18:4(9*c*,11*t*,13*t*,15*c*)	*Impatiens edgeworthii* (48 %)
	4-oxo-18:4(9*c*,11*t*,13*t*,15*c*)	*Chrysobalanus icaco* (18 %)
	18-hydroxy-18:3(9*c*,11*t*,13*t*)	*Mallotus phillipensis*
cis-12	D-9-hydroxy-18:2(10*t*,12*c*)	*Xeranthemum annuum*
		Calendula officinalis
	9-hydroxy-18:3(3*t*,10*t*,12*c*)	*Stenachaenium macrocephalum*
	18:3(8*t*,10*t*,12*c*)	*Calendula officinalis* (56 %)
		Osteospermum hyoseroides (36 %)
	18:3(8*c*,10*t*,12*c*)	*Jacaranda mimosifolia*
trans-9	13-oxo-18:2(9*t*,11*t*)	*Monnina emarginata*
	18:3(9*t*,11*t*,13*c*)	*Catalpa ovata*
trans-12	9-hydroxy-18:2(10*t*,12*t*)	*Dimorphotheca sinuata* (65 %)
	9-oxo-18:2(10*t*,12*t*)	*Dimorphotheca sinuata* (2 %)
	18:2(10*t*,12*t*)	*Chilopsis linearis* (9 %)

[a] After Hitchcock and Nichols (1971) where references are given; see also Hopkins (1972) and Smith (1970).
[b] The percentage weight of the unusual acid (relative to the total weight of acids present in a sample of the seed oil) is given if available.
[c] For abbreviations, see p. 6.

(Table V). Spencer *et al.* (1973) have shown that levels of oxygenated fatty acids in seeds can increase during storage by enzymic oxygenation of unsaturated acids. Auricolate (14-hydroxy-*cis*-11, *cis*-17-eicosadienoate) is a principal component of *Lesquerella auriculata* oil (Kleiman *et al.*, 1972b), while ricinoleate (12-hydroxy-*cis*-9-octadecenoate) has now been found in flax seed oil by Kleiman and Spencer (1971). The lactones of several long-chain 4-hydroxyacids have been discovered in rose flower wax (Hadjieva *et al.*, 1974).

The epoxyacids of soyabean oil have been isolated and characterized as the *cis*-9,10-epoxyderivatives of oleic, linoleic and linolenic acids by Sen Gupta (1972). Wheat bran contains 5,8,12-trihydroxy-*trans*-9-octadecenoate

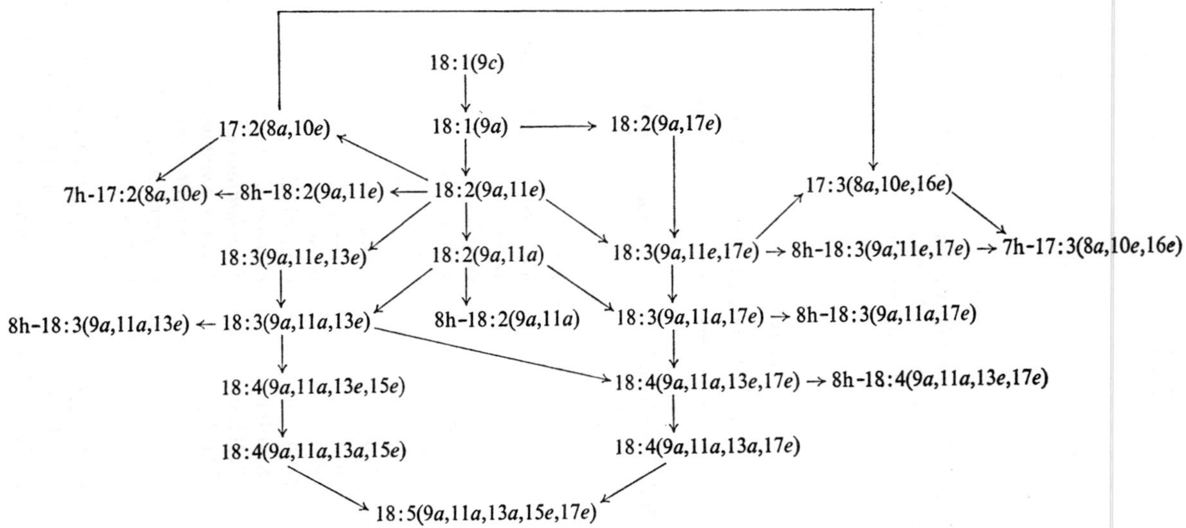

FIG. 1. The stearolic family of acetylenic fatty acids. Arrows denote possible pathways of biogenesis. After Hitchcock and Nichols (1971). [h = hydroxy; for other abbreviations, see p. 6].

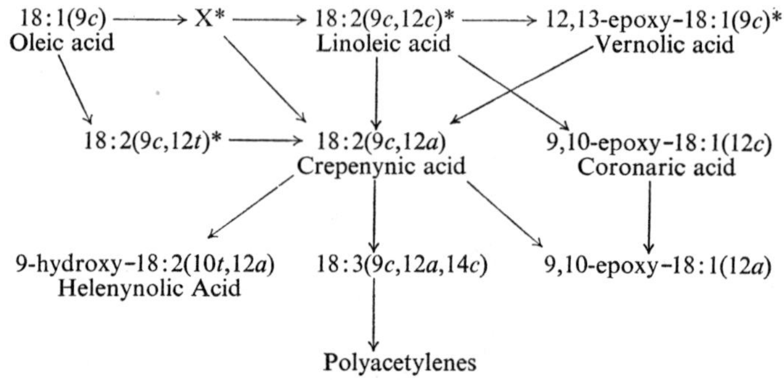

FIG. 2. The crepenynic family of acetylenic fatty acids. Arrows denote possible pathways of biogenesis, and asterisks postulated enzyme-bound intermediates. After Hitchcock and Nichols (1971). [for abbreviations, see p. 6].

TABLE V

Structure and distribution of unusual oxygenated fatty acids[a, b]

Position and nature of oxygenated function (h = hydroxy)	Chain length and unsaturation[d]	Occurrence
2h-acids		leaf, seed sphingolipids
3h-acids		yeast
4h-acids		rose flower wax[c]
5h,8h,12h-	18:1(9*t*)	wheat bran[c]
7h-	10:0	Convolvulaceae
11h-	14:0	Convolvulaceae
8h-	16:0	spores
8-oxo-	16:0	spores
9h-	16:0	yeast
11h-	16:0	Convolvulaceae
12h-	16:1(9*c*)	*Lesquerella densipila*
9h-	18:1(12*c*)	*Strophanthus sarmentosus* (7%)
12h-	18:1(9*c*)	*Ricinus communis* (90%)
		Linum usitatissimum[c] (15%)
		Lesquerella auriculata[c] (5%)
		fungus
		cutin[c]
12h-	18:2(9*c*,15*c*)	*Lesquerella* (2 spp.) (2–38%)
14h-	20:1(11*c*)	*Lesquerella* (3 spp.) (10–72%)
14h-	20:2(11*c*,17*c*)	*Lesquerella auriculata* (32%)
15h-	16:0	yeast
16h-	16:0	yeast
17h-	18:0	yeast
17h-	18:1(9*c*)	yeast

TABLE V *continued*

Position and nature of oxygenated function (h = hydroxy)	Chain length and unsaturation[d]	Occurrence
15h-,16h-	16:0	corn smut
2h,15h,16h-	16:0	corn smut
9h-	18:0	yeast
9h,10h-	18:0	*Onguekoa gore* *Ricinus* sp. *Cardamine impatiens* (1%) spores
11h,12h-	20:0	*Cardamine impatiens* (1%)
13h,14h-	22:0	*Cardamine impatiens* (16%)
15h,16h-	24:0	*Cardamine impatiens* (6%)
9h,10h,18h-	18:0	*Chamaepeuce afra* (9%)
9h,10h,18h-	18:1(12*c*)	*Chamaepeuce afra* (14%)
9,10-epoxy-	18:0	*Xeranthemum annuum* (3%) *Onguekoa gore* (1%) spores wheat stem rust *Soja hispida*[c]
9,10-epoxy-	18:1(12*c*)	*Xeranthemum annuum* (8%) *Chrysanthemum coronarium* *Aster alpinus* *Soja hispida*[c] cutin
9,10-epoxy-	18:1(12*a*)	*Helichrysum brachteatum*
12,13-epoxy-	18:1(9*c*)	*Vernonia anthelmintica* (72%) *Crepis* spp. (0–68%) Malvaceae (10 spp.) (0–7%)
15,16-epoxy-	18:2(9*c*,12*c*)	*Camelina sativa*
9,10-epoxy-	18:2(12*c*,15*c*)	*Soja hispida*[c]
8h,9h,13h-	22:0	yeast
9h,10h,12h,13h-	22:0	lichen
15-oxo-	24:1(18*c*)	*Cuspidaria pterocarpa* (5%)
17-oxo-	26:1(20*c*)	*Cuspidaria pterocarpa* (13%)
19-oxo-	28:1(22*c*)	*Cuspidaria pterocarpa* (3%)

[a] After Hitchcock and Nichols (1971) where references are given; see also Smith (1970) and Galliard (1974).
[b] Where the source is a seed oil, the percentage weight of the unusual acid (relative to the total weight of acids present in a sample) is given if available.
[c] Recent references are given in the text.
[d] For abbreviations, see p. 6.

(Albro and Fishbein, 1970) while 9,10,13-trihydroxystearate occurs in kidney bean roots (Takasugi *et al.*, 1973).

Depolymerized cutin has been analysed extensively over the past few years, and several oxygenated fatty acids isolated from various sources (Brieskorn and Kabelitz, 1971; Croteau and Fagerson, 1972; Holloway and

Deas, 1973; Hunneman and Eglinton, 1972; Kolattukudy and Walton, 1972; Kolattukudy, 1972; Kolattukudy, 1973; see also Chapter 7). These include derivatives of palmitic and stearic acids carrying combinations of hydroxy, epoxy and keto-groups at the 9,10 and ω-positions as well as related dibasic acids (Table VI).

5. *Branched-chain Acids*

While most natural fatty acids have a normal ("straight") carbon skeleton, a few plant seeds contain acids with cyclopropene or cyclopentene ring systems which are listed in Table VII: recent analytical reports are by Madrigal and Smith (1973) and by Spener (1972) respectively. *Anteiso*-branched acids

TABLE VI

Structure of cutin and suberin acids[a]

Position and nature of oxygenated function (h = hydroxy)	Chain length and un-saturation[b]	Position and nature of oxygenated function (h = hydroxy)	Chain length and un-saturation[c]
Monobasic acids		Monobasic acids *continued*	
12h-	12:0	12h,18h-	18:1(9)*
14h-	14:0	9h,10h,18h-	18:0
10h-	14:1	9,10-epoxy-18h-	18:0
9h-	15:0	9,10-epoxy-18-oxo-	18:0
9h,15h-	15:0	9,10-epoxy-18h-	18:1(12)
16h-	16:0	9h,10h,13h-	18:0
8h,16h-	16:0*	12h,18h-9,10-epoxy-	18:0*
9h,16h-	16:0	9h,10h,18h-12,13-epoxy-	18:0*
9h,16-oxo-	16:0	9h,10h,12h,13h,18h-	18:0
10h,16h-	16:0	20h-	20:0
10h,16-oxo-	16:0	22h-	22:0
6h,7h,16h-	16:0	Dibasic acids	
16h-	16:1(9)	7h-	15:0
9h,10h,17h-	17:0	6h-	15:0
9h,10h,17h-	17:1		16:0
10h-	18:0		16:1(9)
18h-	18:0	7h-	16:0
18h-	18:1(9)	8h-	16:0
18h-	18:2(9,12)	8h,9h-	17:0
9h,18h-	18:0	9h,10h-	18:0
10h,18h-	18:0	9h-	18:0
10h,18h-	18:1(12)		18:0
			18:1(9)
			18:2(9,12)
			20:0
		9,10-epoxy-	18:0

[a] References given in this text except for structures* which are described by Kolattukudy in Chapter 7.

[b] For abbreviations, see p. 6.

TABLE VII

Structures and distribution of unusual carbocyclic acids[a,b]

Position and nature of cyclic function	Chain length and unsaturation[d]	Common name	Occurrence
9,10-methylene-	18:0	dihydrosterculate	*Euphoria longana* (17%) *Hibiscus syriacus* (1%) *Sterculia foetida* (0·4%) *Bombacopsis glabra* (3%)
9,10-methylene-	18:1(9*c*)	sterculate	*Hibiscus syriacus* (2%) *Sterculia foetida* (53%) *Bombacopsis glabra* (34%) *Pavonia sepium*[c] (7%)
9,10-methylene-	18:2(9*c*,17*a*)	sterculynate	*Sterculia alata* (8%)
2-hydroxy-9,10-methylene-	18:1(9*c*)	2-hydroxysterculate	*Pachira insignis* (20%) *Bombacopsis glabra* (10%)
8,9-methylene-	17:0	dihydromalvalate	*Brachychiton acerifolium* (1%)
8,9-methylene-	17:1(8)	malvalate	*Hibiscus syriacus* (14%) *Bombacopsis glabra* (3%) *Brachychiton acerifolium* (20%) *Pavonia sepium*[c] (4%)
11-(2-cyclopentenyl)	11:0	hydnocarpate	*Hydnocarpus wightiana* (49%) *Oncoba echinata*[c] (1%)
11-(2-cyclopentenyl)	11:1	manoa-ate	*Oncoba echinata*[c]
13-(2-cyclopentenyl)	13:0	chaulmoograte	*Hydnocarpus wightiana* (27%) *Oncoba echinata*[c]
13-(2-cyclopentenyl)	13:1(6*c*)	gorlate	*Hydnocarpus wightiana* (12%) *Oncoba echinata*[c] (25%)
15-(2-cyclopentenyl)	15:1	hormelate	*Oncoba echinata*[c]
15-(2-cyclopentenyl)	15:2	oncobate	*Oncoba echinata*[c] (0·5%)

[a] After Hitchcock and Nichols (1971) where references are given.
[b] The percentage weight of the unusual acid (relative to the total weight of acids present in a sample of seed oil) is given if available.
[c] Recent references are given in the text. [d] For abbreviations see p. 6.

have been found in *Chlorella vulgaris* (Matucha *et al.*, 1972), though they are characteristic of Gram-positive bacteria. On the other hand, cyclopropane fatty acids are usually associated with Gram-negative bacteria, but Kuiper and Stuiver (1972) report they also accumulate in the sulpholipids of early spring plants and in the phosphatidylcholine of a drought-tolerant plant.

D. DISTRIBUTION WITHIN LIPID CLASSES

Fatty acids are seldom distributed at random between the different classes of lipid present, but frequently exhibit clear tendencies to associate with individual lipids at specific positions.

The 1- and 3-positions of the triglyceride molecule are not biochemically equivalent, but both tend to accumulate the saturated fatty acids available during biosynthesis. The secondary hydroxyl group displays a correspondingly marked preference for unsaturated acids, so that in many vegetable oils 95–100% of the fatty acids in the 2-position are oleate, linoleate and linolenate, even when the total concentration of these acids in the oil is as low as 37%. In Cruciferae oils, erucate (22:1 13*c*) behaves like a saturated acid, and other unsaturated acids will fill the 1- and 3-positions if there is not enough saturated acid available.

In phospholipids, saturated fatty acids tend to accumulate at the 1-position, and the more highly unsaturated acids at the 2-position.

Galactosylglycerides display the same selectivity if the component acids are primarily of a single chain length; if both C_{16} and C_{18} acids are present, the latter are located preferentially at the 1-position.

IV. DISTRIBUTION IN PLANTS

A. ACYL LIPIDS

In general, polar acyl lipids (phospholipids and glycolipids) are associated with membraneous structures in plant cells, while triglycerides and waxes are not strongly membrane-bound. In photosynthetic tissue, lipid composition is variable (Roughan and Batt, 1969), but phosphatidylglycerol, mono- and di-galactosyldiglycerides and sulphoquinovosyldiglyceride tend to predominate: indeed some photosynthetic blue-green algae (Cyanophyceae) contain only these four classes.

The envelope membranes of chloroplasts have been isolated by Poincelot (1973) who reports that they contain more galactosyldiglycerides and phosphatidylcholine, and less sulpholipid and phosphatidylglycerol than the lamellae, and that small amounts of cerebroside, sterylglycoside, acylated sterylglycoside, phosphatidylethanolamine, sterol and steryl esters are exclusive to the envelopes.

Phosphoglycerides form the major lipids of non-photosynthetic membranes. For instance, over half the lipid of plant mitochondria is phospholipid, but the additional presence of the glycosyldiglycerides is characteristic when compared with animal mitochondria (Schwertner and Biale, 1973). On the other hand, seed oils are rich in triglyceride, the small proportion of phospholipid in the seed being associated with membranes such as those enclosing the oil bodies. Similarly, in those fruit containing an oily exocarp (e.g. olive, oil-palm), the fruit-coat lipid is mainly triglyceride; watery exocarp (e.g. apple pulp) contains little acyl lipid, most of which is polar. Notable exceptions include seed oil of some individual *Briza* species in which galactosyl diglycerides accumulate (Smith and Wolff, 1966) and of *Simmondsia californica* (Green *et al.*, 1936) and *Murraya koenigii* (Kartha and Singh, 1969) in which the major lipid components are wax ester and hydrocarbon. The latter non-polar lipids are normally found with cutin in the cuticles of leaves and fruits (Chapter 7).

B. FATTY ACIDS

The relative proportions of fatty acids isolated from plant sources depend on the identity and growth history of the plant, the part of the plant sampled, and the identity and function of the acyl lipid examined. The seven "major" acids often tend to predominate, but their relative quantitative importance is variable. For instance, growth at lower temperatures leads to an increase in the degree of unsaturation of the fatty acid mixtures. Photoautotrophic cultures of green algae exhibit similar fatty acid compositions to many leaves (though the proportions of linolenic acid are somewhat lower); but when grown on organic nutrients, the algae accumulate less polyenoic acid, reflecting the production of fewer chloroplasts. Plant leaves all have a similar function, and this is reflected in the similarity of the fatty acids present; in particular, *trans*-3-hexadecenoic acid occurs exclusively esterified at the 2-position of phosphatidylglycerol. Similarly, 2-hydroxyacids are located almost exclusively in the cerebrosides. Small quantities of these unusual acids are widely distributed in these capacities, but they are also found as triglyceride in a few specific seed-oils. In general, however, when unusual acids accumulate in seeds they are present as triglyceride and not phospholipid (Vijayalkshmi and Rao, 1972) and are seldom found in the rest of the plant; but notable exceptions include the presence of unusual acids in the leaves as well as the seeds of species of Malvaceae and Boraginaceae.

Within the fruit, the composition of the exocarp oil is usually significantly different from the corresponding seed. For example, palm fruit coat oil (palm oil) contains mainly palmitic, oleic and linoleic acids, while the oil from the nut of the same fruit (palm kernel oil) is rich in lauric, myristic, palmitic and oleic acids.

C. TAXONOMY

The variety of the unusual fatty acids in plants suggests that plants could be classified on the basis of qualitative or quantitative composition of their seed-oils. Attempts have been made by Hilditch and Williams (1964) Shorland (1963) and Hitchcock and Nichols (1971). The usefulness of this system is limited, for the large number of seeds in which no unusual acid occurs makes it necessary to group them according to proportions of the "major" acids, which is not distinctive and open to uncertainty; for instance, a large group of linoleate-rich seed oils cannot be sub-classified. Secondly, some species that are closely related botanically give seed oils unrelated chemically and vice versa. For instance, the Lauraceae and Palmae are unrelated families whose species have similar seed fats rich in laurate; conversely some rapeseed oils contain no erucate, which is typical of others. These limitations may well become more significant as new strains of plants are developed for novel seed-oils and the taxonomic significance of fatty acid composition must be treated with reserve. However, sometimes the presence of an unusual fatty acid is closely correlated with accepted botanical classification, as with the cyclopentenoids of the Flacourtaceae.

The taxonomic possibilities of the triglyceride structure of fats having similar overall fatty acid compositions have been reviewed by Litchfield (1973). Others have suggested that cutin acid composition may be a useful taxonomic criterion (Mazliak, 1968; Hunneman and Eglinton, 1972).

REFERENCES

Albro, P. W. and Fishbein, L. (1970). *Phytochemistry* **10**, 631–636.

Bergelson, L. D. (1969). *In* "Progress in the Chemistry of Fats and Other Lipids" (R. T. Holman, ed.), Vol. 10 (Part 3), pp. 241–286. Pergamon Press, Oxford.

Bohlmann, F., Burkhardt, T. and Zdero, C. (1973). "Naturally Occurring Acetylenes". Academic Press, London and New York.

Brieskorn, C. H. and Kabelitz, L. (1971). *Phytochemistry* **10**, 3195–3204.

Bryce, T. A., Welti, D., Walsby, A. E. and Nichols, B. W. (1972). *Phytochemistry* **11**, 295–302.

Carter, H. E., Strobach, D. R. and Hawthorne, J. N. (1969). *Biochemistry, Easton* **8**, 383–388.

Croteau, R. and Fagerson, I. S. (1972). *Phytochemistry* **11**, 353–363.

Galliard, T. (1974). *In* "Metabolism and Regulation of Secondary Plant Products" (V. C. Runeckles and E. E. Conn, eds.)., pp. 209–241. Academic Press, New York and London.

Gardner, H. W., Christianson, D. D. and Kleiman, R. (1973). *Lipids* **8**, 271–276.

Green, T. G., Hilditch, T. P. and Stainsby, W. J. (1936). *J. chem. Soc.* 1750–1755.

Hadjieva, P., Stoianova-Ivanova, B. and Danieli, B. (1974). *Chem. Phys. Lipids* **12**, 60–63.

Hilditch, T. P. and Williams, P. N. (1964). "The Chemical Constitution of the Natural Fats", 4th edition. Chapman and Hall, London.

Hitchcock, C. and Nichols, B. W. (1971). "Plant Lipid Biochemistry". Academic Press, London and New York.

Holloway, P. J. and Deas, A. H. B. (1973). *Phytochemistry* **12**, 1721–1735.

Hopkins, C. Y. (1972). *In* "Topics in Lipid Chemistry" (F. D. Gunstone, ed.), Vol. 3, pp. 37–87. Paul Elek, London.

Hunneman, D. H. and Eglinton, G. (1972). *Phytochemistry* **11**, 1989–2001.

Jamieson, G. R. and Reid, E. H. (1972). *Phytochemistry* **11**, 269–275.

Kartha, A. R. S. and Singh, S. P. (1969). *Chemy Ind.* 1342–1343.

Kleiman, R. and Spencer, G. F. (1971). *Lipids* **6**, 962–963.

Kleiman, R., Rawls, M. H. and Earle, F. R. (1972a). *Lipids* **7**, 494–496.

Kleiman, R., Spencer, G. F., Earle, F. R., Nieschlag, H. J. and Barclay, A. S. (1972b). *Lipids* **7**, 660–665.

Kolattukudy, P. E. (1972). *Biochem. Biophys. Res. Commun.* **49**, 1040–1046.

Kolattukudy, P. E. (1973). *Lipids* **8**, 90–92.

Kolattukudy, P. E. and Walton, T. J. (1972). *Biochemistry, Easton* **11**, 1897–1907.

Kuiper, P. J. C. and Stuiver, B. (1972). *Pl. Physiol., Lancaster* **49**, 307–309.

Litchfield, C. (1973). *Fette, Seifen, Anstrichm.* **75**, 223–231.

Madrigal, R. V. and Smith, C. R. (1973). *Lipids* **8**, 407–409.

Matucha, M., Zilka, L. and Svihel, K. (1972). *J. Chromatog.* **65**, 371–376.

Mazliak, P. (1968). *In* "Progress in Phytochemistry" (L. Reinhold and Y. Liwschitz, eds), Vol. 1, pp. 49–111. Interscience, New York.

Mikolajczak, K. L. and Smith, C. R. (1971). *Lipids* **6**, 349–350.

Pearl, M. B., Kleiman, R. and Earle, F. R. (1973). *Lipids* **8**, 627–630.

Phillips, B. E. and Smith, C. R. (1970). *Biochim. biophys. Acta* **218**, 71–82.

Poincelot, R. P. (1973). *Archs Biochem. Biophys.* **159**, 134–142.

Rankov, D., Popov, A., Panov, P. and Daleva, M. (1971). *J. Am. Oil Chem. Soc.* **48**, 700–701.

Roughan, P. G. and Batt, R. D. (1969). *Phytochemistry* **8**, 363–369.

Schwertner, H. A. and Biale, J. B. (1973). *J. Lipid Res.* **14**, 235–242.

Sen Gupta, A. (1972). *Chem. Ind.* 257–258.

Shorland, F. B. (1963). *In* "Chemical Plant Taxonomy" (T. Swain, ed.), pp. 253–303. Academic Press, London and New York.

Smith, C. R. (1970). *In* "Progress in the Chemistry of Fats and Other Lipids" (R. T. Holman, ed.), Vol. 11, Part 1, pp. 139–177. Pergamon Press, Oxford.

Smith, C. R. and Wolff, I. A. (1966). *Lipids* **1**, 123–127.

Spencer, G. F. and Earle, F. R. (1972). *Lipids* **7**, 435–436.

Spencer, G. F., Kleiman, R., Miller, R. W. and Earle, F. R. (1971). *Lipids* **6**, 712–714.

Spencer, G. F., Earle, F. R., Wolff, I. A. and Tallent, W. H. (1973). *Chem. Phys. Lipids* **10**, 191–202.

Spener, F. (1972). Abstracts of papers, 11th World Congress I.S.F., Göteborg 18–22 June 1972, pp. 28–29.

Sprecher, H. W., Maier, R., Barber, M. and Holman, R. T. (1965). *Biochemistry, Easton* **4**, 1856–1863.

Takasugi, M., Kobayashi, K., Anetai, M., Veno, S., Katsui, N. and Masamune T. (1973). *Chem. Lett.* 445–446.

Vijayalkshmi, B. and Rao, S. V. (1972). *Chem. Phys. Lipids* **9**, 82–86.

Wren, J. J. and Merryfield, D. S. (1970). *J. Sci. Fd Agric.* **21**, 254–257.

CHAPTER 2

Determination of the Structure of Fatty Acids

F. D. GUNSTONE

Department of Chemistry,
The University, St. Andrews, Scotland

I. Introduction

This article is concerned with methods in current use for determining the structure of a fatty acid. As in other fields of organic chemistry in recent years there has been a movement away from classical procedures of degradation to an increasing reliance on physical procedures which are mainly spectroscopic.

If the acid under investigation is of natural origin then it is likely to fall into the general pattern of structure now recognized as a result of two converging studies. The structure determination of a large number of compounds has led to the recognition of certain preferred structural features among natural compounds and biosynthetic studies have revealed the ground rules of biosynthesis and hence of structure. Nevertheless, there are still occasional surprises arising from unusual combinations of more familiar structural features, especially as a result of enzymic reactions of polyenoic acids in both the plant and animal kingdom. The structures of the allenic ester acid (**1**, Sprecher *et al.*, 1965) and the ether acid (**2**, Galliard *et al.*, 1973)

$$CH_3(CH_2)_4CH\overset{c}{=}CHCH\overset{t}{=}CHCOOCH_2CH{=}C{=}CH(CH_2)_3COOH \qquad (1)$$

$$CH_3CH_2CH\overset{c}{=}CHCH_2CH\overset{c}{=}CHCH\overset{t}{=}CHOCH\overset{t}{=}CH(CH_2)_6COOH \qquad (2)$$

would have been impossible to predict. Acids resulting from synthetic procedures or from natural acids by chemical reactions may also display novel structural features. In this review the emphasis is on natural acids though the procedures described are applicable to other long-chain acids of unknown or uncertain structure.

It will be helpful to preface the detailed discussion which follows with a brief summary of the more common structural features which are likely to be encountered.

(a) Natural acids, whether saturated or unsaturated, are usually straight-chain compounds with an even number of carbon atoms but acids with branched chains, cyclic systems, and an odd number of carbon atoms are also known.

(b) Unsaturated acids may contain up to six double bonds which are usually of *cis* (*Z*) configuration and are separated from each other by a single methylene group as, for example, in arachidonic acid (**3**). In addition, however, *trans* (*E*) olefinic, allenic, and acetylenic acids are known along with others that have conjugated unsaturation. (Several of these unusual features are present in structures **1** and **2**.)

$$CH_3(CH_2)_4CH\overset{c}{=}CHCH_2CH\overset{c}{=}CHCH_2CH\overset{c}{=}CHCH_2CH\overset{c}{=}CH(CH_2)_3COOH \qquad (3)$$

(c) Unsaturation usually occurs in certain preferred positions along the carbon chain. This is sometimes correlated with the carboxyl group and sometimes with the end methyl group.

(d) Other functional groups are less common, but natural hydroxy, epoxy, methoxy, oxo and fluoro acids are known.

(e) The rarer acids with unusual unsaturation and/or additional functional groups are most commonly C_{18} compounds of plant origin.

A fatty acid of unusual structure often reveals itself through its unexpected behaviour during adsorption chromatography (usually t.l.c.) or partition chromatography (usually g.l.c.). Sometimes it is possible to identify the acid without isolating it but generally it will be purified, or at least concentrated, by some chromatographic procedure. Its behaviour on t.l.c. (with and without silver nitrate) and g.l.c. will probably give some indication of its chain length, its degree of unsaturation, and the presence of unusual chain-branching or of additional polar groups (see Section II. F). Procedure thereafter will depend on the amount of acid available, on the nature of the spectroscopic equipment which can be called on, and on the ability of the experimenter to interpret this information. Sometimes the structure can be completely recognized by the use of appropriate spectroscopic techniques with a choice between ultraviolet, infrared, nuclear magnetic resonance (^{1}H with or without shift reagents and ^{13}C) and mass spectroscopy. On other occasions it is desirable to hydrogenate the acid first, to identify the perhydro derivative, and then to tackle the problems related to unsaturation—its nature, configuration, and position. Classical degradation procedures are now so refined that, with the assistance of g.l.c. to identify the products, these can be satisfactorily effected on a few milligrams of material or less and, if necessary, can be applied to mixtures of isomeric acids on a quantitative or semi-quantitative basis. Special difficulties which sometimes arise with polyunsaturated compounds can be solved using partial hydrogenation or partial oxidation techniques (see Section III.C).

The several physical and chemical procedures employed will be reviewed systematically and in each case recent examples of their use will be cited. This topic was previously reviewed in 1966 (Holman and Rahm, 1966; Privett, 1966).

II. Physical Methods

A. ULTRAVIOLET SPECTROSCOPY AND ISOMERIZATION PROCESSES

Compounds like oleic acid (18:1 9)* with one double bond absorb ultraviolet light only at wavelengths too low for convenient study, and methylene-interrupted polyenoic acids, such as linoleic (18:2 9,12), linolenic (18:3 9,12,15) and arachidonic (20:4 5,8,11,14 3), with several isolated double bonds, absorb in the same region but with increased intensity. Ultraviolet spectroscopy is thus of limited value except for the small, but growing, group of acids with conjugated unsaturation. Differences in ultraviolet absorption will distinguish various chromophores but *cis* and *trans* isomers are not usually differentiated. Some examples are given in Table I and many more are cited by Hopkins (1972) in a review of fatty acids with conjugated unsaturation.

Natural polyenoic acids are commonly methylene-interrupted and procedures have been developed by which they may be converted, under the influence of base, to conjugated isomers. Though no longer used as a quantitative analytical procedure, this technique is still employed, on occasions, to determine the number of methylene-interrupted double bonds.

* Abbreviated notation for fatty acid structures is explained on p. 6.

TABLE I

Absorption maxima (nm) of some ultraviolet chromophores present in natural fatty acids

Chromophore	Fatty acid[1]	$\lambda_{max}(\log \varepsilon_{max})$[2]
Diene	18:2(9*t*,11*t*)[a]	227 (infl), 231 (4·52), 239 (infl)
Enyne	18:2(9*a*,11*t*)[a]	229 (4·22), 240 (infl)
Diyne	16:2(8*a*,10*a*)[b]	215 (2·55), 226 (2·62), 239 (2·62), 254 (2·38)
Dienoic acid	10:2(2*t*,4*c*)[a]	263 (4·58)
Triene	18:3(9*c*,11*t*,13*t*)[a]	261 (4·56), 271 (4·67), 281 (4·58)
Dienyne	18:3(9*a*,11*t*,13*t*)[e]	267 (4·59), 277 (4·48)
Enediyne	18:3(9*a*,11*a*,13*c*)[c]	214 (?), 229 (3·58), 240 (3·84), 253 (4·19), 268 (4·35), 284 (4·26)
Tetraene	18:4(9*t*,11*t*,13*t*,15*t*)[d]	286 (4·77), 299 (4·97), 313 (4·94)
Enetriyne	18:4(9*a*,11*a*,13*a*,15*e*)[e]	212, 223, 232, 244, 258, 274, 291

[1] Solvent: *a* ethanol, *b* cyclohexane, *c* hexane, *d* methanol, *e* unknown.
[2] infl = inflexion.

Two C_{18} tetraenes are known (6,9,12,15 and 3,9,12,15): when submitted to alkali isomerization the former gives conjugated dienes, trienes, and tetraenes which are easily recognized by ultraviolet spectroscopy (Craig and Bhatty, 1964) whilst the latter gives only conjugated dienes and trienes (Hopkins and Chisholm, 1965). This is evidence that, though both acids are tetraenoic, one contains four and the other three methylene-interrupted double bonds. This procedure is particularly useful in examining those acids which are not entirely methylene-interrupted (Bagby *et al.*, 1965: Wolff and Miwa, 1965).

The polyene acids of human testicular tissue were recently recognized as completely methylene-interrupted polyenes with four (315 nm), five (346 nm), and six (375 nm) double bonds because ultraviolet maxima were observed, after isomerization, at the positions indicated (Grogan *et al.*, 1973).

As a modification of this procedure, lipoxygenase-catalysed peroxidation of polyene acids (see Chapter 11) gives conjugated diene hydroperoxides which can be detected by ultraviolet spectroscopy. This is a highly specific procedure which cannot be applied to all 1,4-dienes independently of their position and configuration but the complete specificity of the various enzyme preparations is not yet clearly defined (Bagby *et al.*, 1965; Wolff and Miwa, 1965).

$$\text{—CH}\overset{c}{=}\text{CHCH}_2\text{CH}\overset{c}{=}\text{CH—} \xrightarrow[\text{lipoxygenase}]{O_2} \text{—CH(OOH)CH}\overset{t}{=}\text{CHCH}\overset{c}{=}\text{CH—} + \text{—CH}\overset{c}{=}\text{CHCH}\overset{t}{=}\text{CHCH(OOH)—}$$

B. INFRARED AND RAMAN SPECTROSCOPY

Infrared spectroscopy is most useful in the recognition of unusual functional groups and of *trans* unsaturation. Workers at the Northern

Regional Research Laboratory in Peoria (Wolff and Miwa, 1965), having examined a large number of seed oils, described how infrared spectroscopy can be used to recognize the following groups: hydroxyl (absorption at 3450 cm^{-1}), oxo (1725 cm^{-1}), cyclopropene (1850 and 1010 cm^{-1}), epoxide (850 and 825 cm^{-1}), allene (2220 and 1960 cm^{-1}), vinyl (990 and 910 cm^{-1}), and conjugated enyne (950 cm^{-1}).

Infrared spectroscopy is most commonly used to examine acids suspected of containing *trans* unsaturation. One *trans* double bond leads to characteristic absorption at 968 cm^{-1}. For non-conjugated *trans* polyenes there is an increase in intensity but not in the position of absorption. Conjugated polyene systems give more complex absorption patterns with small shifts in the absorption maxima which may be of value for identification purposes (Table II).

TABLE II

Infrared absorption maxima in compounds containing *trans* unsaturation in conjugated polyenoic systems

Chromophore	Configuration	Absorption maxima (cm^{-1})
Monoene	*t*	968
Diene	*tt*	988
	tc	985, 950
Triene	*ttt*	994
	ttc	993, 965
	ctc	988, 937
Tetraene	*tttt*	997
	cttc	993, 952

Infrared spectroscopy does not provide any assistance in detecting unsaturation which is *cis*-olefinic or acetylenic but the recent study of a range of monounsaturated C_{18} acids has shown that *cis* and *trans* olefinic and acetylenic compounds all have characteristic absorption bands in their Raman spectra at 1656 ± 1 cm^{-1} (s) for *cis* alkenes, 1670 ± 1 cm^{-1} (s) for *trans* alkenes, and at 2232 ± 1 (s) and 2291 ± 2 cm^{-1} (m) for alkynoic acids; slightly different results were reported for the Δ^2 and Δ^{17} (vinyl) isomers (Davies *et al.*, 1972). Another group independently used this procedure to recognize the acetylenic group in a trienynoic acid (Andersson *et al.*, 1974).

C. NUCLEAR MAGNETIC RESONANCE (n.m.r.) SPECTROSCOPY

At first n.m.r. spectroscopy was of limited value in determining the structure of fatty acids because the spectra obtained with low resolution instru-

TABLE III

Nuclear magnetic resonance deshielding effects of functional groups along an alkyl chain (in ppm). Basic δ values: CH_2 1·253, CH_3 0·883. [From Frost and Barzilay (1971), Frost and Gunstone (unpublished observations)]

Functional group	Proton position (relative to functional group) α	β	γ	δ	ε	ζ
COOH	1·037	0·362	0·094	0·055	0·029	0·005
COOMe	0·959	0·322	0·061	0·036	0·021	0·005
—C≡C— (X)[a]	0·821	0·162	0·120	0·040	0·023	0·013
—XCH_2X—[a]	0·863	0·212	0·145	0·047	0·027	0·013
—CH$\overset{c}{=}$CH— (Y)	0·733	0·063	0·019	0·015	0·007	0
—YCH_2Y—	0·768	0·073	0·025	0·024	0·009	0
—CH$\overset{t}{=}$CH—	0·683	0·052	0·003	0	0	0

[a] These values are applicable for CH_2, different values are reported for CH_3.

ments were not very informative. Fatty acid molecules contain too many protons in a similar environment. Even so it was possible to determine the number of double bonds and their relative positions and to detect unusual functional groups. Important chemical shifts were listed in a review by Hopkins (1965) which was updated by Gunstone and Inglis (1971). A simple but effective illustration of the use of n.m.r. spectroscopy is given by Hinchcliffe and Riley (1971) who identified a sample of 22:6 from cod liver oil on the basis of its n.m.r. spectrum. This showed the molecule to contain six double bonds in a completely methylene-interrupted pattern. Of the five structures which would satisfy this requirement only the 4,7,10,13,16,19-hexaene was consistent with the observed spectrum.

Most of the early results were obtained with 60 or 100 MHz instruments. Developments in this field include (i) the use of higher resolving spectrometers (220 MHz), (ii) the use of chemical shift reagents, and (iii) the study of ^{13}C n.m.r. spectra. D. J. Frost has made particularly effective use of the 220 MHz spectrometer and, having examined a wide range of natural and synthetic acids, has given results which supercede many earlier reports. He claims that long-range deshielding effects may be apparent up to six carbon atoms from their source and that these effects are additive. Some of his values are presented in Table III. Similar studies have been conducted on epoxy esters ((**4**, X=O) and cyclopropane esters (**4**, $X{=}CH_2$) (D. J. Frost and F. D. Gunstone, unpublished observations). The distinctive signals

$n + m = 14$ **(4)**

associated with the cyclopropane ring protons are particularly useful in recognizing this system (Christie, 1970; Gunstone and Perera, 1973). Using this approach Frost is able to distinguish all the sixteen methyl *cis*-octadecenoates except the Δ^{10} and Δ^{11} pair. He has exploited his understanding of these spectra to correct the structure of bongkrekic acid (de Bruijn *et al.*, 1973) and to determine the configuration of the unsaturated centres in the unusual ether acids produced by enzymes in the potato from linoleic and linolenic acid (Galliard *et al.*, 1973) (Table IV).

Derivatives of transition metals such as europium and praseodymium interact with atoms having lone pairs—such as oxygen—and induce considerable shifts in the n.m.r. signals of protons in the vicinity of this inter-

TABLE IV

Chemical shifts (δ) for each proton in the n.m.r. spectra of ether acids formed by enzymic oxidation of linoleic and linolenic acids by an enzyme from potatoes. [From Galliard *et al.* (1973)]

		δ (ppm)			δ (ppm)
	$COOCH_3$	3·59		$COOCH_3$	3·61
	CH_2	2·21		CH_2	2·22
	CH_2	1·57		CH_2	1·58
	$(CH_2)_3$	1·32		$(CH_2)_3$	1·34
	CH_2	1·93		CH_2	1·95
	CH	5·02		CH	5·02
t	‖		*t*	‖	
	CH	6·16		CH	6·16
	O			O	
	CH	6·37		CH	6·40
t	‖		*t*	‖	
	CH	5·83		CH	5·86
	CH	5·71		CH	5·73
c	‖		*c*	‖	
	CH	5·16		CH	~5·15
	CH_2	2·07		CH_2	2·82
	$(CH_2)_3$	1·32		CH	5·28
			c	‖	
	CH_3	0·89		CH	5·28
				CH_2	2·07
				CH_3	0·98

action. Alkanols, for example, normally show only three signals in their n.m.r. spectrum but in the presence of chemical shift reagents separate signals are observed for the first seven CH_2 groups (Smentowski and Stipanovic,

c b a signals without shift reagent

$CH_3CH_2CH_2CH_2CH_2CH_2CH_2CH_2CH_2CH_2OH$

i h g f e d c b a signals with shift reagent

1972). The application of this technique to fatty acids is being pursued by Wineburg and Swern (1972, 1973) who have outlined the careful study required for the complete and accurate interpretation of these spectra. Using only n.m.r. (with shift reagents) they conclude that (i) in methyl petroselinate (18:1 6*c*) the double bond is at the C-6 position and that there is no branching in the carbon chain before C-9, (ii) in methyl oleate (18:1, 9*c*) the double bond is at C-9 or beyond and that there is no chain branching before C-6, and (iii) in methyl ricinoleate (12-hydroxy-18:1 9*c*), with two centres to react with the shift reagent, all the individual proton signals can be distinguished except those associated with C-5, C-6, and C-7.

The application of ^{13}C n.m.r. spectroscopy to fatty acid structure determination lies in the future though some useful introductory reports are available (de Haan and van de Ven, 1971; Tulloch and Mazurek, 1973).

D. MASS SPECTROMETRY

The use of mass spectrometry for determining the structure of fatty acids—now increasingly used with profit—has been reviewed by McCloskey (1970) and by Zeman and Scharmann (1972, 1973). This is a very effective procedure and when combined with the separating efficiency of gas chromatography it provides one of the most powerful methods of structure determination and requires only small amounts of material. Gas chromatography-mass spectrometry (GC/MS) becomes even more efficient when linked with a computer.

The determination of an oxygenated function in a carbon chain is a fairly simple matter by mass spectrometry and a rather more difficult problem by classical degradation processes (Section III.E). Such groups—either present in natural products or produced by chemical manipulation—include hydroxy, alkoxy, oxo, and epoxy acids. Hydroxy esters are best investigated as their trimethylsilyl ethers (which also have superior g.l.c. properties) and epoxy esters as hydroxy-methoxy or methoxy-trimethylsiloxy derivatives (Kleiman and Spencer, 1973).

$$—CH(OH)— \longrightarrow —CH(OSiMe_3)—$$

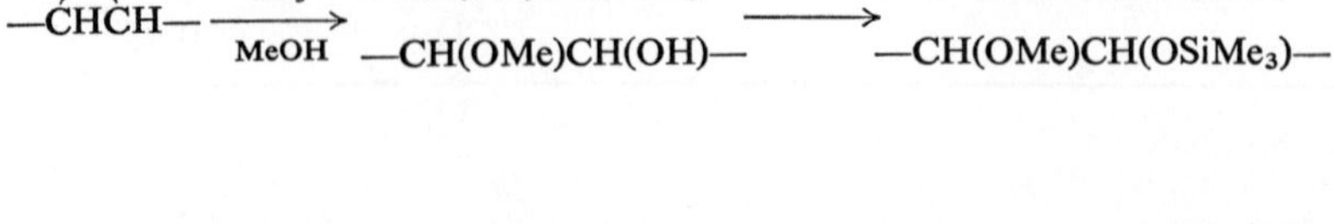

It is sometimes possible with these derivatives to locate the position of unsaturated centres which accompany the oxygenated function.

The position of branched methyl and cyclic groups in otherwise unbranched molecules is not readily deduced by degradation methods (Section III.E) but is more easily determined by appropriate mass spectroscopic procedures (Abrahamsson *et al.*, 1964; McCloskey, 1970; Lough, 1973). Acids containing cyclopropane groups have attracted considerable attention (Christie, 1970; Christie and Holman, 1966; Christie *et al.*, 1968; Gunstone *et al.*, 1971; Gunstone and Perera, 1973). Like the unsaturated compounds to be discussed later cyclopropane compounds are generally unstable under electron bombardment and they isomerize to a mixture of isomeric alkenoates which are not readily distinguishable from each other. Campbell and Naworal (1969) claim that cyclopropane and isomeric alkene esters can be distinguished by the relative intensity of their M (molecular ion) and M − 32 peaks.

More usefully, the cyclopropane system may be "fixed" by some chemical reaction leading to a product, or more usually a mixture of products, which are identified by mass spectrometry. These reactions include hydrogenolysis (McCloskey and Law, 1967), oxidation (Promé, 1968), and reaction with methanolic boron trifluoride (Minnikin, 1972).

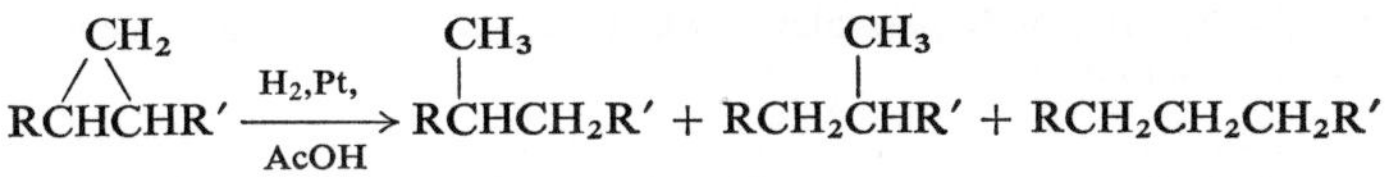

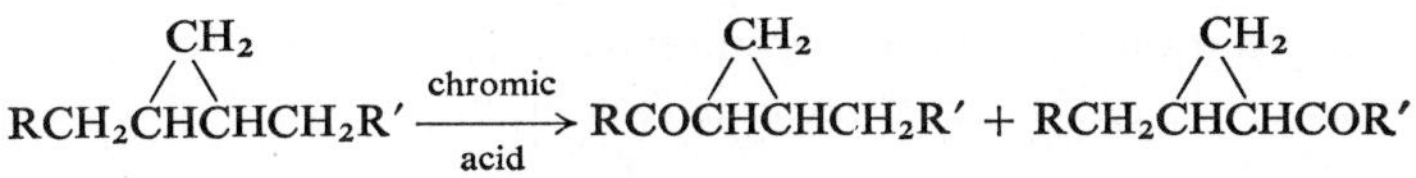

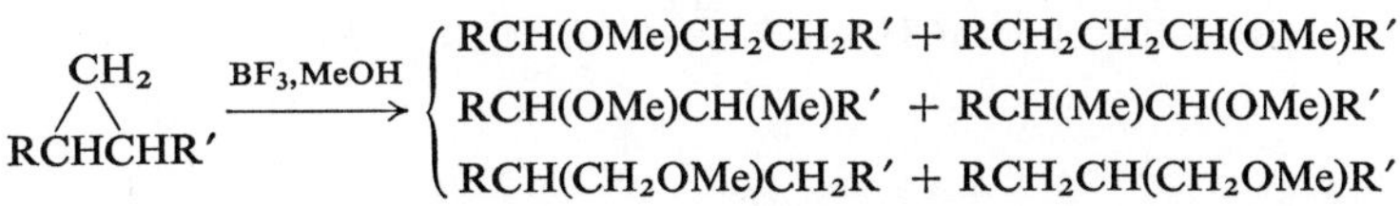

Structure determination involves, more often, the determination of double bond position and it is unfortunate that mass spectrometry does not provide an immediate solution to that problem. The double bond system is labile on electron bombardment and isomeric esters give almost indistinguishable mass spectra of little diagnostic value. Since 1960 various procedures for overcoming this difficulty have been proposed. Each procedure involves a chemical modification of the alkenoate to "fix" the double bond. Some reactions give mixed products and this complicates the interpretation a little. This is not too serious with monoenoates but the situation may become confused with polyunsaturated esters and although mass spectrometry has been applied to such systems no single widely applied procedure has been evolved. The following reactions have been reported for the mass spectrometric study of alkenoates:

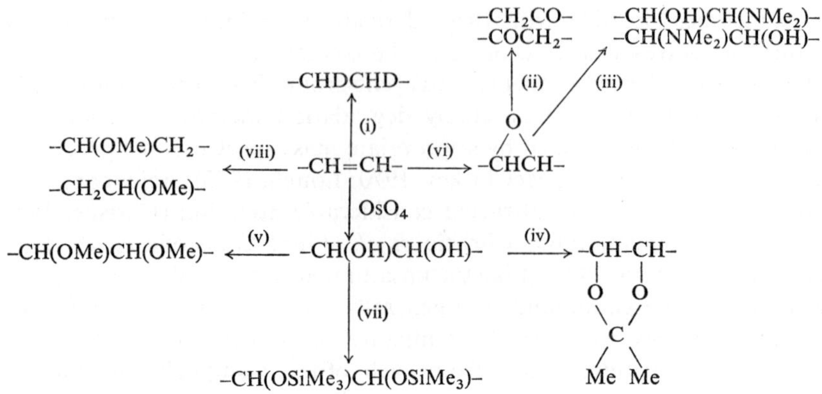

(i) D_2, catalyst (Dinh-Nguyen *et al.*, 1960); D_2 tris(triphenylphosphine)rhodium^{I} chloride (Morandi and Jensen, 1969)
(ii) NaI (Kenner and Stenhagen, 1964)
(iii) Me_2NH (Audier *et al.*, 1964)
(iv) $COMe_2$ (McCloskey and McClelland, 1965)
(v) MeI (Niehaus and Ryhage, 1967, 1968)
(vi) Perlauric acid $C_{11}H_{23}CO_3H$ (Aplin and Coles, 1967)
(vii) $(Me_3Si)_2NH$ and Me_3SiCl or $(Me_3Si)NAc$ (Capella and Zorzut, 1968; Eglinton and Hunneman, 1968; Eglinton *et al.*, 1968; Argoudelis and Perkins, 1968; Perkins and Argoudelis, 1969)
(viii) $Hg(OAc)_2$, MeOH, $NaBH_4$ (Abley *et al.*, 1970; Gunstone and Tallent, unpublished observations).

Acetylenic groups can be placed by the methods discussed above after partial hydrogenation to alkenes or directly after conversion to ethylene ketals (Audier *et al.*, 1967; Gunstone and Tallent, unpublished observations):

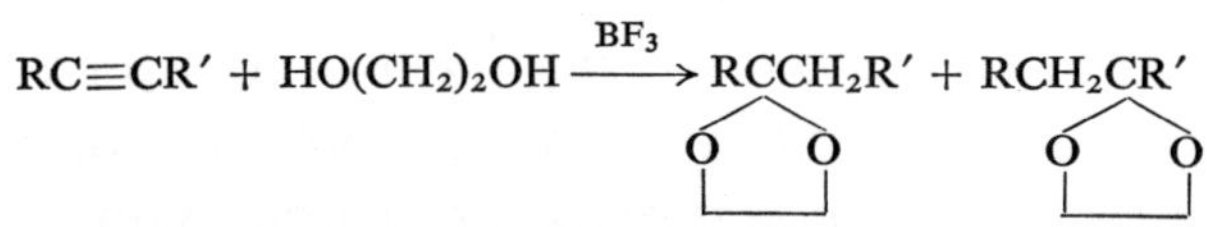

Andersson and Holman (1974) have recently reported a more satisfactory solution to the problem of determining double bond position. In concentrating on the problem of double bond movement occurring during mass spectroscopic study of unsaturated *esters*, the possibility of using other acyl derivatives in which this problem is absent or minimized has been overlooked. Following a proposal by Vetter *et al.* (1971) it has now been demonstrated that more useful results are obtained with amides—and particularly pyrrolidides—from the mass spectra of which double bond position can be deduced directly. The spectra of the octadecenylpyrollidides contain clusters of peaks corresponding to fragments with 18, 17, 16, 15 etc. carbon atoms attached to the pyrrolidide moiety. The major peaks in each cluster differ by 14 mass units except around the double bond where one major peak differs by only 12 mass units from its nearest neighbour. In the 18:1 series this holds for the members with the double bond between C-5 and C-15 and isomers outside this range are readily recognized in other ways.

The figures shown for methyl oleate (Table V) differ by 14 mass units except between the C_8 (196) and C_9 (208) fragments. Results of this kind led Andersson and Holman (1974) to conclude, "If an interval of 12 mass units, instead of the regular 14, is observed between the most intense peak of each cluster of fragments containing n and $n-1$ carbon atoms in the acid moiety then a double bond occurs between carbon atoms n and $n+1$ in that acid".

TABLE V

Major ions (m/e) in each cluster of fragments in the mass spectrum of *N*-oleylpyrrolidide. [From Andersson and Holman (1974)]

$CH_3CH_2CH_2CH_2CH_2CH_2CH_2CH_2CH{=}CHCH_2CH_2CH_2CH_2CH_2CH_2CH_2CON$

355 306 278 250 222 196 168 140 113

320 292 264 236 208 182 154 126 98

E. X-RAY CRYSTALLOGRAPHY

Though used extensively in other areas of natural product chemistry, X-ray crystallography has not been widely employed in the determination of fatty acid structure (Chapman, 1965).

F. GAS–LIQUID CHROMATOGRAPHY (g.l.c.) DATA

Retention data for long-chain esters are now usually recorded in the form of equivalent chain length (Miwa *et al.*, 1960) and so many have been reported that certain generalizations have been derived and useful structural information may be obtained from retention data on one or more stationary phases.

Along with t.l.c. data, g.l.c. behaviour will usually indicate the nature of additional functional groups which may be present and suspicions can often be confirmed by further study of g.l.c. behaviour after the preparation of appropriate derivatives. For example, hydroxyl groups are easily recognized by their g.l.c. behaviour before and after conversion to a derivative such as the acetate, trifluoroacetate, methyl ether, or trimethylsilyl ether. The presence of unsaturated groups (alkenoic or alkynoic) and of branched and cyclic groups is often indicated by retention times and careful work with a capillary or support-coated open tubular (SCOT) column may even suggest the position of the unsaturated centre or functional group. Valuable studies by Ackman, Gunstone, Haken, Holman, and Tulloch have been reviewed by Jamieson (1970) and more recent publications have been made by Ackman (1969); Ackman and Hooper (1973); Scholfield and Dutton (1971); Gunstone and Lie Ken Jie (1970); Gunstone *et al.* (1971); Barve *et al.* (1972); Gunstone and Jacobsberg (1972): Gunstone and Perera (1973); and Gunstone (1973).

The g.l.c. separation of diastereoisomers has been used to identify enantiomeric compounds (see Section II.G).

G. DETERMINATION OF ABSOLUTE CONFIGURATION

Natural compounds with one or more chiral centres usually exist in an enantiomeric form and complete structural identification requires an assignment of absolute configuration to each chiral centre. This is never an easy problem and the difficulty is increased with many long-chain compounds because of their very weak optical activity. Absolute configuration is usually assigned on the basis of correlation with some other compound of known configuration, the correlation being made on the basis of optical rotatory dispersions curves, g.l.c., or some other physical property. In such studies it is frequently necessary to make some chemical modification of the enantiomeric molecule and it is then important to ensure that the stereochemical consequences of any such change are fully appreciated and allowed for. This topic has been fully reviewed by Smith (1970).

Among oxygenated acids the synthetic enantiomers of 12-L(*S*)-hydroxystearic (Serck-Hanssen, 1958) and 9-D(*R*)-hydroxystearic acid (Baker and Gunstone, 1963) are key compounds. Each was prepared from the appropriate enantiomer of methyl hydrogen 3-acetoxypentanedioate. These have

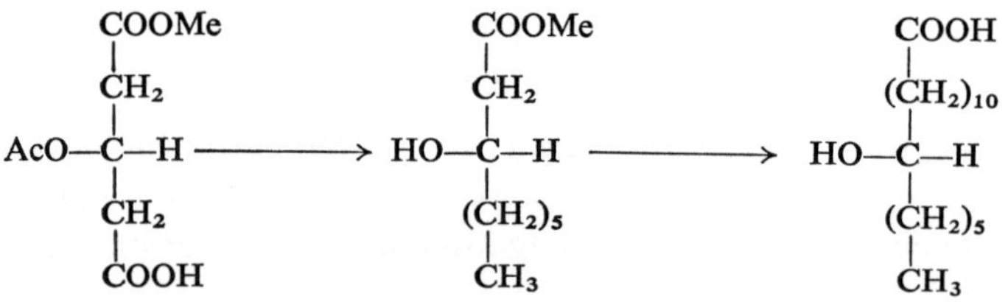

methyl hydrogen 3-L(*S*)-acetoxy-pentanedioate — methyl 3-L(*S*)-hydroxynonanoate — methyl 12-L(*S*)-hydroxystearate

been used to assign absolute configuration to hydroxy, hydroperoxy, and epoxy acids after appropriate reduction procedures. For example, since hydrogenation of ricinoleic acid gives 12-hydroxystearic acid which is the enantiomer of the synthetic L(*S*) acid it is concluded that the natural product has the D(*R*) configuration.

The synthesis of enantiomeric allenic acids has also been employed in determining the absolute configuration of natural allenic acids (Cowie *et al.*, 1972).

Gas–liquid chromatography is used increasingly as a method of comparing compounds in order to determine absolute configuration. This is based on the possibility of separation after conversion to diastereoisomeric derivatives. Two recent examples are cited: Brooks *et al.* (1973) reported the separation on SE30 liquid phase of enantiomeric alcohols (such as methyl 13-hydroxystearate) after conversion into diastereoisomeric esters by reaction with drimanoyl chloride. This information makes it possible to determine the absolute configuration of the 13-hydroperoxy and related compounds produced from linoleic acid under the influence of enzymes (Chapter 11). As an alternative approach to a similar problem, Hamberg (1971) described a procedure for detecting and determining the 9- and 13-hydroperoxyoctadecadienoates in both D and L forms. This involves reduction of the hydroperoxide to hydroxyl, acylation of the hydroxy with (—)-menthyl chloroformate, and oxidative ozonolysis of the unsaturated ester to give diastereoisomeric esters of methyl 2-hydroxydecanedioate (**6**, from the 9-hydroperoxy compounds) and of methyl 2-hydroxyheptanoate (**5**, from the 13-hydroxyperoxy compounds). The product obtained from linoleic acid in the presence of soyabean lipoxygenase was shown to be mainly the 13-L(*S*)-hydroperoxy acid.

$$\text{RCH(OOH)CH{=}CHCH{=}CHR}' \longrightarrow \text{RCH(OZ)CH{=}CHCH{=}CHR}' \longrightarrow \text{RCH(OZ)COOMe} \quad (\mathbf{5})$$

$$\text{RCH{=}CHCH{=}CHCH(OOH)R}' \longrightarrow \text{RCH{=}CHCH{=}CHCH(OZ)R}' \longrightarrow \text{MeOOCCH(OZ)R}' \quad (\mathbf{6})$$

$R = CH_3(CH_2)_4$ $\quad R' = -(CH_2)_7COOH$ $\quad Z =$

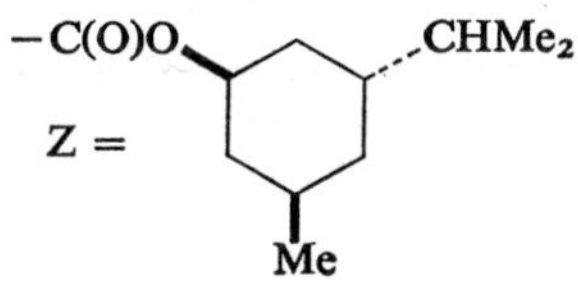

III. Chemical Methods

A. INTRODUCTION TO CHEMICAL METHODS

The procedures described thus far are essentially physical (and instrumental) methods which, as indicated in Section I, are increasingly important. Nevertheless the classical chemical procedures, updated to take account of the more refined methods (chromatographic) now available for examining the products, continue to play a useful role.

Complete hydrogenation to the perhydro derivative produces a structurally simpler product which is usually a straight-chain acid, a branched-chain acid, or a hydroxy acid. These are readily identified and the main carbon skeleton is thus quickly identified. There remain, however, such details as the position and nature of unsaturated centres (*cis*-alkenes, *trans*-alkenes, alkynes) and the nature and position of oxygenated functions and of chain-branching.

The reaction of polyenoic acids with appropriate dienophiles such as maleic anhydride or tetracyanoethylene, which is useful only for those acids having conjugated unsaturation, has been reviewed by Hopkins (1972) and does not warrant further discussion. The following sections will deal with methods of oxidative fission (Section III.B), the need for and use of partial hydrogenation or partial oxidation procedures (III.C), a special procedure for Δ3, 4, and 5 double bonds (III.D), methods of placing functional groups (III.E), the degradation of labelled acids (III.F), and synthetic procedures (III.G).

B. OXIDATIVE FISSION OF UNSATURATED ACIDS

Alkyne unsaturation can usually be distinguished from alkene unsaturation by g.l.c. and by n.m.r. and Raman spectroscopy: *cis* and *trans* alkenes are usually distinguished by infrared spectroscopy though Raman spectra have some advantages for this purpose. Alkynes are partially reduced to *cis* alkenes by catalytic hydrogenation in the presence of Lindlar's catalyst: alkenes are not further reduced under these conditions. Methods of locating alkene and alkyne centres by mass spectrometry have already been discussed. The position of unsaturated centres is more usually determined by oxidative degradation followed by recognition of the fission fragments by g.l.c. Oxidation with potassium permanganate or with ozone are the only commonly employed procedures.

The classical vigorous reaction of alkenoic acids with potassium permanganate in boiling acetone or acetic acid has been largely replaced by the milder oxidation technique of von Rudloff (1956) which involves reaction at room temperature with potassium periodate and potassium permanganate in a molar ratio of 39:1. Under these conditions there is never a high concentration of permanganate but this reagent is continually regenerated by periodate. Over-oxidation is thereby reduced to the minimum. The experimental procedure described by Tulloch and Craig (1964) is widely employed and Downing and Greene (1968) have reported some improvements.

$$\mathrm{RCH{=}CHR'} \xrightarrow{\mathrm{KMnO_4}} \mathrm{RCH(OH)CH(OH)R'} \xrightarrow{\mathrm{KIO_4}} \mathrm{RCHO + R'CHO}$$

$$\xrightarrow{\mathrm{KMnO_4}} \mathrm{RCO_2H + R'CO_2H}$$

Ozonolysis is more widely employed for structure determination. Ozonization is effected in a non-participating solvent, such as pentane or methylene chloride, at low temperature (−65 to −75°C) or in a participating solvent, such as methanol, at 0 to 20°C. The ozonized product may be decomposed in various ways to give alcohols, aldehydes or acids. The choice between these depends on the ease with which they are identified and aldehydes seem to find greatest favour at the present time (for a review see Pryde and Cowan, 1971). Alcohols result from the ozonide by reduction with lithium aluminium hydride (Ramachandran *et al.*, 1968b) or with hydrogen in the presence of palladium charcoal (Nickell and Privett, 1969); aldehydes from reaction with triphenylphosphine (Stein and Nicolaides, 1962; Beroza and Bierl, 1967; Kleiman *et al.*, 1969), tetracyanoethylene (Conacher and Page, 1972); Conacher *et al.*, 1972), dimethyl sulphide (Ramachandran *et al.*, 1968a), hydrogen and Lindlar's catalyst (Privett and Nickell, 1962), or on pyrolysis (Nickell and Privett, 1966; Spencer *et al.*, 1969); and acids by oxidative fission with silver oxide or potassium permanganate (Keppler, 1957) or hydrogen peroxide (Ackman *et al.*, 1961; Stoffel, 1965). These procedures can easily be adapted to a milligram (Stein and Nicolaides, 1962) or microgram (Privett *et al.*, 1963; Privett and Nickell, 1966) scale and the combined operation of ozonolysis has been conducted in a micro-reactor attached to a g.l.c. instrument (Davison and Dutton, 1966; Bitner *et al.*, 1969; Johnston and Dutton, 1972).

C. PARTIAL HYDROGENATION AND PARTIAL OXIDATION

The positions of unsaturated centres in a long-chain molecule are frequently identified by some form of oxidative fission (Section III.B) but oxidation alone does not always lead to an unequivocal solution. For example, a tetra-ene C_{20} acid oxidized to hexanoic, malonic, and glutaric acids can have several structures, viz. the 5,8,11,14; 3,8,11,14; 3,6,11,14; and 3,6,9,14 isomers. If these unsaturated centres are partly *cis* and partly *trans* then the complete identification of the tetra-ene is even more difficult.

This problem is now circumvented by partial hydrogenation followed by isolation of the monoene fraction (or, if necessary, of the *cis* and *trans* monoenoates) by silver ion t.l.c. (Morris, 1966) from the saturated and polyene esters. For example, after partial hydrogenation of a 20:4 acid, oxidative fission of the monoene fraction(s) will indicate the position of each double bond thus:

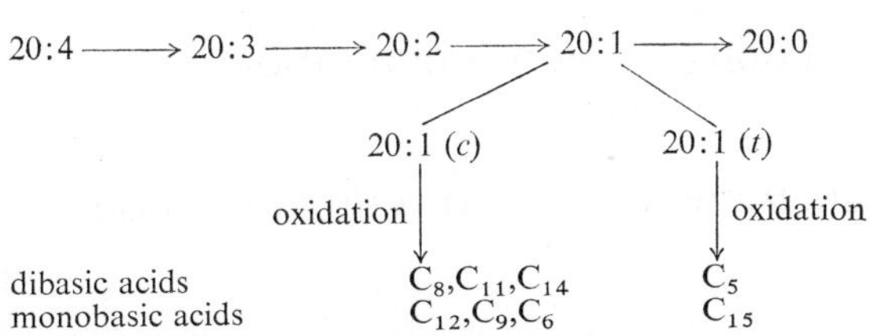

These results would provide complete proof that the tetraene was the 5*t*,8*c*,11*c*,14*c* isomer.

For this procedure to be successful it is imperative that partial hydrogenation should occur without stereomutation and without double bond migration. This is successfully achieved by reaction with hydrazine but *not* by catalytic hydrogenation. This procedure has been detailed by Privett and Nickell (1966) and by Roehm and Privett (1969) and has been employed, for example, in the identification of 18:3(3*t*,9*c*,12*c*) (Bagby *et al.*, 1965), 18:4(3*t*,9*c*,12*c*,15*c*) (Hopkins and Chisholm, 1965), and 20:3(5*c*,11*c*,14*c*) (Smith *et al.*, 1968).

Some investigators have tackled this problem by partial oxidation with performic acid (Gunstone and Sykes, 1962), osmium tetroxide (de Jong, 1967), or ozone (Kleiman *et al.*, 1967) but these procedures are less convenient and sometimes less informative than the partial hydrogenation process.

D. IDENTIFICATION OF DOUBLE BONDS IN POSITIONS 3, 4 AND 5

It would be useful to have simple and unique procedures for detecting unsaturation in all possible positions in a long-chain. A recently-discovered procedure, which makes it easy to recognize double bonds which are 3*t*, 4*c*/*t* or 5*c*/*t* with respect to the carboxyl group, is based on the intramolecular oxymercuration of the alcohols derived from these acids. This cyclization is confined to compounds having a double bond and a hydroxyl group in certain relative positions.

Alkenes react with mercuric acetate (and some other Hg^{II} compounds) to give intermediates which then react with hydroxy compounds such as methanol used as a solvent for the reaction. The mercury-containing product furnishes an ether when reduced with sodium borohydride; methyl oleate, for example, furnishes the methyl 9(10)-methoxystearates in high yield.

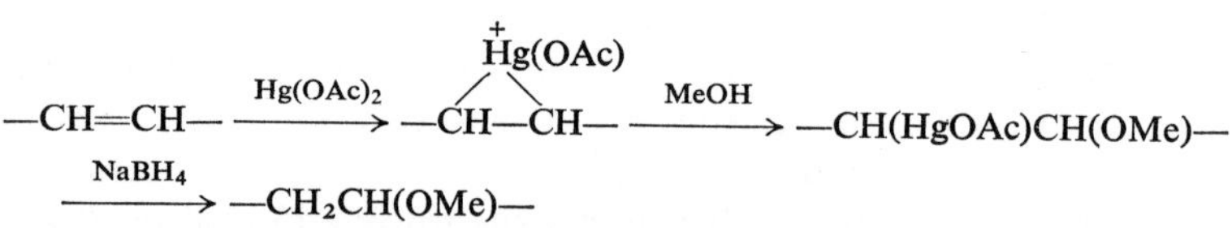

Intramolecular reaction occurs when the double bond and hydroxyl group are suitably disposed and these furnish substituted tetrahydrofurans

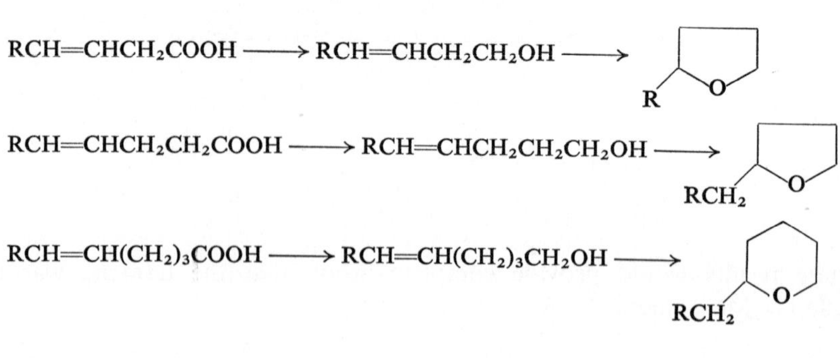

(1,4-epoxides) from β- and γ-hydroxyalkenes and substituted tetrahydropyrans (1,5-epoxides) from δ-hydroxyalkenes. Reactions occur in high yields and cyclic ethers are easily separated and recognized by their chromatographic behaviour.

With monoene alcohols of this type intramolecular reaction predominates over intermolecular reaction with methanol so the latter can be used as a solvent but with polyene alcohols or with mixtures of monoene alcohols it is desirable to use dimethylformamide as solvent so that reaction is confined to that occurring in an intramolecular fashion as with arachidonic acid:

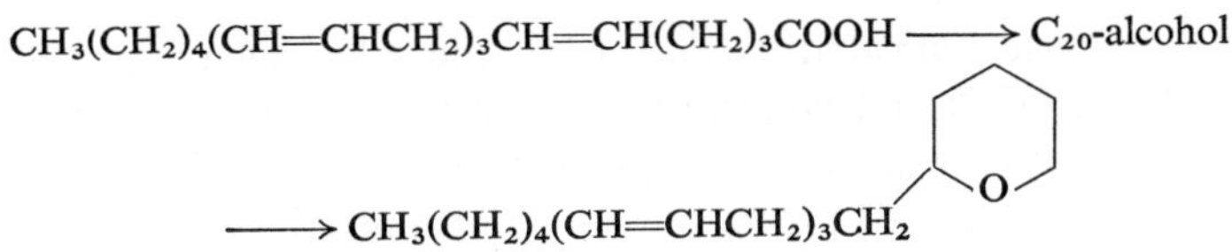

This reaction, which has been applied as an analytical technique and for the isolation of alcohols having unsaturation in these positions, should prove useful as an identification procedure (Gunstone and Inglis, 1973).

E. IDENTIFICATION OF FUNCTIONAL GROUPS

Although functional groups other than the various types of carbon to carbon unsaturation are not common, acids with branched methyl and cyclic groups (cyclopropane, cyclopropene, cyclopentene) are known as are those containing oxo, hydroxy, and epoxy groups (see Chapter 1). Many of these groups can be recognized and their positions determined by a combination of chromatographic and spectroscopic procedures but chemical degradation techniques still prove useful on occasion.

The presence of one or more hydroxyl groups would be indicated by t.l.c. and g.l.c. behaviour, particularly after conversion of the hydroxy compound to a more volatile derivative (*O*-methyl ether, *O*-trimethylsilyl ether, acetate, trifluoro-acetate) (Tulloch, 1964). The position of the hydroxy group can usually be determined by mass spectrometry after conversion to the trimethylsilyl ether or methyl ether or after oxidation to the oxo ester (McCloskey, 1970). The best degradative procedure involves oxidation to the oxo ester and determination of the position of the oxo group as described below.

Oxo esters are also indicated by t.l.c. and g.l.c. behaviour in their original form or after reduction by sodium borohydride to hydroxy esters (Tulloch, 1964) and are easily confirmed by infrared or n.m.r. spectroscopy. The position of the oxo function can usually be fixed by mass spectrometry since ketones undergo characteristic α and β (McLafferty) cleavage (McCloskey, 1970). The chemical procedure most commonly employed involves (i) hydrogenation to a saturated oxo ester, (ii) preparation of its oxime, (iii) Beckmann rearrangement of the stereoisomeric oximes to a mixture of two amides, and (iv) hydrolysis of the amide mixture to give an amine, an amino acid, and a mono- and dibasic acid followed by identification of at least one of these fragments. A recent use of this procedure has been reported (Cobern *et al.*, 1966).

$$\mathrm{RCOR'} \longrightarrow \mathrm{RC({=}NOH)R'} \longrightarrow \begin{matrix} \mathrm{RCONHR'} \\ \mathrm{RNHCOR'} \end{matrix} \longrightarrow \begin{matrix} \mathrm{RCOOH + R'NH_2} \\ \mathrm{RNH_2 + R'COOH} \end{matrix}$$

$$\mathrm{R = CH_3(CH_2)_n} \qquad \mathrm{R' = (CH_2)_mCOOMe\ or\ (CH_2)_mCOOH}$$

Epoxides are also indicated by their t.l.c. and g.l.c. behaviour (Gunstone and Jacobsberg, 1972) and confirmed by infrared and n.m.r. spectroscopy. The latter procedure will also distinguish between *cis* and *trans* epoxides. The position of an epoxide group can be determined by mass spectrometry of the epoxy ester or after conversion to a number of derivatives including the *O*-methyl *O*-trimethylsilyl ether (see Section II.D). The readiness with which epoxides undergo cleavage directly with periodic acid (Maerker and Haeberer, 1966; Kusamran and Polgar, 1971) or after ring opening to the diol, is the basis of a simple degradation procedure.

$$\mathrm{R\overset{\overset{O}{\frown}}{CHCH}R'} \xrightarrow{\mathrm{HIO_4}} \mathrm{RCHO + R'CHO}$$

Branched methyl groups may be recognized by a careful study of g.l.c. retention data (Ackman, 1969) and by n.m.r. spectroscopy. They usually occur in a limited number of positions along the carbon chain (often *iso* or *anteiso*) and are often recognizable by direct comparison with authentic standards. In difficult cases mass spectrometry may be useful and a vigorous oxidation technique coupled with gas–liquid chromatography of the complex mixture of branched-chain acids, straight-chain acids, and a ketone has also been employed (Murray, 1959, 1962; Dinh-Nguyen and Raal, 1969).

$$\mathrm{CH_3(CH_2)_m\overset{\overset{CH_3}{|}}{C}H(CH_2)_nCOOH} \xrightarrow{\mathrm{KMnO_4/acetone}}$$

$$\left\{\begin{array}{lll} \text{branched-chain acids} & \mathrm{CH_3(CH_2)_m\overset{\overset{CH_3}{|}}{C}H(CH_2)_pCOOH} & \\ & & p = 0 \text{ to } n \\ \text{ketone} & \mathrm{CH_3(CH_2)_mCOCH_3} & \\ \text{straight-chain acids} & \mathrm{CH_3(CH_2)_qCOOH} & q = 0 \text{ to } m \end{array}\right.$$

Chromic acid oxidation is also useful on occasions (Cason *et al.*, 1959).

$$\mathrm{CH_3(CH_2)_m\overset{\overset{CH_3}{|}}{C}H(CH_2)_nCOOH} \xrightarrow{\mathrm{CrO_3}} \left\{\begin{array}{l} \mathrm{CH_3(CH_2)_mCOCH_3}\ (1\%) \\ \mathrm{CH_3(CH_2)_{m-1}COOH}\ (10\%) \\ \mathrm{CH_3CO(CH_2)_nCOOH}\ (9\%) \\ \mathrm{HOOC(CH_2)_{n-1}COOH}\ (22\%) \end{array}\right.$$

Cyclopropane rings are most readily recognized by their characteristic n.m.r. signals (Christie, 1970; Gunstone and Perera, 1973) and are located by mass spectrometry after oxidation (Promé and Asselineua, 1966; Promé, 1968) or reaction with methanolic boron trifluoride (Minnikin, 1972) or by oxidative degradation (Conacher and Gunstone, 1969). Cyclopropene systems can be detected by their positive Halphen reaction and identified by mass spectrometry after chemical modification (Christie, 1970).

F. DEGRADATION OF LABELLED ACIDS

Labelled acids which contain unsaturated centres or oxygen-containing functional groups can be degraded by the methods already described (Sections III.B and E) but special procedures are sometimes required to determine the activity of a particular carbon atom. Such reactions should occur in high yield, without complicating side reactions, and it is also desirable that the reaction sequence can be repeated to give information about several carbon atoms. In addition to the familiar Barbier-Wieland, Hunsdiecker, and Schmidt reactions two others (summarized below), have been employed (Hunter and Popják, 1951; Dauben *et al.*, 1953 and Steinberg *et al.*, 1956).

$$RCH_2CH_2COOH \xrightarrow[\text{(ii) } Br_2]{\text{(i) } SOCl_2} RCH_2CHBrCOCl \xrightarrow[R'=Me_3CCH_2]{R'OH} RCH_2CHBrCOOR'$$

$$\xrightarrow{PhNEt_2} RCH{=}CHCOOR' \xrightarrow[300\text{–}350^\circ]{KOH} RCOOH + \overset{2}{C}H_3\overset{1}{C}OOH \xrightarrow[380^\circ]{\text{Li salt}}$$

$$\overset{2}{C}H_3\overset{1}{C}O\overset{2}{C}H_3 \xrightarrow{NaOI} \overset{2}{C}HI_3 + CH_3COONa + Li_2\overset{1}{C}O_3$$

$$RCH_2COOH \xrightarrow[\text{(ii) } C_6H_6,AlCl_3]{\text{(i) } SOCl_2} RCH_2COPh \xrightarrow{HNO_2} RC(:NOH)COPh$$

$$\xrightarrow{TsCl,NaOH} Ph\overset{1}{C}OOH + RCN$$

G. SYNTHETIC PROCEDURES

In classical terms synthesis has always been considered as a means of confirming structure. The increasing facility of structure determination means that synthesis has often lagged behind, though a natural product which has not been synthesized is a continual challenge to the synthetic organic chemist. The preparation of hydroxy acids of known absolute configuration has already been discussed (Section II.G). The synthesis of unsaturated acids—mainly via acetylenic intermediates—is being exploited continually (Barve and Gunstone, 1971; Beerthuis *et al.*, 1971) and has been reviewed (Osbond, 1966; Kunau, 1973).

REFERENCES

Abley, P., McQuillin, F. J., Minnikin, D. E., Kusamran, K., Maskens, K. and Polgar, N. (1970). *Chem. Commun.* 348–349.

Abrahamsson, S., Ställberg-Stenhagen, S. and Stenhagen, E. (1964). *Progr. Chem. Fats Lipids* **7**, 1–164.

Ackman, R. G. (1969). *J. Chromatogr.* **42**, 170–175.

Ackman, R. G. and Hooper, S. N. (1973). *J. Chromatogr.* **86**, 73–81 and 83–88.

Ackman, R. G., Retson, M. E., Gallay, L. R. and Vendenheuvel, F. A. (1961). *Can. J. Chem.* **39**, 1956–1963.

Andersson, B. A. and Holman, R. T. (1974) *Lipids*, **9**, 185–90; with Heimermann, W. H., *ibid* pp. 443–449.

Åndersson, B., Anderson, W. H., Chipault, J. R., Ellison, E. C., Fenton, S. W., Gellerman, J. L., Hawkins, J. M., and Schlenk, H. (1974) *Lipids*, **9**, 506–511.

Aplin, R. T. and Coles, L. (1967). *Chem. Commun.* 858–859.

Argoudelis, C. J. and Perkins, E. G. (1968). *Lipids* **3**, 379–381.

Audier, H. E., Bégué, J. P., Cadiot, P., and Fétizon, M. (1967). *Chem. Commun.* 200–201.

Audier, H. E., Bory, S., Fétizon, M., Longevialle, P. and Toubiana, R. (1964). *Bull. Soc. chim. fr.* 3034–3035.

Bagby, M. O., Siegl, W. O. and Wolff, I. A. (1965). *J. Am. Oil Chem. Soc.* **42**, 50–53.

Baker, C. D. and Gunstone, F. D. (1963). *J. chem. Soc.* 759–760.

Barve, J. A. and Gunstone, F. D. (1971). *Chem. Phys. Lipids* **7**, 311–323.

Barve, J. A., Gunstone, F. D., Jacobsberg, F. R. and Winlow, P. (1972). *Chem. Phys. Lipids* **8**, 117–126.

Beerthuis, R. K., Nugteren, D. H., Pabon, H. J. J., Steenhoek, A. and van Dorp, D. A. (1971). *Recl. Trav. chim. Pays-Bas Belg.* **90**, 943–960.

Beroza, M. and Bierl, B. A. (1967). *Analyt. Chem.* **39**, 1131–1135.

Bitner, E. D., Davison, V. L. and Dutton, H. J. (1969). *J. Am. Oil. Chem. Soc.* **46**, 113–117.

Brooks, C. J. W., Gilbert, M. T. and Gilbert, J. D. (1973). *Analyt. Chem.* **45**, 896–902.

Campbell, I. M. and Naworal, J. (1969). *J. Lipid Res.* **10**, 589–592.

Capella, P. and Zorzut, C. M. (1968). *Analyt. Chem.* **40**, 1458–1463.

Cason, J., Fessenden, J. S. and Agre, C. L. (1959). *Tetrahedron* **7**, 289–298.

Chapman, D. (1965). "The Structure of Lipids by Spectroscopic and X-ray Techniques". Methuen, London.

Christie, W. W. (1970). *In* "Topics in Lipid Chemistry" (F. D. Gunstone, ed.), Vol. 1, pp. 1–49. Logos Press, London.

Christie, W. W., Gunstone, F. D., Ismail, I. A. and Wade, L. (1968). *Chem. Phys. Lipids* **2**, 196–202.

Christie, W. W. and Holman, R. T. (1966). *Lipids* **1**, 176–182.

Cobern, D., Hobbs, J. S., Lucas, R. A. and Mackenzie, D. J. (1966). *J. chem. Soc.* (*C*) 1897–1902.

Conacher, H. B. S. and Gunstone, F. D. (1969). *Chem. Phys. Lipids* **3**, 203–220.

Conacher, H. B. S. and Page, B. D. (1972). *J. Am. Oil Chem. Soc.* **49**, 283–286.

Conacher, H. B. S., Page, B. D. and Chadha, R. K. (1972). *J. Am. Oil Chem. Soc.* **49**, 520–523.

Cowie, J. S., Landor, P. D., Landor, S. R. and Punja N. (1972). *J. Chem. Soc.* (*Perkin Trans I*) 2197–2201.

Craig, B. M. and Bhatty, M. K. (1964). *J. Am. Oil Chem. Soc.* **41**, 209–211.

Davies, J. E. D., Hodge, P., Barve, J. A., Gunstone, F. D. and Ismail, I. A. (1972). *J. chem. Soc.* (*Perkin Trans. II*) 1557–1561.
Davison, V. L. and Dutton, H. J. (1966). *Analyt. Chem.* **38**, 1302–1305.
Dauben, W. G., Hoerger, E. and Peterson, J. W. (1953). *J. Am. chem. Soc.* **75**, 2347–2351.
de Bruijn, J., Frost, D. J., Nugteren, D. H., Gaudemer, A., Lijmbach, G. W. M., Cox, H. C. and Berends, W. (1973). *Tetrahedron* **29**, 1541–1547.
de Haan, J. W. and van de Ven, L. J. M. (1971). *Tetrahedron Lett.* 3965–3968.
de Jong, K. (1967). *Fette, Seifen, Anstrichmittel* **69**, 277–279.
Dinh-Nguyen, N. and Raal, A. (1969). *Acta chem. scand.* **23**, 1442–1443.
Dinh-Nguyen, N., Ryhage, R. and Ställberg-Stenhagen, S. (1960). *Arkiv. Kemi.* **15**, 433–438.
Downing, D. T. and Greene, R. S. (1968). *Lipids* **3**, 96–100.
Eglinton, G. and Hunneman, D. H. (1968). *Phytochemistry* **7**, 313–322.
Eglinton, G., Hunneman, D. H. and McCormick, A. (1968). *Org. Mass Spectrometry* **1**, 593–611.
Frost, D. J. and Barzilay, J. (1971). *Analyt. Chem.* **43**, 1316–1318.
Galliard, T., Phillips, D. R. and Frost, D. J. (1973). *Chem. Phys. Lipids* **11**, 173–180.
Grogan, W. M. Jr., Coniglio, J. G. and Rhamy, R. K. (1973). *Lipids* **8**, 480–482.
Gunstone, F. D. (1973). *J. Am. Oil Chem. Soc.* **50**, 486A–490A.
Gunstone, F. D. and Inglis, R. P. (1971). *In* "Topics in Lipid Chemistry" (F. D. Gunstone, ed.), Vol. **2**, pp. 287–307. Paul Elek (Scientific Books), London.
Gunstone, F. D. and Inglis, R. P. (1973). *Chem. Phys. Lipids* **10**, 73–88; 89–104; 105–113.
Gunstone, F. D. and Jacobsberg, F. R. (1972). *Chem. Phys. Lipids* **9**, 26–34.
Gunstone, F. D. and Lie Ken Jie, M. (1970). *Chem. Phys. Lipids* **4**, 131–138 and 139–146.
Gunstone, F. D., Lie Ken Jie, M. and Wall, R. T. (1971). *Chem. Phys. Lipids* **6**, 147–151.
Gunstone, F. D. and Perera, B. S. (1973). *Chem. Phys. Lipids* **10**, 303–308.
Gunstone, F. D. and Sykes, P. J. (1962). *J. chem. Soc.* 3055–3058.
Hamberg, M. (1971). *Analyt. Biochem.* **43**, 515–526.
Hinchcliffe, P. R. and Riley, J. P. (1971). *J. Am. Oil Chem. Soc.* **48**, 514.
Holman, R. T. and Rahm, J. J. (1966). *Progr. Chem. Fats Lipids* **9**, 15–90.
Hopkins, C. Y. (1965). *Progr. Chem. Fats Lipids* **8**, 213–252
Hopkins, C. Y. (1972). *In* "Topics in Lipid Chemistry" (F. D. Gunstone, ed.), Vol. **3**, pp. 37–87. Paul Elek (Scientific Books), London.
Hopkins, C. Y. and Chisholm, M. J. (1965). *J. chem. Soc.* 907–910.
Hunter, G. D. and Popják, G. (1951). *Biochem. J.* **50**, 163–168.
Jamieson, G. R. (1970). *In* "Topics in Lipid Chemistry" (F. D. Gunstone, ed.), Vol. **1**, pp. 107–159. Logos Press, London.
Johnston, A. E. and Dutton, H. J. (1972). *J. Am. Oil Chem. Soc.* **49**, 98–100.
Kenner, G. W. and Stenhagen, E. (1964). *Acta chem. scand.* **18**, 1551–1552.
Keppler, J. G. (1957). *Recl. Trav. chim. Pays-Bas Belg.* **76**, 49–57.
Kleiman, R. and Spencer, G. F. (1973). *J. Am. Oil Chem. Soc.* **50**, 31–38.
Kleiman, R., Spencer, G. F., Earle, F. R. and Wolff, I. A. (1967). *Chemy. Ind.* 1326–1327.
Kleiman, R., Spencer, G. F., Earle, F. R. and Wolff, I. A. (1969). *Lipids* **4**, 135–141.
Kunau, W-H. (1973). *Chem. Phys. Lipids* **11**, 254–269.
Kusamran, K. and Polgar, N. (1971). *Lipids* **6**, 961–962.
Lough, A. K. (1973). *Progr. Chem. Fats Lipids* **14**, 1–48.
McCloskey, J. A. (1970). *In* "Topics in Lipid Chemistry" (F. D. Gunstone, ed.), Vol. **1**, pp. 369–440. Logos Press, London.

McCloskey, J. A. and Law, J. H. (1967). *Lipids* **2**, 225–230.
McCloskey, J. A. and McClelland, M. J. (1965). *J. Am. chem. Soc.* **87**, 5090–5093.
Maerker, G. and Haeberer, E. T. (1966). *J. Am. Oil Chem. Soc.* **43**, 97–100.
Minnikin, D. E. (1972). *Lipids* **7**, 398–403.
Miwa, T. K., Mikolajczak, K. L., Earle, F. R., and Wolff, I. A. (1960). *Analyt. Chem.* **32**, 1739–1742.
Morandi, J. R. and Jensen, H. B. (1969). *J. org. Chem.* **34**, 1889–1891.
Morris, L. J. (1966). *J. Lipid Res.* **7**, 717–732.
Murray, K. E. (1959). *Aust. J. Chem.* **12**, 657–670.
Murray, K E. (1962) *Aust. J. Chem.* **15**, 510–520.
Nickell, E. C. and Privett, O. S. (1966). *Lipids* **1**, 166–170.
Nickell, E. C. and Privett, O. S. (1969). Paper presented to the American Oil, Chemists' Society, Minneapolis. *J. Am. Oil Chem. Soc.* **46**, Abstract 122.
Niehaus, W. G. Jr. and Ryhage, R. (1967). *Tetrahedron Lett.* 5021–5026.
Niehaus, W. G. Jr. and Ryhage, R. (1968). *Analyt. Chem.* **40**, 1840–1847.
Osbond, J. M. (1966). *Progr. Chem. Fats Lipids* **9**, 121–157.
Perkins, E. G. and Argoudelis, C. J. (1969). *Lipids* **4**, 619–621.
Privett, O. S. (1966). *Progr. Chem. Fats Lipids* **9**, 91–117.
Privett, O. S., Blank, M. L. and Romanus, O. (1963). *J. Lipid Res.* **4**, 260–265.
Privett, O. S. and Nickell, E. C. (1962). *J. Am. Oil Chem. Soc.* **39**, 414–419.
Privett, O. S. and Nickell, E. C. (1966). *Lipids* **1**, 98–103.
Promé, J-C. (1968). *Bull. Soc. Chim. Fr.* 655–660.
Promé, J-C. and Asselineau, C. (1966). *Bull. Soc. Chim. Fr.* 2114–2115.
Pryde, E. H. and Cowan, J. C. (1971). *In* "Topics in Lipid Chemistry" (F. D. Gunstone, ed.), Vol. **2**, pp. 1–98. Logos Press, London.
Ramachandran, S., Rao, P. V. and Cornwell, D. G. (1968a). *J. Lipid Res.* **9**, 137–139.
Ramachandran, S., Sprecher, H. W. and Cornwell, D. G. (1968b). *Lipids* **3**, 511–618.
Roehm, J. N. and Privett, O. S. (1969). *J. Lipid Res.* **10**, 245–246.
Scholfield, C. R. and Dutton, H. J. (1971). *J. Am. Oil Chem. Soc.* **48**, 228–231.
Serck-Hanssen, K. (1958). *Chemy Ind.* 1554.
Smentowski, F. J. and Stipanovic, R. D. (1972). *J. Am. Oil Chem. Soc.* **49**, 48–51.
Smith, C. R. Jr. (1970). *In* "Topics in Lipid Chemistry" (F. D. Gunstone, ed.), Vol. **1**, pp. 277–368. Logos Press, London.
Smith, C. R. Jr., Kleiman, R. and Wolff, I. A. (1968). *Lipids* **3**, 37–42.
Spencer, G. F., Kleiman, R., Earle, F. R. and Wolff, I. A. (1969). *Analyt. Chem.* **41**, 1874–1876.
Sprecher, H. W., Maier, R., Barber, M. and Holman, R. T. (1965). *Biochemistry* **4**, 1856–1863.
Stein, R. A. and Nicolaides, N. (1962). *J. Lipid Res.* **3**, 476–478.
Steinberg, G., Slaton, W. H., Howton, D. R. and Mead, J. F. (1956). *J. biol. Chem.* **220**, 257.
Stoffel, W. (1965). *J. Am. Oil Chem. Soc.* **42**, 583–587.
Tulloch, A. P. (1964). *J. Am. Oil Chem. Soc.* **41**, 833–836.
Tulloch, A. P. and Craig, B. M. (1964). *J. Am. Oil Chem. Soc.* **41**, 322–326.
Tulloch, A. P. and Mazurek, M. (1973). *Chem. Commun.* 692–693.
Vetter, W., Walther, W. and Becchi, M. (1971). *Helv. chim. Acta* **54**, 1599–1605.
Von Rudloff, E. (1956). *J. Amer. Oil Chem. Soc.* **33**, 126–128.
Wineburg, J. P. and Swern, D. (1972). *J. Am. Oil Chem. Soc.* **49**, 267–273.
Wineburg, J. P. and Swern D. (1973). *J. Am. Oil Chem. Soc.* **50**, 142–146.
Wolff, I. A. and Miwa, T. K. (1965). *J. Am. Oil Chem. Soc.* **42**, 208–215.
Zeman, A. and Scharmann, H. (1972). *Fette, Seifen Anstrichmittel* **74**, 509–519.
Zeman, A. and Scharmann, H. (1973). *Fette, Seifen Anstrichmittel* **75**, 32–44.

CHAPTER 3

Fatty Acid Biosynthesis

J. L. HARWOOD

Department of Biochemistry, University College, Cardiff, Wales

I. Introduction

Interest in plant fatty acid biosynthesis has expanded rapidly in recent years as a result both of technical advances and the application to plants of discoveries from bacterial or mammalian systems. We now know of several pathways for fatty acid formation and the whole picture is much more complex than was first surmised.

This review will attempt to describe the major pathways but with little emphasis on the specific role of the chloroplast, a subject reviewed separately in Chapter 4 of this volume. Several other relevant articles, to which I am indebted, have appeared recently (Hitchcock and Nichols, 1971; Mazliak, 1973; Sedgwick, 1973; Stumpf, 1969, 1970).

During germination and growth, different plants have many features of fatty acid metabolism in common. For example, the first acids into which radioactive acetate is incorporated by germinating seeds are always saturated; isolated chloroplasts generally synthesize palmitate and oleate readily; and *trans*-3-hexadecenoic acid is a ubiquitous fatty acid of higher plant chloroplasts. It is only during fruit and seed development that marked differences between species appear and the neutral lipids of seed oils are the source of the more unusual fatty acid types (see Chapter 1). This point is emphasized in Table I which shows the fatty acid composition of leaf tissue and seeds from a number of different plants.

TABLE I

Comparison of fatty acid composition of some common leaves and seeds [Adapted from Hitchcock and Nichols (1971) and Stumpf (1969)]

Plant tissue	Fatty acid composition (% total)						
	16:0	16:1[a]	18:0	18:1	18:2	18:3	Others
Leaves							
pasture grass	16	3	2	3	13	61	2
stinging nettle	12	2	1	2	15	66	2
spinach	13	3	tr	7	16	56	5
broad bean	12	7	3	3	14	56	5
holly	22	tr	tr	3	14	60	1
Seeds							
safflower	—	—	7	1	16	76	—
castor bean	—	—	—	2	7	1	87[b]
rapeseed	4	tr	1	13	7	7	59[c]
wheat	22	—	1	10	63	3	1

[a] 16:1(9*c*) + 16:1(3*t*).
[b] Ricinoleic acid.
[c] Mainly erucic acid.

II. Acyl CoA Synthesis

A. ACETYL CoA

It was observed many years ago that (compared with formate, glucose or butyrate) acetate was the most effective precursor of fatty acids in germinating or developing seeds of peanuts (Newcomb and Stumpf, 1953). The C_3 homologue, propionate, is poorly incorporated into lipids (Hatch and Stumpf, 1962).

Although acetyl CoA is the end product of β-oxidation of fatty acids (p. 332), this process obviously cannot be regarded as a normally important source of acetate for synthesis. One key enzyme for this is acetyl thiokinase which was shown to be present in the soluble fraction of homogenates from germinating peanut cotyledons (Rebeiz *et al.*, 1965). Recently, a thiokinase from potato tubers has been partially purified (Huang and Stumpf, 1970). The initial homogenate was subjected to $MnCl_2$ and $(NH_4)_2SO_4$ precipitations, gel filtration, DEAE-cellulose chromatography and hydroxyapatite column chromatography to yield up to 40-fold purification with high recovery (38%). Five peaks of acetyl CoA synthetase activity were isolated after DEAE-cellulose chromatography. All had molecular weights close to 59 500. The multiple acetyl CoA synthetases were similar in substrate specificity, pH optimum and response to AMP inhibition. The major form of the enzyme was studied further by Cleland kinetics using initial velocities and product inhibition. The mechanism was found, like the yeast enzyme (Berg, 1956), to follow Iso-Bi-Uni-Uni-Bi Ping-Pong kinetics. The reaction can be simply written thus:

$$\text{acetate} + \text{ATP} \longrightarrow \text{acetyl AMP} + \text{PP}_i$$

$$\text{acetyl AMP} + \text{CoA} \longrightarrow \text{acetyl CoA} + \text{AMP}$$

B. MALONYL CoA

Malonyl CoA, the key intermediate in fatty acid biosynthesis, is formed from acetyl CoA by carboxylation. When isolated from liver (Numa *et al.*, 1966) or yeast (Matsuhashi *et al.*, 1964), the carboxylating enzyme is a tight complex which cannot be dissociated into active subunits. The enzyme from *E. coli*, on the other hand, easily separates into three protein fractions: one sub-unit (biotin carboxylase, E_1) catalyses the ATP-dependent carboxylation of added (+)biotin; another smaller sub-unit (biotin carboxyl-carrier protein, BCCP, E_2) contains covalently bound biotin which is the normal substrate for biotin carboxylase: the third (transcarboxylase, E_3), free of biotin, catalyses the transfer of CO_2 from biotin to acetyl CoA (see Volpe and Vagelos, 1973):

$$ATP + HCO_3^- + BCCP \xrightarrow{\textit{biotin carboxylase}} CO_2^- \text{-}BCCP + ADP + P_i$$

$$CO_2^- \text{-}BCCP + \text{acetyl CoA} \xrightarrow{\textit{transcarboxylase}} BCCP + \text{malonyl CoA}$$

Carboxylation of acetyl CoA has been studied in a number of plant tissues, principally by Stumpf and his co-workers. Hatch and Stumpf (1961) described the distribution, isolation and purification of the enzyme. Using wheat germ, they purified a preparation some 150-fold. It was sensitive to avidin and required free -SH groups for activity. Although acetyl CoA was the preferred substrate, propionyl CoA and butyryl CoA were also carboxylated. Barron *et al.* (1961) demonstrated the carboxylase in avocado mesocarp and it has also been shown in seeds of bean, castor bean, pea, peanut, safflower and soyabean, leaves of barley, alfalfa, bean, clover and spinach and in potato tubers (Hatch and Stumpf, 1962).

The wheat germ carboxylase has now been purified 1000-fold (Heinstein and Stumpf, 1969) and two protein components, of 7·4S and 9·5S, revealed by analytical ultracentrifugation. The biotin carboxylation and transcarboxylase steps appear to be catalysed by the 9·5S and 7·4S components respectively. Avidin was found to bind to the 9·5S protein, increasing its sedimentation coefficient to 11·5S. A single peak of active enzyme obtained on gel filtration revealed five bands after disc gel electrophoresis. Two of the bands contained biotin and catalysed ATP–^{32}Pi exchange and these were deduced to represent monomers and dimers of a carboxylase+BCCP complex. The transcarboxylase was present in two further bands as a monomer or a tetramer and the fifth band represented the active enzyme complex. The total reaction is summarized in Fig. 1.

Because it is only partly dissociable (to two components of similar charge) the wheat germ acetyl CoA carboxylase appears to occupy an intermediate position between those of animals and bacteria.

Isolated chloroplasts, which incorporate acetate efficiently into fatty acids, must contain an active acetyl CoA carboxylase. However, attempts to study the enzyme from this source were thwarted at first due to the release of an inhibitor when the organelles were disrupted (Burton and Stumpf, 1966). The heat-stable inhibitor markedly depressed the carboxylase activity without affecting fatty acid synthesis from malonyl CoA. The inhibitor was present

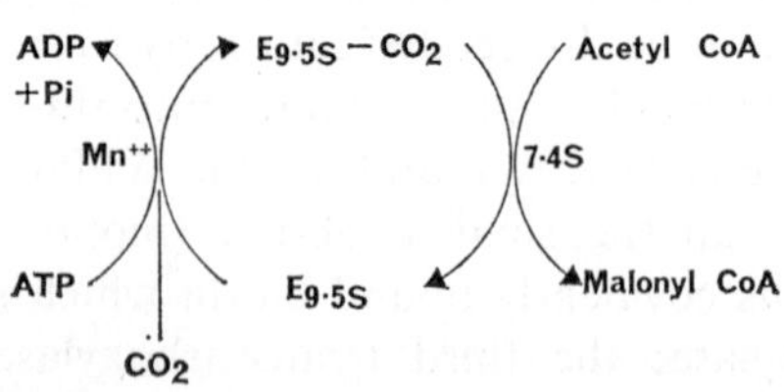

FIG. 1. The action of acetyl CoA carboxylase in plants. [Adapted from Hitchcock and Nichols (1971) and Stumpf (1970).]

in stromal and lamellar fractions of lettuce chloroplasts and was also found in avocado mesocarp (Harwood and Stumpf, 1972a).

Leaf chloroplasts, which have some procaryotic characteristics contain an acetyl-CoA carboxylase system similar to that described for *E. coli*. Using spinach chloroplasts, Kannangara and Stumpf (1972b) found that, whereas $^{14}CO_2$ was bound to a protein by intact chloroplasts, no such binding to isolated stromal or lamellar proteins occurred. Further work revealed that BCCP was present in the lamellar membranes but that biotin carboxylase and the transcarboxylase were in the stroma. Thus, in contrast to wheat germ, three protein components could be separated from chloroplasts (see Chapter 4).

The biotinyl protein (BCCP) has been studied in a number of eucaryotic organisms (Kannangara and Stumpf, 1973). In higher plant chloroplasts and in *Chlamydomonas reinhardii* it is membrane-bound, in the blue-green alga *Anacystis nidulans* it showed an intermediate condition, while the filamentous blue-green alga *Anabaena flos-aquae* resembled higher plant chloroplasts. On a protein basis, BCCP was evenly distributed between stroma and grana membranes. In dormant embryos of barley and wheat germ, acetyl CoA carboxylase was a soluble protein localized in the proplastids but on germination, the BCCP component became associated with the lamellar membranes.

Although wheat germ carboxylase is intermediate in structure between bacterial and animal carboxylases, it is not under allosteric control. Burton and Stumpf (1966) tested over thirty compounds without obtaining any increase in activity. In fact, concentrations of isocitrate above 4 mM were actually inhibitory (Heinstein and Stumpf, 1969) and only sulphydryl compounds caused any activation. Thus, in their lack of stimulation by citrate and isocitrate, the two types of plant acetyl CoA carboxylase resemble that from *E. coli* (Alberts and Vagelos, 1968) and differ from the mammalian enzyme.

Malonyl CoA may also be formed directly from malonate by a thiokinase which has been demonstrated in a variety of plant tissues including seeds, fruits, leaves and roots (Hatch and Stumpf, 1962). The enzyme from avocado mesocarp is probably soluble (Yang and Stumpf, 1965a). Malonate is formed by oxidative carboxylation of oxaloacetate in bush bean roots (Shannon *et al.*, 1963) and the various other potential sources of this and acetyl CoA in chloroplasts have been discussed by Stumpf *et al.* (1967).

C. LONG CHAIN ACYL CoA

The thiokinases responsible for activation of long chain acids have been reported in a number of animal, plant and microbial systems. In animals, several thiokinases have been differentiated on the basis of their substrate specificities and another utilizes GTP rather than ATP as energy source (Rossi *et al.*, 1969).

In plants, long chain fatty acid thiokinase is associated with microsomal membranes and has been reported in avocado mesocarp (Barron and Stumpf, 1962) and castor bean endosperm (Yamada and Stumpf, 1965). A method for synthesizing palmityl CoA using the plant enzyme has also been reported by Galliard and Stumpf (1968).

III. *DE NOVO* Fatty Acid Synthesis

A. FATTY ACID SYNTHETASE

Synthesis of saturated fatty acids *de novo* from malonyl CoA is catalysed by fatty acid synthetase. The stoichiometry of the reaction is as follows:

$$CH_3CO\text{-}SCoA + 7\ HOOC\text{-}CH_2CO\text{-}SCoA + 14\ NADPH + 14\ H^+$$
$$CH_3(CH_2)_{14}COOH + 7\ CO_2 + 14\ NADP^+ + 8\ CoAS\text{-}H + 6\ H_2O$$

Lynen (1961) proposed that the intermediates in fatty acid biosynthesis were bound via a thioester linkage to the individual enzymes of the yeast complex. Experiments with the easily dissociable bacterial synthetase from *E. coli* confirmed this and shows that a central component was acyl carrier protein (ACP—see below). The individual reactions were also demonstrated separately (Majerus and Vagelos, 1967) and shown to represent seven enzyme steps thus (E_{cond} = condensing enzyme):

1. $CH_3CO\text{-}SCoA + HS\text{-}ACP \rightleftharpoons CH_3CO\text{-}S\text{-}ACP + CoA\text{-}SH$
   acetyl CoA—ACP transacylase
2. $CH_3CO\text{-}S\text{-}ACP + HS\text{-}E_{cond} \rightleftharpoons CH_3CO\text{-}S\text{-}E_{cond} + ACP\text{-}SH$
   acetyl transferase
3. $HOOC\text{-}CH_2CO\text{-}S\text{-}CoA + HS\text{-}ACP \rightleftharpoons HOOC\text{-}CH_2CO\text{-}S\text{-}ACP + CoA\text{-}SH$
   malonyl CoA-ACP transacylase
4. $HOOC\text{-}CH_2CO\text{-}S\text{-}ACP + CH_3CO\text{-}S\text{-}E_{cond} \longrightarrow CO_2 + HS\text{-}E_{cond} + CH_3COCH_2CO\text{-}S\text{-}ACP$
   condensing enzyme
5. $CH_3COCH_2CO\text{-}S\text{-}ACP + NADPH + H^+ \rightleftharpoons NADP^+ + CH_3CHOHCH_2\text{-}S\text{-}ACP$
   β-ketoacyl-ACP reductase
6. $CH_3CHOHCH_2CO\text{-}S\text{-}ACP \rightleftharpoons H_2O + CH_3CH{=}CHCO\text{-}S\text{-}ACP$
   β-hydroxyacyl-ACP dehydrase
7. $CH_3CH{=}CH\text{-}CO\text{-}S\text{-}ACP + NADPH + H^+ \rightleftharpoons CH_3CH_2CH_2CO\text{-}S\text{-}ACP + NADP^+$
   enoyl-ACP reductase

Steps 4–7 are repeated, further malonyl ACP being incorporated, until palmitic acid (16:0) has been synthesized. Termination then occurs under the action of palmityl ACP thioesterase. The actual mechanism of termination is not established at present and the matter is complicated by the varying

TABLE II

Properties of fatty acid synthetase systems

Source	Molecular weight	Products		Reference
Type I				
yeast	$2 \cdot 3 \times 10^4$	14:0–18:0	CoA ester	adapted from Brindley *et al.* (1969)
pigeon liver	$4 \cdot 5 \times 10^5$	16:0	free acid	adapted from Brindley *et al.* (1969)
rat liver	$5 \cdot 4 \times 10^5$	16:0	free acid	adapted from Brindley *et al.* (1969)
adipose tissue	—	14:0–16:0	free acid	adapted from Brindley *et al.* (1969)
mammary gland	—	8:0–18:0	free acid	adapted from Brindley *et al.* (1969)
Euglena gracilis (etiolated)	1×10^6	16:0	CoA ester	adapted from Brindley *et al.* (1969)
Mycobacterium phlei	$1 \cdot 7 \times 10^6$	14:0–26:0	esters	adapted from Brindley *et al.* (1969)
Type II				
E. coli	—	16:0, 18:0, 16:1	ACP esters	adapted from Brindley *et al.* (1969)
Clotridium sp.	—	18:1, 3-OH-10:0	ACP esters	adapted from Brindley *et al.* (1969)
Pseudomonas sp.	—	3-OH-12:0, 3-OH-14:0	ACP esters	adapted from Brindley *et al.* (1969)
Bacillus subtelis	—	C_{15} and C_{17} *iso*acids and *anteiso*acids		adapted from Brindley *et al.* (1969)
avocado mesocarp	—	16:0	ACP ester	Harwood and Stumpf (1972b)
potato tuber	—	8:0–16:0	free acids + ACP esters	Huang and Stumpf (1971)
pea seed	—	16:0	—	Harwood and Stumpf (1971)
Euglena gracilis (photoauxotrophic)	$3 \cdot 6 \times 10^5$	18:0–22:0	ACP ester	Ernst-Fonberg (1973)

products of different synthetases (Table II) and the possibility of direct utilization of palmityl-*S*-ACP for phospholipid biosynthesis, at least in bacteria (Goldfine *et al.*, 1967). More detailed discussion of fatty acid synthetase and its termination reactions was given by Hitchcock and Nichols (1971), Wakil (1970) and Volpe and Vagelos (1973).

B. ACYL CARRIER PROTEIN

Acyl carrier protein (ACP), which occupies a key position in fatty acid synthesis, has been purified from avocado mesocarp and spinach plastids (Brooks and Stumpf, 1966; Simoni *et al.*, 1967; Simoni and Stumpf, 1969).

The first demonstration that ACP was required for fatty acid synthesis in plants was the observation by Overath and Stumpf (1964) that fatty acid biosynthesis, catalysed by a soluble extract of avocado mesocarp, was greatly stimulated by the addition of boiled, crude extracts. The heat-stable substance was destroyed by papain and could be replaced by a similar fraction from *E. coli*, later shown to be ACP.

Purification of plant ACP has been achieved by methods analogous to those used for *E. coli* (Simoni *et al.*, 1967). The molecular weights were 11 400 and 10 300 for avocado and spinach ACP respectively. Amino acid compositions for these two ACPs were compared to those of the bacteria, *E. coli* and *Arthrobacter viscosus*. The two plant ACPs were very similar to the bacterial proteins except that spinach ACP lacked arginine and tyrosine and both contained twice as many lysine residues. Now the complete amino acid sequence for *E. coli* ACP is known (Vanaman *et al.*, 1968) the comparative results become even more interesting. In cross-over studies using fatty acid synthetases and ACP from *E. coli* and avocado it was found that the bacterial ACP was more effective in both systems. In addition, when plant ACP was used with *E. coli* synthetase, a whole series of intermediates accumulated, indicating that β-hydroxyacyl-(plant)ACP derivatives were ineffective substrates for the β, γ-dehydrase and the α,β-dehydrase enzymes (see Table III). Stumpf (1969) has suggested that tyrosine and arginine residues are not essential but the lysine content may have a modifying effect on the activities of plant ACPs. In support of this view, nitration of the single tyrosine at position 71 of *E. coli* ACP did not alter fatty acid synthesis in this organism (Abita *et al.*, 1971). However, replacement of lysine residues by norleucine did not affect activity in *E. coli* (Volpe and Vagelos, 1973). The importance of arginine at position 6 in *E. coli* ACP is difficult to gauge since the amino-end of the ACP chain seems to be particularly important in the conformation of the protein (Hancock *et al.*, 1973).

Matsumura and Stumpf (1968) compared the amino acid sequence of spinach ACP to those of *E. coli* and *Arthobacter*. The six to seven residues surrounding the prosthetic group (attached to serine at position 36 in *E. coli* ACP) were identical but the amino terminal residue was serine in both

TABLE III

Comparison of products of synthesis by spinach or *E. coli* fatty acid synthetases
[From Simoni *et al.* (1967)]

Synthetase source	Products	% Fatty acid formed (% of total products)	
		APC source: spinach	*E. coli*
Spinach			
	16:0	20	0
	18:0	80	100
E. coli			
	3-OH-12:0	10	0
	3-OH-12:1(5)	6	0
	3-OH-14:0	8	0
	3-OH-14:1(7)	5	0
	3-OH-16:0	10	0
	3-OH-16:1(9)	2	0
	3-OH-18:0	12	0
	3-OH-18:1(11)	6	0
	18:0	33	25
	18:1(11)	9	75

bacterial ACPs compared to alanine in spinach ACP. Because of the importance of the first six residues in the activity of *E. coli* ACP (Majerus, 1968), differences in this region may account for much of the alteration in fatty acid patterns yielded by mixing plant ACPs with *E. coli* fatty acid synthetase (Table III). It would be most profitable to determine the amino acid sequence of a plant ACP to define these interesting differences at a molecular level.

C. *De novo* SYNTHESIS IN PLANTS

In many fatty acid synthesizing systems from plants, ACP is a necessary addition *in vitro*. Table IV gives some examples of the stimulation of fatty acid biosynthesis observed on adding *E. coli* ACP to various subcellular fractions from plants. Such results indicate that in these cases the synthetase is of the bacterial type. Although individual reactions of fatty acid synthesis have not been demonstrated in plants, there is no reason to doubt that the usual condensation–reduction–dehydration–reduction sequence occurs. In view of the possible vital role of the C_{10}-acyl intermediates in the formation of unsaturated fatty acids (see later), examination of individual enzymes of plant fatty acid synthetase is an important and pressing problem.

Demonstration of separate fatty acid synthetase and elongation systems for saturated fatty acid synthesis in germinating pea seed (Harwood and

TABLE IV

Stimulation of fatty acid synthesis *in vitro* by ACP
[From Harwood and Stumpf (1972a), Hawke and Stumpf (1965b), Macey and Stumpf (1968), Overath and Stumpf (1964)]

System used		Substrate	Stimulation factor
castor bean seed	fat fraction	malonyl CoA	7·5
	particle-free supernatant	malonyl CoA	4·4
pea seed	microsomes	malonyl CoA	1·7
	particle-free supernatant	malonyl CoA	9·8
avocado mesocarp	homogenate	malonyl CoA	37
	particle-free supernatant	malonyl CoA	87
barley seedling	chloroplast	malonyl CoA	2
barley seedling	chloroplast	acetate	1·2
broad bean	chloroplast	acetate	1·3

Stumpf, 1971) and avocado mesocarp (Harwood and Stumpf, 1972b) has served to highlight much earlier work with plant systems. Thus, most preparations used produce a mixture of fatty acid products because they contain elongation or desaturation systems as well as fatty acid synthetase. This point is most important when examining results obtained with plant preparations and palmitic acid should be regarded as the end product of higher plant fatty acid synthetase. Labelling studies have frequently confirmed

TABLE V

Fatty acids produced from ^{14}C-acetate or ^{14}C-malonyl-CoA with different preparations from plants
[From Harwood and Stumpf (1971, 1972a), Hawke and Stumpf (1965b) and Macey and Stumpf (1968)]

System	Precursor	Incorporation (% total radio-activity in fatty acids)			
		16:0	18:0	18:1	others
barley seedling (chloroplast)	acetate	36	31	27	6
	malonyl CoA	46	54	—	—
pea seed (slices)	acetate	16	3	31	50
	malonyl CoA	19	0	19	62
pea seed (supernatant)	acetate	43	57	—	2
	acetyl CoA	45	53	—	2
	malonyl CoA	20	75	—	5
castor bean seed (fat fraction)	acetate	7	22	71	0
	malonyl CoA	3	48	44	5
castor bean seed (homogenate)	acetate	17	18	65	0
	malonyl CoA	5	53	37	5

that palmitate is made *de novo* whereas stearate is made by elongation (e.g. Harwood and Stumpf, 1972b; Hawke and Stumpf, 1965a).

Subcellular organelles have often been used in these studies with ^{14}C-acetate as substrate. Although "physiological", this approach has the disadvantage that synthesis from acetate is dependent on the activity of acetyl CoA carboxylase, which may be absent, rate-limiting or subject to inhibition (see p. 46). ^{14}C-malonyl CoA generally gives synthesis rates for unsaturated fatty acids at least as good as with ^{14}C-acetate but, because of better incorporation overall, may give a higher pecentage yield of saturated fatty acids. In a recent study of the subcellular distribution of biosynthetic enzymes in a number of plants, Harwood and Stumpf (1972a) used both precursors thus attempting to minimize these problems. Comparison of the fatty acid patterns obtained with acetate or malonyl CoA substrates by different plant preparations is shown in Table V.

It is a common feature of plant systems that tissue extracts seldom produce a pattern of fatty acids related to the composition *in vivo*. Typical results are shown in Table VI. The lack of polyunsaturated fatty acid synthesis is a noticeable feature of plant extracts and few systems have been described where significant polyenoate formation takes place *in vitro*.

Two approaches have been used to study fatty acid biosynthesis in plants. In the first, changes in composition are followed in the whole plant, seed or leaf either with reference to total fatty acid composition or by following the incorporation of labelled precursors. Such an approach has yielded much

TABLE VI

Comparison of fatty acids synthesized by various preparations of plant tissue with the endogenous fatty acids

[From Stumpf (1970)]

Plant tissue	Preparation	Fatty acids (% total)					
		14:0	16:0	18:0	18:1	18:2	18:3
avocado mesocarp	*endogenous fatty acids*	—	*20*	*1*	*60*	*18*	—
acetate incorporation	slices	—	48	10	31	3	—
	10 000 *g* particles	—	50	30	20	1	—
	acetone powder from particles	—	40	50	0	0	—
	10 000 *g* supernatant	—	82	18	0	0	—
lettuce leaves	*endogenous fatty acids*	*1·7*	*14·0*	*0·8*	*0·5*	*6·6*	*71·2*
acetate incorporation	leaf tissue	4·8	41·5	3·9	41·8	8·0	—
	whole chloroplasts	1·0	24·0	—	73·0	—	—
	soluble fraction from chloroplasts	—	—	98	—	—	—

useful data on metabolic pathways and control but, naturally, can contribute little information on the properties of fatty acid synthetases. The second approach involves homogenization and subcellular fractionation. The active fractions are particularly deficient in unsaturated fatty acid biosynthesis (see Table VI) but fatty acid synthetase (i.e. the ability to synthesize palmitic acid *de novo*) seems to be relatively stable.

Following the observation that the crude mitochondrial fraction from avocado mesocarp would incorporate acetate into long chain fatty acids (Stumpf and Barber, 1957), Squires *et al.* (1958) prepared a water-soluble system which was sensitive to avidin (Squires and Stumpf, 1959). Barron *et al.* (1961) prepared acetone-dried powders of the same fraction to give a source of stable "synthetase" which converted acetate into palmitate and stearate in the presence of carbon dioxide, ATP, CoA, Mn^{2+} and NADPH. Only NADPH was required when acetyl CoA or malonyl CoA were used as substrates and the system was then avidin-insensitive. Whereas the particles could synthesize oleic acid, the solubilized enzymes could neither elongate added palmitate nor desaturate stearate. The mechanism of synthesis in these preparations, therefore, resembled the mammalian fatty acid synthetase except that the plant enzyme was derived from a particulate fraction.

Yang and Stumpf (1965a) studied the avocado system further and found that both the particulate and supernatant fractions could synthesize long-chain fatty acids from suitable substrates. There were differences, however, in the relative incorporation of different precursors (Table VII). From the data it appeared that:

(a) Acetyl CoA carboxylase activity was absent from the supernatant (see Section II.B).

(b) Malonate: CoA ligase was absent from the particles.

The supernatant fraction gave rise to palmitic and stearic acids and unsaturated fatty acid synthesis was associated with the particulate fraction. The proportion of stearate synthesized was increased when there was

TABLE VII

Synthesis of fatty acids by particulate and supernatant fractions of avocado mesocarp from ^{14}C-labelled precursors

[From Yang and Stumpf (1965a)]

Precursor	Incorporation into fatty acids (nmol)	
	Particulate	Supernatant
acetate	17·0	0·1
acetyl CoA	35·0	0·1
malonate	1·0	11·6
malonyl CoA	37·5	30·0

TABLE VIII

Effect of antibodies on fatty acid synthesis by subcellular fractions from avocado mesocarp

[From Harwood and Stumpf (1972a)]

Fraction	Fatty acid synthesis activity (% of serum-free controls)		
Serum added:	Normal	Anti-lamellae	Anti-stroma
chloroplast lamellae	104	25	114
chloroplast stroma	103	98	0
105 000 *g* supernatant	105	102	52

a high malonyl CoA:acetyl CoA ratio. This agrees with results in mammalian and yeast systems where the malonyl CoA concentration can alter the pattern of fatty acids (Sumper *et al.*, 1969; Carey *et al.*, 1970). Requirement for NADPH and stimulation of synthesis by an added boiled fraction (shown later to be ACP) were demonstrated by Overath and Stumpf (1964). They also showed that avocado ACP and *E. coli* ACP were interchangeable in the system with little alteration in fatty acids produced although better incorporation was obtained using the bacterial ACP.

More recently, the avocado particulate and supernatant fractions have been studied further. Harwood and Stumpf (1972a) demonstrated the existence of immunologically distinct synthesizing systems in the chloroplast lamellar and supernatant fractions. The latter contained activity derived both from the cytosol and the proplastid stroma. Weaire and Kekwick (1973) examined the synthesis by stroma alone and found stearic acid to be the sole product, in keeping with many previous observations using chloroplast soluble fractions (e.g. Brooks and Stumpf, 1966; Stumpf *et al.*, 1967). In contrast, the 105 000 *g* supernatant which contains stromal and cytoplasmic proteins, synthesizes palmitic as well as stearic acid (Overath and Stumpf, 1964; Yang and Stumpf, 1965a; Harwood and Stumpf, 1972a). It would appear that fatty acid synthetase of avocado mesocarp originates in the cytoplasm. Immunological experiments indicated that the stromal system differed from that of the cytoplasm. These results are shown in Table VIII. Therefore three antigenically distinct fatty acid synthesizing systems are present in avocado.

Elongation of palmitate to stearate has been found to be sensitive to arsenite inhibition, whereas fatty acid synthetase is unaffected (Harwood and Stumpf, 1971, 1972b; Huang and Stumpf, 1971). Whereas arsenite almost completely inhibits the avocado chloroplast stroma fraction, it only partially inhibits the 105 000 *g* supernatant. Similar results are obtained with antiserum (to the stroma) when it is added to the stromal and supernatant fractions (Table IX).

TABLE IX

Effect of arsenite and antisera on the pattern of fatty acids formed from ^{14}C-malonyl-CoA by avocado supernatant and plastid stroma fractions
[From Harwood (1974b)]

Fraction	c.p.m.	Fatty acids formed (% total)		
		16:0	18:0	Others
plastid stroma	10 130	3	97	tr
+ arsenite (3 mM)	3 755	5	95	tr
+ anti-stroma serum	0	0	0	0
supernatant	6 240	46	51	3
+ arsenite (3 mM)	3 860	84	10	6
+ anti-stroma serum	3 250	95	tr	5

In summary, the fatty acid synthetase of the avocado pear originates in the cytoplasm, uses malonyl CoA and NADPH to make palmitate and needs ACP for full activity. The synthetase can be separated from the elongating system by gel filtration (Harwood and Stumpf, 1972b). Fig. 2 shows the separation of fatty acid reaction products, protein-bound ^{3}H-ACP and

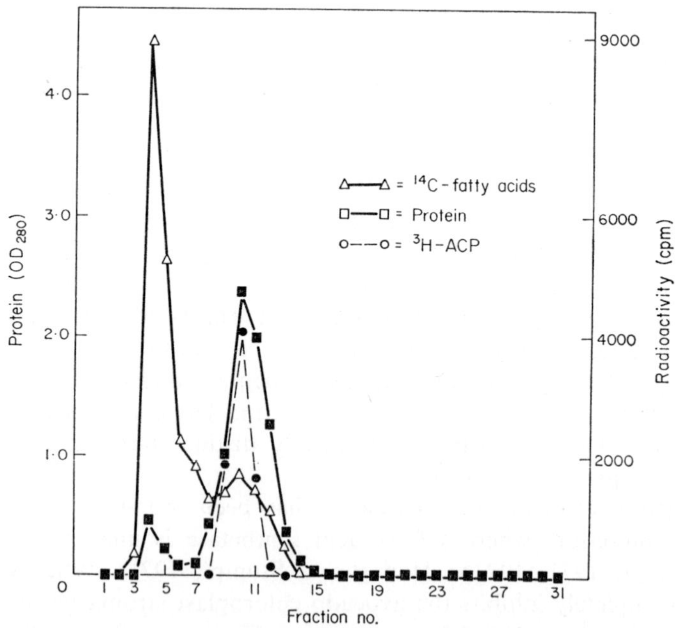

FIG. 2. Sephadex G-75 separation of the reaction products formed from ^{14}C-malonyl CoA by a supernatant fraction from avocado in the presence of ^{3}H-ACP. [From Harwood and Stumpf (1972b).]

protein. Two protein peaks were rapidly eluted from the column. The first, and smaller one, contained more of the fatty acid products, mainly stearate (65–80%). The second peak bound ^{3}H-ACP and contained palmitate (40–70%) and fatty acid thioesters (30%); it also retained the fatty acid synthetase activity of the original supernatant with palmitate as its sole product.

A purified soluble synthetase has been prepared from potato (Huang and Stumpf, 1971). The enzyme synthesizes saturated even-chain acids in the C_8–C_{16} range with 50–70% of the products present as free acids. ACP and CoA thioesters of the shorter chain acids also occurred (but no C_{16} thioesters), reflecting the specificity of a palmityl thiosterase contaminant of the preparation. Increasing the concentration of malonyl CoA or NADPH in the incubation increased the average chain length of product while variations of acetyl CoA concentration had no effect. At very high ACP concentrations more of the medium chain-length acids were produced. This was probably due to:

(a) competition between acetyl ACP and other acyl ACPs for the condensing enzyme.
(b) an increase in transferase activity converting fatty acyl ACP→fatty acyl CoA (favoured by high fatty acyl-ACP concentrations).

Although *de novo* synthesis of palmitate has been observed in many plant systems, no other synthetase has been isolated. Chloroplast preparations from various sources such as lettuce (Brooks and Stumpf, 1966) spinach (Givan and Stumpf, 1971; Kannangara and Stumpf, 1972a) pea (Panter and Boardman, 1973) and broad bean (Harwood, 1974a) synthesize palmitic acid from ^{14}C-acetate. Degradation studies on the product and cofactor requirements indicate that the reactions involved are those of fatty acid synthetase. However, the chloroplast system is more complex since oleic acid is also a typical major product. Other subcellular fractions which have been used include mitochondria (Mazliak and Ben Abdelkader, 1970; Mazliak *et al.*, 1968, 1972) and microsomes (Macey and Stumpf, 1968).

IV. Elongation

A. VERY LONG CHAIN FATTY ACIDS

Hawke and Stumpf (1965a) observed the incorporation of [1-^{14}C] acetate into the very long-chain acids of seedlings of the family Gramineae. About half of the radioactivity was present in the C_{20}–C_{24} acids even though these only represented minor component fatty acids. The labelling pattern indicated that, whereas palmitate was made *de novo*, stearate and the very long chain acids (in this case C_{20}–C_{26}) were synthesized by elongation of a C_{16} precursor. Macey and Stumpf (1968) observed similar results with slices from pea seedlings and presented evidence that the site of the elongation by malonyl CoA was the microsomal fraction. Germinating seeds incorporated significant amounts of ^{14}C-acetate or ^{3}H-H_2O into the very long chain fatty acids in spite of a low tissue concentration of these substances (Table X). Permanganate oxidation again indicated elongation at the C_{16} level (Harwood and Stumpf, 1970).

The pea system was examined further and results from inhibitor and heat-lability studies indicated at least two separate elongating enzymes thus:

$$C_{16} \xrightarrow[\text{arsenite-inhibited}]{\text{System I}} C_{18} \xrightarrow[\text{F}^{-}\text{inhibited, temperature-sensitive}]{\text{System II}} C_{20} \text{ etc.}$$

The former was active in the supernatant fraction whereas the latter was located in the microsomes (Macey and Stumpf, 1968; Harwood and Stumpf, 1971). The inhibition of microsomal formation of very long chain fatty acids at 35°C explained their lack of synthesis when seeds were germinated at this temperature. A simultaneous defective formation of waxes by the same seeds provided evidence in favour of Kolattukudy's proposals for elongation/decarboxylation as a mechanism for hydrocarbon formation (Kolattukudy, 1967; see also p. 212). Kolattukudy and Buckner (1972) have also used microsomes, in this case from pea leaves, to produce the very long chain fatty acids.

TABLE X

Fatty acids formed from ^{14}C-acetate by different seeds during germination at 30°C

[From Harwood and Stumpf (1970)]

Seed		Fatty acids labelled (% total)									
		16:0	18:0	18:1	18:2	18:3	20:0	22:0	24:0	26:0	others
Low lipid content											
pea	mass	17	3	15	56	5	—	—	—	—	4
	counts	26	62	—	—	—	—	3	6	3	tr
rice	mass	43	—	25	32	—	—	—	—	—	—
	counts	—		73 (18:0–18:2)		—	5	2	2	—	—
High lipid content											
broccoli	mass	5	tr	10	12	13	—	—	1	—	59
	counts	32	30	31	—	—	4	1	2	—	—
cotton	mass	32	tr	10	54	—	—	—	—	—	3
	counts	48	33	—	—	—	2	9	3	—	5
safflower	mass	16	1	7	76	—	—	—	—	—	—
	counts	25	tr	16	46	—	2	4	5	2	—
castor bean	mass	4	tr	6	8	—	—	—	—	—	82
	counts	32	—	17	33	—	tr	3	14	tr	1

TABLE XI

The effect of 10^{-3}M sodium trichloroacetate on ^{14}C-acetate incorporation into seed lipids

[From Harwood (1974a), Harwood and Stumpf (1971)]

Seed	Lipid fraction	Incorporation into lipid (% of trichloroacetate control)	
		Fatty acids	Total lipid
pea	total	105 ± 7	—
	surface	—	18
	non-surface	—	130
runner bean	surface	—	42
	non-surface	—	108

According to Kolattukudy (see Chapter 7, p. 212), the palmitic acid precursor is elongated by a trichloroacetate-sensitive system to a C_{30} fatty acid which is then decarboxylated to a C_{29} hydrocarbon. It was noted, however, that trichloroacetate did not interfere with palmitate formation. Harwood and Stumpf (1971) could find no effect of trichloroacetate on the incorporation of ^{14}C-acetate into fatty acid by germinating pea seeds even though significant amounts of the very long chain (C_{20}–C_{24}) fatty acids were formed. However, trichloroacetate does actually inhibit acetate incorporation into surface lipids as shown in Table XI, but this effect is not seen in the total fatty acids of germinating seeds since they represent only a minor portion. The lack of effect of trichloroacetate on non-surface lipids may be related to permeability problems.

B. STEARATE SYNTHESIS

Palmitate elongation activity has been found in a number of tissues and has been partially characterized in a soluble fraction from avocado (Harwood and Stumpf, 1972b). Inhibition by arsenite (reversible by dithiothreitol) indicated the presence of vicinal sulphydryl groups at or near the active centre (Harwood and Stumpf, 1971). Stearate was formed apparently by addition of malonyl CoA to a C_{16} precursor. Palmityl ACP was a far more effective precursor than palmitic acid or palmityl CoA. Phospholipids had little effect on the amount of stearate formed and phospholipid-bound fatty acids are probably not effective substrates. The involvement of malonyl CoA in stearate synthesis by soluble preparations of lettuce chloroplasts was also indicated by the work of Brooks and Stumpf (1966) and Burton and Stumpf (1966). In all these cases, the utilization of acetyl CoA for elongation was poor.

Recently, Weaire and Kekwick (1973) have partially purified the stromal elongation enzyme from avocado mesocarp. Their results agree with those of Harwood and Stumpf (1972b) in that malonyl CoA is the source of the C_2-unit. The incoporation of ^{14}C-malonyl CoA into stearate, catalysed by a 35%–65% $(NH4)_2SO_4$ fraction of the soluble enzyme, was stimulated by unlabelled acetyl CoA, ACP, $NADP^+$ and NADH. Harwood (1974b) measured stearate synthesis from palmityl CoA and reduced pyridine nucleotides, in a similar fraction and found that ^{14}C-malonyl CoA was a much better substrate than ^{14}C-acetyl CoA. Since carboxylation of acetyl CoA is inhibited in soluble chloroplast fractions from avocado (Harwood and Stumpf, 1972a), these results show that malonyl CoA is the likely C_2 donor.

A palmityl-ACP elongation system has recently been prepared from safflower seed (Jaworski *et al.*, 1974). Like the avocado enzyme, elongation required palmityl-ACP, malonyl CoA and NADPH. Again, acetyl CoA could not serve as a C_2 donor. The elongation system clearly differed from the safflower fatty acid synthetase in that it was more heat sensitive, had a different pH optimum, NADH was not required and the enzyme was relatively insensitive to cerulenin. By employing a protein-bound palmityl substrate Jaworski and Stumpf (1974a) were able to measure palmitate elongation directly. Under these conditions, exogenous ACP was not necessary, presumably because of sufficient endogenous ACP for transfer of the malonyl residue. In this system, arsenite did not inhibit—probably because only a single sulphydryl residue (that of the malonyl binding site) was present. Addition of ACP to the reaction caused arsenite sensitivity—perhaps because the inhibitor could now react with endogenous ACP at the malonyl binding site and with exogenous ACP.

Euglena gracilis, with its potential for both animal and plant characteristics, offers unique opportunities for comparing fatty acid synthesis systems. When the etiolated organism is grown heterotrophically an "animal-type" fatty acid synthetase is present which, *in vitro*, does not require exogenous ACP, has a high molecular weight and produces large amounts of medium chain-length acids (Delo *et al.*, 1971; Khan and Kolattukudy, 1973). Cells grown auxotrophically in the light contain a second enzyme system which is associated with the chloroplasts and which forms stearate as its principal product. Although synthesis of the enzyme was found to be coupled to protein synthesis in the chloroplast it was not dependent on an active organelle. This suggests that a light receptor other than protochlorophyll is involved (Ernst-Fonberg and Bloch, 1971). The enzyme *in vitro* is dependent on ACP addition and appears to be similar to the stromal palmitate elongation enzyme described above although Ernst-Fonberg and Bloch (1971) did not examine the labelling pattern of the products. The products of the reaction are bound to ACP which also affects association of the enzyme (Ernst-Fonberg, 1973).

Cheniae and Kerr (1965) found that *Euglena* synthesized stearate from acetate *de novo*. The extracted enzymes did not require bicarbonate, were not inhibited by avidin and could not utilize malonate. However, Khan and Kolattukudy (1973) were able to get good rates of synthesis with malonyl CoA as well as acetate so the exact nature of the pathway indicated by Cheniae and Kerr (1965) is unknown. The fatty acid synthesizing systems of *Chlamydomonas* spp. are somewhat similar to *Euglena* in that light causes production of a chloroplast synthetase responsible for 18:0 and 20:0 synthesis (Sirevag and Levine, 1972).

C. UNSATURATED FATTY ACIDS

Elongation of unsaturated fatty acids has been noted in a number of cases. For example, Downey and Craig (1964) suggested, on the basis of rapeseed oil analyses, that a genetically controlled elongating system operated to form 20:1(11) and 22:1(13) (erucic acid) sequentially from 18:1(9) (oleic acid). Distribution of label in the fatty acids synthesized from ^{14}C-acetate by immature rapeseed pods confirmed this. The seed oil *Crambe abyssinica* also contains rather high levels of erucic acid (60%) the formation of which from ^{14}C-acetate is rather sensitive to trichloroacetate inhibition (Appleby *et al.*, 1974). Isolated plant mitochondria synthesize 18:1(11) (vaccenic acid) presumably by elongation of 13:1(9) (palmitoleic acid) (Mazliak *et al.*, 1972). Yang and Stumpf (1965b) showed that the elongation of ricinoleic acid to 14-hydroxy-20:1(11) used acetyl CoA but not malonyl CoA. Cofactors included ATP, NADPH and NADH. The enzyme appeared to be quite specific as 18:0, 16:0, 14:0, 3-hydroxy-12:0 and 3-hydroxy-10:0 were all inactive substrates; 12-hydroxy-18:0 was elongated at about half the rate of ricinoleate (12-hydroxy-18:1 9).

Recently, while studying the formation of linolenic acid, Jacobson *et al.* (1973b) observed the elongation of medium chain length trienoic fatty acids to α-linolenic acid by a spinach stromal enzyme. Acetyl CoA or malonyl CoA were effective as donors but the elongation system would not use medium chain and long chain saturated or monounsaturated fatty acids. In contrast to the stromal palmitate elongation system (see p. 59), ACP was not required.

V. DESATURATION

A. THE AEROBIC PATHWAY

Aerobic desaturation of fatty acids has been demonstrated in a whole variety of organisms ranging from bacteria (such as *Bacillus megaterium* or *Mycobacterium phlei*), through algae, euglenoids, yeasts and plants to animals. The alternative formation of monoenoic acids by the anaerobic pathway (β,γ-elimination at the C_{10} stage of fatty acid synthetase) has not been found in plants.

Originally, the aerobic pathway was described in a yeast where CoA esters of saturated acids were desaturated when NADPH and O_2 were also available (Bloomfield and Bloch, 1960).

$$CH_3\text{-}(CH_2)_{16}\text{-CO-SCoA} + O_2 + \text{NADPH} + H^+ \longleftrightarrow$$
$$CH_3\text{-}(CH_2)_7\text{-CH}{=}\text{CH}(CH_2)_7\text{-CO-SCoA} + \text{NADP}^+ + 2H_2O$$

Formation of polyunsaturated fatty acids by another yeast also required NADPH and O_2 cofactors (Yuan and Bloch, 1961). Cell-free preparations from *M. phlei* required a Fe^{2+} salt and either FMN or FAD in addition to the above cofactors, in order to desaturate palmityl or stearyl-CoA (Fulco and Bloch, 1964). The requirement for Fe^{2+} in desaturation is disputed by Kashiwabara and Sato (1973). In general, desaturation enzymes are membrane-bound and this fact has hindred progress towards elucidating the reaction mechanism.

B. MECHANISM OF DESATURATION

Certain requirements have to be fulfilled in any hypothesis for fatty acid desaturation. The enzyme concerned, which requires O_2 and NADPH, must be able to insert a double bond into a relatively unreactive part of the acyl chain. The cofactor requirements strongly suggest similarity with a mixed-function oxygenase reaction in which one atom of the O_2 molecule is reduced by NADPH and one is used to form a hydroxyl group on the substrate. Unfortunately, direct experimental evidence has not been forthcoming in spite of numerous attempts. Even with soluble desaturases, conversion of hydroxy-derivatives into monoenoic acids has not been shown (Gurr and Bloch, 1966; Nagai and Bloch, 1968). Label from hydroxy fatty acids only enters monoenoic acids after prior conversion to saturated components (Marsh and James, 1962; Elovson, 1964). Therefore, if oxygenated intermediates are involved they must have a transient existence and be tightly bound to the enzyme. Isotopic experiments have been interpreted as showing that hydrogen removal during desaturation is by a concerted mechanism (Morris, 1970) and, if this is so, no hydroxy-intermediate can be involved. Bloch (1969) suggested, on the other hand, that hydrogen removal is sequential, in which case intermediates could be formed. He has also proposed the involvement of an electron-transport chain thus:

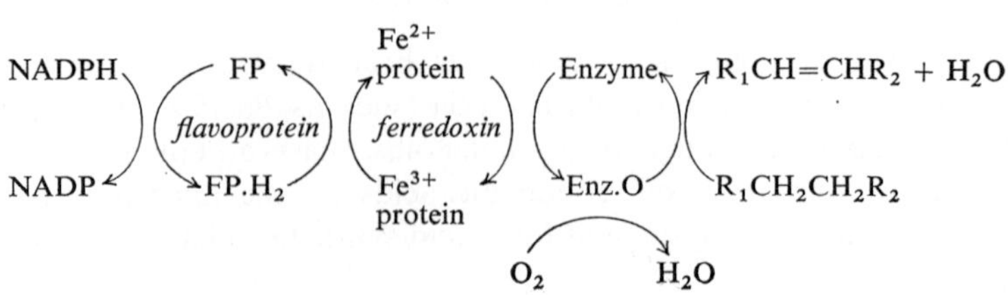

After studying the $\Delta 9$ desaturase enzyme in a number of systems, Brett *et al.* (1971) proposed an alternative scheme based, in part, on the isotope experiments (mentioned above) showing concerted hydrogen removal (Morris, 1970). They suggested that the fatty acyl substrate is enfolded by the enzyme following a conformational change as the two interact. This would bring the two D-hydrogens at C-9 and C-10 into close contact with a single active site. The involvement of NADPH and O_2 cofactors could, in this case, be in production of a conformational change in the enzyme-substrate complex. A possible sequence is shown below:

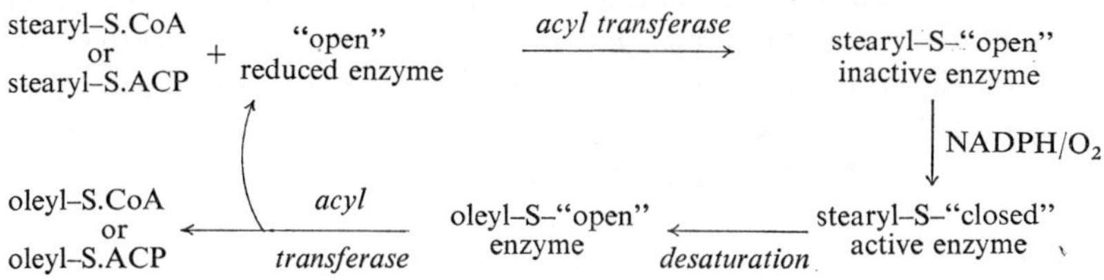

C. OLEIC ACID FORMATION

Rapid synthesis of oleic acid has been shown in many plant systems. Two general approaches have been used (Stumpf 1970). Firstly, ^{14}C-labelled precursors have been utilized to label fatty acids in intact, chopped or sliced tissues. Secondly, experiments have been conducted using isolated sub-cellular fractions and, recently, attempts have been made to purify individual desaturases.

Smirnov (1960) first described the synthesis of fatty acids by isolated chloroplasts. These organelles required O_2 in order to make oleate (Stumpf and James, 1963) as did intact leaves (James 1963). Similarly, synthesis of oleate from acetate by avocado mitochondria was absolutely dependent on O_2 (Mudd and Stumpf, 1961). A detailed discussion of experiments using photosynthetic tissues up to 1966 has been made (Stumpf *et al.*, 1967). There was no evidence for the anaerobic desaturation pathway in plants, but labelling patterns of stearate and oleate formed from ^{14}C-acetate were different. Most stearate was not converted to oleate, the labelling of which indicated *de novo* synthesis. Suggestions were made that a "plant pathway" could be operating which utilized shorter chain fatty acids as oleate pre-cursors. This subject has been well discussed (Hitchcock and Nichols, 1971). Failure to desaturate exogenous palmitate and stearate was possibly due to a lack of the long-chain fatty acyl CoA-ACP acyl transferase, or, at least, failure to form the appropriate acyl–desaturase complex.

The first successful demonstration of stearate desaturation was with *Euglena gracilis*. Nagai and Bloch (1968) used cell-free particulate fractions to desaturate palmityl and stearyl CoA thioesters; ACP-derivatives were inactive. They also examined a soluble fraction which would use stearyl

ACP and could be split into three components: NADPH-oxidoreductase, the desaturase and a ferredoxin. Conversion of stearyl ACP to oleate was also demonstrated in spinach chloroplasts.

Chlorella vulgaris appears to contain two desaturases (Howling *et al.*, 1968): one inserting a double bond at the Δ7 position and is effective in the substrate range C_{14}–C_{16}, and another which inserts a Δ9 double bond in substrates of 14–19 carbon atoms. Furthermore the Δ9 desaturase had two chain-length maxima, at C_{15} and C_{18}. Methyl substitution at the 5–15 positions of stearic acid prevented desaturation (Brett *et al.*, 1971). Sterculic acid is a potent inhibitor of stearate desaturation but does not prevent oleate formation from acetate (James *et al.*, 1968); palmitoleate synthesis from acetate, on the other hand, was inhibited by sterculate. The reason for this difference could be because palmitate (the substrate for palmitoleate formation) is formed outside the chloroplast. Inhibition by sterculate would then be at the point of acyl transfer into the chloroplast as suggested by James *et al.* (1968) since this would have to take place before the chloroplast Δ9 desaturase could operate.

Efforts to study the desaturase have been hampered by the lack of a soluble preparation. When chloroplast preparations, which synthesize predominantly palmitate and oleate (Givan and Stumpf, 1971), are fractionated into soluble and membrane portions, the desaturation activity is confined to the particulate fraction (Yang and Stumpf, 1965a; Weaire and Kekwick, 1973). The biosynthesis of oleic acid has also been demonstrated in non-photosynthetic tissue. Isolated mitochondria converted ^{14}C-acetate to oleic acid in high yield (Mazliak *et al.*, 1968, 1972). These authors, who concluded that double bonds were probably introduced at the Δ9 position of saturated acids, have also demonstrated desaturation by a microsomal fraction of apple parenchyma (Thibaudin *et al.*, 1968).

The fatty acids of castor bean have been examined by a number of workers. Canvin's original analysis of endogenous lipids (Canvin, 1963) indicated that different acids were synthesized sequentially during seed development and this was confirmed by experiments with tissue slices (James *et al.*, 1965). Drennan and Canvin (1969) found the highest fatty acid synthesis activity in a particulate fraction, later identified as proplastids (Zilkey and Canvin, 1969, 1972), which synthesized high amounts of oleate from acetyl CoA. NADPH and O_2 were essential but ACP was not necessary. Unfortunately, in their initial subcellular fractionation studies, these workers did not examine the oil droplets (fat fraction) nor assay the supernatant with ACP. Harwood and Stumpf (1972a) showed that, although rates of fatty acid synthesis in the mitochondrial and microbody fractions were similar to those of Drennan and Canvin (1969), the fat fraction of castor bean was the most active particulate preparation (Fig. 3). When preparations made from seeds 25 DAF (days after flowering) were used, a high proportion of oleate was made from either acetate or malonyl CoA. Furthermore, with

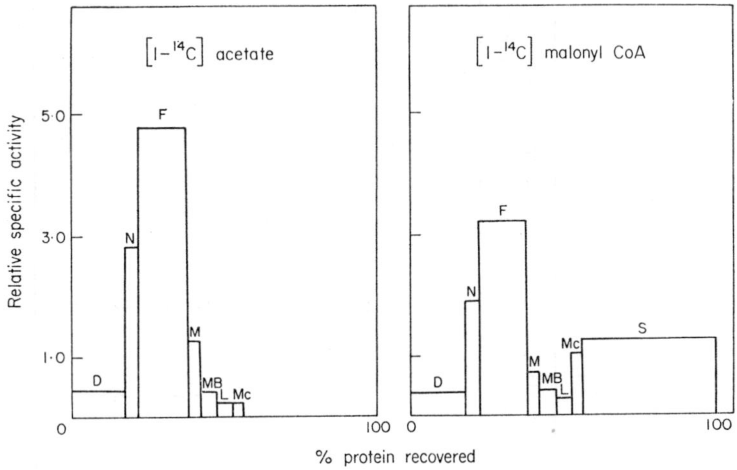

FIG. 3. Fatty acid synthesis by subcellular fractions of developing castor bean. The total homogenate incorporated 41 pmol of acetate (or 295 pmol of malonyl CoA)/min/mg protein. Fractions analysed: debris (D), nuclei (N), fat fraction (F), mitochondria (M) microbodies (MB), lyosomes (L), microsomes (Mc) and particle-free supernatant (S). [From Harwood and Stumpf (1972a).]

the 105 000 *g* supernatant from 25 DAF seeds, oleic acid represented 50% of the fatty acids formed from malonyl CoA (Table XII).

Another high-lipid plant source is the developing safflower seed. Harwood and Stumpf (1972a) found that, in this tissue also, the fat fraction had a very active fatty acid synthesis and that a soluble desaturase was present (Table XII). This enzyme was examined further (Stumpf *et al.*, 1973) and it was found that, while malonyl CoA was a good precursor of oleic acid, stearic acid, stearyl CoA and stearyl ACP were all inactive. The desaturation was aerobic and the possibility of a tightly bound stearyl-protein intermediate has now been examined (Jaworski *et al.*, 1974). Jaworski and Stumpf (1974b) showed that the reaction required stearyl-ACP, O_2, ferredoxin and NADPH. ANDPH was probably only needed to reduce ferredoxin and could be replaced by a lamellae-ferredoxin reducing system. The desaturation was inhibited by 1 mM KCN but was insensitive to CO.

Soluble desaturation has also been found in soyabean cotyledons (Inkpen and Quackenbush, 1969; Rinne, 1969). The enzyme was a Δ9 desaturase, active with C_{16} and C_{18} acids.

D. LINOLEIC AND LINOLENIC ACID FORMATION

Desaturation of an acyl chain between an existing double bond and the methyl end is a characteristic of plant tissues. Although leaf tissues will efficiently utilize acetate or medium chain-length acids to form linoleic acid,

TABLE XII

Fatty acids produced from ^{14}C-malonyl CoA by developing seeds of castor bean and safflower
[From Harwood and Stumpf (1972a) and Stumpf *et al.* (1973)]

Seed		Fatty acids formed (% total)						Specific activity
		14:0	16:0	16:1	18:0	18:1	20:0	pM min^{-1} (mg protein)$^{-1}$
castor bean								
25 days after flowering	homogenate	—	5	4	53	36	2	295
	fat fraction	—	3	—	48	44	5	942
	105 000 *g* supernatant	—	12	2	35	51	—	383
40 days after flowering	homogenate	—	1	—	92	—	7	217
	fat fraction	—	—	—	97	—	3	660
	105 000 *g* supernatant	—	—	—	88	7	5	221
safflower								
	microsomal fraction	1	57	—	16	26	0	72 ± 11
	105 000 *g* supernatant	1	19	—	49	30	1	127 ± 24
	105 000 *g* supernatant (anaerobic)	3	26	—	71	0	0	114 ± 13
	acetone powder of supernatant	0	34	—	61	0	5	141 ± 34

isolated chloroplasts rarely produce significant quantities of the polyenoic fatty acids (James, 1962a; Stumpf and James, 1962, 1963; Mudd and McManus, 1962). From a time study of [1-^{14}C]oleate incorporation into lineoleic and linolenic acids in leaves of castor bean or lettuce and *Chlorella vulgaris*, Harris and James (1965) found evidence for a precursor–product relationship during sequential desaturation. The label was retained in the carboxyl group. Because isolated chloroplasts from higher plants were inactive, James and his co-workers used the much more active *Chlorella* for most of their later work. Cofactors for oleate desaturation were NADPH or NADH and O_2 (Harris *et al.*, 1967) but sterculic acid caused much less inhibition than with stearate desaturation (Gurr, 1971). Again the stereochemistry of hydrogen removal was *cis* and of D-absolute configuration (Morris *et al.*, 1968).

Isolated chloroplasts from *Chlorella* desaturated oleyl CoA but high labelling of phosphatidylcholine was also noted (Harris *et al.*, 1967). These studies were later extended by Gurr and his co-workers (Gurr *et al.*, 1969; Gurr and Brawn, 1970; Gurr, 1971) who suggested that the actual substrate for desaturation might be enzyme-bound oleoyl phosphatidylcholine. Willemot and Verret (1973) have presented evidence that another photosynthetic tissue, alfalfa leaflets, may utilize phosphatidylcholine for desaturation of linoleic acid. Recently, Kates and his colleagues (Kates and Paradis, 1973; Pugh and Kates, 1973) found oleate desaturation in *Candida*

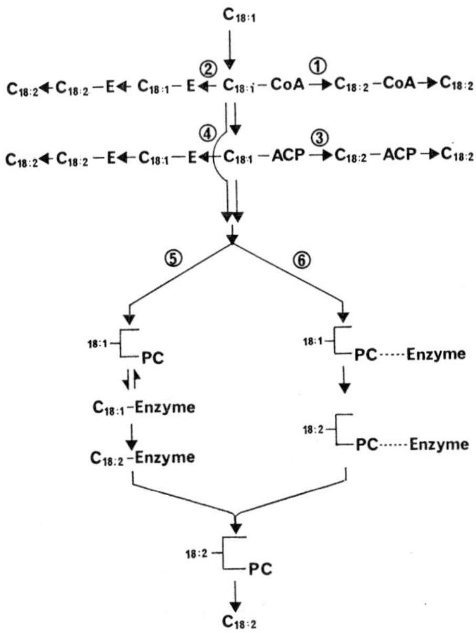

FIG. 4. Hypothetical pathways for the conversion of oleate (C_{18}:1) into linoleate (C_{18}:2) by *Chlorella vulgaris*. E = enzyme; APC = acyl carrier protein; PC = phosphatidylcholine. [From Gurr *et al.* (1969).]

lipolytica with two systems, one using oleoyl CoA as substrate the other using phosphatidylcholine. In a similar way, monogalactosyldiglyceride may be used by the blue-green alga, *Anabaena variabilis* (Appleby *et al.*, 1971). Possible pathways for conversion of oleate to linoleate by *Chlorella* are shown in Fig. 4.

Using safflower seeds harvested at particular stages of development McMahon and Stumpf (1964, 1966) showed that oleate→linoleate was rapid in the seed and that oleoyl CoA would act as a substrate in a particulate system provided that O_2 and a NADPH-generating system were present. The nature of the substrate was further examined by Vijay and Stumpf (1971) who showed that oleoyl CoA, but not acyl-ACP or phospholipid, was the substrate. Both the product and the substrate were rapidly transacylated into phosphatidylcholine. These results are shown in Fig. 5. In further work Vijay and Stumpf (1972) found that the position and configuration of the substrate double bond were critical for desaturation. Any change in the bond position or chain shortening caused complete loss of activity. In *Chlorella* and castor bean seed, however, chain length was much less critical for linoleic acid synthesis provided that the *cis*-Δ9 or *cis*-ω9 configuration was retained (Howling *et al.*, 1972). The safflower desaturase was strongly inhibited by a number of thiol reagents and metal chelators. The requirement for NADH could be replaced by photochemically reduced ferredoxin suggesting that NADH did not interact directly with O_2 and that the de-

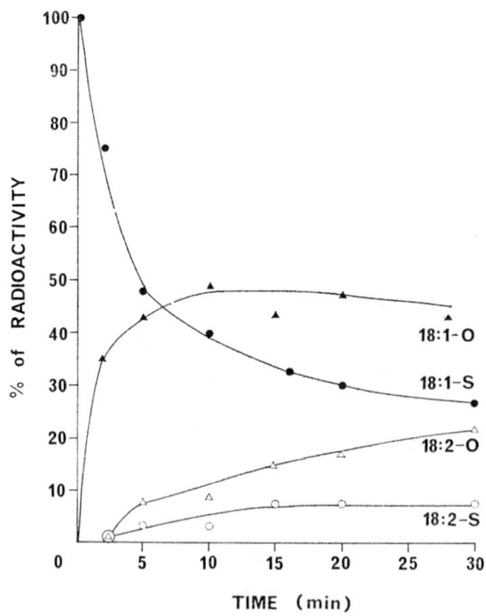

FIG. 5. Time course study of safflower oleyl CoA desaturase reaction. The levels of oxygen esters (O) and sulphur esters (S) of oleic (18:1) and linoleic (18:2) acids are shown. [From Vijay and Stumpf (1971).]

saturase merely provided a source of electrons. In *Chlorella*, in which linoleic acid can also be made from 18:1(12) as substrate, the desaturase also acts on saturated fatty acids (of chain length around C_{14}) and, in comparison to the oleate desaturase, is much more sensitive to sterculate (Gurr, 1971; Gurr *et al.*, 1972b).

Another experimental system which has proved useful is the "aged" slice from potato tuber. Willemot and Stumpf (1967b) observed that, during the first hours of ageing, fatty acid synthesis increased 6–12-fold. Particularly remarkable was the increased formation of linoleic and the very long-chain fatty acids. The increase was apparently related to a temporary rise in the rate of protein and RNA synthesis (Willemot and Stumpf, 1967a). Ben Abdelkader, Mazliak and their associates have followed up this work with a detailed examination of the oleoyl CoA desaturase (Ben Abdelkader *et al.*, 1969, 1973). As in safflower, the enzyme was microsomal and it required O_2 and either NADH or NADPH. Both substrate and product were rapidly incorporated into phospholipids, particularly phosphatidylcholine. The desaturase could not operate with palmityl CoA or stearyl CoA as substrates and had a temperature optimum at a physiological value (25°C).

Evidence from a number of different types of experiment has accumulated over the years for a sequential desaturation:

$$\begin{array}{ccccccc} \text{stearate} & \rightarrow & \text{oleate} & \rightarrow & \text{linoleate} & \rightarrow & \text{linolenate} \\ 18{:}0 & \rightarrow & 18{:}1(9) & \rightarrow & 18{:}2(9,12) & \rightarrow & 18{:}3(9,12,15) \end{array}$$

This includes the use of leaves (James, 1963), developing seeds (Dutton and Mounts, 1966) isolated mitochondria (Mazliak *et al.*, 1968) and *Penicillium chrysogenum* (Bennett and Quackenbush, 1969). However, isolated chloroplasts, which contain high levels of linolenic acid, have proved remarkably poor systems for demonstrating the insertion of the third double bond. Kannangara and Stumpf (1972a) were able to boost incorporation of acetate into linoleic and linolenic acids by using immature spinach chloroplasts. Cyanide inhibition of linoleic acid, but not oleic or linolenic acid synthesis, suggested that sequential desaturation might not represent the whole picture in spinach. Subsequent work (Jacobson *et al.*, 1973a, b; Kannangara *et al.*, 1973) has revealed that in a number of systems, linolenic acid is formed by a series of reactions involving desaturation at the C_{12} level followed by elongation of the 12:3 acid. This interesting new pathway is described in detail by Stumpf in Chapter 4 of this volume. Spinach and the other systems studied by Stumpf's group contain significant amounts of, e.g., 16:3 acid and it remains to be seen if this alternative pathway is ubiquitous. In contrast, recent evidence has been presented for sequential desaturation in leaves (Trémolières, 1973) possibly involving desaturation on a phosphatidylcholine species containing linoleic acid (Roughan, 1970; Willemot and Verret, 1973). It seems quite possible that both pathways may occur in higher plants, their relative importance varying from species to species.

TABLE XIII

Changes in the fatty acid composition of broad bean leaves on exposure of the plant to light/dark periods

Treatment	Fatty acid composition (% total)						
	16:0	16:1(3*t*)	18:0	18:1	18:2	18:3	Others
etiolated	16·1	n.d.	1·8	4·4	30·0	47·2	0·5
normal	12·7	3·2	2·4	6·2	14·3	60·7	0·0
7 days dark	10·6	0·3	2·5	4·4	15·8	64·0	2·4
7 days dark +7 days light	14·5	4·8	2·1	3·9	18·0	55·2	1·5

E. OTHER UNSATURATED ACIDS

Many minor unsaturated fatty acid plant constituents have been reported and their biosynthesis, in a few instances, studied. Details can be found in Hitchcock and Nichols (1971).

One fatty acid, *trans*-3-hexadecenoic acid, appears to be associated with chloroplasts of higher plants and is only found in the phosphatidylglycerol of leaf tissue (Bartels *et al.*, 1967). However, in seed oils it may also be found in other lipid fractions. The unusual structure, specific location and its light-dependent synthesis, make the function of this monoenoic acid an intriguing question. It has been noted by a number of workers that *trans*-3-hexadecenoic acid is absent from etiolated tissue (e.g. Nichols *et al.*, 1965a) and increases on exposure to light (Nichols *et al.*, 1965b). Palmitate seems to be the direct precursor and, although direct evidence has not been obtained,

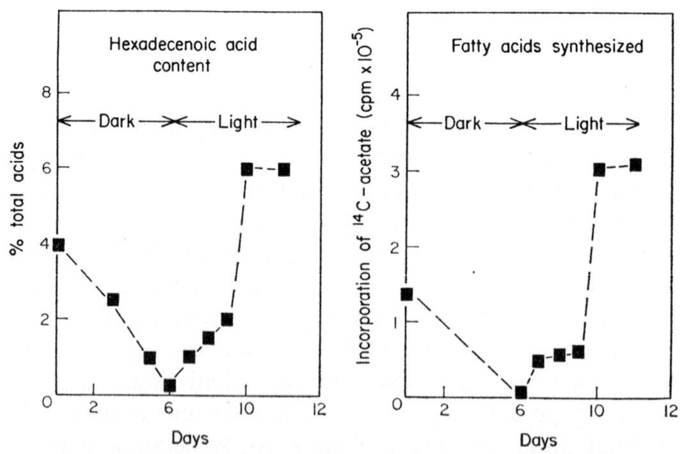

FIG. 6. Effects of light on endogenous *trans*-3-hexadecenoic acid and on fatty acid synthesis from ^{14}C-acetate by isolated chloroplasts from broad bean.

desaturation may take place on the phosphatidylglycerol molecule (Bartels, 1969). Harwood and James (1975) have observed that the level of the fatty acid is reduced in green tissue by placing the plants in the dark (Table XIII). After 5–7 days the level is negligible but can be increased again on exposure to light. The ability of chloroplast to synthesize fatty acids parallels the rise and fall of *trans*-3-hexadecenoate (Fig. 6). It should be noted that normal dark periods are not sufficient to affect its concentration. The observation that the percentage content of the acid increases only when the fully mature chloroplast is formed (Leech *et al.*, 1973) is in keeping with proposals that *trans*-3-hexadecenoic acid may play a functional role in the higher plant chloroplast.

VI. Hydroxy Acids

Hydroxy acids are formed by two routes in plants (Hitchcock and Nichols, 1971). These are:

(a) by direct oxygenation of a saturated carbon atom;
(b) by hydration of an unsaturated bond.

Molecular O_2 is the source of oxygen in the former while water serves in the latter.

A. 2-Hydroxy Acids

The 2-hydroxy acids are widespread but minor components of plants. Their biosynthesis involves an α-hydroxylation mechanism as in α-oxidation. These processes are discussed in Chapter 11 (p. 330).

B. 3-Hydroxy Acids

Davidoff and Korn (1964) noted that long chain fatty acids are converted to their D(−)-3-hydroxy derivatives by subcellular fractions of the slime mould, *Dictyostelium discoideum*, or by guinea pig liver mitochondria. The configuration of the intermediates ruled out formation during conventional β-oxidation of fatty acids in which L(+)-3-hydroxyacids are intermediates. The accumulation might, therefore, involve cleavage of intermediates from fatty acid synthetase.

A soluble preparation from maturing avocado mesocarp was found to be highly active in forming 3-hydroxy derivatives of medium chain fatty acids (Harwood *et al.*, 1972). CoA, ATP and O_2 were required and addition of ACP gave some stimulation. Use of acyl-CoAs as substrates eliminated the requirements for CoA and ATP. Inhibitor studies indicated that thiol and heavy metal groups were involved but no requirement for reduced nucleotide could be found. Marked chain-length specificity of the enzyme was observed

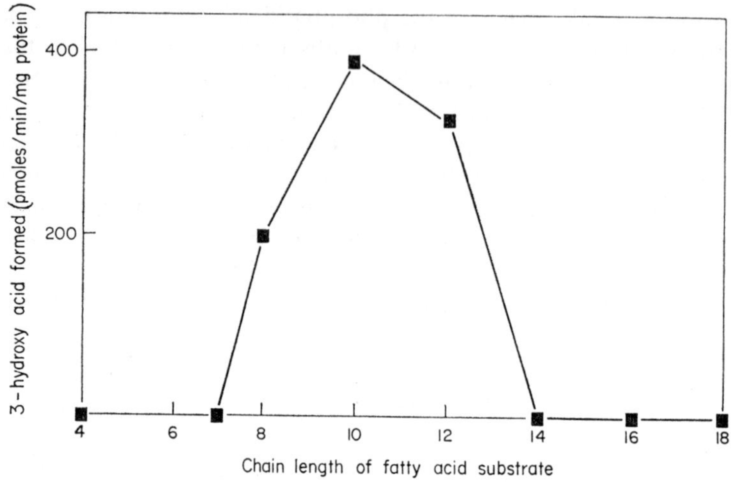

FIG. 7. Chain length specificity of the β-hydroxylating system of avocado. [From Harwood *et al.* (1972).]

with good activity in the C_8–C_{12} range, but none with shorter or longer chain acids (Fig. 7).

The importance of this very active hydroxylating system is unknown but the chain-length specificity tempts speculation on the possible formation of unsaturated fatty acids by this pathway. Proposals for desaturation at the C_{12} stage have been made in the past and a so-called "plant pathway" was invoked at one time (see Hitchcock and Nichols, 1971). Labelling studies in the rat also led Raju and Reiser (1969) to propose this alternative route for oleate formation. Harwood *et al.* (1972) were not able to demonstrate any conversion of the hydroxy-derivatives to unsaturated acids. However, the observations that the accumulating acid contains the D(−) isomer (Sodja and Stumpf, 1973) and that linolenic acid may be synthesized following desaturation at the C_{12} stage (see above) have given the hypothesis a new lease of life.

C. RICINOLEIC ACID

The biosynthesis of ricinoleic acid, 12-D-hydroxy-18:1(9*c*)—the major fatty acid component in the seed oil of castor beans—has been thoroughly studied by the groups of James and Stumpf. Canvin (1963) found that ricinoleic acid first appeared in the seed 12 days after flowering and represented 90% of the total oil fatty acids after 36 days. The composition of the oil then remained constant during final maturation. Isotope studies indicated that oleic acid was the precursor but that linoleate was not an intermediate (James, 1962b; Canvin, 1965; James *et al.*, 1965). Yamada and Stumpf (1964) demonstrated a soluble system which would convert oleoyl-CoA into a hydroxy-acid derivative but further work (Galliard and Stumpf, 1966) indicated that the product of the soluble hydroxylation system was 3-hydroxy-oleate and that the hydroxylating system that produced ricinoleate was microsomal. Oxygen and NADH were required and a marked substrate specificity was noted. Although oleic acid is the best substrate, the castor bean seed system also hydroxylates other monoenoic acids but the hydroxyl group is always inserted in the same position relative to the existing double bond (Howling *et al.*, 1972).

By the use of tritiated substrates, Morris (1967) confirmed that overall retention of configuration in the 12-position was obtained and excluded the possibility of hydration of a double bond during formation. In fungal ergot oil, however, ricinoleic acid appears to be formed by a specific anaerobic hydration of linoleate (Morris *et al.*, 1966).

D. OTHER HYDROXY ACIDS

Long-chain hydroxy acids are important components of cutin. Biosynthesis of 10,16-dihydroxy palmitate via a mixed function oxidase reaction has been indicated by experiments of Kolattukudy and his associates (see p. 230). Hitchcock and Nichols (1971) have also discussed the formation of various other hydroxy acids.

VII. SUBCELLULAR SITE OF FATTY ACID BIOSYNTHESIS

Fatty acid biosynthesis has been demonstrated in a number of subcellular organelles. The picture is further confused by the fact that, in most cases, individual enzymes have not been considered but only the incorporation of precursors (usually acetate) into long chain fatty acids has been studied. Further work on more clearly defined reaction steps is needed.

Localization of the site of mammalian fatty acid synthesis has now been achieved with a number of tissues. Generally, the fatty acid synthetase complex is soluble, while the mitochondria contain elongation enzymes and the microsomes contain the desaturases (Olsen, 1966; Stumpf, 1969). In contrast, studies on fatty acid biosynthesis in plants have been less well defined. Several workers have used isolated fractions, but detailed analysis of intracellular distribution has rarely been undertaken. A list of some of the diverse organelles, or their subfractions, which have been found to synthesize fatty acids is given in Table XIV.

Early work with isolated plant organelles was, not surprisingly, with chloroplasts. Smirnov (1960) reported rapid incorporation of ^{14}C-acetate into long chain acids by spinach chloroplasts and the rate of acetate uptake has been studied (Jacobsen and Stumpf, 1972). The organelle was indicated to be the major site of fatty acid biosynthesis in the leaf (Stumpf and James,

TABLE XIV

Synthesis of fatty acids by subcellular fractions

Fraction	Plant	Precursor	Major acids formed	Reference
mitochondrial	apple	acetate	16:0, 18:1	Mazliak *et al.* (1968)
proplastid	castor bean	malonyl CoA	18:1	Zilkey and Canvin (1969)
plastid	avocado	acetate	16:0, 18:0, 18:1	Mudd and Stumpf (1961)
plastid stroma	avocado	malonyl CoA	18:0	Harwood (1974b)
chloroplast	spinach	acetate	16:0, 18:1	Kannangara and Stumpf (1972a)
chloroplast	barley	acetate	16:0, 18:0, 18:1	Hawke and Stumpf (1965b)
chloroplast	lettuce	acetate	16:0, 18:1	Brooks and Stumpf (1966)
chloroplast stroma	lettuce	malonyl CoA	18:0	Brooks and Stumpf (1966)
oil droplet	castor bean	acetate, malonyl CoA	18:0, 18:1	Harwood and Stumpf (1972a)
oil droplet	crambé	malonyl CoA	20:0, 20:1, 22:1	Appleby *et al.* (1974)
microsomal	castor bean	oleyl CoA	12-OH-18:1	Galliard and Stumpf (1966)
microsomal	potato (aged)	oleyl CoA	18:2	Ben Abdelkader *et al.* (1973)
microsomal	pea	malonyl CoA	20:0, 22:0, 24:0, 26:0	Macey and Stumpf (1968)
soluble	crambé	malonyl CoA	16:0	Appleby *et al.* (1974)
soluble	avocado	malonyl CoA	16:0, 18:0	Harwood and Stumpf (1972b)
soluble	avocado	C_8–C_{12} acids	3-OH-C_8–C_{12} acids	Harwood *et al.* (1972)
soluble	pea	malonyl CoA	16:0, 18:0	Macey and Stumpf (1968)
soluble	castor bean	malonyl CoA	16:0, 18:0, 18:1	Harwood and Stumpf (1972a)

1963; Mudd and McManus, 1964). The major advances have been reviewed by Stumpf *et al.* (1967) and Stumpf (Chapter 4 of this volume).

The avocado mesocarp has been studied in some depth (see Section III). In summary, this tissue shows fatty acid synthesis activity in the plastids (two systems) and a cytoplasmic fatty acid synthetase. Extrapolation of this system to leaves may provide an explanation for a number of perplexing observations made over the years. Thus in any discussion of chloroplast biosynthesis one has to explain:

1. The ability to incorporate medium-chain but not long-chain free fatty acids into oleate.
2. The difference in labelling patterns between stearate and oleate.
3. The inhibition by sterculate of palmitoleic acid synthesis from acetate but not of oleic acid synthesis.

I would propose the following possible explanations. The chloroplast lamellar membranes are capable of synthesizing palmitate and oleate (Harwood and Stumpf, 1972b; Weaire and Kekwick, 1970). They do this by the operation of a relatively integrated fatty acid synthetase—elongase—Δ9 desaturase system, in which stearate does not normally accumulate. The end product of the synthetase is palmityl-ACP, which is the substrate for palmitate elongation (Harwood and Stumpf, 1972b) and can also serve as a donor of lipid acyl groups (Kannangara and Stumpf, 1971). The elongation system, which is very easily solubilized (Weaire and Kekwick, 1973), can under abnormal circumstances also accept endogenous fatty acids (from complex lipids) resulting in the synthesis of stearic acid with heavy labelling at its carboxyl end. The desaturase also uses ACP-derivatives which can be supplied endogenously (Nagai and Bloch, 1968).

When activated as the ACP derivatives, medium chain-length fatty acids, which are effectively incorporated into oleate (Stumpf *et al.*, 1967; James *et al.*, 1968), are substrates for the condensation step in fatty acid synthetase. Palmitate and stearate, on the other hand, are neither used by fatty acid synthetase nor activated to their ACP derivatives in higher plants. Clear evidence from a number of sources indicates that stearate is formed by elongation (see Mazliak, 1973) but, under normal circumstances the elongation system is tightly coupled in the chloroplast to newly-synthesized palmitate. Stearyl-ACP formed *de novo* from labelled C_2 precursors and used as substrate for desaturation will thus tend to be uniformally labelled as is oleate in typical incorporation studies. Under conditions when desaturation is prevented and stearate accumulates e.g. with viologen dyes (Givan and Stumpf, 1971) or in isolated stroma (Weaire and Kekwick, 1973) the tightly coupled system of oleate synthesis is perturbed and stearate can be formed by addition to a non-radioactive C_{16} primer. The inhibition of palmitoleate synthesis by sterculate (James *et al.*, 1968) may be explained because the desaturase accepts ACP-derivatives (Nagai and Bloch, 1968). In the isolated

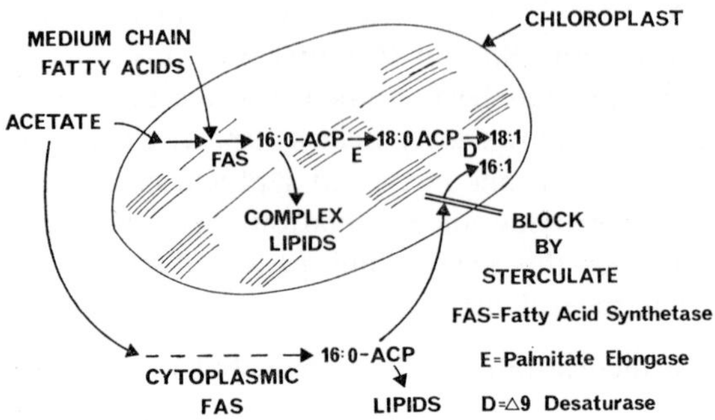

FIG. 8. Diagrammatic representation of a proposal for monoenoate formation in leaves.

chloroplast the radioactive ACP-derivative is C_{18} (from elongation) but in the leaf, palmityl-ACP is also available. The cytoplasmic fatty acid synthetase liberates palmityl-ACP as end product (Harwood and Stumpf, 1972b) and this is then available for the Δ9 desaturase. However, sterculate probably acts by preventing transfer of exogenous acid to the desaturase and hence prevents acetate incorporation into palmitoleate. A schematic diagram of the system is shown in Fig. 8.

Apart from the possibility of cytoplasmic fatty acid synthetase, leaves have also been shown to contain synthesizing enzymes in the cuticle (Kolattukudy, 1967). Formation of palmitate in the leaf cytoplasm has an obvious role in the system for cutin formation (Kolattukudy, p. 207).

Harwood and Stumpf (1972a) made a study of fatty acid synthesis by subcellular fractions from a number of different plants. They examined nuclear, mitochondrial, plastid, microbody, lysosomal, fat fraction, microsomal and 105 000 *g* supernatant fractions (depending on the tissue used). This work revealed a surprising diversity in pattern from species to species. All the 105 000 *g* supernatant fractions were able to synthesize saturated fatty acids from malonyl CoA but not from acetate (see Section II.A). Active particulate fractions included pea seed microsomes, avocado mesocarp plastids and the fat fraction of castor bean and safflower seeds. Figure 3 (p. 65) shows the results for castor bean.

Although, with hindsight, it seems logical that the oil droplets formed in plants should have the capacity for lipid synthesis, the finding of very active incorporation of acetate by the fat fraction was unexpected at the time. Most workers had neglected this fraction during analysis (e.g. Yang and Stumpf, 1965a; Zilkey and Canvin, 1969). A detailed examination of its properties (Harwood *et al.*, 1971) revealed that at least 80% of the total fatty acid and triglyceride synthesis in the maturing castor bean was provided by the fat droplets. The enzymes appeared to be membranous and could be separated from the stored triglyceride by detergents. Although no bounding membrane could be seen, membranous inclusions were shown by electron microscopy of the isolated fractions and of the droplets *in situ*. Malonyl CoA was also rapidly used by the fat fraction, which represents one of the most active fatty acid synthesizing systems yet studied in plants.

Erucic acid, the principle storage acid of *Crambe abyssinica* seeds, is most actively synthesized by the oil body fraction (Appleby *et al.*, 1974). Again, as in the maturing castor bean, the isolated fat fraction is capable of synthesis of its own triglyceride from malonyl CoA and glycerol-3-phosphate (Gurr *et al.*, 1974). The oil droplets in this plant differ from those of castor bean in that a bounding membrane is clearly visible and there are no particulate inclusions.

The study by Harwood and Stumpf (1972a), and numerous publications by others, have shown that many plant organelles are capable of fatty acid synthesis. The relative importance of these fractions depends, not only on the species, but also on the stage of physiological development. Thus fatty acid synthesis in castor bean fat fractions is negligible in the mature seed (Harwood *et al.*, 1971), the developing safflower microsomes only synthesize linoleate at certain times (Vijay and Stumpf, 1971) and ricinoleic acid formation by castor bean microsomes is also time-dependent (Galliard and Stumpf, 1966).

VIII. Control of Fatty Acid Synthesis

A. EXPERIMENTAL SYSTEMS USED

A number of experimental systems can be used to follow environmental influence on fatty acid synthesis. These include:

(a) *In vivo* systems. Changes in the fatty acid patterns in growing or senescing plant tissue can be readily analysed. Labelling of fatty acids from $^{14}CO_2$ can be followed, though this is so indirect as to make conclusions difficult (Eberhardt and Kates, 1957).

(b) Droplet uptake by leaves. Microdroplets of, e.g. acetate can be used to label leaf fatty acids (Trémolières, 1973).

(c) Uptake through petiole. This technique has been used to introduce $^{32}P_i$, ^{35}S-sulphate and ^{14}C-acetate (Kates, 1960) and is fairly rapid though suffering from similar interpretational problems as (a), especially in the case of acetate.

(d) Leaf slices. Appelqvist *et al.* (1968) examined a number of techniques for labelling fatty acids in barley leaves. They concluded that leaf slices were a very effective method but pointed out that, since chlorophyll synthesis was inhibited during incubation, incorporation might be different from the *in vivo* situation.

(e) Germinating seeds. Since seeds contain a very small amount of water, quantities of which they imbibe during germination, they can be very effectively labelled by using ^3H_2O (Harwood and Stumpf, 1970). Various seeds were studied and the fatty acid labelling pattern was the same as with ^{14}C-acetate.

(f) Cell cultures. Plant cell cultures have now been used in a number of metabolic studies including lipid biosynthesis (Fletcher and Beevers, 1970). This method offers somewhat simpler "*in vivo*" systems but has the usual limitations of tissue culture experiments.

(g) Acetone powders. Acetone powders can be made of organelles or tissues. These have the advantage of stability but may only make a limited range of fatty acids.

(h) Subcellular fractions. It has been pointed out (Stumpf, 1970) that fractions rarely duplicate incorporation *in vivo*. However, in the study of individual enzyme steps, this is clearly the method of choice, since interpretation and results are usually clear-cut.

B. CONDITIONS AFFECTING BIOSYNTHESIS

Fatty acid biosynthesis in plants can be altered at several levels. These may be roughly divided into factors controlling enzymes, the supply of necessary substrates and cofactors, or factors indirectly altering fatty acid synthesis by regulating plant growth or metabolism generally. Since relatively few isolated enzymes have been described, control at this molecular level has not been examined in much detail. It therefore seems more logical to divide this section into a number of basic conditions which result in altered fatty acid metabolism *in vivo*.

1. *Germination*

There have been several studies on the changes in lipids and fatty acids which accompany the germination process but these, since they represent total analysis, reflect both catabolic and anabolic processes (see Hitchcock and Nichols, 1971). A simple, yet effective, way of following biosynthesis is to allow the dry seed to imbibe ^3H-labelled water (see (e) above).

Synthesis of membrane lipids is common to both high- and low-lipid seeds. Storage oils often contain fatty acids not present in membrane lipids. Thus the fatty acid compositions of flax (Zimmerman and Klosterman, 1965) and oil palm seed (Boatman and Crombie, 1958; Hardman and Crombie, 1958), for example, undergo large changes during germination with an increase of the typical polyunsaturated membrane components.

Harwood and Stumpf (1970) studied a number of different seeds, both high- and low-lipid, of which pea was studied in some detail. The biosynthesis of very long-chain saturated fatty acids, apparently as precursors of wax has already been mentioned (p. 58). Total fatty acid analysis of germinating flax seed also indicated the appearance of these acids (Zimmerman and Klosterman, 1965). The enzymes required for fatty acid synthesis are, apparently,

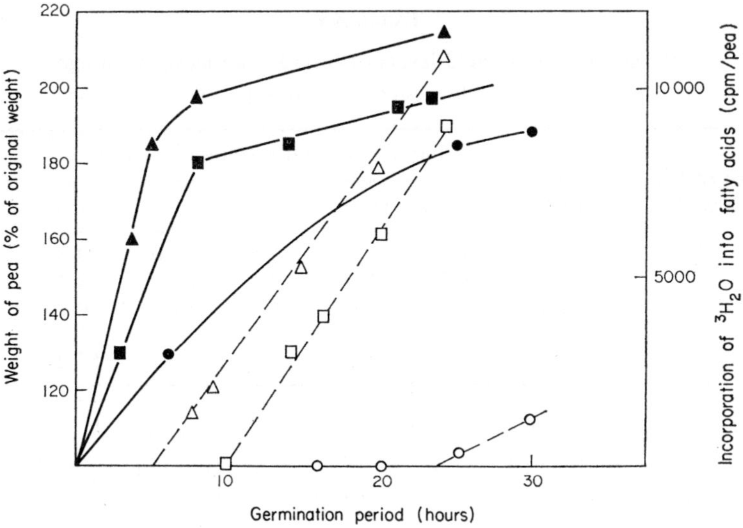

FIG. 9. Correlation of water uptake with the onset of fatty acid synthesis in germinating peas. Uptake of ^{3}H-labelled H_2O into the pea (solid symbols) and into fatty acids (open symbols) was measured during germination at 5° (●, ○, 20° (■, □) and 35° (▲, △). [From Harwood and Stumpf (1970).]

already present in the seed (Harwood and Stumpf, 1970). The onset of synthesis corresponds very accurately to a specific water content (Fig. 9) and it is possible that water either activates the enzymes or makes cofactors available. Interestingly, relatively little unsaturated acid was made by any of the seeds studied (Table X). In hazel seeds also, saturated acids were the major groups synthesized (Shewry *et al.*, 1972). Raising the germination temperature to 35°C or adding fluoride blocked the elongation of stearate while arsenite prevented palmitate elongation in peas (Harwood and Stumpf, 1971). A number of carbamate weed killers also severely reduced the very long chain fatty acids and, hence, the leaf waxes (Still *et al.*, 1970).

2. *Chloroplast development*

Since fatty acid metabolism in the chloroplast is reviewed separately (Chapter 4) I will deal only briefly with fatty acid changes in this organelle. It is well known that light stimulates acetate incorporation into lipids of greening chloroplasts (Appelqvist *et al.*, 1968; Kannangara *et al.*, 1971). Apparently this stimulation in pea leaves is due to non-cyclic phosphorylation by the chloroplast (Panter and Boardman, 1973). The principal changes accompanying "greening" or chloroplast development can be summarized as:

(a) An increase in the levels of typical chloroplast complex lipids such as the glycosyldiglycerides, sulpholipid and phosphatidylglycerol.

(b) A change in fatty acid composition, most noticeably with increases in linolenic acid and the appearance of *trans*-3-hexadecenoic acid.

TABLE XV

Variation of fatty acid levels of broad bean leaves with age
[From Harwood and James (1975)]

Age of leaf (days)	Fatty acid composition (% total)						
	16:0	16:1(3*t*)	18:0	18:1	18:2	18:3	Others
1	21·2	0·6	2·6	7·5	26·0	39·9	2·2
3	15·3	2·2	1·9	7·4	20·2	49·4	3·6
5	12·4	2·8	2·4	6·2	14·3	60·7	2·2
7	12·1	3·5	1·6	7·4	14·6	55·8	5·0
11	13·0	3·9	1·8	7·8	15·5	53·2	4·8

These ubiquitous and specific alterations lead one to believe that certain, or all, of these lipids must play a role in the functional chloroplast. It is not clear whether lipid changes create, or are a result of, initiation of photosynthesis.

Because *trans*-3-hexadecenoic acid (16:1 3*t*) is found specifically in the chloroplasts of higher plants (James and Nichols, 1966) only acylated to phosphatidylglycerol, its role there has intrigued biochemists. Light was indicated as being absolutely necessary for its synthesis in *Chlorella* (Nichols *et al.*, 1965b). The changes induced in 16:1(3*t*) levels and chloroplast fatty acid synthesis by broad bean have been described earlier (p. 70). This plant contains rather high levels of the *trans*-3-hexadecenoate (Crombie, 1958) and also has chloroplasts capable of good levels of acetate incorporation. It will be seen from Table XIII that levels of the fatty acid are reduced in 5–7 days to values corresponding to those of etiolated tissue. "Re-greened" leaves contain levels of the acid and synthesis significantly higher than controls. Leech *et al.* (1973) observed that, in maize, the levels of 16:1(3*t*) increased with the maturation of the leaf and parallel changes are also seen in the broad bean (Table XV). Since synthesis of 16:1(3*t*) lags behind many other typical chloroplast components and, reduction and increase during "re-greening" is not paralleled by the other lipids, light may indirectly or directly be the primary control for its formation.

3. *Growth*

There have been few studies on the maintenance of lipid levels in mature or slowly-growing plants. As mentioned previously, the chloroplast can be regarded as a major site of fatty acid biosynthesis in leaves and proplastids have even been isolated from roots (Davey *et al.*, 1973). There are considerable differences, however, in chloroplasts prepared from different parts of maize (Poincelot, 1973).

Transacylation reactions are common in plants and major changes in the fatty acid composition of individual lipids can occur subsequent to their *de novo* synthesis. ^{14}C-acetate incorporation into *Chlorella* revealed that many of the newly synthesized fatty acids were rapidly transferred into phosphatidylglycerol and monogalactosyldiglyceride (Nichols *et al.*, 1967). More recently Devor and Mudd (1971) have described transacylation reactions for phosphatidylcholine in spinach leaves.

4. *Ageing*

Physical damage to a tissue initiates wound responses which, in storage organs, are associated with the development of new enzyme activities and a large increase in membraneous components of which lipids constitute a major part.

The experiments with aged potato slices have already been mentioned on p. 69. The rise in lipid synthesis appears simultaneously with changes in protein and RNA synthesis and inhibitors such as cycloheximide will prevent the increase in fatty acid synthesis (Willemot and Stumpf, 1967a, b). Similar results are also observed with peel discs from pre-climateric apples (Galliard *et al.*, 1968). The increase in oleoyl CoA desaturase in ageing discs of potato tuber is also subject to inhibition by cycloheximide and actinomycin D (Ben Abdelkader *et al.*, 1973). The control mechanisms involved are unknown at present but Willemot and Stumpf (1967a) have speculated that slicing causes de-repression of short-lived RNAs which can then direct synthesis of proteins, including lipid synthetases.

5. *Seed and fruit maturation*

Notable qualitative and quantitative changes in the fatty acids occur during maturation of fruit. The major results up to 1970 have been covered by Hitchcock and Nichols (1971). Since that time there have been quite a number of relevant studies including those on crambé (Appleby *et al.*, 1974; Gurr *et al.*, 1972b, 1974) castor bean (Drennan and Canvin, 1969; Harwood and Stumpf, 1972a; Harwood *et al.*, 1971; Zilkey and Canvin, 1969, 1972) avocado pear (Harwood and Stumpf, 1972a, b; Harwood *et al.*, 1972; Weaire and Kekwick, 1970, 1973) and safflower (Vijay and Stumpf, 1971, 1972).

The sudden and temporary appearance of many enzymes involved in fatty acid synthesis is a notable feature of this developmental period. This point is emphasized in Table XVI where formation of particular fatty acids is tabulated with age of the fruit. The control of these systems has been studied little and is ill-understood. Some evidence on environmental regulation is described on p. 83.

6. *Senescence*

Senescence in leaves is apparently accompanied by very rapid losses of those lipids that are components of chloroplasts. Thus the glycosyldiglycerides are quickly deacylated (Draper, 1969) and the liberated linolenic

TABLE XVI

Fatty acids synthesized by some fruits at particular phases of maturation

Plant source	Fatty acid synthesized		Age when active (days after flowering)	Reference
white mustard	22:1(13)	(*erucic*)	35	Dasgupta and Friend (1973)
ivy	18:1(6)	(*petroselenic*)	30–60	Grosbois (1971)
Ceiba pentandra	14:0	(*myristic*)	38–65	Kaimal and Lakshminarayana (1972)
safflower	18:2(9,12)	(*linoleic*)	14–18	Vijay and Stumpf (1971)
castor bean	12-OH-18:1(9)	(*ricinoleic*)	20–25	Yamada and Stumpf (1964)
castor bean	18:1(9)	(*oleic*)	20–40	Harwood and Stumpf (1972a)
crambé	22:1(13)	(*erucic*)	8–16	Gurr *et al.* (1972)
corn	18:1(9)	(*oleic*)	10–30	Weber (1969)

acid is particularly susceptible to lipoxygenase-catalysed attack (Newman, 1966; Holden, 1970). However, incorporation of ^{14}C-acetate indicates that there is some synthesis in senescing cocklebur leaves (Newman *et al.*, 1973) or cucumber cotyledons (Draper and Simon, 1970). The incorporation, which decreases with age, falls first in phosphatidylethanolamine and, later, in the galactolipids. Once more, there are obviously important control mechanisms here but little information is available.

7. *Control by the environment*

The influence of light on plant growth is important at several stages including germination, development of chloroplasts, steady-state photosynthesis and flowering. It follows that light also influences fatty acid synthesis. Much of this work has already been mentioned but a few examples are:

(a) dependence on light for fatty acid synthesis in pea etioplasts (Panter and Boardman, 1973).
(b) stimulation by light of acetate incorporation into mature or young chloroplasts (Stumpf *et al.*, 1967)
(c) induction of specific fatty acid synthesizing systems by light (e.g. Delo *et al.*, 1971)
(d) changes in the pattern of acetate incorporation which may not be directly related to (c) (Trémolières and Lepage, 1971). This is particularly true for linolenic acid (Trémolières, 1972).

The stimulation of chloroplast development by light can be inhibited by chloramphenicol which prevents formation of chlorophyll, chloroplast protein and galactosyldiglyceride whereas synthesis of phospholipids is unaffected. Cycloheximide, on the other hand, prevents phospholipid syn thesis but, after an initial lag phase, chlorophyll and galactosyldiglycerides accumulate at an accelerated rate (Bishop and Smillie, 1970). Clearly, control of fatty acid synthesis by light is complex. The use of specific inhibitors *in vivo* may offer the best chance for further understanding of these effects.

The influence of temperature on plant growth and fatty acid synthesis has been studied by a number of workers, and some of the observed effects have been rationalized. An increase in unsaturation at lower temperatures has been attributed to an increase in dissolved O_2 (Harris and James, 1969a, b) and to induction of specific desaturase (Fulco, 1970). Conversely, as the growth temperature is raised, more saturated fatty acids are synthesized; a typical example is *Cyanidium caldarium* (Table XVII). Apparently the organism attempts to adjust for changes in membrane packing caused by temperature through alterations in its fatty acid composition (Wakil and Esfahani, 1972). In *Chlorella sorokiniana*, however, although the proportion of unsaturated acids was altered by growth temperature, the average chain length was also changed such that the average melting point of the constituent fatty acids did not fall as the temperature was lowered (Patterson, 1970).

TABLE XVII

Changes in the fatty acids of *Cyanidium caldarum* with growth temperature [From Kleinschmidt and McMahon (1970)]

Lipid fraction	Growth temperature (°C)	Fatty acid composition (% total)				
		16:0	18:0	18:1	18:2	18:3
total	20	18	7	34	11	30
	55	53	5	21	21	0
monogalactosyl-diglyceride	20	26	0	5	12	57
	55	54	2	20	23	0
digalactosyl-diglyceride	20	30	0	6	17	47
	55	46	2	16	34	0

Short- and long-term exposure of tobacco leaves to high temperature (47–49°C) caused alterations in lipid synthesis and a tendency for chloride to easily leak from the tissue (Benzioni and Itai, 1973). In pea seedlings grown at 35°C a marked change in the wax composition was noted (Harwood and Stumpf, 1971) and similar results are also obtained with runner bean (Harwood, 1974a). Typical wax components such as hydrocarbons and long chain alcohols and esters are reduced, apparently because of a decrease in formation of the very long chain fatty acids at 35°C (Harwood and Stumpf, 1971) (Table XVIII). The *in vitro* formation of these acids by microsomes is severely reduced at 35°C apparently because the elongation system is rather heat labile.

Several groups have studied the changes in lipids with relation to cold hardiness. De la Roche *et al.* (1972) and Yoshida and Sakai (1973) have

TABLE XVIII

Effect of inhibitors and temperature on the synthesis of fatty acids by germinating pea seeds

[From Harwood and Stumpf (1971)]

	Fatty acids synthesized (% total)				
	16:0	18:0	20:0	22:0	24:0
control (25°C)	27	55	5	5	8
+ fluoride (10^{-3}M)	31	66	1	2	tr
+ [a]EPTC (10^{-5}M)	33	66	1	0	0
+ arsenite (10^{-3}M)	87	13	0	0	0
germination at 35°C	28	70	1	1	tr

Seeds were germinated in the presence of ^{14}C-acetate

[a] [EPTC = *S*-ethyl dipropylthiocarbonate (a carbamate weed killer)].

reported that lower temperatures increase the content of phospholipids in germinating wheat seedlings and poplar stem respectively. A marked increase in linolenic acid was also noted in wheat seeds grown at 2°C in comparison to those at 24°C. For seedlings at comparable morphological stages of growth, their phospholipid composition was similar (De la Roche *et al.*, 1973). Changes in the growth temperature of alfalfa, on the other hand, did not alter the fatty acid composition of individual lipids—variations were due to increases in the relatively more saturated lipids (phosphatidylinositol, phosphatidylglycerol and sulpholipid) at high temperatures or in the more unsaturated lipids (galactosyldiglycerides, phosphatidylcholine and phosphatidylethanolamine) at lower temperatures (Kuiper, 1970). In a study of drought tolerance and earliness in spring, Kuiper and Stuiver (1972) noted the presence of long-chain cyclopropane fatty acids in resistant plants. Up to 80% of the sulpholipid fatty acids in early spring cow parsley and 70% of the phosphatidyl choline fatty acids in drought-tolerant *Corynephorus canescans* were cyclopropane types. By May, or after a period of rain, these were reduced to 0 and 20% respectively.

A number of simple chemicals will cause pronounced changes in fatty acid metabolism *in vivo*. Fluoride and various carbonate weed killers inhibit stearate elongation in pea whilst arsenite prevents palmitate elongation (Harwood and Stumpf, 1971). These effects have been ascribed to inhibition of specific enzymes (Harwood and Stumpf, 1971, 1972b). Typical results are shown in Table XVIII. Appleby *et al.* (1974) have shown that trichloroacetate will inhibit the elongation of oleic acid by *Crambe* enzymes. This reagent has already been mentioned as preventing very long fatty acid synthesis in the epidermis (p.59).

Deficiencies of nitrogen and minerals will lead to a decline in nutritional status, metabolism and, hence, lipid synthesis in plants. Thus lack of Fe, Mn, and N each caused depression in the levels of polyunsaturated fatty acids in leaves but these elements and S, P, K, Ca and Mg were all without marked effect on the development of seed oils (Hitchcock and Nichols, 1971; Sedgwick, 1973). Newman (1966) found that nitrogen deficiency caused a decline in the unsaturates:saturates ratio and also decreased acetate incorporation into the linolenic acid of isolated chloroplasts. This suggests a direct control by nitrogen. In *Chlorella*, very high concentrations of CO_2 increase fatty acid synthesis, especially palmitate and oleate (Dickson *et al.*, 1969).

8. *Genetic Control*

This is of particular importance commercially where species which synthesize storage oils of desired composition can be selected. Breeding trials have included those for flax, rape, safflower and sunflower. In safflower there are three genes at one chromosome locus which determine the proportions of linoleic and oleic acids while two genes at a second locus control stearic

TABLE XIX

Oil content and fatty acid composition in varieties of rapeseed.
[From Downey and Craig (1964)]

Plant genotype	Oil content (%)	Fatty acid composition (% total)								
		16:0	16:1	18:0	18:1	18:2	18:3	20:0	20:1	22:1[a]
+ + + + (high erucic)	39·0	3·8	0·4	0·8	18·4	15·8	9·8	0·8	11·0	39·2
− + + +	45·5	3·8	0·5	1·3	24·3	15·7	9·1	1·0	13·9	30·4
− − + +	38·9	4·1	0·4	1·4	34·2	16·3	7·9	0·7	14·1	20·9
− − − +	39·4	4·8	0·7	0·8	41·8	18·7	7·8	1·0	12·1	12·8
− − − − (zero erucic)	43·4	4·8	0·9	1·2	63·1	20·0	8·2	0·8	1·0	0·0

[a] 22:1(13*c*) is erucic acid

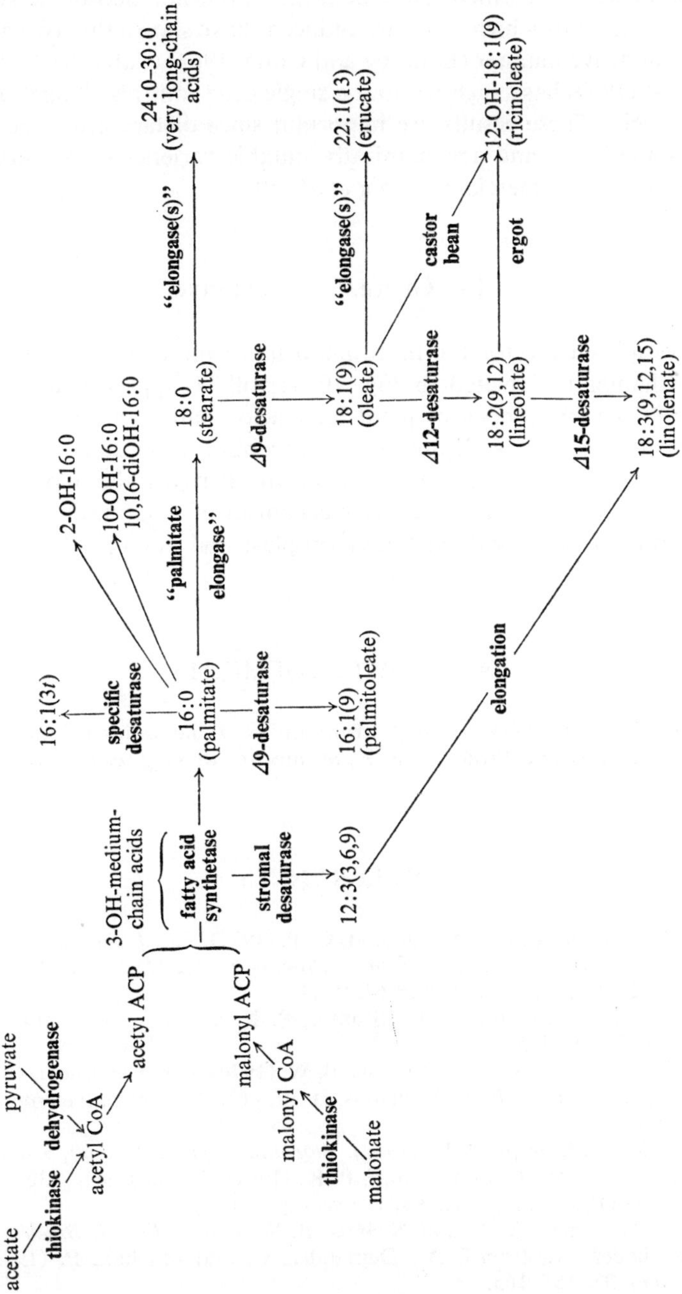

FIG. 10. Summary of the known enzymic pathways for fatty acid biosynthesis in plants. (Unless otherwise shown, all unsaturated bonds are *cis*.)

acid levels (Krzymanski and Downey, 1969; Knowles, 1969). Evidence with rapeseed varieties, in which erucic acid (22:1) but not eicosenoic acid (20:1) decline, or those in which both are reduced, have shown that two genes may act in an additive manner (Downey and Craig, 1964) (Table XIX). However, Appelqvist (1969) has suggested that a single gene controls elongation of oleic to erucic acid. These results are important since dietary erucic acid is considered undesirable and agronomically suitable varieties (2–5% erucic acid) are now grown commercially (Craig *et al.*, 1973).

IX. General Conclusions

An overall scheme for the more important synthetic pathways described in this Chapter is depicted in Fig. 10. It will be appreciated that a large increase in our knowledge of plant fatty acid anabolism has been made in recent years but that, with one or two exceptions, the details of individual enzyme steps are still relatively unknown. Perhaps the most important problems are those concerning the mechanism of desaturation, the specific (if any) role of fatty acids in the chloroplast and attempts to control biosynthesis.

ACKNOWLEDGEMENTS

I would like to thank Drs J. M. Basford and A. T. James for helpful discussions and Dr M. I. Gurr and Professor P. K. Stumpf for making results available to me prior to publication.

REFERENCES

Abita, J. P., Lazdunski, M. and Ailhaud, G. P. (1971). *Eur. J. Biochem.* **23**, 412–420.

Alberts, R. W. and Vagelos, P. R. (1968). *Proc. natn. Acad. Sci. U.S.A.* **59**, 561–568.

Appelqvist, L. A. (1969). *Hereditas* **61**, 9–44.

Appelqvist, L. A., Boynton, J. E., Stumpf, P. K. and Von Wettstein, D. (1968). *J. Lipid Res.* **9**, 425–436.

Appelby, R. S., Gurr, M. I. and Nichols, B. W. (1974). *Eur. J. Biochem.* **48**, 209–216.

Appleby, R. S., Safford, R. and Nichols, B. W. (1971). *Biochim. biophys. Acta* **248**, 205–211.

Barron, E. J. and Stumpf, P. K. (1962). *Biochim. biophys. Acta* **60**, 329–337.

Barron, E. J., Squires, C. and Stumpf, P. K. (1961). *J. biol. Chem.* **236**, 2610–2614.

Bartels, C. T. (1969). Ph.D. Thesis, University of Utrecht.

Bartels, C. T., James, A. T. and Nichols, B. W. (1967). *Eur. J. Biochem.* **3**, 7–10.

Ben Abdelkader, A., Cherif, A., Demandre, C. and Mazliak, P. (1973) *Eur. J. Biochem.* **32**, 155–165.

Ben Abdelkader, A., Mazliak, P. and Catesson, A. M. (1969). *Phytochemistry* **8**, 1121–1137.

Bennett, A. S. and Quackenbush, F. W. (1969). *Archs Biochem. Biophys.* **130**, 567–572.

Benzioni, A. and Itai, C. (1973). *Physiologia Pl.* **28**, 493–497.

Berg, P. (1956). *J. biol. Chem.* **222**, 991–1013.

Bishop, D. G. and Smillie, R. M. (1970). *Archs Biochem. Biophys.* **139**, 179–189.

Bloch, K. (1969). *Acc. Chem. Res.* **2**, 193–202.

Bloomfield, D. K. and Bloch, K. (1960). *J. biol. Chem.* **235**, 337–345.

Boatman, S. G. and Crombie, W M. (1958). *J. exp. Bot.* **9**, 52–74.

Brett, D., Howling, D., Morris, L. J. and James, A. T. (1971). *Archs Biochem. Biophys.* **143**, 535–547.

Brindley, D. W., Matsumura, S. and Bloch, K. (1969). *Nature, Lond.* **224**, 666–669.

Brooks, J. L. and Stumpf, P. K. (1966). *Archs Biochem. Biophys.* **116**, 108–116.

Burton, D and Stumpf, P. K. (1966). *Archs Biochem. Biophys.* **117**, 604–614.

Canvin, D. T. (1963). *Can. J. Biochem. Physiol.* **41**, 1879–1885.

Canvin, D. T. (1965). *Can. J. Bot.* **43**, 49–62.

Carey, E. M., Dils, R. and Hansen, H. J. M. (1970). *Biochem. J.* **117**, 633–635.

Cheniae, G. M. and Kerr, P. C. (1965). *Pl. Physiol., Lancaster* **40**, 452–457.

Craig, B. M., Mallard, T. M., Wight, R. E., Irvine, G. N. and Reynolds, J. R. (1973). *J. Am. Oil Chem. Soc.* **50**, 395–403.

Crombie, W. M. (1958). *J. exp. Bot.* **9**, 254–261.

Dasgupta, S. K. and Friend, J. (1973). *J. Sci. Fd Agric.* **24**, 463–470.

Davey, M. R., Cocking, E. C. and Bush, E. (1973). *Nature, Lond.* **244**, 460.

Davidoff, F. and Korn, E. D. (1964). *J. biol. Chem.* **239**, 2496–2506.

De la Roche, I. A., Andrews, C. J., Pomeroy, M. K., Weinberger, P. and Kates, M. (1972). *Can. J. Bot.* **50**, 2401–2410.

De la Roche, I. A., Andrews, C. J. and Kates, M. (1973). *Pl. Physiol., Lancaster* **51**, 468–473.

Delo, J., Ernst-Fonberg, M. L. and Bloch K. (1971). *Archs Biochem. Biophys.* **143**, 384–391.

Devor, K. A. and Mudd, J. B. (1971). *J. Lipid Res.* **12**, 412–419.

Dickson, L. G., Galloway, R. A. and Patterson, G. W. (1969). *Pl. Physiol., Lancaster* **44**, 1413–1416.

Downey, R. K. and Craig, B. W. (1964). *J. Am. Oil Chem. Soc.* **41**, 475–478.

Draper, S. R. (1969). *Phytochemistry* **8**, 1641–1647.

Draper, S. R. and Simon, E. W. (1970). *Phytochemistry* **9**, 1997–2002.

Drennan, C. H. and Canvin, D. T. (1969). *Biochim. biophys. Acta* **187**, 193–200.

Dutton, H. J. and Mounts, T. L. (1966). *J. Lipid Res.* **7**, 221–225.

Eberhardt, F. M. and Kates, M. (1957). *Can. J. Bot.* **35**, 907–921.

Elovson, J. (1964). *Biochim. biophys. Acta* **84**, 275–293.

Ernst-Fonberg, M. L. (1973). *Biochemistry, Easton* **12**, 2449–2455.

Ernst-Fonberg, M. L. and Bloch, K. (1971). *Archs Biochem. Biophys.* **143**, 392–400.

Fletcher, J. S. and Beevers, H. (1970). *Pl. Physiol., Lancaster* **45**, 765–772.

Fulco, A. J. (1970). *J. biol. Chem.* **245**, 2985–2990.

Fulco, A. J. and Bloch, K. (1964). *J. biol. Chem.* **239**, 993–997.

Galliard, T. and Stumpf, P. K. (1966). *J. biol. Chem.* **241**, 5806–5812.

Galliard, T. and Stumpf, P. K. (1968). *In* "Biochemical Preparations" (W. E. M. Lands, ed.), Vol. 12, p. 66. Wiley, New York.

Galliard, T., Rhodes, M. J. C., Wooltorton, L. S. C. and Hulme, A. C. (1968). *Phytochemistry* **7**, 1453–1463.

Givan, C. V. and Stumpf, P. K. (1971). *Pl. Physiol., Lancaster* **47**, 510–516.

Goldfine, H., Ailhaud, G. P. and Vagelos, P. R. (1967). *J. biol. Chem.* **242**, 4466–4475.

Gresbios, M. (1971). *Phytochemistry* **10**, 1261–1273.

Gurr, M. I. (1971). *Lipids* **6**, 266–273.
Gurr, M. I. and Bloch, K. (1966). *Biochem. J.* **99**, 16c–18c.
Gurr, M. I. and Brawn, P. (1970). *Eur. J. Biochem.* **17**, 19–22.
Gurr, M. I., Robinson, M. P. and James, A. T. (1969). *Eur. J. Biochem.* **9**, 70–78.
Gurr, M. I., Blades, J. and Appleby, R. S. (1972a). *Eur. J. Biochem.* **29**, 362–368.
Gurr, M. I., Robinson, M. P., James, A. T., Morris, L. J. and Howling, D. (1972b). *Biochim. biophys. Acta* **280**, 415–421.
Gurr, M. I., Blades, J., Appleby, R. S., Smith, C. G., Robinson, M. P. and Nichols, B. W. (1974). *Eur. J. Biochem.* **43**, 281–290.
Hancock, W. S., Marshall, G. R. and Vagelos, P. R. (1973). *J. biol. Chem.* **248**, 2424–2434.
Hardman, E. E. and Crombie, W. M. (1958). *J. exp. Bot.* **9**, 239–246.
Harris, P. and James, A. T. (1969a). *Biochem. J.* **112**, 325–330.
Harris, P. and James, A. T. (1969b). *Biochim. biophys. Acta* **187**, 13–18.
Harris, R. V. and James, A. T. (1965). *Biochim. biophys. Acta* **106**, 456–464.
Harris, R. V., James, A. T. and Harris, P. (1967). *In* "Biochemistry of the Chloroplasts" (T. W. Goodwin, ed.), Vol. 2, pp. 239–253. Academic Press, New York and London.
Harwood, J. L. (1974a). Unpublished observations.
Harwood, J. L. (1974b). *Lipids* **9**, 850–854.
Harwood, J. L. and James, A. T. (1975). *Eur. J. Biochem.* **50**, 325–334.
Harwood, J. L. and Stumpf, P. K. (1970). *Pl. Physiol. Lancaster* **46**, 500–508.
Harwood, J. L. and Stumpf, P. K. (1971). *Archs Biochem. Biophys.* **142**, 281–291.
Harwood, J. L. and Stumpf, P. K. (1972a). *Lipids* **7**, 8–19.
Harwood, J. L. and Stumpf, P. K. (1972b). *Archs Biochem. Biophys.* **148**, 282–290.
Harwood, J. L., Sodja, A., Stumpf, P. K. and Spurr, A. R. (1971). *Lipids* **6**, 851–854.
Harwood, J. L., Sodja, A. and Stumpf, P. K. (1972). *Biochem. J.* **130**, 1013–1018.
Hatch, M. D. and Stumpf, P. K. (1961). *J. biol. Chem.* **236**, 2879–2885.
Hatch, M. D. and Stumpf, P. K. (1962). *Pl. Physiol., Lancaster* **37**, 121–126.
Hawke, J. C. and Stumpf, P. K. (1965a). *Pl. Physiol., Lancaster* **40**, 1023–1032.
Hawke, J. C. and Stumpf, P. K. (1965b). *J. biol. Chem.* **240**, 4746–4752.
Heinstein, P. F. and Stumpf, P. K. (1969). *J. biol. Chem.* **244**, 5374–5381.
Hitchcock, C., Morris, L. J. and James, A. T. (1968). *Eur. J. Biochem.* **3**, 419–423.
Hitchcock, C. and Nichols, B. W. (1971). "Plant Lipid Biochemistry". Academic Press, London and New York.
Holden, M. (1970). *Phytochemistry* **9**, 507–512.
Howling, D., Morris, L. J. and James, A. T. (1968). *Biochim. biophys. Acta* **152**, 224–226.
Howling, D., Morris, L. J., Gurr, M. I. and James, A. T. (1972). *Biochim. biophys. Acta* **260**, 10–19.
Huang, K. P. and Stumpf, P. K. (1970). *Archs Biochem. Biophys.* **140**, 158–173.
Huang, K. P. and Stumpf, P. K. (1971). *Archs Biochem. Biophys.* **193**, 412–427.
Inkpen, J. A. and Quackenbush, F. W. (1969). *Lipids* **4**, 539–543.
Jacobsen, B. S., Kannangara, C. G. and Stumpf, P. K. (1973a). *Biochem. biophys. Res. Commun.* **51**, 487–493.
Jacobsen, B. S., Kannangara, C. G. and Stumpf, P. K. (1973b). *Biochem. biophys. Res. Commun.* **52**, 1190–1198.
Jacobsen, B. S. and Stumpf, P. K. (1972). *Archs Biochem. Biophys.* **153**, 656–663.
James, A. T. (1962a). *Biochim. biophys. Acta* **57**, 167–169.
James, A. T. (1962b). *Bull. Soc. Chim. Biol.* **44**, 951–963.
James, A. T. (1963). *Biochim. biophys. Acta* **70**, 9–19.
James, A. T. and Nichols, B. W. (1966). *Nature, Lond.* **210**, 372–375.
James, A. T., Harris, P. and Bezard, J. (1968). *Eur. J. Biochem.* **3**, 318–325.

James, A. T., Hadaway, H. C. and Webb, J. P. W. (1965). *Biochem. J.* **95**, 448–452.
Jaworski, J. G. and Stumpf, P. K. (1974a). *Archs Biochem. Biophys.* **162**, 166–173.
Jaworski, J. G. and Stumpf, P. K. (174b). *Archs Biochem. Biophys.* **162**, 158–165.
Jaworski, J. G., Goldschmidt, E. E. and Stumpf, P. K. (1974). *Archs Biochem. Biophys.* **163**, 769–776.
Kaimal, T. N. B. and Lakshminarayana, G. (1972). *Phytochemistry* **11**, 1617–1622.
Kalra, S. K. and Brooks, J. L. (1973). *Phytochemistry* **12**, 487–492.
Kannangara, C. G. and Stumpf, P. K. (1971). *Biochem. biophys. Res. Commun.* **44**, 1544–1551.
Kannangara, C. G. and Stumpf, P. K. (1972a). *Archs Biochem. Biophys.* **148**, 414–424.
Kannangara, C. G. and Stumpf, P. K. (1972b). *Archs Biochem. Biophys.* **152**, 83–91.
Kannangara, C. G. and Stumpf, P. K. (1973). *Archs Biochem. Biophys.* **155**, 361–369.
Kannangara, C. G., Jacobson, B. S. and Stumpf, P. K. (1973). *Biochem. biophys. Res. Commun.* **52**, 648–655.
Kannangara, C. G., Henningsen, K. W., Stumpf, P. K., Appelqvist, L. A. and Von Wettstein, D. (1971). *Pl. Physiol., Lancaster* **48**, 526–531.
Kashiwabara, Y. and Sato, R. (1973). *J. Biochem., Tokyo* **74**, 405–414.
Kates, M. (1960). *Biochim. biophys. Acta* **41**, 315–328.
Kates, M. and Paradis, M. (1973). *Can. J. Biochem.* **51**, 184–197.
Khan, A. A. and Kolattukudy, P. E. (1973). *Biochemistry, Easton* **12**, 1939–1947.
Kleinschmidt, M. G. and McMahon, V. A. (1970). *Pl. Physiol., Lancaster* **46**, 290–293.
Knowles, P. F. (1969). *J. Am. Oil Chem. Soc.* **46**, 130–132.
Kolattukudy, P. E (1967). *Phytochemistry* **6**, 963–975.
Kolattukudy, P. E. and Buckner, J. S. (1972). *Biochem. biophys. Res. Commun.* **46**, 801–807.
Krzymanski, J. and Downey, R. K. (1969). *Can. J. Pl. Sci.* **49**, 313–319.
Kuiper, P. J. C. (1970). *Pl. Physiol., Lancaster* **45**, 684–686.
Kuiper, P. J. C. and Stuiver, B. (1972). *Pl. Physiol., Lancaster* **49**, 307–309.
Leech, R. M., Rumsby, M. G. and Thomson, W. W. (1973). *Pl. Physiol., Lancaster* **52**, 240–245.
Lynen, F. (1961). *Fed. Proc. Fedn Am. Socs exp. Biol.* **20**, 941–951.
Macey, M. J. K. and Stumpf, P. K. (1968). *Pl. Physiol., Lancaster* **43**, 1637–1647.
Majerus, P. W. (1968). *Science, N.Y.* **159**, 428–430.
Majerus, P. W. and Vagelos, P. R. (1967). *Adv. Lipid Res.* **5**, 1–30.
Marsh, J. B. and James, A. T. (1962). *Biochim. biophys. Acta* **60**, 320–328.
Matsuhashi, M., Matsuhashi, S. and Lynen F. (1964). *Biochem. Z.* **340**, 263–289.
Matsumura, S. and Stumpf, P. K. (1968). *Archs Biochem. Biophys.* **125**, 932–941.
Mazliak, P. (1973). *A. Rev. Pl. Physiol.* **24**, 287–310.
Mazliak, P. and Ben Abdelkader, A. (1970). *Rev. gen. Bot.* **77**, 53–71.
Mazliak, P., Stoll, U. and Ben Abdelkader, A. (1968). *Biochim. biophys. Acta* **152**, 414–417.
Mazliak, P., Oursel, A., Ben Abdelkader, A. and Grosbois, M. (1972). *Eur. J. Biochem.* **28**, 399–411.
McMahon, V. and Stumpf, P. K. (1964). *Biochim. biophys. Acta* **84**, 359–361.
McMahon, V. and Stumpf, P. K. (1966). *Pl. Physiol., Lancaster* **41**, 148–156.
Morris, L. J. (1967). *Biochem. biophys. Res. Commun.* **29**, 311–315.
Morris, L. J. (1970). *Biochem. J.* **118**, 681–693.
Morris, L. J., Hall, S. W. and James, A. T. (1966). *Biochem. J.* **100**, 29c–30c.
Morris, L. J., Harris, R. V., Kelly, W. and James, A. T. (1968). *Biochem. J.* **109**, 673–678.

Mudd, J. B. and McManus, T. T. (1962). *J. biol. Chem.* **237**, 2057–2063.
Mudd, J. B. and McManus, T. T. (1964). *Pl. Physiol., Lancaster* **39**, 115–119.
Mudd, J. B. and Stumpf, P. K. (1961). *J. biol. Chem.* **236**, 2602–2609.
Nagai, J. and Bloch, K. (1968). *J. biol. Chem.* **243**, 4626–4633.
Newcomb, E. H. and Stumpf, P. K. (1953). *J. biol. Chem.* **200**, 233–239.
Newman, D. W. (1966). *Pl. Physiol., Lancaster* **41**, 328–334.
Newman, D. W., Rowell, B. W. and Byrd, K. (1973). *Pl. Physiol., Lancaster* **51**, 229–233.
Nichols, B. W., Wood, J. B. and James, A. T. (1965a). *Biochem. J.* **95**, 6P.
Nichols, B. W., Harris, P. and James, A. T. (1965b). *Biochem. biophys. Res. Commun.* **21**, 473–479.
Nichols, B W., James, A. T. and Breuer, J. (1967). *Biochem. J.* **104**, 486–496.
Numa, S., Ringelmann, E. and Reidel, B. (1966). *Biochem. biophys. Res. Commun.* **24**, 750–757.
Olsen, J. A. (1966). *A. Rev. Biochem.* **35**, 559–598.
Overath, P. and Stumpf, P. K. (1964). *J. biol. Chem.* **239**, 4103–4110
Panter, R A and Boardman, N. K (1973). *J. Lipid Res.* **14**, 664–671.
Patterson, G. W. (1970). *Lipids* **5**, 597–600.
Poincelot, R. P. (1973). *Pl. Physiol., Lancaster* **51**, 802–804.
Pugh, E. L. and Kates, M. (1973). *Biochim. biophys. Acta* **316**, 305–316.
Raju, P. K. and Reiser, R. (1969). *Biochim. biophys. Acta* **176**, 48–53.
Rebeiz, C. A., Castelfranco, P. and Breidenbach, R. W. (1965). *Pl. Physiol., Lancaster* **40**, 286–289.
Rinne, R. W. (1969). *Pl. Physiol., Lancaster* **44**, 89–94.
Rossi, C. R., Galzigna, L. and Gibson, D. M. (1969). *In* "Methods in Enzymology" (J. M. Lowenstein, ed.), Vol. XIV, pp. 91–95. Academic Press, New York and London.
Roughan, P. G. (1970). *Biochem. J.* **117**, 1–8.
Sedgwick, B. (1973). *In* "Biosynthesis and its Control in Plants" (B. V. Milbarrow, ed.), pp. 179–217. Academic Press, London and New York.
Shannon, L. M., de Vellis, J. and Law, J. Y. (1963). *Pl. Physiol., Lancaster* **38**, 691–697.
Shewry, P. R., Pinfield, N. J. and Stobart, A. K. (1972). *Phytochemistry* **11**, 2149–2157.
Simoni, R. D. and Stumpf, P. K. (1969). *In* "Methods in Enzymology" (J. M. Lowenstein, ed.), Vol. XIV, pp. 84–88. Academic Press, New York and London.
Simoni, R. D., Criddle, R. S. and Stumpf, P. K. (1967). *J. biol. Chem.* **242**, 573–581.
Sirevag, R. and Levine, R. P. (1972). *J. biol. Chem.* **247**, 2586–2591.
Smirnov, B. P. (1960). *Biokhimiya* **25**, 419–423.
Sodja, A. and Stumpf, P. K. (1973). Unpublished results.
Squires, C. L. and Stumpf, P. K. (1959). *Fed. Proc. Fedn Am. Socs exp. Biol.* **18**, 328.
Squires, C. L., Stumpf, P. K. and Schmid, C. (1958). *Pl. Physiol., Lancaster* **33**, 365–366.
Still, G. G., Davis, D. G. and Zander, G. (1970). *Pl. Physiol., Lancaster* **46**, 307–314.
Stumpf, P. K. (1969). *A. Rev. Biochem.* **38**, 159–207.
Stumpf, P. K. (1970). *In* "Comprehensive Biochemistry" (M. Florkin and E. M. Stotz, eds), Vol. 18, pp. 265–292. Elsevier, Amsterdam.
Stumpf, P. K. and Barber, G. A. (1957). *J. biol. Chem.* **227**, 407–417.
Stumpf, P. K. and James, A. T. (1962). *Biochim. biophys. Acta* **57**, 400–402.
Stumpf, P. K. and James, A. T. (1963). *Biochim. biophys. Acta* **70**, 20–32.
Stumpf, P. K., Brooks, J., Galliard, T., Hawke, J. C. and Simoni, R. (1967). *In*

"Biochemistry of Chloroplasts" (T. W. Goodwin, ed.), Vol. 11, pp. 213–239. Academic Press, London and New York.

Stumpf, P. K., Vijay, I. K. and Harwood, J. L. (1973). *In* "Current Trends in the Biochemistry of Lipids" (J. Ganguly and R. M. S. Smellie, eds), pp. 57–63. Academic Press, London and New York.

Sumper, M., Oesterhelt, D., Riepertinger, C. and Lynen, F. (1969). *Eur. J. Biochem.* **10**, 377–387.

Thibaudin, A., Mazliak, P. and Catesson, A. M. (1968). *Seances Acad. Sci. Series D., Nature (Paris)* **266**, 784–787.

Trémolières, A. (1972). *Phytochemistry* **11**, 3453–3460.

Trémolières, A. (1973). *Proc. 16th Int. Conf. on Biochem. of Lipids*, Nottingham, p. 21.

Tremolieres, A. and Lepage M. (1971). *Pl. Physiol., Lancaster* **47**, 329–334.

Vanaman, T. C., Wakil, S. J. and Hill, R. L. (1968). *J. biol. Chem.* **243**, 6420–6431.

Vijay, I. K. and Stumpf, P. K. (1971). *J. biol. Chem.* **246**, 2910–2917.

Vijay, I. K. and Stumpf, P. K. (1972). *J. biol. Chem.* **247**, 360–366.

Volpe, J. J. and Vagelos, P. R. (1973). *A. Rev. Biochem.* **42**, 21–60.

Wakil, S. J. (1970). "Lipid Metabolism" (S. J. Wakil, ed.). Academic Press, New York.

Wakil, S. J. and Esfahani, M. (1972). *In* "Current Trends in the Biochemistry of Lipids" (J. Ganguly and R. M. S. Smellie, eds), pp. 395–406. Academic Press, London and New York.

Weber, E. J. (1969). *J. Am. Oil Chem. Soc.* **46**, 485–488.

Weaire, P. J. and Kekwick, R. G. O. (1970). *Biochem. J.* **119**, 48–49P.

Weaire, P. J. and Kekwick, R. G. O. (1973). *Biochem. Soc. Transactions* **1**, 899–900.

Willemot, C. and Stumpf, P. K. (1967a). *Pl. Physiol., Lancaster* **42**, 391–397.

Willemot, C. and Stumpf, P. K. (1967b). *Can. J. Bot.* **45**, 579–584.

Willemot, C. and Verret, G. (1973). *Lipids* **8**, 588–591.

Yamada, M. and Stumpf, P. K. (1964). *Biochem. biophys. Res. Commun.* **14**, 165–171.

Yamada, M. and Stumpf, P. K. (1965). *Pl. Physiol., Lancaster* **40**, 653–658.

Yang, S. F. and Stumpf, P. K. (1965a). *Biochim. biophys. Acta* **98**, 19–26.

Yang, S. F. and Stumpf, P. K. (1965b). *Biochim. biophys. Acta* **98**, 27–35.

Yoshida, S. and Sakai, A. (1973). *Pl. Cell Physiol.* **14**, 353–360.

Yuan, C. and Bloch, K. (1961). *J. biol. Chem.* **236**, 1277–1279.

Zilkey, B. and Canvin, D. T. (1969). *Biochem. biophys. Res. Commun.* **34**, 646–653.

Zilkey, B. and Canvin, D. T. (1972). *Can. J. Bot.* **50**, 323–327.

Zimmerman, D. C. and Klosterman, H. J. (1965). *J. Am. Oil Chem. Soc.* **42**, 58–62.

CHAPTER 4

Biosynthesis of Fatty Acids in Spinach Chloroplasts*

P. K. STUMPF

Department of Biochemistry and Biophysics
University of California, Davis, California, U.S.A.

I. General Introduction

The chloroplast is a unique organelle in the plant cell since its stroma and lamellar phases house the enzyme systems responsible for CO_2 fixation and oxygen evolution, two key processes essential for the maintenance of life on this planet.

The stroma phase contains the complete enzymatic machinery for the conversion of CO_2 to starch and sucrose and for the transport of high energy bonds and reducing potential to the cytosolic compartments. In addition it contains all the enzymes responsible for *de novo* synthesis of fatty acids. The lamellar membrane, on the other hand, consisting of about 50% protein and 50% lipids, is the site for oxygen evolution and the generation of high energy phosphate and reducing potential. Indeed, the products of the membrane-localized non-cyclic photophosphorylation, namely ATP, NADPH and molecular oxygen, are essential components for fatty acid synthesis and desaturation.

* The research discussed here was supported by grants NIH GM 19213–02 and NSF GB 37880X.

Clearly a fuller understanding of the mechanisms of lipid biosynthesis and of the regulation of these synthetic functions in chloroplasts would be relevant for a deeper insight into the biochemical events occurring in this organelle.

By 1967, the general outline of the biosynthetic events leading to the incorporation of acetate into long chain fatty acids was already known in chloroplasts (Stumpf *et al.*, 1967). Since then a marked refinement of the basic observations has been made. Thus, considerable information is now available for chloroplasts concerning acetyl CoA carboxylase, acyl carrier protein, the *de novo* as well as the elongation systems for long chain fatty acid synthesis and the several desaturation systems which form the monoenoic, dienoic and trienoic fatty acids. The previous chapter in this volume has reviewed the general aspects of fatty acid biosynthesis in plants; the following discussion will now examine in more detail these systems in chloroplasts.

II. Initial Phases of Fatty Acid Synthesis

A. Movement of Acetate across Chloroplast Membranes

The specificity of permeation of substrates into chloroplasts is well established. Thus, the nicotinamide nucleotides and a number of phosphorylated sugars do not pass through the outer envelope of chloroplasts.

In studies on the biosynthesis of fatty acids with isolated intact chloroplasts, the conventional substrate is ^{14}C-acetate. Previous workers had provided evidence that the acetate anion penetrated slowly into the stroma phase (Heldt and Rapley, 1970). Recently, Jacobson, in our laboratory, clearly demonstrated that external acetate moved rapidly into the stroma phase (Jacobson and Stumpf, 1972). The $t_{\frac{1}{2}}$ for equilibration of 1 mM acetate across the envelope membrane in the dark was less than 5 s while the penetration across the lamellar membrane was much longer, over 30 min at pH 7·9. Thus, the movement of free acetate into the stroma phase of intact chloroplasts is rapid and is certainly not rate-limiting.

The origin of acetate in leaf tissue has not as yet been fully clarified although one could invoke either the generation of acetate by leaf mitochondria and its diffusion out into the cytosol with a subsequent movement into the intact chloroplasts, or the generation of acetyl CoA *in situ* in the chloroplast. That the latter situation might exist is suggested by recent evidence in our laboratory that pyruvic dehydrogenase activity is associated with lamellar membranes (J. Lina and P. K. Stumpf, unpublished results). When either [1-^{14}C]pyruvate or [2-^{14}C]pyruvate is incubated with intact chloroplasts, a release of $^{14}CO_2$ and the formation of ^{14}C-fatty acids respectively, identical in pattern to those formed from added ^{14}C-acetate, is consistently observed. Further investigation of this system is merited to answer the question of the nature of endogenous precursor for fatty acid biosynthesis.

B. ACTIVATION AND CONVERSION OF ACETYL CoA TO MALONYL CoA

While no attempts have been made in our laboratory to characterize the acetyl CoA synthetase in chloroplasts, our evidence clearly indicates a stable soluble enzyme which occurs in the stroma phase and which converts acetate to acetyl CoA in the presence of ATP and Mg^{2+}. However, the acetyl CoA synthetase of potato tubers has been thoroughly examined (Huang and Stumpf, 1970).

Brooks and Stumpf (1966) demonstrated the presence of a very labile acetyl CoA carboxylase in disrupted chloroplasts. In the same year, Burton and Stumpf (1966) showed that an endogenous inhibitor in chloroplasts markedly decreased the activity not only of the chloroplast carboxylase but also of a well defined wheat germ acetyl CoA carboxylase. Nevertheless, the problem concerning the low activity of chloroplast acetyl CoA carboxylase remained a puzzle for a number of years. In the meantime Alberts and Vagelos (1968) made the important observation that in *E. coli*, a soluble acetyl CoA carboxylase could be readily separated into three distinct proteins (biotin carboxyl carrier protein (BCCP) biotin carboxylase and transcarboxylase) which catalyzed the following reactions (see also p. 45):

$$HCO_3^- + ATP + BCCP \xrightarrow[\text{biotin carboxylase}]{Mg^{2+}} CO_2^- \sim BCCP + ADP + P_i \qquad (1)$$

$$CO_2^- \sim BCCP + \text{acetyl CoA} \xrightarrow{\text{transcarboxylase}} \text{malonyl CoA} + BCCP \qquad (2)$$

Since chloroplasts have some procaryotic-like properties, a reexamination of the chloroplast acetyl CoA carboxylase was initiated with the *E. coli* carboxylase as a working model (Kannangara and Stumpf, 1972a). It was necessary to show: (a) a component of the chloroplast capable of binding CO_2 which could then be isolated and employed as a substrate for the formation of malonyl CoA from acetyl CoA, namely a BCCP component; and (b) the presence of biotin carboxylase and acetyl CoA transcarboxylase.

When disrupted chloroplasts were incubated with $^{14}CO_2$, ATP, Mn^{2+} and Mg^{2+} and then placed on a Sephadex G-50 column, ^{14}C was associated with a green protein fraction. The ^{14}C was readily lost when the protein fraction was treated with acid or heated. Table I demonstrates that photophosphorylation (or ATP) and biotin (avidin-sensitivity) were requirements for $^{14}CO_2$ binding. To localize the chloroplast site of $^{14}CO_2$ binding, three experiments were conducted. In the first experiment, disrupted chloroplasts were allowed to fix $^{14}CO_2$ in the presence of ATP, $^{14}CO_2$ and Mn^{2+} and then the suspension was separated into the lamellar membrane fraction and

TABLE I

Cofactor requirements for formation of bound-$^{14}CO_2$ by disrupted spinach chloroplasts [From Kannangara and Stumpf (1972a)]

Experimental conditions	Bound $^{14}CO_2$ (cpm)
complete ($^{14}CO_2$, Mn^{2+}, ATP), light	287 000
complete, dark	232 840
complete + avidin, light	26 000
−ATP, light	182 110
−ATP, dark	4 650
−ATP + DCMU, light	5 616
complete + DCMU, light	260 120

the stroma fraction. Each fraction was then tested for the capacity to transfer the bound $^{14}CO_2$ to acetyl CoA in the presence of *E. coli* transcarboxylase. As indicated in Table II, it was clearly demonstrated that only the $^{14}CO_2$ associated with the lamellar membrane was incorporated into malonyl CoA while the stroma phase proteins were essentially inactive. Thus the BCCP component was membrane bound. In the second experiment, it could be shown that all the biotin associated covalently with protein in the chloroplast was bound to the lamellar membranes (Table III). In the third experiment,

TABLE II

Localization of bound $^{14}CO_2$ in disrupted spinach chloroplasts [From Kannangara and Stumpf (1972a)]

Chloroplast fraction (previously charged with $^{14}CO_2$)	*E. coli* Trans-carboxylase	Acetyl CoA	^{14}C-Malonyl CoA formed (cpm)
lamellar membrane (84 000 cpm)	+	+	51 220
	+	−	0
	−	+	0
stroma phase (121 000 cpm)	+	+	524
	+	−	0
	−	+	0

TABLE III

Biotin content of chloroplast fractions [From Kannangara and Stumpf (1972a)]

	Lamellar membranes (nmol/mg protein)	Stroma (nmol/mg protein)
free biotin	0·005	0·80
protein-bound biotin	1·56	0·04

we showed that bound $^{14}CO_2$ was all associated with a biotinyl protein by carrying out a methylation-stabilization of the bound $^{14}CO_2$, pronase digestion and then chromatography of the stabilized ^{14}C fragment. This fragment co-chromatographed with authentic 1′*N*-methoxycarbonyl biocytin. Therefore, we could definitely assign the localization of chloroplast BCCP as a firmly bound component of the lamellar membrane. Since all attempts to solubilize the membraneous BCCP met with no success, it can be probably classified as an integral protein firmly associated with the lamellar membranes.

To localize the position of biotin carboxylase we carried out the experiment summarized in Table IV. While disrupted chloroplasts readily fixed $^{14}CO_2$

TABLE IV

Requirement of stroma fraction for formation of bound $^{14}CO_2$
[From Kannangara and Stumpf (1972a)]

Chloroplast fraction	Protein-bound $^{14}CO_2$ (cpm)
1. broken chloroplasts	78 680
2. lamellar membranes	540
3. stroma	102
4. fractions 2+3	29 780

to a protein, the membrane fraction was totally inactive as were the stroma fractions. With membrane and stroma fractions together, considerable fixation occurred. Thus, the biotin carboxylase could be assigned to the stroma phase.

The localization of the transcarboxylase was somewhat more difficult to assign. As indicated in Table V, disrupted chloroplasts were unable to

TABLE V

Activation of chloroplast acetyl CoA carboxylase by *E. coli* components
[From Kannangara and Stumpf (1972a)]

Enzyme System	Malonyl CoA formed (cpm)
1. *E. coli* biotin carboxylase+BCCP	316
2. *E. coli* transcarboxylase	584
3. 1+2	29 264
4. disrupted chloroplasts	0
5. 4+1	0
6. 4+2	12 640

TABLE VI

Localization of transcarboxylase in disrupted chloroplasts prepared in the presence of 0·06 M HCO_3^-
[From Kannangara and Stumpf (1972a)]

System	^{14}C-Malonyl CoA (cpm) formed from ^{14}C-Acetyl CoA
1. *E. coli* biotin carboxylase + BCCP	599
2. *E. coli* transcarboxylase	820
3. stroma	202
4. lamellar membranes	660
5. 1 + stroma	17 760
6. 1 + lamellar membranes	1 006
7. 2 + stroma	908
8. 2 + lamellar membranes	13 213
9. 2 + lamellar membranes − ATP	11 157

catalyze the formation of malonyl CoA from $^{14}CO_2$. However, supplementation of the system with *E. coli* transcarboxylase restored considerable activity, thereby indicating that the reason for the low activity of chloroplast carboxylase was not related either to an absence or instability of biotin carboxylase or BCCP but to a loss of transcarboxylase activity. However, we observed that when chloroplasts were isolated in a 0·06 M bicarbonate medium, the transcarboxylase could be considerably stabilized. When disrupted chloroplasts were prepared with bicarbonate included in the isolation medium, we could demonstrate transcarboxylase activity in the stroma (Table VI). In addition, the data in Table VI clearly show that the BCCP was already charged with CO_2 since ^{14}C-acetyl CoA could be converted to ^{14}C-malonyl CoA in the presence of lamellar membrane fragments and transcarboxylase but in the absence of ATP. Additional data also indicated that Mg^{2+} was required for the transcarboxylation reaction. We can now summarize our current thoughts concerning chloroplast acetyl CoA carboxylase in Fig. 1. Thus the biotin carboxylase and the transcarboxylase are

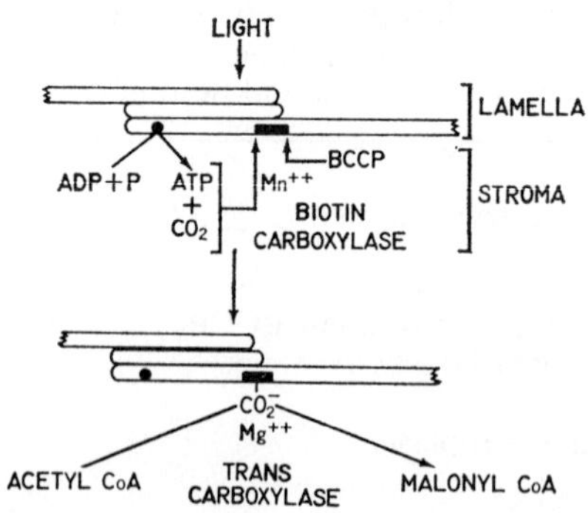

FIG. 1. A proposed scheme for acetyl CoA carboxylase activity in spinach chloroplast.

localized in the stroma phase and BCCP in the lamellar membranes, presumably in juxtaposition to a photophosphorylation site which generates ATP from ADP+P_i. Thus the ATP formed at the surface of the lamellar membranes is presumably available for the charging of membrane-bound BCCP to form $CO_2^- \sim$ BCCP.

Interestingly, the chloroplast acetyl CoA carboxylase has many of the properties of the procaryotic type described by Alberts and Vagelos (1968), with the exception that, rather than having a freely dissociable BCCP, the chloroplast BCCP is firmly buried in the lamellar membrane system. Similar systems have been described in chloroplast preparations from other leaf tissues (Kannangara and Stumpf, 1973).

C. NATURE OF THE CHLOROPLAST ACYL CARRIER PROTEIN

The key role of acyl carrier protein (ACP) in procaryotic fatty acid synthetase systems is well defined (Vagelos, 1973). We first observed the presence and requirement of ACP in an avocado mesocarp preparation which readily synthesized fatty acids from malonyl CoA providing ACP was added to the system (Overath and Stumpf, 1964). Brooks and Stumpf (1965) first described a similar requirement in both a lettuce and spinach chloroplast system. Simoni *et al.* (1967), isolated, purified and characterized spinach ACP. Matsumura and Stumpf (1968) partially sequenced spinach ACP and showed that the core amino acids surrounding the 4′-phosphopantetheine were identical in *E. coli*, *Arthrobacter*, avocado and spinach ACPs. A summary of these results is presented in Table VII.

That ACP is an essential component of the *de novo* fatty acid synthetase system is now well documented. We have observed this requirement in synthetases prepared from avocado mesocarp (Overath and Stumpf, 1964), safflower cotyledons (Jaworski and Stumpf, 1974), potato tubers (Huang and Stumpf, 1971), spinach and lettuce chloroplasts (Brooks and Stumpf, 1965), barley chloroplasts and castor bean cotyledons (P. K. Stumpf, unpublished observations). Not only is this protein essential for *de novo* synthesis but, as we shall see, acyl ACPs serve as specific substrates in elongation and in desaturation reactions.

III. *De Novo* Fatty Acid Synthesis

We shall define the *de novo* fatty acid synthetase system as a collection of soluble enzymes and ACP which together convert acetyl CoA and malonyl CoA to 16:0 ACP. This system has been now demonstrated in potato tuber (Huang and Stumpf, 1971), avocado mesocarp (Overath and Stumpf, 1964), safflower seedlings (Jaworski and Stumpf, 1974), and spinach chloroplasts (Brooks and Stumpf, 1965; Kannangara *et al.*, 1973a). A more detailed

TABLE VII

Properties of ACPs from different sources

Source of ACP	NH_2 terminus	No. of amino acids	Molecular weight	Partial structure
E. coli	serine	77	8 847	4′-phosphopantetheine (attached to Ser) —Gly—Ala—Asp—Ser—Leu—Asp— No. 36
Arthrobacter	serine	81	9 235	same
spinach	alanine	88	9 560	same
avocado	—	117	11 900	—

discussion of this subject is covered in the monograph of Hitchcock and Nichols (1971).

Smirnov (1960) described a spinach chloroplast preparation which readily incorporated ^{14}C-acetate into long chain fatty acids. Mudd and McManus (1962) and Stumpf and James (1963) extended this work by establishing that ATP, CoA Mg^{2+}, HCO_3^- and light were necessary components for the intact chloroplast system. Under aerobic conditions, the synthesized fatty acids were mainly palmitic and oleic acids while under anaerobic conditions, stearate accumulated instead of oleic acid. Whereas intact chloroplasts were very effective in incorporating ^{14}C-acetate into long chain fatty acids, their limited permeability prevented a deeper probing into the mechanism of fatty acid biosynthesis. However, with disrupted chloroplasts, NADH and NADPH and ACP became additional components required for maximal fatty acid synthesis.

A comparison between chloroplasts isolated from immature and mature spinach leaves has been made in our laboratory (Kannangara *et al.*, 1973a). Chloroplast preparations obtained from immature leaves were 1·5–3-fold higher in protein/mg chlorophyll than those from mature spinach leaves. Over 50% of the Class I chloroplasts from immature leaves contained prolamellar bodies with poorly developed grana stacks while fully formed grama and stroma lamellae were present in chloroplasts from mature chloroplasts. Table VIII summarizes the results obtained using different

TABLE VIII

Comparison of fatty acid synthetase activities of disrupted chloroplasts from immature and mature spinach leaves [From Kannangara *et al.* (1973a)]

Substrate	Substrate incorporated into fatty acids (nmol)	
	Immature leaves	Mature leaves
acetate	5·9	1·7
acetyl CoA	8·8	1·2
malonate	7·2	0·94
malonyl CoA	8·4	1·4

substrates for fatty acid synthesis with disrupted chloroplasts from immature and mature leaves and clearly shows that the synthetic capacity of the maturing chloroplasts is 3–6-fold more active on a chlorophyll basis than that of the fully developed chloroplast. Similarly, four different enzyme systems show major differences in activities as illustrated in Table IX and their different activities would clearly reflect the differences between mature and immature leaves noted in Table VIII. Table X summarizes the different fatty acids formed by the two different types of chloroplasts with ^{14}C-acetate as substrate.

TABLE IX

Comparison of some enzyme activities in chloroplast preparations from immature and mature leaves
[From Kannangara *et al.* (1973a)]

Enzymes	Immature leaves	Mature leaves
	nmol converted h^{-1}(mg protein)$^{-1}$	
fatty acid synthetase (^{14}C-acetate as substrate)	0·104	0·07
acetyl CoA synthetase	24·0	14·0
malonyl CoA synthetase	2·1	0·68
acetyl CoA carboxylase	102·0	63·0

In contrast to yeast and animal synthetases but like most procaryotic systems, the *de dovo* system in chloroplasts consists of soluble enzymes, i.e., prolonged centrifugation at 100 000 *g* does not yield a pellet containing a synthetase complex; rather, all the activity remains in the soluble protein fraction. Very little work has been done in isolating and characterizing the individual enzymes involved in the formation of long chain fatty acids (see Chapter 3, p. 51). It is assumed, of course, that because of the ACP requirement, the acetyl CoA:ACP transacylase and the malonyl CoA-ACP transacylase are present in adequate concentrations in chloroplast stroma. Presumably the β-keto acyl ACP synthetase, the β-keto acyl ACP reductase, the β-keto acyl ACP dehydrase and the enoyl ACP reductase are present in adequate concentrations although investigations are necessary to determine the kinetic properties as well as the chain specificities of these enzymes, i.e., whether one enzyme with broad specificity or several enzymes with narrow specificities participates in *de novo* synthesis of 16:0 ACP.

Recently, Vagelos's group (D'Agnolo *et al.*, 1973) have observed that cerulenin, an antibiotic (which at 30 μM concentration, completely and irreversibly inhibits fatty acid synthesis in a number of procaryotic and eucaryotic organisms) exerts its action by binding irreversibly with the two binding sites of β-keto acyl ACP synthetase of *E. coli.* We have tested cerulenin with chloroplasts as well as with a *de novo* system from safflower and have observed

TABLE X

^{14}C-Acetate incorporation into fatty acids by chloroplasts prepared from immature and mature spinach leaves

Source of chloroplasts	^{14}C Incorporation into lipids (cpm)	Distribution of ^{14}C in fatty acids (%)					
		14:0	16:0	18:0	18:1	18:2	18:3
immature leaves	389 000	1·1	27·0	—	64·1	2·8	5·0
mature leaves	140 000	tr	14·8	7·5	77·7	0·0	0·0

a similar irreversible inhibition of *de novo* synthesis. These results would strongly suggest similar binding sites for the β-keto acyl ACP synthetase protein in chloroplasts (Jaworski *et al.*, 1974b).

IV. The Elongation Reaction

Although the products of acetate incorporation consist of C_{16} and C_{18} fatty acids (see Table X), information has gradually accumulated in our laboratory to support the idea that a separate elongation system, other than the *de novo* system, was operating in chloroplasts as well as in cytosolic systems (see also Chapter 3, p. 59).

The first indication that two distinct systems function in plant tissues was the observation in our laboratory (Huang and Stumpf, 1971) that a partially purified system from potato tuber synthesized a mixture of saturated fatty acids consisting of C_8, C_{10}, C_{12}, C_{14}, C_{16} but no C_{18} fatty acids, while the original extract also had a capacity for C_{18} fatty acid synthesis. Clearly, in the purification steps the capacity for C_{18} fatty acid synthesis had been lost. Harwood and Stumpf (1972) extended these observations by noting that ^{14}C-16:0 ACP was elongated to 18:0 by an acetone preparation of avocado mesocarp with malonyl CoA as the cosubstrate. At the same time Kannangara and Stumpf (1972c) noted that, in the presence of 5mM arsenite, (an inhibitor previously used by Harwood and Stumpf (1971) to inhibit C_{18} synthesis in seed tissue) a 3-fold increase in 16:0 synthesis and a significant drop in 18:0+ 18:1 synthesis occurred with spinach chloroplasts using ^{14}C-acetate as substrate.

A direct attack on the elongation problem was finally made when Jaworski *et al.* (1974) enzymically prepared pure palmityl ACP, which served very effectively as a substrate for elongation to stearyl ACP. The reaction is summarized:

$$\text{Palmityl ACP} + \text{malonyl ACP} \xrightarrow[\text{elongation enzymes}]{\text{NADPH}} \text{stearyl ACP} \quad (3)$$

In brief, 16:0 ACP was elongated specifically to 18:0 ACP by a system which required specifically 16:0 ACP, malonyl ACP and NADPH. The elongation system was found in chloroplast stroma, in extracts of avocado mesocarp and in maturing safflower seedlings. The system differed from the *de novo* system in that: (1) it was inactivated rapidly by 37°C whereas the *de novo* system remained fully active; (2) the pH optimum for elongation was 7·8–8·6 whereas for *de novo* synthesis, pH 7·0 was optimal; (3) NADPH was specifically required; and (4) the elongation system was relatively insensitive to cerulenin whereas the *de novo* system was highly sensitive. Acetyl CoA did not serve as a C_2 donor; only malonyl CoA functioned in this respect, presumably by being first converted to malonyl ACP prior to the condensation reaction. 18:0 ACP, 12:0 CoA, 14:0 CoA and 16:0 CoA were all inactive. Since these substrates were inactive, it would appear that the long chain acyl CoA:ACP transacylases were missing in plant tissues and hence added palmitic and stearic acids or their CoA derivatives would be biosynthetically inert in the chloroplasts. This observation would readily explain the results obtained by many investigators who have repeatedly shown that stearic acid given to intact plant tissues did not serve as a substrate for conversion to oleate (Hitchcock and Nichols, 1971).

V. Desaturation Reactions

For a number of years it was well known that the fatty acids of chloroplast lipids were unique in their composition and relative constancy (Hitchcock and Nichols, 1971; see also Chapter 3, p. 44). A typical fatty acid composition of spinach lamella is presented in Table XI. The polyunsaturated fatty acids are of considerable interest since they constitute the major acids in lamellar lipids. For several years, investigators have attempted to resolve the mechanisms of biosynthesis of these acids but with mixed success. As techniques of chloroplast isolation have improved and information about the biochemistry of chloroplasts has increased, the opportunities for a successful attack on the problems of desaturation have become more attainable. We shall then summarize here our recent results pertaining to these problems.

A. Biosynthesis of Monoenoic Fatty Acids

Nagai and Bloch (1966, 1968) were the first to demonstrate a stearyl ACP desaturase in isolated spinach chloroplasts. Employing chemically prepared stearyl ACP, they showed that the substrate was converted to oleic acid in the presence of NADPH:ferredoxin reductase, NADPH, O_2 and disrupted chloroplasts. Stearyl CoA was inactive. We have recently confirmed and extended these observations in both photosynthetic and non-photosynthetic tissues. Jaworski and Stumpf (1974) obtained a soluble extract from maturing safflower seeds which rapidly desaturated ^{14}C-stearyl ACP to free oleic acid in the presence of NADPH, ferredoxin and O_2. Stearyl CoA, palmityl ACP and palmityl CoA were inactive. Ferredoxin was an essential component although flavodoxin would substitute partially for ferredoxin. The enzyme was insensitive to CO but was 84% inhibited by 1 mM CN^-. NADH was ineffective. While a reduced nicotinamide nucleotide has always been considered a component in desaturation systems, we have been able to show that its only function was as an electron donor (Jaworski and Stumpf, 1974). Thus, when an ascorbate–dichlorophenolindophenol photosystem I (grana) reducing system was substituted for NADPH, rapid desaturation of stearyl

TABLE XI

Fatty acids of spinach lamellae
[From Allen *et al.* (1966)]

Fatty acid:	14:0	14:3	15:0	16:0	16:1(3t)	16:2	16:3	18:0	18:1	18:2	18:3
% of total	0·1	0·4	0·2	13·5	1·8	tr	10·6	0·1	4·2	9·4	55·8

ACP occurred in the light in the complete absence of NADPH. Thus, a reducing system of a sufficiently negative redox potential was required rather than NADPH or NADH. The properties of the stearyl ACP desaturase in chloroplasts are essentially identical to those of the non-photosynthetic safflower system (Jacobson *et al.*, 1974). In the chloroplast, as well as in the maturing safflower, the desaturase occurred as a stable, completely soluble, non-membraneous protein. Purification of this desaturase is currently in progress. Thus, we can now conclude that the formation of oleic acid occurs by a direct desaturation of stearyl ACP formed by a specific elongation system from palmityl ACP which in turn is synthesized by the *de novo* system from acetyl and malonyl ACPs (Fig. 2). Since the long-chain acyl CoA:ACP transacylases are missing in higher plants, the observation made by many investigators that free stearic acid and stearyl CoA were not desaturated can now be readily explained.

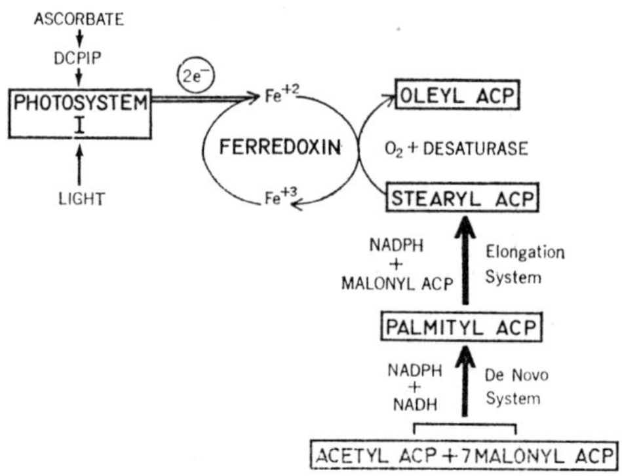

FIG. 2. A summary of the biosynthesis of palmityl, stearyl and oleyl ACPs by chloroplast systems.

Recently in our laboratory A. Sodja and P. K. Stumpf (unpublished observations) have been able to observe the formation of 12:1(2*t*), 12:1(3*c*) and DL-3-hydroxy-12:0 when a leaf homogenate was incubated with 12:0 CoA. The system required molecular O_2 although ferredoxin could substitute as an electron acceptor for O_2 under anaerobic conditions. Since the system did not occur in isolated spinach chloroplasts, but only in whole homogenates or supernatant fractions, we have suggested that this desaturation system is localized in the cytosol and may be the first desaturation step in the formation of a C_{12} trienoic acid.

B. BIOSYNTHESIS OF DIENOIC FATTY ACIDS

The principal dienoic acid in chloroplast lipids is linoleic acid (18:2 9,12). A number of investigators have examined the biosynthesis of this acid, each group providing evidence for different mechanisms of desaturation. Briefly,

Gurr *et al.* (1969), and later Talamo *et al.* (1973), Pugh and Kates (1973) as well as Baker and Lynen (1971) have provided substantial evidence that desaturation of oleic acid to linoleic acid occurs with 2-oleyl phosphatidylcholine as the preferred substrate for conversion to 2-linoleyl phosphatidylcholine. Each of these investigators has also observed that oleyl CoA was rapidly desaturated to linoleic acid. Both substrates required a microsomal preparation, NADPH, and O_2. In contrast, Vijay and Stumpf (1971, 1972), employing a microsomal system from maturing safflower and Abdelkader *et al.* (1973) working with microsomes from potato tubers, showed that oleyl CoA was rapidly converted to linoleyl CoA in the presence of NADPH and O_2. Vijay and Stumpf (1971) further showed that an ascrobate–dichlorophenolindophenol–photosystemI–ferredoxin–light system could completely replace the NADPH requirement; their results are summarized in Fig. 3. They also showed that acyl CoA transferases were present in their microsomal preparations which rapidly transferred ^{14}C-oleyl CoA and ^{14}C-linoleyl CoA to a phosphatidylcholine acceptor to form 2-oleyl and 2-linoleyl phosphatidylcholine. Recently Trémolières and Mazliak (1974) have employed pea leaf microsomes to catalyze a similar reaction.

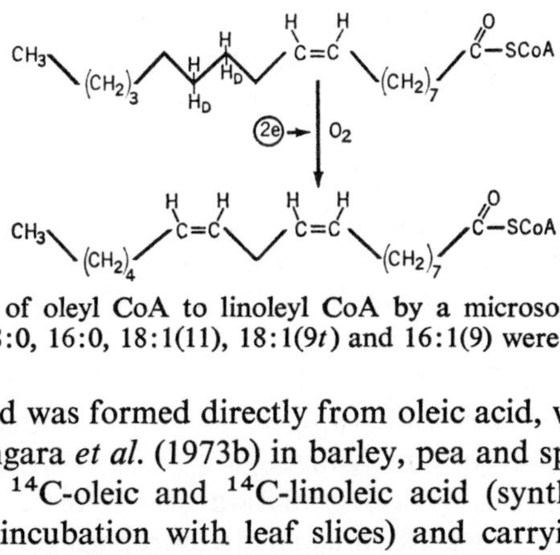

FIG. 3. Desaturation of oleyl CoA to linoleyl CoA by a microsomal safflower system. CoA derivatives of 18:0, 16:0, 18:1(11), 18:1(9*t*) and 16:1(9) were ineffective substrates.

That linoleic acid was formed directly from oleic acid, was clearly demonstrated by Kannangara *et al.* (1973b) in barley, pea and spinach leaf systems by isolating both ^{14}C-oleic and ^{14}C-linoleic acid (synthesized from ^{14}C-acetate by a 4 h incubation with leaf slices) and carrying out ozonolysis degradations. As shown in Table XII, since the degradation patterns were almost identical for 18:1 and 18:2, a direct sequential desaturation of 18:1 to 18:2 occurred. Furthermore, from both *in vitro* and *in vivo* data, the conversion of oleic to linoleic acid was O_2-dependent and CN^--sensitive. We can, therefore, conclude that the reaction:

$$\text{oleyl-X} \xrightarrow[O_2]{(2e)} \text{linoleyl-X}$$

where X = phosphatidylcholine or CoA

is occurring although the clearcut nature of X is still under active, if not controversial, examination.

TABLE XII

Distribution of ^{14}C in unsaturated fatty acids synthesized in whole leaf tissue for 4 h from ^{14}C-acetate as determined by reductive ozonolysis [From Kannangara *et al.* (1973b)]

Tissue	Fatty acid	Distribution of radioactivity in each half of the carbon chain (%) C_1–C_9	C_{10}–C_{18}
spinach	18:3	100	0
	18:2	51	49
	18:1	50·1	49·9
barley	18:3	95	5
	18:2	60	40
	18:1	52	48
Chlorella	18:3	71	29
	18:2	52	48
	18:1	51	49

C. BIOSYNTHESIS OF TRIENOIC ACIDS

While α-linolenic acid is the major polyunsaturated fatty acid in chloroplasts, 16:3(7,10,13) is present in substantial amounts in spinach chloroplast lipids while 14:3(5,8,11) is a trace acid (Allen *et al.*, 1966).

For a number of years now, investigators have had difficulties in clearly understanding the biosynthesis of trienoic acids in chloroplasts. Kannangara and Stumpf (1972b) showed that chloroplasts isolated from immature spinach leaves synthesized small amounts of 18:3 from ^{14}C-acetate. The surprising observation was that the synthesis of 18:3 was O_2-independent and CN^--insensitive whereas the formation of 18:1 and 18:2 had the opposite properties. They therefore suggested a CN^--sensitive and CN^--insensitive pathway for the synthesis of 18:3 from 18:2. Further investigation by Jacobson *et al.* (1973a, b) and Kannangara *et al.* (1973b) revealed the surprising results that ozonolysis degradation of ^{14}C-18:3, ^{14}C-18:2 and ^{14}C-18:1, formed from ^{14}C-acetate in isolated chloroplast or in intact leaf tissue, revealed strikingly different degradation patterns as illustrated in Table XII. Furthermore, when spinach chloroplasts were maintained at 4°C for 6 h and were then assayed for 18:3 synthesis with ^{14}C-acetate, a marked increase in synthesis of 18:3 occurred. On chemical α-oxidation of ^{14}C-stearate, obtained by Pt/H_2 reduction of 18:3, it could be readily postulated that the origin of 18:3 was 16:3. When 16:3 was added to a stroma preparation of chloroplasts supplemented with ATP, CoA, Mg^{2+} and ^{14}C-acetyl CoA or ^{14}C-malonyl CoA under anaerobic conditions, synthesis of ^{14}C-linolenic acid readily occurred (Jacobson *et al.*, 1973b). Finally when a number of [$^{14}C-1$] fatty acids were added to intact leaf tissues, [$^{14}C-1$]12:0 acid was the most effective precursor of 18:3 whereas ^{14}C-14:0, ^{14}C-16:0 and ^{14}C-18:0

were completely inactive (Kannangara *et al.*, 1973b). On the basis of these experiments and those described earlier, we concluded that 18:3 was synthesized by a unique pathway, as illustrated in Fig. 4, which is completely separate from the sequential pathway of oleate and linoleate synthesis (18:0 → 18:1 → 18:2) described above.

These results differed rather dramatically from those previously described by James and his colleagues who proposed a sequential desaturation pathway of 18:0 → 18:1 → 18:2 → 18:3 (Harris and James, 1965) and by the data recently published by Trémolières and Mazliak (1974) who propose a 18:1 → 18:2 synthesis in leaf microsomes with subsequent transport into chloroplasts for the formation of 18:3 directly from 18:2.

We conclude that, in spinach chloroplasts, the complete machinery for *de novo* synthesis of 16:0 acid, the elongation to 18:0 ACP, and a subsequent desaturation to 18:1 ACP takes place. Preliminary results suggest a further desaturation of 18:1 (via its CoA ester?) to 18:2 CoA which is then rapidly inserted into the complex lipids making up the lamellar membranes. From $^{14}CO_2$-labelling data and from recent data concerning the origin of 12:1(3*c*) we conclude that the formation of 12:3(3,6,9) occurs in a cytosolic compartment and this precursor is then transported into the chloroplast stroma compartment for elongation to 18:3. With this explanation at hand, we can explain the complete lack of ^{14}C-labelling of 16:3 when isolated chloroplasts, incubated with ^{14}C-acetate, readily form 16:0, 18:0, 18:1, 18:2 and 18:3. We suggest that with isolated chloroplasts, since the potential for synthesizing 12:3 had been deleted by the removal of the cytosolic components,

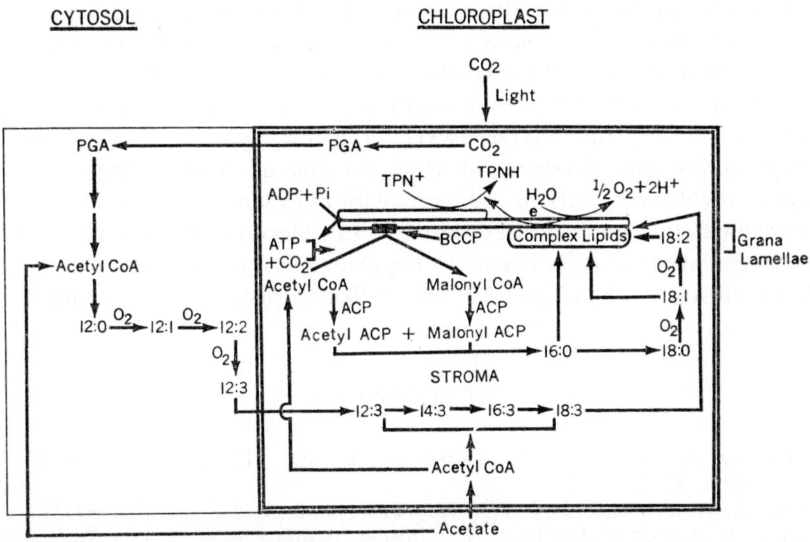

FIG. 4. A summary of the biochemical events related to the biosynthesis of chloroplast fatty acids.

16:3 cannot be formed. Since only a trace of 14:3 occurs in isolated chloroplast lipids, a sufficient concentration of 14:3 may not be available for elongation to 16:3, etc.

As to the question of transport of 12:3 from a cytosolic compartment to the chloroplast stroma across a membrane barrier, a recent paper of interest (Crotty and Ledbetter, 1973) describes observations in which plant cell membranes were examined in thin sections by electron microscopy. Numerous membrane continuities appeared between chloroplasts and other organelles such as endoplasmic reticula and mitochondria in leaves of fern and tomato. These results would suggest the possibility of a direct transfer of substrate from one organelle to the next without any elaborate transfer mechanisms being required and they would also imply that many of the transport studies carried out with isolated organelles, in which these continuities between organelles had been severed by isolation procedures, may indeed be artifacts.

VI. Regulation of Lipid Synthesis in Chloroplasts

A cursory examination of an electron micrograph of a normal chloroplast in plant cells would show an organelle with well defined membranes, perhaps with a few osmiophilic globules and starch granules interspersed between the membranes. Certainly one never observes a chloroplast engorged with fat droplets such as is normally found in cotyledons of high lipid-containing seeds. Obviously, the rate of lipid synthesis is carefully regulated so that normal construction and maintenance of lamellar membranes occurs in order to allow maximal functioning of the total photosynthetic process.

At present we know very little about the control of lipid synthesis in the chloroplast. The obvious factors such as the flow of acetyl CoA precursors into the chloroplast or their synthesis in the chloroplast itself, the kinetic relationship between CO_2 concentrations, ATP levels, ratios of S-S/SH, $NADPH/NADP^+$ and $NADH/NAD^+$ and lipid synthesis, the interaction of thioesterase with thioester substrate, possible allosteric properties of the enzymes involved directly or indirectly with lipid synthesis—all these factors require a careful and critical examination in order to describe what mechanism of regulation actually occurs in the chloroplast *in situ*, rather than what a biochemist's vivid imagination would like to picture as occurring in the chloroplast.

REFERENCES

Abdelkader, A. B., Cherif, A., Demandre, C. and Mazliak, P. (1973). *Eur. J. Biochem.* **32**, 155–165.

Alberts, A. W. and Vagelos, P. R. (1968). *Proc. natn. Acad. Sci. U.S.A.* **59**, 561–564.

Allen, C. F., Good, P., David, H. F., Chisum, P. and Fowler, S. D. (1966). *J. Am. Oil Chem. Soc.* **43**, 223–231.

Baker, N. and Lynen, F. (1971). *Eur. J. Biochem.* **19**, 200–210

Brooks, J. L. and Stumpf, P. K. (1965). *Biochim. biophys. Acta* **98**, 213–216

Brooks, J. L. and Stumpf, P. K. (1966). *Archs Biochem. Biophys.* **116**, 108–116

Burton, D. and Stumpf, P. K. (1966). *Archs Biochem. Biophys.* **117**, 604–614.

Crotty, W. J. and Ledbetter, M. C. (1973). *Science, N.Y.* **182**, 839–841.

D'Agnolo, G., Rosenfeld, I. S., Awaya, J., Omura, S. and Vagelos, P. R. (1973). *Biochim. biophys. Acta* **326**, 155–166.

Gurr, M. L., Robinson, M. P. and James, A. T. (1969). *Eur. J. Biochem.* **9**, 70–78.

Harris, R. V. and James, A. T. (1965). *Biochim. biophys. Acta* **106**, 456–464.

Harwood, J. L. and Stumpf, P. K. (1971). *Archs Biochem. Biophys.* **142**, 281–291.

Harwood, J. L. and Stumpf, P. K. (1972). *Archs Biochem. Biophys.* **148**, 282–290.

Heldt, H. W. and Rapley, L. (1970). *FEBS Lett.* **10**, 143–148.

Hitchcock, C. and Nichols, B. W. (1971). "Plant Lipid Biochemistry", pp. 1–387. Academic Press, London and New York.

Huang, K. P. and Stumpf, P. K. (1970). *Archs Biochem. Biophys.* **140**, 158–173.

Huang, K. P. and Stumpf, P. K. (1971). *Archs Biochem. Biophys.* **143**, 412–427.

Jacobson, B. S. and Stumpf, P. K. (1972). *Archs Biochem. Biophys.* **153**, 656–663.

Jacobson, B. S., Kannangara, C. G. and Stumpf, P. K. (1973a). *Biochem. biophys. Res. Commun.* **51**, 487–493.

Jacobson, B. S., Kannangara, C. G. and Stumpf, P. K. (1973b). *Biochem. biophys. Res. Commun.* **52**, 1190–1197.

Jacobson, B. S., Jaworski, J. G. and Stumpf, P. K. (1974). *Pl. Physiol. Lancaster* **54**, 484–486.

Jaworski, J. G. and Stumpf, P. K. (1974). *Archs Biochem. Biophys.* **162**, 158–165.

Jaworski, J. G., Goldschmidt, E. E. and Stumpf, P. K. (1974). *Archs Biochem. Biophys.* **163**, 769–776.

Kannangara, C. G. and Stumpf, P. K. (1972a). *Archs Biochem. Biophys.* **152**, 83–91.

Kannangara, C. G. and Stumpf, P. K. (1972b). *Archs Biochem. Biophys.* **148**, 414–424.

Kannangara, C. G. and Stumpf, P. K. (1972c). *Pl. Physiol., Lancaster* **49**, 497–501.

Kannangara, C. G. and Stumpf, P. K. (1973). *Archs Biochem. Biophys.* **155**, 391–399.

Kannangara, C. G., Jacobson, B. S. and Stumpf, P. K. (1973a). *Pl. Physiol., Lancaster* **52**, 156–161.

Kannangara, C. G., Jacobson, B. S. and Stumpf, P. K. (1973b). *Biochem. biophys. Res. Commun.* **52**, 648–655.

Matsumura, S. and Stumpf, P. K. (1968). *Archs Biochem. Biophys.* **125**, 932–941.

Mudd, J. B. and McManus, T. T. (1962). *J. biol. Chem.* **237**, 2057–2063.

Nagai, J. and Bloch, K. (1966). *J. biol. Chem.* **241**, 1927.

Nagai, J. and Bloch, K. (1968). *J. biol. Chem.* **243**, 4626–4633.

Overath, P. and Stumpf, P. K. (1964). *J. biol. Chem.* **239**, 4103–4110.

Pugh, E. L. and Kates, M. (1973). *Biochim. biophys. Acta* **316**, 305–316.

Simoni, R. D., Criddle, R. S. and Stumpf, P. K. (1967). *J. biol. Chem.* **242**, 573–581.

Smirnov, B. P. (1960). *Biokhimiya* **25**, 419–423.

Stumpf, P. K. and James, A. T. (1963). *Biochim. biophys. Acta* **70**, 260–270.

Stumpf, P. K., Brooks, J. L., Galliard, T., Hawke, J. C. and Simoni, R. (1967). *In* "Biochemistry of Chloroplasts" (Goodwin, T. W., ed.), Vol. 2, pp. 213–239. Academic Press, London and New York.

Talamo, B., Chang, N. and Bloch, K. (1973). *J. biol. Chem.* **248**, 2738–2742.

Trémolières, A. and Mazliak, P. (1974). *Plant Science Lett.* **2**, 193–201.

Vagelos, P. R. (1973). *In* "The Enzymes" (Boyer, P. D., ed.), 3rd edition, Vol. 8, pp. 155–199. Academic Press, New York and London.

Vijay, I. K. and Stumpf, P. K. (1971). *J. biol. Chem.* **246**, 2910–2917.

Vijay, I. K. and Stumpf, P. K. (1972). *J. biol. Chem.* **247**, 360–366.

CHAPTER 5

Biosynthesis of Phosphoglycerides in Plants

M. KATES AND M. O. MARSHALL

Department of Biochemistry, University of Ottawa, Ottawa, Canada

I. Introduction

Biological membranes not only provide a physical barrier separating cells and subcellular organelles from their respective environments, but are intimately involved in essential life processes. In both procaryotic and eucaryotic cells they facilitate and control the transport of metabolites between the two environments they separate, and are also involved in the respiratory chain, oxidative phosphorylation and protein synthesis, as well as in many other cell functions. Among eucaryots, photosynthetic organisms are of special interest in that the molecular machinery for photosynthesis is localized in the lamellae membranes of chloroplasts (Branton, 1969; Korn, 1969; Stoeckenius and Engelman, 1969).

The study of membranes is one of the most exciting fields of modern biochemistry, and of particular interest is the problem of membrane assembly from the individual lipid and protein constituents. Clearly this problem can only be approached when detailed knowledge has been gained concerning: (a) the pathways in the biosynthesis of lipid and protein; and (b) the sub-cellular localization of individual lipids and protein components and of the enzymes involved in their biosynthesis. Lipids generally comprise 20–40% by weight of membranes. A major class of membrane lipids are the phospholipids, and during the last 20 years knowledge of their structure and biosynthesis has grown rapidly (Kennedy, 1961, 1962; Cronan and Vagelos, 1972; Mangnall and Getz, 1973). However, studies on the biosynthesis of phospholipids has largely been confined to animals and bacteria, while very few detailed studies have been made with plant tissues and even less with photosynthetic tissue of higher plants (Kates, 1970; Hitchcock and Nichols, 1971). In this paper we shall present our current studies on phospholipid biosynthesis in spinach leaves against the background of present knowledge concerning the biosynthetic pathways for phospholipids in animals and microorganisms.

II. Glycerolipids of Plant Tissues

A. COMPOSITION

In most photosynthetic tissues, phospholipids amount to about 20%, and glycolipids about 40% of total lipid. In leaves and green algae, phospholipids consist mainly of phosphatidylcholine (PC), phosphatidylglycerol (PG), phosphatidylethanolamine (PE) and phosphatidylinositol (PI), of which PC is the major component (Table I). However, in photosynthetic bacteria, PC is replaced by PE as the major component, and in blue-green algae PC, PE and PI are completely absent (see Table I, and Kates, 1970). Minor phospholipids occasionally reported to be present in photosynthetic and other plant tissues include phosphatidylserine (PS), diphosphatidylglycerol (cardiolipin) and phosphatidic acid (PA). In regard to phosphatidylserine, Hutt and co-workers (1950) and Kates and Eberhardt (1957) isolated small amounts of serine from the hydrolysate of peanut and runner bean leaf phospholipids, respectively, and Benson and Strickland (1960) detected [^{14}C]glycerylphosphorylserine among the deacylated products of ^{14}C-labelled *Chlorella* lipids. Though this evidence is not conclusive, the presence of PS in plants is also suggested by the finding of PS labelled from [U-^{14}C]-serine in tomato root (Willemot and Boll, 1967). Other minor phospholipids are *N*-methyl and *N*-acyl derivatives of PE; the former has only been detected in diatoms (Kates and Volcani, 1966) and the latter in seeds (Aneja *et al.*, 1969; Dawson *et al.*, 1969). Also, Debuch and Rotsch (1966) have isolated a novel polyglycerophosphatide from spinach leaves, the precise structure of which is not known.

TABLE I

Phospholipid composition of photosynthetic tissue[a]

Tissue	Phosphatidyl-glycerol	Phosphatidyl-choline	Phosphatidyl-ethanolamine	Phosphatidyl-serine	Phosphatidyl-inositol	Diphosphatidyl-glycerol
LEAVES						
Spinach, total[b]	26	47	17	—	9	tr
Chloroplast[c]	58	27	tr	—	11	tr
Lamellae[d]	65	24	0	—	11	—
Chloroplast envelope[e]	19	62	9	—	7	2
Cauliflower chloroplasts[f]	44	19	13	9	12	3
Runner bean[g]	22	45	17	—	9	—
Tomato[h]	19	50	20	—	4	4
Maidenhair tree[h]	22	48	14	—	6	7
Vicia faba						
Chloroplast envelope[i]	23	77	0	—	—	—
Lamellae	66	34	0	—	—	—
GREEN ALGAE						
Chlorella vulgaris[j]	27	46	9	—	14	1
Scenedesmus obliquus[k]	41	36	6	1	15	—
PHOTOSYNTHETIC BACTERIA						
Rhodospirillum rubrum[l]	29	6	57	—	—	8
Rhodopseudomonas spheroides[l]	54	4	36	—	—	6

[a] Values are expressed as percentage of total phospholipids. Dash (—) indicates data unavailable.
[b] Wintermans (1960).
[c] Douce *et al.* (1968).
[d] Allen *et al.* (1966).
[e] Poincelot (1973)
[f] Schwertner and Biale (1973).
[g] Kates (1960).
[h] Roughan and Batt (1969).
[i] Mackender and Leech (1974).
[j] Sastry and Kates (1965).
[k] Benson and Maruo (1958); Benson and Strickland (1960).
[l] Wood *et al.* (1965); Haverkate and van Deenen (1965b).

A unique characteristic of all plant photosynthetic tissues, except photosynthetic bacteria, is the high content of glycolipids. These consist chiefly of monogalactosyl and digalactosyl diglycerides and sulfoquinovosyl diglyceride, of which monogalactosyl diglyceride is the major single component. In leaves, minor glycolipids include glucocerebrosides (Sastry and Kates, 1964; Carter and Koob, 1969), sterol and acylsterol glycosides (Lepage, 1964; Eichenberger and Menke, 1966; Kiribuchi *et al.*, 1966 and 1967) and a complex sphingolipid called phytoglycolipid (Carter and Koob, 1969).

In non-photosynthetic plant tissues, the lipid composition is qualitatively similar to that of photosynthetic tissues, but the proportions of PG and glycolipids are much lower (see Hitchcock and Nichols, 1971; Kates, 1970). Even more striking differences are apparent between the lipid composition of plant photosynthetic tissue, and that of animals and bacteria. Thus, in all animal tissues except brain, the glycolipid content is very low, and sulfoquinovosyl diglyceride is completely absent. A major phospholipid of photosynthetic tissues, PG is only a trace component in animal tissues, while the latter tissue contains much higher amounts of cardiolipin, and also contains sphingomyelin and phosphatidylinositol phosphates which are absent from photosynthetic tissues (Gray, 1964; Wells and Dittmer, 1966; and Fleischer and Rouser, 1965). In bacteria, the major phospholipids are PE and PG, but PC is generally absent except as a minor component in *Thiobacillus novellus* and in certain species of soil bacteria including the *Agrobacteria*, the nitrifying bacterium *Nitrosocystis oceanus*, the nitrogen-fixing bacterium *Azotobacter agilis* and in some of the photosynthetic bacteria (see Goldfine, 1972).

The fatty acid composition of the glycerolipids of all photosynthetic tissues, except photosynthetic bacteria and certain species of blue-green algae, is characterized by a high content of polyunsaturated species of which α-linolenic acid is the major component (see Kates, 1970). A unique fatty acid is *trans*-3-hexadecenoic acid, which is found in all photosynthetic tissues except blue-green algae and photosynthetic bacteria, but is absent in etiolated and non-photosynthetic tissues (see reviews by Kates, 1970, and Hitchcock and Nichols, 1971). This acid is located specifically at the 2-position of the phosphatidylglycerol component in the chloroplast lamellae of leaves, green algae and *Euglena gracilis* (Haverkate and van Deenen, 1965a, b). The major molecular species of the phosphatidylglycerol component of leaves is 1-linolenoyl-2-(*trans*-3′-hexadecenoyl)-*sn*-3-glycerophosphoryl-1″-*sn*-glycerol (Haverkate and van Deenen, 1965a).

B. SUBCELLULAR LOCALIZATION

In the leaves of spinach and tobacco, about 50% and 75%, respectively, of the total cell content of phosphatidylglycerol has been found localized in the chloroplasts, in which it is the major phospholipid component (see Table I; also Douce *et al.*, 1968; Ongun *et al.*, 1968). Phosphatidylcholine and phosphatidylethanolamine, however, are the major phospholipid components of purified mitochondria from non-photosynthetic plant tissue (Schwertner and Biale, 1973), and are also probably the major phospholipid components of mitochondria from photosynthetic tissue. The question whether phosphatidylcholine and phosphatidylethanolamine are also present in leaf chloroplasts has only recently been answered by lipid analysis of chloroplasts freed from mitochondrial contamination and of chloroplast envelope membranes. Douce *et al.* (1968) showed that in intact spinach chloroplasts judged to be free of mitochondria by electron microscopic examination, phosphatidylcholine accounted for 27% of phospholipid while only small amounts of phosphatidylethanolamine were detected. Douce's results are consistent with the finding by Allen *et al.* (1966) of 24% phosphatidylcholine, but no phosphatidylethanolamine, in the phospholipids of purified spinach chloroplast lamellae. Schwertner and Biale (1973) found a similar concentration of PC in purified cauliflower chloroplasts, but in contrast a much higher concentration of phosphatidylethanolamine than was found by Douce *et al.* (1968) and Allen *et al.* (1966) for spinach chloroplasts. Recently, Poincelot (1973) has shown conclusively that both phosphatidylcholine and phosphatidylethanolamine are indeed chloroplast constituents and are associated with the chloroplast envelope membrane, the latter exclusively so (Table I). These studies thus establish that phosphatidylcholine is a component of leaf chloroplasts, localized partly in the lamellae membrane but mostly in the envelope membranes, while phosphatidylethanolamine is localized exclusively in the chloroplast envelope membranes; the variable reports on the contents of these two phospholipids in chloroplasts is probably due to the ease with which the envelope membrane is lost during isolation of chloroplasts.

In regard to the intracellular localization of phosphatidylinositol and cardiolipin, the former has been found in both leaf chloroplasts (Table I) and in mitochondria of nonphotosynthetic tissue (10% of phospholipid), while the latter has been found in mitochondria from non-photosynthetic tissue (10% of phospholipid) but appears to be only a trace component in leaf chloroplasts (see Table I; Douce *et al.*, 1968; Schwertner and Biale, 1973).

III. General Outline of the Biosynthetic Pathways of Phospholipids

Biosynthetic pathways for the major phospholipids in animal, fungal and bacterial cells are now known in considerable detail (see Fig. 1). The key intermediate in these systems is *sn*-3-phosphatidic acid (derived from *sn*-3-glycerophosphate; reactions **1** and **2**, Fig. 1), in its metabolically active form, cytidine diphosphate diglyceride (CDP-diglyceride, reaction **3**, Fig. 1) which

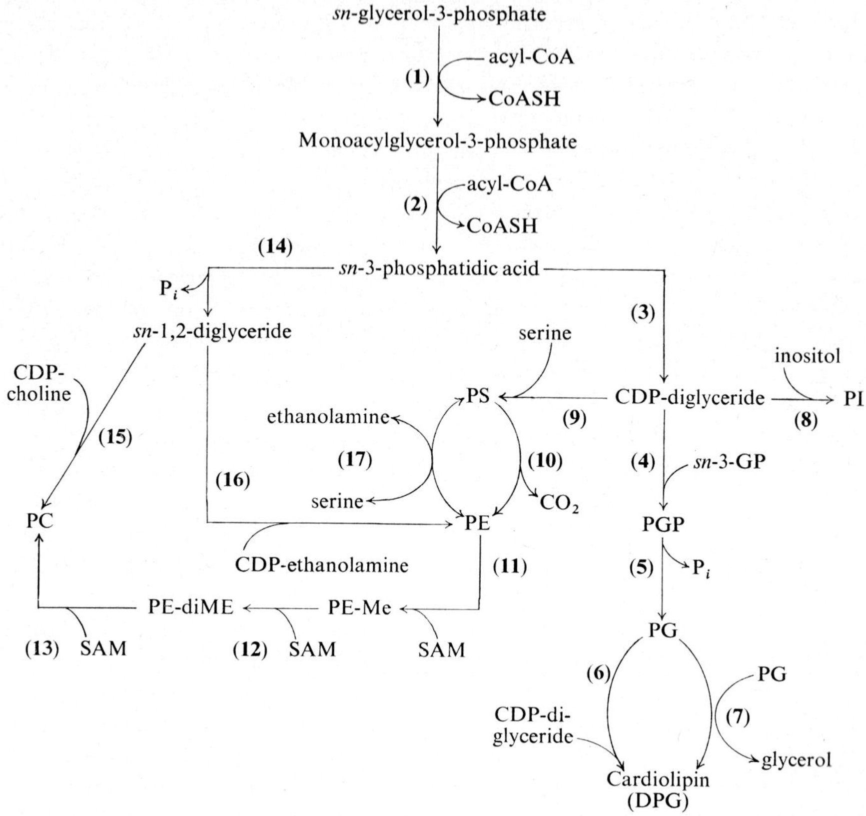

FIG. 1. Pathways for biosynthesis of phospholipids (see Kennedy, 1961, 1962; Mangnal and Getz, 1973).

Abbreviations: PG, phosphatidylglycerol; PGP, phosphatidylglycerophosphate; DPG, diphosphatidylglycerol(cardiolipin); PI, phosphatidylinositol; PS, phosphatidylserine; PE, phosphatidylethanolamine; PE-Me, phosphatidylmonomethylethanolamine; PE-diMe, phosphatidyldimethylethanolamine; PC, phosphatidylcholine(lecithin); SAM, *S*-adenosylmethionine.

serves as a common intermediate for the synthesis of phosphatidylglycerol (reactions **4** and **5**, Fig. 1), cardiolipin (reactions **6** and **7**, Fig. 1), phosphatidylinositol (reaction **8**, Fig. 1), and phosphatidylserine (reaction **9**, Fig. 1). Decarboxylation of phosphatidylserine (reaction **10**, Fig. 1) gives rise to phosphatidylethanolamine, which is converted by stepwise methylation with *S*-adenosylmethionine as methyl donor (reactions **11**, **12** and **13**) to phosphatidylcholine. Phosphatidylcholine and phosphatidylethanolamine may also be synthesized from diglyceride, which is derived from phosphatidic acid (reaction **14**, Fig. 1), by condensation with CDP-choline (reaction **15**) and CDP-ethanolamine (reaction **16**), respectively.

Interconversion of phospholipids by exchange of polar head groups also occurs, for example, the exchange of the ethanolamine moiety of phosphatidylethanolamine for free serine to form phosphatidylserine, and vice versa (reaction **17**), and the exchange of the glycerol moiety of phosphatidylglycerol for another molecule of phosphatidylglycerol to form cardiolipin (reaction **7**).

Although until recently relatively little was known about the pathways of phospholipid biosynthesis in plants, the central role of phosphatidic acid was clearly established by the finding of high specific activity [^{32}P]phosphatidic acid in leaves supplied with [^{32}P]orthophosphate (Kates, 1960; Eberhardt and Kates, 1957) and in *Chlorella* cells supplied with [^{32}P]glycerophosphate (Sastry and Kates, 1965) during photosynthesis.

Information on biosynthetic pathways for phospholipids in plants will now be presented in detail.

IV. Biosynthetic Pathways of Individual Phospholipids in Plants

A. Phosphatidic Acid (PA)

Four pathways are known for the biosynthesis of PA, as follows:

1. Stepwise acylation of *sn*-glycerol-3-phosphate by acyl CoA:

$$\textit{sn}\text{-glycerol-3-phosphate} \xrightarrow{\text{acyl-CoA}} \text{monoacyl-}\textit{sn}\text{-glycero-3-phosphate} \qquad \textbf{(1)}$$

$$\text{monoacylglycerophosphate} \xrightarrow{\text{acyl-CoA}} \text{phosphatidic acid} \qquad \textbf{(2)}$$

These reactions have been shown to be catalyzed by separate enzymes in liver (Lands and Hart, 1964) and in *E. coli* (Ray *et al.*, 1970; Hechemy and Goldfine, 1971). The biosynthesis of PA in spinach leaves has been studied by Cheniae (1965) and Sastry and Kates (1966). They demonstrated the *in vitro* incorporation by microsomal-bound enzymes of *sn*-glycerol-3-phosphate into PA and lyso PA, either by direct acylation with acyl-CoA or under conditions in which acyl-CoA was being actively synthesized. Thus, it is most probable that phosphatidic acid is synthesized in leaves by the above pathway. Although acyl-CoA clearly can act as acyl donor, it is possible that acyl-*S*-ACP might also be an acyl donor, as has been found in bacteria (Ailhaud and Vagelos, 1966). However, in green cells of *Euglena*, acyl-*S*-ACPs are acyl donors for monogalactosyl diglyceride synthesis but not for phospholipid or glyceride synthesis (Renkonen and Bloch, 1969).

2. Acylation of dihydroxyacetone phosphate, followed by reduction and reacylation:

$$\text{dihydroxyacetone phosphate} + \text{acyl-CoA} \rightarrow \text{monoacyldihydroxyacetone phosphate} \xrightarrow{\text{NADPH}} \text{monoacylglycerophosphate} \xrightarrow{\text{acyl-CoA}} \text{phosphatidic acid} \quad (3)$$

These reactions have been demonstrated in rat liver microsomes (Hajra and Agranoff, 1968; Hajra, 1968; and Puleo *et al.*, 1970), but have so far not been detected in bacteria or plants.

3. Phosphorylation of *sn*-1,2-diglyceride with ATP, catalyzed by diglyceride kinase:

$$sn\text{-1,2-diglyceride} \xrightarrow{\text{ATP}} \text{phosphatidic acid} \quad (4)$$

This enzyme is present in brain (Hokin and Hokin, 1959) and *E. coli* (Pieringer and Kunnes, 1965). It has been found in only one plant tissue so far: peanut cotyledons (Bradbeer and Stumpf, 1960).

4. Phosphorylation of monoglyceride with ATP to lyso PA, followed by acylation with acyl CoA to PA:

$$\text{monoglyceride} \xrightarrow{\text{ATP}} \text{monoacylglycerophosphate} \xrightarrow{\text{acyl-CoA}} \text{phosphatidic acid} \quad (5)$$

These reactions have been demonstrated in brain (Pieringer and Hokin, 1962a, b) and *E. coli* (Pieringer and Kunnes, 1965), but has so far not been demonstrated in plants.

B. CDP-DIGLYCERIDE

The only known pathway for the biosynthesis of CDP-diglyceride is that catalyzed by CTP:phosphatidic acid cytidyltransferase:

$$\text{CTP} + \text{phosphatidic acid} \rightarrow \text{CDP-diglyceride} + \text{PP}_i \quad (6)$$

The rat liver enzyme system has been described in detail by Carter and Kennedy (1966). The same enzyme system can also synthesize dCDP-diglyceride from dCTP and PA (Ter Schegget *et al.*, 1971). Synthesis of CDP-diglyceride by the above pathway has also been demonstrated in brain (Petzold and Agranoff, 1967), in bacteria (Carter, 1968; Patterson and Lennarz, 1971; McCaman and Finnerty, 1968; and Bell *et al.*, 1971) and in yeast (Hutchison and Cronan, 1968).

The central role of CDP-diglyceride in the biosynthesis of bacterial (Cronan and Vagelos, 1972) and yeast (Steiner and Lester, 1972) phospholipids has been established beyond doubt. However, direct proof of the natural occurrence of CDP-diglyceride was only recently presented by Raetz and Kennedy (1973) who isolated small amounts of ^{32}P-labelled CDP-diglyceride from *E. coli* cells grown in glycerol-3-[^{32}P]phosphate. Interestingly, equivalent amounts of ^{32}P-labelled dCDP-diglyceride were also isolated.

The very low level of total labelled cytosine liponucleotide relative to labelled PA led Raetz and Kennedy (1973) to suggest that the conversion of PA to liponucleotide may be the rate determining step in the overall process of phospholipid biosynthesis in *E. coli*.

Recent evidence indicates that CDP-diglyceride is also biosynthesized by the above pathway in plants. Thus, Douce (1968), Bahl *et al.* (1970) and Sumida and Mudd (1970a) have demonstrated the incorporation of ^{32}P from [α-^{32}P]CTP into CDP-diglyceride by cell-free preparations from spinach leaves, maize and cauliflower inflorescence. However, these workers were unable to show a dependence of CDP-diglyceride synthesis on added PA, presumably because of high levels of endogenous PA in the cell free preparations.

C. PHOSPHATIDYLINOSITOL (PI)

The currently accepted pathway for the *de novo* biosynthesis of PI in animal tissues is as follows:

$$\text{CDP - diglyceride} + \textit{myo}\text{-inositol} \leftrightarrows \text{phosphatidyl inositol} + \text{CMP} \qquad (7)$$

Direct proof that CDP-diglyceride was a substrate in *de novo* PI biosynthesis was first provided by Paulus and Kennedy (1960) who showed that incorporation of inositol by chicken liver microsomes into PI in the presence of Mn^{2+} was accompanied by release of [^{32}P]CMP from [α-^{32}P]CDP-diglyceride. These authors also found that inositol, in the presence of Mn^{2+}, could be incorporated by liver microsomes into PI in the absence of added CDP-diglyceride, and suggested that this occurred by an exchange reaction with endogenous PI and that the exchange reaction and the CDP-diglyceride reaction were part of the same enzyme system, as follows:

$$\text{CDP-diglyceride} + \text{enzyme} \leftrightarrows \text{phosphatidyl-enzyme} + \text{CMP} \qquad (8)$$

$$\text{phosphatidyl-enzyme} + \text{inositol} \leftrightarrows \text{phosphatidyl inositol} + \text{enzyme} \qquad (9)$$

It should be emphasized that the enzyme involved in the exchange reaction with inositol requires Mn^{2+} and is different from the enzyme described below for exchange of serine, ethanolamine and choline which specifically requires Ca^{2+}. CDP-diglyceride:inositol phosphatidyltransferase has also been found in a wide variety of animal tissues (Benjamins and Agranoff, 1969; Prottery and Hawthorne, 1967) and in yeast (Steiner and Lester, 1972), but little is known about the biosynthesis of PI in the few species of bacteria (e.g. Mycobacteria) in which it is found.

The biosynthesis of PI in non-photosynthetic plant tissue has been studied by Sumida and Mudd (1970b). A mitochondrial preparation from cauliflower inflorescence was found to catalyze the incorporation of inositol into

PI in the presence of Mn^{2+}, and this incorporation could be greatly stimulated by the addition of CDP-diglyceride. CTP was also found to be as effective as CDP-diglyceride, presumably because of its conversion to the latter by reaction with endogenous PA. This evidence suggests that in non-photosynthetic plant tissue PI is biosynthesized by the same pathway as in animals (eqn **7**).

Evidence for the operation of the same pathway in spinach leaves was obtained by Sastry and Kates (1966) who found that ^{32}P from *sn*-glycerol-3-[^{32}P]phosphate could be incorporated into PI, in the presence of CTP and Mn^{2+} by spinach cell fractions actively synthesizing [^{32}P]PA. This observation was recently followed up (M. O. Marshall and M. Kates, unpublished results) using a particulate spinach leaf fraction (1200–100 000 *g* pellet) incubated with [U-^{14}C]*myo*-inositol, Mn^{2+} and Triton X-100, either in the presence or the absence of CDP-diglyceride. The rate of [^{14}C]PI synthesized in the absence of CDP-diglyceride was 0·8 nmol/h, and was stimulated 3-fold by the addition of CDP-diglyceride (Table II). When Triton X-100 was omitted from the assay system PI synthesis in the presence of CDP-diglyceride was decreased, whereas PI synthesis in the absence of CDP-diglyceride was increased (Table II). However, both activities required Mn^{2+} (5 mM) which could not be replaced by Mg^{2+} or Ca^{2+} at the same concentration.

The ^{14}C-labelled product from reactions carried out both in the presence and absence of CDP-diglyceride was identified as PI, as follows: (a) the

TABLE II

Cofactor requirements for phosphatidylinositol synthesis

System[a]	Phosphatidylinositol synthesized (nmol/h)
A. WITH CDP-DIGLYCERIDE	
Complete	2·5
Complete minus Triton	0·6
Complete minus Triton and Mn^{2+}	0·0
Complete minus Triton and Mn^{2+} plus Mg^{2+} (5 mM)	0·0
Complete minus Triton and Mn^{2+} plus Ca^{2+} (5 mM)	0·0
Complete boiled enzyme	0·0
B. WITHOUT CDP-DIGLYCERIDE	
Complete	0·8
Complete minus Triton	1·5
Complete minus Triton and Mn^{2+}	0·0
Complete minus Triton and Mn^{2+} plus Mg^{2+} (5 mM)	0·0
Complete minus Triton and Mn^{2+} plus Ca^{2+} (5 mM)	0·0

[a] Complete assay system (1 ml): A, 0·1 mM [U-^{14}C]-*myo*-inositol (2·4 × 10^5 dpm), 0·064 mM CDP-diglyceride, 0·1 M Tris HCl (pH 8·0), 6 mM $McCl_2$, and 0·15 ml of 1200–80 000 *g* pellet suspension (1·2 mg protein). Incubation, 1 h at 30°C. B, as in A but without CDP-diglyceride.

product in each case migrated on TLC with an R_f identical to that of authentic PI (Fig. 2A); (b) after deacylation of the combined products, the ^{14}C-labelled water-soluble phosphate ester migrated as a single spot with R_f identical to that of authentic glycerophosphorylinositol (Fig. 2B). Thus, the product formed from [U-^{14}C]*myo*-inositol either in the presence or absence of CDP-diglyceride was [^{14}C]phosphatidyl inositol.

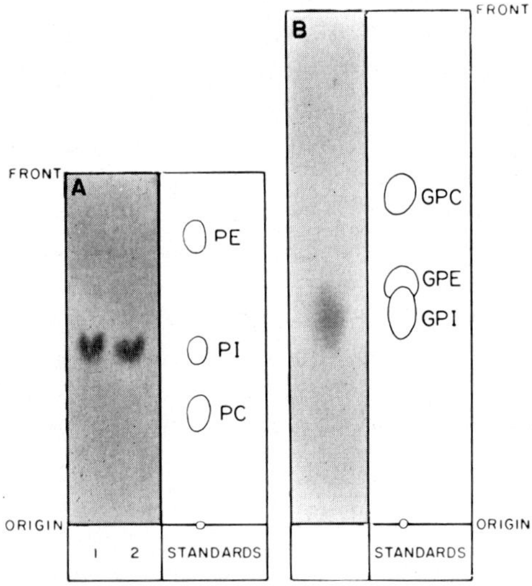

FIG. 2. Autoradiogram of chromatograms of phosphatidyl-[^{14}C] inositol and its deacylation product labelled from [U-^{14}C]*myo*-inositol in the presence and absence of CDP-diglyceride.

A: Phosphatidyl [^{14}C]inositol product from assay systems described in Table II: 1, in the presence of CDP-diglyceride; 2, in the absence of CDP-diglyceride. Chromatography was on silica gel H plates in chloroform–methanol–acetic acid–water (65:25:8:4). B: Deacylation product of combined phosphatidyl [^{14}C]inositol products from assays in presence and absence of CDP-diglyceride. GPI, GPE and GPC were prepared by deacylation of standard PI, PE and PC, respectively. Chromatography was on Whatman No. 1 paper in methanol–98% formic acid–water (80:13:7).

With regard to the pathways of PI biosynthesis (eqns **7–9**), the observed CDP-diglyceride-independent Mn^{2+}-dependent synthesis of PI (Table II) most likely represents an exchange reaction of the type found in liver (Paulus and Kennedy, 1960; eqn **9**), whereas the CDP-diglyceride stimulated synthesis of PI most likely arises from CDP-diglyceride:inositol phosphatidyl transferase activity (eqn **7**).

D. PHOSPHATIDYLGLYCEROL (PG)

The only known pathway for the biosynthesis of PG involves the intermediate formation of phosphatidyl glycerophosphate (PGP) from CDP-diglyceride and *sn*-glycerol-3-phosphate, followed by dephosphorylation to PG:

$$\text{sn-glycerol-3-phosphate} + \text{CDP-diglyceride} \rightarrow \text{3-sn-phosphatidyl-1'-sn-glycerol-3'-phosphate} + \text{CMP} \qquad \mathbf{(10)}$$

$$\text{3-sn-phosphatidyl-1'-sn-glycerol-3'-phosphate} \rightarrow \text{3-sn-phosphatidyl-1'-sn-glycerol} + P_i \qquad \mathbf{(11)}$$

This pathway was first demonstrated by Kiyasu *et al.* (1963) in rat and chicken liver, and by Kanfer and Kennedy (1964) in *E. coli*. Their evidence was based mainly on the CDP-diglyceride dependent incorporation of $[^{14}C]$GP into PG, and of $[^{32}P]$GP into PGP, by cell-free extracts of these tissues. The pathway has since been established in sheep brain (Possmayer *et al.*, 1968; Stanacev *et al.*, 1968; Davidson and Stanacev, 1970), in beef and rat heart (Stanacev *et al.*, 1969; Stuhne-Sekalec and Stanacev, 1970), in rat ventral prostate (Stanacev *et al.*, 1970), in a *Bacillus megaterium* mutant (Patterson and Lennarz, 1971) and in *Salmonella typhimurium* (Bell *et al.*, 1971).

The enzymes catalyzing the above reactions (eqns **10, 11**) have been isolated from the particulate fraction of *E. coli* and partially purified by Chang and Kennedy (1967a, b). The partially purified *sn*-glycerol-3-phosphate: CDP-diglyceride phosphatidyltransferase (PGP synthetase; eqn **10**) requires Mn^{2+} or Mg^{2+} for activity, and does not accept *sn*-glycerol-1-phosphate, glycerol, *myo*-inositol or serine as substrates. The partially purified PGP phosphatase (eqn **11**) was much more active than the PGP synthetase; thus, PGP does not accumulate, which probably explains why it has not been detected in *E. coli* lipids. The enzyme also requires Mg^{2+}, and does not act on GP or PA. Unlike the PGP synthetase it is strongly inhibited by sulfhydryl reagents. In contrast, the PGP phosphatase in a strain of *Bacillus* has been reported to be uninhibited by sulfhydryl reagents (Patterson and Lennarz, 1971).

Recently, the specificity of PGP synthetase towards different liponucleotides has been studied. Raetz and Kennedy (1973) found that the activity of non-cytosine nucleoside diphosphate diglycerides was measurable but very much lower than that of either CDP-diglyceride or its deoxy analogue. The latter was almost twice as effective as CDP-diglyceride at high concentrations, though whether this was also true at low concentrations was not determined. At certain concentrations, dCDP-diglyceride was also found to be more effective than CDP-diglyceride in the synthesis of PS (see below). Since dCDP-diglyceride has been found to accumulate to the same extent as CDP-diglyceride in growing *E. coli* cells (Raetz and Kennedy, 1973), these findings raise the question, at present unresolved, as to which of the two cytosine-containing liponucleotides is functioning *in vivo* in *E. coli* in the biosynthesis of PG and PS.

The reactions shown in eqns **10** and **11** have been proposed for the formation of PG in photosynthetic tissues since the plant PG also has the 3-*sn*-phosphatidyl-1′-*sn*-glycerol configuration (Benson and Miyano, 1961;

Haverkate and van Deenen, 1965a, b). The first evidence for this proposal was presented by Sastry and Kates (1966) who showed that a crude cell-free homogenate from spinach leaves could incorporate ^{32}P from *sn*-glycerol-3-[^{32}P]phosphate into PG as well as PA, in a system containing CTP, ATP, Mg^{2+}, CoA and fatty acids. However, the direct involvement of CDP-diglyceride in the formation of PG was not demonstrated. The above pathway may also function in *Chlorella vulgaris* since PG is rapidly labelled with ^{32}P from DL-α-glycerol[^{32}P]phosphate during photosynthesis (Sastry and Kates, 1965). In non-photosynthetic plant tissue, Douce (1968) and Douce and Dupont (1969) have presented good evidence that in cauliflower inflorescence PG is biosynthesized by the above reactions. Their evidence was as follows: (a) ^{14}C activity from *sn*-[^{14}C]glycerol-3-phosphate was incorporated into [^{14}C]PG in the presence of Mg^{2+} by a crude mitochondrial fraction actively synthesizing CDP-diglyceride *in situ* from CTP and endogenous PA; (b) $HgCl_2$ caused inhibition of PG formation and an accumulation of PGP (Douce and Dupont, 1969).

Though the above evidence is consistent with the biosynthesis of PG in plants by eqns **10** and **11**, knowledge concerning the kinetic properties and cofactors of the PG synthesizing system and of the ability of exogenous CDP-diglyceride to act as substrates was lacking, as well as information concerning intracellular distribution of the enzyme system. Studies bearing on these points have been carried out in this laboratory using spinach leaf preparations (Marshall and Kates, 1972) and will now be described.

1. *Phosphatidylglycerol synthetase in spinach leaves*

To define the biosynthetic pathway for PG synthesis in spinach leaves (eqns **10** and **11**), a 40 000 *g* pellet fraction ("heavy" microsomes) or a 90 000 *g* pellet ("light" microsomes) was incubated at pH 8 in a system containing CDP-diglyceride (fatty acid composition: 56% palmitate, 1% palmitoleate, 4% stearate, and 37% oleate), Mg^{2+}, Triton X-100 and *sn*-[1,3-^{14}C]glycerol-3-[^{32}P]phosphate. With each microsomal pellet fair incorporation of $^{14}C+^{32}P$ into lipid was obtained (2% and 0·7% for "heavy" and "light" microsomes, respectively) and a major radioactive product with mobility corresponding to that of PG both on TLC and on silicic acid-loaded paper and labelled *only* with ^{14}C was formed; a minor radioactive product, labelled both with ^{14}C and ^{32}P and migrating with R_f similar to that of authentic PGP was also detected. The major product was further identified as PG by its formation of [^{14}C]GPG after mild alkaline deacylation.

The formation of [^{14}C]PG was found to be absolutely dependent on the presence of exogenous CDP-diglyceride (Table III), showing that CDP-diglyceride was an obligatory substrate. Maximum synthesis of PG occurred at a CDP-diglyceride concentration of 0·04 mM, above which there was a marked inhibition. A similar inhibition has been observed for the *E. coli* system (Chang and Kennedy, 1967a).

TABLE III

The effect of CDP-diglyceride and divalent ionic concentration on phosphatidylglycerol synthesis

Concentration (mM)	Phosphatidylglycerol formed (nmol/h)
CDP-Diglyceride[a]	
0·0	0·0
0·02	14·3
0·04	14·8
0·08	12·9
0·16	10·2
0·25	9·5
$MgCl_2$[b]	
0·0	0·0
1·0	13·2
2·0	20·3
5·0	28·6
10·0	38·0
30·0	42·9
$MnCl_2$[b]	
10·0	18·8

[a] Assay system (1 ml): 1 mM [^{14}C]GP (8·5 × 10^5 dpm), 30 mM $MgCl_2$, 0·25 M Tris HCl (pH 7·4), 1·5 mg Triton X-100, 0·2 ml (0·33 mg protein) of 40 000 *g* pellet fraction suspended in 0·1 M potassium phosphate (pH 8·0) containing 1 mM β-mercaptoethanol, and varying amounts of CDP-diglyceride as indicated. Incubation, 1 h at 37°C.

[b] Assay system (1 ml): 1 mM [^{14}C]GP (1 × 10^5 dpm), 0·1 mM CDP-diglyceride, 0·25 M Tris-HCl buffer (pH 7·3), 1·5 mg Triton X-100, 40 000 *g* pellet fraction (0·4 mg protein) and varying amounts of $MgCl_2$ or $MnCl_2$ as indicated. Incubation, 1 h at 37°C. (Data from Marshall and Kates, 1972.)

PG formation was also absolutely dependent upon Mg^{2+}, half maximal activity being observed at 2 mM Mg^{2+} (Table III); at all concentrations of Mg^{2+}, PG was the only product detected. Mn^{2+} was also found to be effective as a co-factor, but at a concentration of 10 mM it was only half as effective as Mg^{2+}. The effect of varying Triton X-100 concentration was not studied, though it was observed that in the absence of Triton little or no PG was formed.

The evidence presented shows that the 40 000 *g* and 90 000 *g* pellet fractions from spinach leaves contain an enzyme system catalyzing the biosynthesis of phosphatidylglycerol from CDP-diglyceride and *sn*-glycerol-3-phosphate by the biosynthetic pathways (eqns **10, 11**) involving the intermediate formation of PGP, followed by its dephosphorylation. The enzyme system in spinach leaves appears to be very similar to the bacterial system (Chang and Kennedy, 1967a, b) and was further characterized as follows.

The pH optimum of PG formation in Tris-HCl was 7·3; over the pH range (6·8–8·5) covered, only small amounts of PGP were detected (2–10% of total ^{14}C products). The temperature optimum of PG formation measured by the incorporation of ^{14}C into the total lipid extract was found to be 37°C. The thermal stability of the PG synthesizing system at the optimum temperature was determined by preincubating the 40 000 *g* fraction at 37°C for varying lengths of time up to 3 h, and then assaying its activity. The rate of PG formation was constant for preincubation times up to 1 h, but decreased thereafter to 70% of the initial activity after a 3 h preincubation. At no time was any appreciable PGP formation observed, indicating that the PGP phosphatase is not more heat labile than the PGP synthetase. In some experiments it was observed that storage of the 40 000 *g* fraction at – 10°C for 3 days resulted in a 10% loss of activity. The PG synthetase in spinach is thus fairly heat-labile.

Under optimal conditions of pH and Mg^{2+}, CDP-diglyceride and GP concentrations the formation of PG by the 40 000 *g* fraction was linear with time of incubation up to 1 h (after a brief lag phase of about 2 min) and with protein concentration up to 0·35 mg/ml. Under conditions of linearity with time, the initial rate of formation of PG showed typical Michaelis-Menten kinetics as a function of glycerophosphate concentration and reached a maximum at a GP concentation of 1·3 mM (Fig. 3A). The PGP intermediate accumulated with increasing GP concentration and reached a steady state above 0·53 mM GP (Fig. 3A), but at all concentrations of GP, the amount of PGP formed accounted for only a small proportion of total phosphatide synthesized (8–14%, Fig. 3A). Thus, the rate determining step in PG synthesis from GP must be the first step, that catalyzed by PGP synthetase:

$$\text{CDP-dig} + \text{GP} \xrightarrow[\text{slow}]{} \text{PGP} + \text{CMP} \qquad \text{step 1} \quad \textbf{(12)}$$

$$\text{PGP} \xrightarrow[\text{fast}]{} \text{PG} + \text{P}_i \qquad \text{step 2} \quad \textbf{(13)}$$

The Michaelis-Menten constant for GP in the first step may be derived as follows (Dixon and Webb, 1964): The rate equation for the first step, assuming there is no effect of one substrate on the affinity of the enzyme for the other, is

$$\frac{-d[\text{GP}]}{dt} = \frac{V_{max}[\text{GP}]\,[\text{CDP-dig}]}{(K_a + [\text{GP}])\,(K_b + [\text{CDP-dig}])} \qquad \textbf{(14)}$$

where K_a and K_b are the Michaelis-Menten constants for GP and CDP-diglyceride, respectively. Under conditions of saturation by CDP-diglyceride, $[\text{CDP-dig}] \gg K_b$, and the equation reduces to

$$\frac{-d[\text{GP}]}{dt} = \frac{V_{max}[\text{GP}]}{K_a + [\text{GP}]} \qquad \textbf{(15)}$$

Since GP forms the products PG and PGP, it follows that

$$\frac{-d[GP]}{dt} = \frac{d(PG + PGP)}{dt} \tag{16}$$

Thus,

$$\frac{d(PG + PGP)}{dt} = \frac{V_{max}[GP]}{K_a + [GP]} \tag{17}$$

The plot of rate of (PG+PGP) formation *vs* GP concentration showed typical Michaelis-Menten kinetics (Fig. 3A), and from the Lineweaver-Burke plot (Fig. 3B) the Michaelis-Menten constant for GP was calculated to be 0·25 mM. For comparison, the corresponding K_a obtained from the Lineweaver-Burke plot of PG formation alone, was 0·20 mM. These values are in excellent agreement with the corresponding apparent K_m (0·25 mM) reported

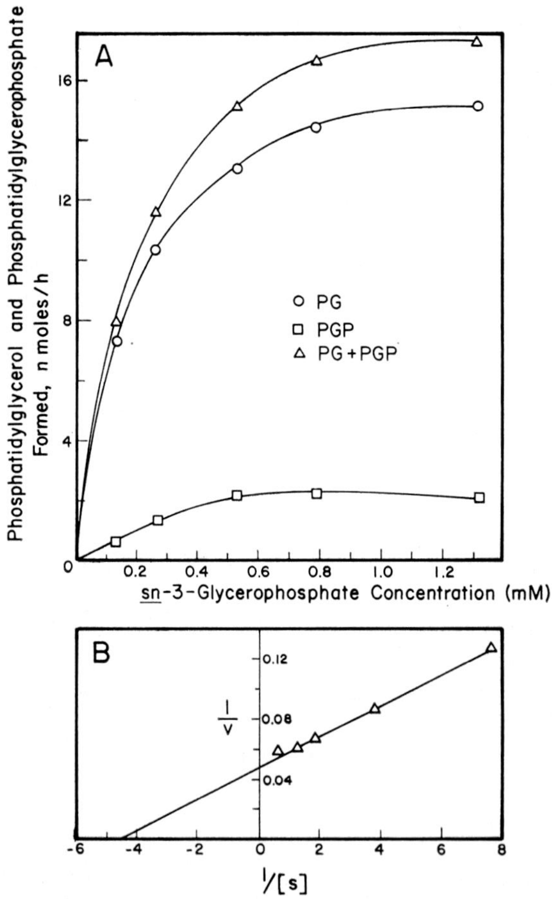

FIG. 3 The effect of *sn*-glycerol-3-phosphate concentration on the synthesis of phosphatidylglycerol and phosphatidylglycerophosphate by the 40 000 *g* pellet fraction.

A: Rate of lipid formation as a function of GP concentration. B: Plot of 1/V *vs* 1/S, where V is the rate of (PG + PGP) formation and S is GP concentration.

for the isolated PGP-synthetase from *E. coli* (Chang and Kennedy, 1967a).

Overall PG synthesis by the 40 000 *g* fraction was not significantly inhibited (8%) by $HgCl_2$ at a concentration (10 mM) reported to be completely inhibitory for PG synthesis in cell free preparations from cauliflower inflorescence (Douce and Dupont, 1969), chicken liver (Kiyasu *et al.*, 1963) and *E. coli* (Chang and Kennedy, 1967a). Furthermore, only 17% inhibition was observed on increasing the $HgCl_2$ concentration to 40 mM. The most pronounced inhibition was observed with *p*-chloromercuribenzoate (PCMB), but this still amounted to only 21%. No PGP was detected in any of the inhibition assay systems. It should be mentioned that a small amount of β-mercaptoethanol derived from the 40 000 *g* suspension was present in the reaction mixtures, but at a concentration (0·25 mM) too low to react with significant amounts of sulfhydryl reagent. Furthermore, it was found that prior dialysis of the 40 000 *g* fraction against phosphate buffer did not affect the results. Since none of the sulfhydryl reagents inhibited PG synthesis to any great extent, nor resulted in accumulation of PGP, it was concluded that neither PGP synthetase nor PGP phosphatase contained sulfhydryl groups actively taking part in their catalytic action. Absence of inhibition by sulfhydryl reagents has also been reported for the *Bacillus megaterium* phosphatidylglycerol synthetase system (Patterson and Lennarz, 1971).

2. *Subcellular distribution of phosphatidylglycerol synthetase*

The total cell-free homogenate of spinach leaves was fractionated (Marshall and Kates, 1972) by differential centrifugation into chloroplast (1200 *g*), mitochondrial (15 000 *g*), "heavy" microsomal (40 000 *g*) and "light" microsomal (90 000–100 000 *g*) fractions. The microsomal fractions were virtually free of chloroplast fragments, mitochondria and peroxisomes, but the mitochondrial fraction, which accounted for most of the cytochrome oxidase activity in the homogenate, was however contaminated with chloroplast fragments and some peroxisomes (Table IV). These fractions were assayed for PG synthetase activity under conditions of linearity of enzyme activity with respect to protein concentration.

The results (Table IV) showed that the 40 000 *g* microsomal fraction contained a high proportion (45%) of phosphatidylglycerol-synthesizing activity, and the highest specific activity. The 40 000 *g* supernatant also contained a high proportion of activity (44%), but with a low specific activity, and about 30% of this activity could be sedimented by further centrifugation at 90 000 *g*. A small (15%) but significant proportion of activity was associated with the 15 000 *g* mitochondrial fraction, but little activity (4%) was found in the 1200 *g* choroplast fraction. Essentially the same distribution pattern was obtained when the cell fractions were ultrasonicated prior to assaying for phosphatidylglycerol-synthesizing activity.

The phosphatidylglycerol-synthesizing system thus appears to be localized primarily in the 40 000 *g* fraction which, being essentially free of chloroplasts,

TABLE IV

Sub-cellular distribution of phosphatidylglycerol synthesizing activity[a]

Cell fraction	Protein	Chloro-phyll	Cytochrome oxidase	Catalase	PG synthesis[b]		
	%	%	%	%	Activity nmol/h	%	Specific activity
Cell-free homogenate (chloroplast)	100	100	100	100	306	100	0·7
1200 *g* (chloroplast) pellet	23	63	19	5	13	4	0·1
15 000 *g* (mitochondrial) pellet	12	41	71	20	46	15	0·9
40 000 *g* ("heavy" microsomal) pellet	2	3	3	0	139	45	16·0
40 000 *g* supernatant	71	2	6	34	133	44	0·4
90 000 *g* ("light" microsomal) pellet	—	—	—	—	40	13	—
% recovery	108	109	99	59	—	108	—

[a] Sub-cellular fractionation was carried out in homogenizing medium containing 0·1 M potassium phosphate (pH 8·0) and marker enzymes, chlorophyll and protein were assayed as described elsewhere (Marshall and Kates, 1972).

[b] Assay mixture (1 ml): 0·36 mM [^{14}C]GP ($1\cdot3 \times 10^6$ dpm), 0·04 mM CDP-diglyceride, 30 mM $MgCl_2$, 0·25 M Tris HCl (pH 7·3), 1·5 mg Triton X-100, and either 0·04 ml of cell-free homogenate or 0·2 ml of remaining fractions suspended in homogenizing medium. Incubation, 1 h at 37°C. (Data from Marshall and Kates, 1972).

peroxisomes and mitochondria, is most probably a microsomal membrane fraction. Biosynthesis of other phospholipids in spinach leaves is also localized in the microsome fraction, e.g. phosphatidic acid (Cheniae, 1965; Sastry and Kates, 1966), CDP-diglyceride (Bahl *et al.*, 1970) and phosphatidycholine (Devor and Mudd, 1971). However, PG synthesis in a non-photosynthetic tissue, cauliflower inflorescence, has been demonstrated in mitochondria (Douce, 1968; Douce and Dupont, 1969), (but no other cell fractions were tested) and is exclusively localized in mitochondria of chicken liver (Kiyasu *et al.*, 1963), rat liver (Dennis and Kennedy, 1972) and sheep brain (Davidson and Stanacev, 1970).

The question then arises whether there are other subcellular sites for PG-synthesis in spinach leaves. The considerable activity remaining in the 40 000 *g* supernatant even after further centrifugation at 90 000 *g*, does not necessarily indicate a cytosol site, since the observed activity in this fraction could have arisen from solubilization or fragmentation of the microsomal system during cell-fractionation. The possibility of a mitochondrial site for PG-synthesis was not ruled by our results since activity was found in the 15 000 *g* fraction (Table IV). Indeed, a mitochondrial site is a likely possibility in view of the presence in spinach leaf mitochondria of the CDP-diglyceride synthesizing system (Bahl *et al.*, 1970). However, the PG-synthesizing activity which we observed most probably arose from contamination by the microsomal fraction. This question should be further investigated using sub-cellular fractions purified by density gradient centrifugation and specific microsomal marker enzymes other than glucose-6-phosphatase which is not specific for plant microsomes (Thompson, 1969; Harwood and Stumpf, 1972).

The very low PG-synthesizing activity found in the chloroplast fraction (Table IV) appears to rule out chloroplasts as another sub-cellular site for PG synthesis. Though the presence of a permeability barrier to the substrate CDP-diglyceride might account for the low activity, this factor does not apply since prior ultrasonication of the chloroplast fraction did not increase its activity. The absence of PG-synthesizing activity in the chloroplast fraction is consistent with the absence of a CDP-diglyceride synthesizing system in this organelle (Bahl *et al.*, 1970). It is surprising, however, that an organelle so rich in phosphatidylglycerol (Table I) should lack the enzymes necessary for synthesis of this lipid. Our results suggest that the PG in chloroplasts may be synthesized by the microsomal system and then incorporated into chloroplast membranes.

3. *Exchange Pathway*

An alternative pathway for PG synthesis has been suggested (Yang *et al.*, 1967; Dawson, 1967), namely, the transfer of a phosphatidyl unit from phosphatidylcholine to glycerol, catalyzed by phospholipase D:

$$\text{phosphatidylcholine} + \text{glycerol} \underset{}{\overset{\text{phospholipase D}}{\rightleftharpoons}} \text{phosphatidyl glycerol} + \text{choline} \quad \mathbf{(18)}$$

However, this pathway cannot possibly be operating *in vivo* since the PG formed by this transphosphatidylation has been shown to be phosphatidyl-*rac*-1′-glycerol (Yang *et al.*, 1967), a mixture of diastercoisomers rather than the one stereoisomer known to be present in leaves. Also, Schantz, *et al.* (1972) showed that in *Euglena gracilis* growing auxotrophically, no ^{32}P was incorporated into PG from PC and PE pulse-labelled from [^{32}P]orthophosphate, even though PG was being rapidly synthesized. This clearly demonstrated that in this organism at least, the phosphatidyl moiety of PG is not derived from PC or PE by an exchange reaction catalyzed either by phospholipase D or by a separate exchange enzyme. In our studies on PG synthesis in spinach leaves, the possibility existed that the labelled glycerophosphate precursor was first dephosphorylated to glycerol which was then incorporated into PG by the exchange reaction (eqn **18**). This was, however, ruled out by the results of an experiment comparing the relative incorporation of [^{14}C]glycerol into PG by a microsomal fraction under the same assay conditions. ^{14}C was found to be incorporated into PG only from [^{14}C]glycerophosphate and not from [^{14}C]glycerol. Furthermore, no [^{14}C]glycerol could be detected in the methanol-water soluble products from the [^{14}C]glycerophosphate incubation mixture, indicating the absence of glycerophosphatase activity in the microsomal fraction.

E. DIPHOSPHATIDYL GLYCEROL (CARDIOLIPIN)

Two pathways are at present known for the biosynthesis of cardiolipin:

$$\text{phosphatidylglycerol} + \text{CDP-diglyceride} \rightarrow \text{cardiolipin} + \text{CMP} \qquad \textbf{(19)}$$

$$2\ \text{phosphatidylglycerol} \rightarrow \text{cardiolipin} + \text{glycerol} \qquad \textbf{(20)}$$

Both pathways involve the transfer of a phosphatidyl group to PG, but in one pathway the phosphatidyl group is derived from CDP-diglyceride (eqn **19**), while in the other it is derived from another molecule of PG (eqn **20**).

Stanacev *et al.* (1967) proposed that cardiolipin in *E. coli* is biosynthesized by the reaction shown in eqn **19**, on the basis of stimulation of [^{14}C]cardiolipin synthesis from phosphatidyl-[^{14}C]glycerol by high concentration of exogenous unlabelled CDP-diglyceride. At the time of this proposal, cardiolipin biosynthesis by the alternate reaction (eqn **20**) was not known. However, later, Stanacev and Stuhne-Sekalec (1970) showed the feasibility of such a reaction by demonstrating the CDP-diglyceride-independent conversion of small amounts of PG into cardiolipin by a crude phospholipase D preparation from cabbage.

The first indication that this alternate pathway might be functioning in bacteria was the finding by De Siervo and Salton (1971) of an almost quantitative conversion of PG to cardiolipin by a particulate fraction from

M. lysodeikticus in the absence of CDP-diglyceride. These studies prompted a re-examination of the biosynthesis of cardiolipin in *E. coli*. Thus, Hirschberg and Kennedy (1972) found that the $^{3}H/^{32}P$ ratio in cardiolipin formed by an *E. coli* particulate fraction was half that of the corresponding ratio in the [^{32}P]phosphatidyl[^{3}H]glycerol substrate. These workers confirmed the original observation by Stanacev *et al.* (1967) that CDP-diglyceride stimulated cardiolipin biosynthesis, but they proved conclusively that this compound was not functioning as a phosphatidyl donor by the finding that ^{14}C from [^{14}C]PG, but not ^{32}P from [β-^{32}P]CDP-diglyceride, was incorporated into cardiolipin. Hostetler *et al.* (1972) have made similar observations in *E. coli*, and the work of Rampini *et al.* (1971) strongly suggests that the same CDP-diglyceride independent pathway operates in *Staphylococcus aureus*. The evidence so far thus favours the formation of cardiolipin in *E. coli* by the reaction shown in eqn **20**. In animal tissue, the pathway of cardiolipin biosynthesis probably follows eqn **19**. Davidson and Stanacev (1971) showed that *sn*-glycerol-3-phosphate was incorporated into cardiolipin by guinea pig liver mitochondria actively synthesizing CDP-diglyceride *in situ*, and Hostetler *et al.* (1971) showed that glycerophosphate and PG incorporation into cardiolipin by rat liver mitochondria was dependent on the presence of exogenous CDP-diglyceride. Stanacev *et al.* (1973) later demonstrated the substrate role of CDP-diglyceride more directly by showing the incorporation of ^{14}C into cardiolipin from CDP-[^{14}C]diglyceride and Hostetler *et al.* (1972) showed that [2-^{3}H]phosphatidyl-[1-^{14}C]glycerol was converted into cardiolipin with conservation of the $^{3}H/^{14}C$ ratio. These findings were consistent with the CDP-diglyceride dependent pathway (eqn **19**) and at the same time ruled out the possibility that the bacterial CDP-diglyceride independent pathway (eqn **20**) was also operating.

In regard to the properties of the enzyme catalyzing the CDP-diglyceride dependent pathway, Domazet *et al.* (1973) have confirmed earlier reports that Mg^{2+} or Mn^{2+} are required for activity. This is in contrast to the animal PG synthetase system which does not require metal ions (Kiyasu *et al.*, 1963).

Very little is known concerning the biosynthesis of cardiolipin in plants. Douce (1968) and Douce and Dupont (1969) found that ^{14}C from [^{14}C]GP was incorporated into cardiolipin as well as into PG by cauliflower inflorescence mitochondria in the presence of CTP and Mg^{2+}. However, this result does not unambiguously establish which of the pathways discussed above was operating in this system. In view of the finding by Stanacev and Stuhne-Sekalec (1970) that phospholipase D (a plant enzyme) catalyzes formation of cardiolipin *in vitro* by eqn **20**, it might be thought that this enzyme also catalyzes the formation of cardiolipin *in vivo* in plants. However, as pointed out above, phospholipase D can carry out many non-specific exchange reactions (in addition to its hydrolyzing activity), and it is therefore unlikely to have any physiological role in *de novo* synthesis of cardiolipin.

F. PHOSPHATIDYLSERINE (PS)

The importance of CDP-diglyceride as a phosphatidyl donor in the biosynthesis of PG, PI and cardiolipin has already been described above. Kennedy and co-workers have demonstrated that in *E. coli* this compound also serves as a phosphatidyl donor in the biosynthesis of PS, a minor lipid in *E. coli* which serves as the intermediate in the biosynthesis of PE. The reaction for the formation of PS, catalyzed by CDP-diglyceride: L-serine phosphatidyltransferance (PS synthetase), is as follows:

$$\text{CDP-diglyceride} + \text{L-serine} \rightarrow \text{phosphatidyl serine} + \text{CMP} \qquad \textbf{(21)}$$

This reaction was first described by Kanfer and Kennedy (1964), who found that serine incorporation into PS by an *E. coli* cell free extract was completely dependent on exogenous CDP-diglyceride. A striking property of the enzyme was its dependence for optimal activity on both an added surface active agent (either octanol or cutscum), and a sufficient ionic strength (0·1 mM) of the incubation medium. The above pathway has also been demonstrated in a mutant of *Bacillus megaterium* by Patterson and Lennarz (1971) and in *Salmonella typhimurium* by Bell *et al.* (1971).

The biosynthesis of PS by eqn (**21**) is characteristic of bacteria and no such reaction has been detected either in liver (Paulus and Kennedy, 1960) or in the protozoan *Tetrahymena pyriformis* (Dennis and Kennedy, 1970), Instead. the evidence available so far suggests that PS is biosynthesized in these tissues by a Ca^{2+}-stimulated exchange reaction involving free serine and the ethanolamine moiety of phosphatidylethanolamine (see Fig. 1 and Kennedy, 1962).

Few studies on the biosynthesis of PS in plants have been carried out, and most of these have utilized only non-photosynthetic tissue. Willemot and Boll (1967) have demonstrated the incorporation of ^{14}C from [U-^{14}C]serine into the serine moiety of PS by excised tomato root. The biosynthesis of PS in *Chlorella vulgaris* by eqn (**21**) was suggested by Sastry and Kates (1965) from the finding that growing cells of this organism incorporated ^{32}P from [^{32}P]GP into PE. This incorporation most likely arose by the formation of CDP-diglyceride from GP followed by conversion of CDP-diglyceride to PS and decarboxylation to PE (eqns **1**, **2**, **6**, **21**, **22**), since this is the only known route in which the phosphate moeity of PE is derived from that of GP. However, it is not possible from these *in vivo* studies to establish with certainty the pathway of PS biosynthesis. Vandor and Richardson (1968) showed that a microsomal fraction from etiolated pea seedlings catalyzed the incorporation of serine into PS by a Ca^{2+}-stimulated exchange reaction (see Fig. 1), but no results were reported on the presence of PS synthetase in this tissue.

In our biosynthetic studies with spinach leaves (Marshall and Kates, 1974) whole leaves or leaf slices were found to incorporate ^{14}C from [U-^{14}C]L-serine into lipid, and a small but significant proportion of the incorporated ^{14}C was found in PS (Table V; Fig. 4). To test for the presence of CDP-diglyceride:

TABLE V

The incorporation of ^{14}C-labelled substrates into lipid by whole leaves and leaf slices

Substrate[a]	^{14}C in total lipids[b]	Distribution[c] of ^{14}C (% of total lipid ^{14}C)		
	dpm × 10^{-3}	PC	PE+ PE-Me	PS
WHOLE LEAF				
[U-^{14}C]L-serine (2·4 μCi)	116	19	28	3
[1,2-^{14}C]ethanolamine (1·7 μCi)	301	11	78	0
[1,2-^{14}C]ethanolamine (2·2 μCi) + cold L-methionine (10 mM)	615	7	85	0
[Me-^{14}C]L-methionine (1·9 μCi)	167	25	4	0
LEAF SLICES				
[U-^{14}C]L-serine (10 μCi)	51	4	20	6
[1,2-^{14}C]ethanolamine (10 μCi)	401	1	66	0
[Me-^{14}C]L-methionine (10 μCi)	169	8	5	0
[1,2-^{14}C]choline (10 μCi)	102	90	0	0

[a] The figures in parenthesis denote amount of ^{14}C activity taken up by each whole leaf, or supplied to the leaf slices.

[b] Per whole leaf or *g* of leaf slices.

[c] ^{14}C-labelled lipids were separated by t.l.c. (see Fig. 4) and each spot counted; PE and PE-Me were counted as one fraction. Traces of lyso PC and lyso PE were found only in the leaf slices; remainder of counts were mostly in the neutral lipid fraction, with traces in an unidentified component (compound X; Fig. 4). (Data from Marshall and Kates, 1974.)

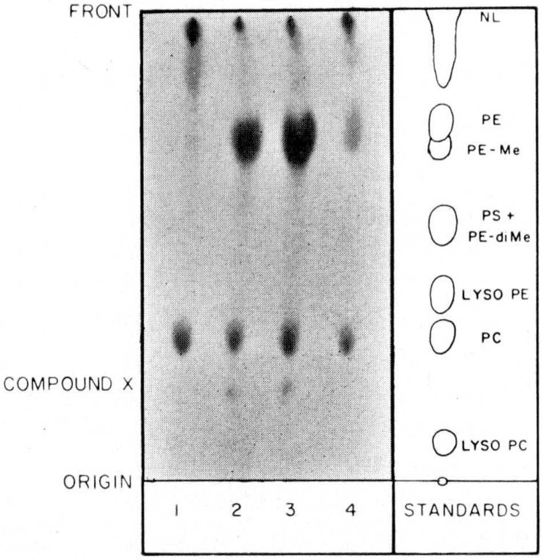

FIG. 4. Autoradiogram of a chromatogram of lipids labelled from [^{14}C]methionine, [^{14}C]ethanolamine and [^{14}C]serine by whole leaves:

1, [Me-^{14}C]methionine; 2, [1,2-^{14}C]ethanolamine; 3, [1,2-^{14}C]ethanolamine + unlabelled methionine; 4, [U-^{14}C]L-serine. Chromatography was on a silica gel H plate in chloroform–methanol–acetic acid–water (65:25:8:4, v/v). Mobility of standard lipids not shown were: PG (R_f 0·60), DGD (R_f 0·75) and MGD (R_f 0·90). (Data from Marshall and Kates, 1974.)

serine phosphatidyltransferase (PS synthetase; eqn **21**), a particulate cell fraction (1200–100 000 *g*) was incubated with [^{14}C]serine in the presence and absence of CDP-diglyceride. Several-fold stimulation of [^{14}C]serine incorporation into PS was observed in the presence of exogenous CDP-diglyceride (Table VI) indicating the existence of PS synthetase. A low level of ^{14}C was also present in PE which was presumably formed from PS by decarboxylation (see below). The PS synthetase in *E. coli* is known to be dependent (Kanfer and Kennedy, 1964) on the presence of cutscum and Na_2SO_4 in the assay system but the addition of these components to the spinach assay system was found almost completely to inhibit PS synthesis (Table VI).

TABLE VI

The effect of CDP-diglyceride on the incorporation of [U-^{14}C]L-serine into phosphatidylserine by a total particulate fraction

Assay system[a]	^{14}C incorporation[b] (dpm)	
	PS	PE
Complete system	5711	126
Complete system, minus CDP-diglyceride	781	0
Complete system + cutscum + Na_2SO_4	194	0

[a] Complete assay system (1 ml): 7·5 μM [U-^{14}C]L-serine (2·1 × 10^6 dpm), 0·13 mM CDP-diglyceride, 50 mM sodium phosphate (pH 8·0), 1·5 mM β-mercaptoethanol and 0·4 ml of 1200–100 000 *g* pellet suspended in water (1·3 mg protein). Cutscum (0·5 mg) and Na_2SO_4 (80 mM) were added as indicated. Incubation, 1 h at 30°C.

[b] ^{14}C-labelled lipids were separated by t.l.c. in chloroform–methanol–acetic acid–water (65:25:8:4 by vol.; R_f values of authentic PS and PE, 0·53 and 0·72, respectively). (Data from Marshall and Kates, 1974.)

The low residual incorporation of [^{14}C]serine into PS in the absence of CDP-diglyceride (Table VI) most probably arose by the action of PS synthetase utilizing endogenous CDP-diglyceride, but might also have arisen from a serine-phospholipid exchange reaction (reaction **17**, Fig. 1). However, this possibility was eliminated by the finding that no ^{14}C-incorporation into PS was observed on incubation of the particulate fraction with [^{14}C]serine in a system designed to test for the exchange pathway either in the absence or presence of Ca^{2+} (see Table VII).

These results thus show that biosynthesis of PS in spinach leaves occurs by the same pathway as that operating in bacteria (eqn **21**), namely by a phosphatidyl transferase reaction between CDP-diglyceride and L-serine, and not by the serine exchange pathway. In contrast, the exchange pathway is the only one known to operate in animal tissues and protozoa. It has also been described in pea seedlings (Vandor and Richardson, 1968), but whether PS synthetase is also present in this tissue is not known.

TABLE VII

Exchange pathway for incorporation of [U-^{14}C]L-serine, [1,2-^{14}C]ethanolamine and [1,2-^{14}C]choline into phospholipids by spinach particulate fraction

Assay system[a]	^{14}C incorporation[b] (dpm)		
	PS	PE	PC
Complete + [U-^{14}C]L-serine + EDTA	0	0	0
Complete + [U-^{14}C]L-serine + Ca^{2+}	0	0	0
Complete + [1,2-^{14}C]ethanolamine + EDTA	0	498	0
Complete + [1,2-^{14}C]ethanolamine + Ca^{2+}	0	4066	0
Complete + [1,2-^{14}C]choline + EDTA	0	0	385
Complete + [1,2-^{14}C]choline + Ca^{2+}	0	0	1622

[a] Complete assay system (1 ml): 50 mM Tris-HCl (pH 8·5), 1·5 mM β-mercaptoethanol, 0·4 ml of 1200–100 000 *g* pellet suspended in H_2O (1·3 mg), and either 7·4 μM [U-^{14}C]L-serine (2·1 × 10^6 dpm), 0·4 mM [1,2-^{14}C]ethanolamine (2·3 × 10^6 dpm) or 0·1 mM [1,2-^{14}C]choline chloride (1·4 × 10^6 dpm). Where applicable, assay systems contained either 2 mM EDTA. Na_2 or 5 mM $CaCl_2$. Incubation, 1 h at 30°C.

[b] ^{14}C-labelled lipids were separated by t.l.c. in chloroform–methanol–acetic acid–water (65:25:8:4, by vol.; R_f values of PE, PS and PC, 0·72, 0·53 and 0·31, respectively. (Data from Marshall and Kates, 1974.)

G. PHOSPHATIDYLETHANOLAMINE (PE)

Two pathways are at present known for the *de novo* biosynthesis of PE, namely the PS decarboxylase pathway (eqns **21**, **22**), present in bacteria, and the nucleotide pathway (eqns **23–25**), which is present in animals.

1. *Phosphatidylserine decarboxylase pathway*

$$\text{CDP-diglyceride} + \text{L-serine} \rightarrow \text{phosphatidylserine} + \text{CMP} \qquad \textbf{(21)}$$

$$\text{phosphatidylserine} \rightarrow \text{phosphatidylethanolamine} + CO_2 \qquad \textbf{(22)}$$

Kanfer and Kennedy (1964) demonstrated the presence of a PS decarboxylase in cell-free extracts from *E. coli* and showed it to be much more active than PS synthetase. PS decarboxylase has recently been isolated from *E. coli* and purified to homogeneity by Wickner and Kennedy (1971). The pure enzyme was inhibited by hydroxylamine and 4-bromo-3-hydroxybenzyloxyamine, indicating the involvement of a pyridoxal cofactor. PS decarboxylase, together with PS synthetase, has also been detected in other bacteria, namely a *Bacillus megaterium* mutant (Patterson and Lennarz, 1971) and in *Salmonella typhimurium* (Bell *et al.*, 1971).

PS decarboxylase has also been detected in rat liver (Borkenhagen *et al.*, 1961) and in the protozoan *Tetrahymena pyriformis* (Dennis and Kennedy, 1970). However, in these tissues, the enzyme is not involved in the *de novo* biosynthesis of PE, since the PS itself is apparently derived from PE by an exchange reaction with free serine (Fig. 1), and not by PS synthetase which is absent in these tissues. Thus, the decarboxylation and exchange reactions in these tissues form a metabolic cycle which, as pointed out by Kennedy (1962), involves the net production of free ethanolamine from free serine.

In our *in vivo* studies with whole leaves (Marshall and Kates, 1974), ^{14}C from [U-^{14}C]L-serine was incorporated not only into PS but also into the glycerylphosphorylethanolamine moiety of PE (Figs 4, 5) (Tables V, VIII). It is unlikely that the glycerol moiety was labelled in view of the lack of ^{14}C-incorporation from [U-^{14}C]serine into PG by whole leaves (see Fig. 4). Thus, assuming that all of the ^{14}C in GPE was localized in the ethanolamine moiety, these findings indicate the existence of PS decarboxylase in spinach leaves.

More conclusive proof of the presence of PS decarboxylase was provided by the *in vitro* demonstration of a direct enzymatic conversion of [^{14}C]PS to [^{14}C]PE (Table IX), using a total cell-free homogenate. A high proportion

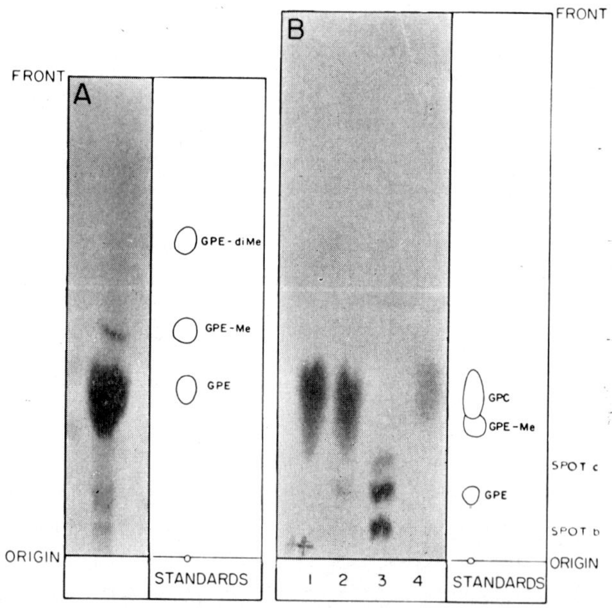

FIG. 5. Autoradiogram of a chromatogram of methanol–water soluble deacylation products of lipids labelled with ^{14}C by whole leaves (see Fig. 4).

A: PE + PE-Me labelled from [1,2-^{14}C]ethanolamine. Chromatography on Whatman No. 1 paper in isopropanol–water–conc. ammonia (7:2:1). B: 1, PC labelled from [Me-^{14}C]-methionine; 2, PC labelled from [1,2-^{14}C]ethanolamine; 3, PE labelled from [U-^{14}C]serine; 4, PC labelled from [U-^{14}C]serine. Chromatography on Whatman No. 1 paper in *n*-butanol–acetic acid–water (5:3:1). Standard GPE, GPE-Me, GPE-diMe and GPC were obtained by deacylation of the corresponding standard phospholipid. Spots b and c are deacylated products of unidentified ^{14}C-labelled lipids chromatographing with the PE fraction labelled from [U-^{14}C]serine (Fig. 4). Spot b may be derived from digalactosyl diglyceride. (Data from Marshall and Kates, 1974).

TABLE VIII

The incorporation of ^{14}C-labelled substrates into deacylated moieties of phosphatidylcholine, phosphatidylethanolamine and phosphatidyl monomethylethanolamine by whole leaves and leaf slices

Substrate	^{14}C incorporation into deacylated lipid[a] dpm $\times 10^{-3}$		
	GPC	GPE	GPE-Me
WHOLE LEAF			
[U-^{14}C]L-serine	5·1 (23)	10 (—)	0
[1,2-^{14}C]ethanolamine	32 (97)	226 (97)	4 (—)
[Me-^{14}C]L-methionine	36 (68)	0	0
LEAF SLICES			
[1,2-^{14}C]ethanolamine	—	245 (92)	—
[Me-^{14}C]methionine	8·1 (67)	—	—
[1,2-^{14}C]choline	91 (99)	—	—

[a] Deacylation was carried out on the individual lipids shown in Fig. 4, and identity of deacylated products was confirmed by paper chromatography, as described in Fig. 5. Numbers in parentheses denote percentage of lipid ^{14}C in deacylated moiety. Dash denotes data unavailable. (Data from Marshall and Kates, 1974.)

(34%) of the total enzyme activity was recovered in the 100 000 *g* supernatant fraction, while of the particulate fractions, the 15 000 *g* pellet contained the most activity (Table IX). The ratio of ^{14}C in the deacylated (water-soluble) moiety to that in the fatty acid (chloroform-soluble) moieties of the [^{14}C] PE formed was 1·08, close to the value 1·11 calculated for loss of one

TABLE IX

Subcellular distribution of phosphatidylserine decarboxylase

Cell fraction	Incorporation of ^{14}C into PE[a]	
	dpm $\times 10^{-3}$ per 20 min	%
Cell-free homogenate	3·71	100
1200 *g* (chloroplast) pellet	0·09	2
15 000 *g* (mitochondrial) pellet	0·25	7
100 000 *g* (microsomal) pellet	0·12	3
100 000 *g* supernatant	1·27	34
Recovery	—	46

[a] Assay system (0·5 ml): [^{14}C]phosphatidylserine (1·7 $\times$ 10^{4} dpm), and 0·1 ml of cell-free homogenate (16·5 mg protein/ml) or equivalent amount of each cell fraction suspended in homogenizing medium (pH 7·0); incubation, 20 min at 30°C. ^{14}C-labelled lipid products were separated by TLC in chloroform–methanol–conc. ammonia (65:25:5, by vol.). R_f values: 0·05; PE, 0·32; lyso PE, 0·11; PC, 0·21. Values of activity are corrected for boiled enzyme control. (Data from Marshall and Kates, 1974.)

TABLE X

Distribution of ^{14}C activity between water-soluble and chloroform-soluble moieties of [^{14}C]phosphatidylserine and [^{14}C]phosphatidylethanolamine

Compound[a]	Initial activity dpm	Distribution[b] of ^{14}C, %		% recovery	Ratio, ^{14}C in H_2O-sol. / ^{14}C in $CHCl_3$-sol.
		Chloroform-soluble	Water-soluble moieties		
[^{14}C]PS substrate	13 000	39	65	104	1·67
[^{14}C]PE product	5 930	46	50	96	1·08[c]

[a] [^{14}C]PS was prepared as described by Raetz and Kennedy (1972). [^{14}C]PE was formed by a 15 000 *g* fraction in a reaction mixture as described in Table IX, and isolated by preparative TLC in chloroform–methanol–28% ammonia (65:25:5, by vol.).

[b] Determined by the mild alkaline hydrolysis procedure (Marshall and Kates, 1972).

[c] Calculated for [^{14}C]PE, 1·11, assuming all ^{14}C activity in GPS is localized in the serine moiety. (Data from Marshall and Kates, 1973, 1974).

^{14}C-carbon from the serine moiety of [^{14}C]PS (initial ratio, 1·67; Table X). This result demonstrates conclusively that PE may be formed in spinach leaves by decarboxylation of PS.

The spinach decarboxylase was found to be strongly inhibited by hydroxylamine (Table XI), as was reported for the *E. coli* enzyme (Raetz and Kennedy, 1972) suggesting that it contains pyridoxal phosphate as coenzyme. The spinach enzyme also resembles the corresponding enzyme in bacteria (Kanfer and Kennedy, 1964; Patterson and Lennarz, 1971), protozoa (Dennis and Kennedy, 1970) and liver (Borkenhagen *et al.*, 1961) in its lack of a requirement for divalent metal ions (Table XI); but unlike the enzyme in these tissues which is stimulated by Triton X-100, the spinach enzyme appears to be strongly inhibited by detergents (Table XI).

Biosynthesis of PE in spinach leaves via decarboxylation of PS, synthesized from CDP-diglyceride and serine (eqns **21**, **22**), thus represents a *de novo* pathway for net synthesis of PE, just as in bacteria. PS decarboxylase and PS synthetase may be distributed generally in plants, as indicated by indirect evidence for their occurrence in *Chlorella vulgaris* (Sastry and Kates, 1965), tomato root (Willemot and Boll, 1967) and pea seedlings (Vandor and Richardson, 1968).

TABLE XI

The incorporation of ^{14}C from [^{14}C]phosphatidylserine into phosphatidylethanolamine by a particulate fraction

Assay system	Incorporation of ^{14}C dpm $\times 10^{-3}$ PE
Expt 1	
Complete[a]	6·0
Complete + sodium lauryl sulphate	1·1
Complete + Triton X-100	3·5
Complete + Triton + Mg^{2+} + Mn^{2+}	3·8
Expt 2	
Complete[b]	1·4
Complete + hydroxylamine	0·3

[a] Complete assay system (vol, 0·5 ml); [^{14}C]PS ($1·1 \times 10^5$ dpm), 50 mM Tris HCl (pH 7·0) and 0·2 ml (3 mg protein) of 1200–100 000 *g* pellet suspended in 2 mM Tris-HCl (pH 7·0). Additions were made as indicated to give the following concentrations: sodium lauryl sulphate (0·5 mg/ml), Triton X-100 (0·5 mg/ml), $MgCl_2$ (10 mM) and $MnCl_2$ (10 mM). Incubation, 1 h at 30°C. Values are corrected for boiled enzyme control.

[b] Complete assay system as described in Table IX, using 15 000 *g* pellet suspension as enzyme source. Hydroxylamine hydrochloride was added where indicated to a final concentration of 30 mM. Values are corrected for boiled enzyme control.

2. *Nucleotide pathway* (*pathway* 2)

$$\text{ethanolamine} + \text{ATP} \rightarrow \text{phosphorylethanolamine} + \text{ADP} \quad (\mathbf{23})$$

$$\text{phosphorylethanolamine} + \text{CTP} \rightarrow \text{CDP-ethanolamine} + \text{PP}_i \quad (\mathbf{24})$$

$$\text{CDP-ethanolamine} + sn\text{-1,2-diglyceride} \rightarrow \text{phosphatidylethanolamine} + \text{CMP} \quad (\mathbf{25})$$

The enzyme catalyzing the first step (eqn **23**), namely ethanolaminekinase, has been detected in a wide variety of animal tissues; it is inhibited by choline and localized specifically in the cytoplasm (Sung and Johnstone, 1967; Weinhold and Rethy, 1972). The enzymes catalyzing the second and third steps (eqns **24** and **25**) are widely distributed in animal tissues (Kennedy and Weiss, 1956) and in yeast (Kennedy and Weiss, 1956; Steiner and Lester, 1972). The CDP-ethanolamine-diglyceride phosphorylethanol-aminetransferase is localized in the microsome fraction in these systems (Dennis and Kennedy, 1972).

In regard to the existence of the nucleotide pathway for PE in plants, Kennedy and Weiss (1956) reported, but presented no detailed evidence for, the presence of the cytidyltransferase (eqn **24**) in carrot root, and Willemot and Boll (1967) demonstrated a very active incorporation of [^{14}C]ethanolamine into PE by excised tomato root. However, the first direct evidence bearing on the existence of this pathway in plants was presented by Macher and Mudd (1974) who demonstrated the synthesis of PE from CDP-ethanolamine by a microsomal fraction of spinach leaves in a system having an optimum pH of 7·5 and requiring Mg^{2+} or Mn^{2+} (see below for further discussion).

In our *in vivo* studies with spinach leaves or slices (Marshall and Kates, 1974), high incorporation of ^{14}C from [1,2-^{14}C]ethanolamine into the GPE moiety of PE was observed (Table VIII; Fig. 5). There are two possible pathways that would give rise to this incorporation, namely the CDP-ethanolamine pathway (eqns **23-25**) and the exchange pathway (see below, Section G3). The strongest evidence in support of the CDP-ethanolamine pathway was the demonstration of a very active diglyceride-stimulated synthesis of [^{14}C]PE from CDP-[1,2-^{14}C]ethanolamine in cell-free systems (Table XII). The 100 000 *g* microsomal fraction had the highest specific activity both in the absence and presence of diglyceride and was essentially free of mitochondrial cytochrome *c* oxidase activity (Table XII). PE synthesis by a dialyzed 100 000 *g* microsomal fraction was found to require Mg^{2+} or Mn^{2+} as co-factors the optimum concentrations being 8 mM and 2 mM, respectively. Both metal ions appeared to inhibit PE synthesis at concentrations above their respective optimum values. The formation of PE at 20 mM Mg^{2+} had an optimum pH in Tris HCl of 7·0.

These results are essentially the same as those reported by Macher and Mudd (1974) and establish the existence of CDP-ethanolamine:*sn*-1,2-diglyceride phosphorylethanolaminetransferase activity in spinach leaf microsomal fractions. The question whether this enzyme is the same as the CDP-

TABLE XII

Subcellular distribution of phosphatidylethanolamine synthesis from CDP-[1,2-^{14}C]ethanolamine

Cell fractions[a]	Phosphatidylethanolamine synthesis[b]					
	Minus diglyceride			Plus diglyceride[c]		
	Activity (nmol/30 min)	%	Specific activity	Activity (nmol/30 min)	%	Specific activity
Cell-free homogenate	35·4	100	0·05	43·2	100	0·06
1200 *g* (chloroplasts) pellet	4·1	12	0·03	4·5	10	0·03
15 000 *g* (mitochondrial) pellet	12·1	34	0·09	16·5	37	0·13
100 000 *g* (microsomal) pellet	10·7	30	0·21	15·2	35	0·30
100 000 *g* supernatant	10·7	30	0·01	8·5	20	0·01
% recovery		106			102	

[a] Cell fractions were prepared and assayed for enzyme markers and chlorophyll as in Table IV.

[b] Assay system (vol. 1 ml): 1·5 μM CDP-[1,2-^{14}C]ethanolamine (2×10^5 dpm), 100 mM Tris-HCl (pH 7·0), 20 mM $MnCl_2$ and an aliquot of cell free homogenate (0·05 ml) or cell fraction (0·1 ml) suspended in homogenizing medium. Prior to their addition to assay mixture, the pH of each fraction was adjusted to 7·0 with 0·2 N HCl. Incubation, 30 min at 30°C. PE synthesis was calculated from the incorporation of ^{14}C into total lipid.

[c] *sn*-1,2-Diglyceride (300 μg prepared from egg lecithin by action of phospholipase C) was suspended in above assay mixture minus CDP-ethanolamine and cell fraction by sonication for 2 min. (Data from Marshall and Kates, 1974.)

F

choline: diglyceride phosphorylcholinetransferase will be discussed below (Section H.1).

The existence of ethanolamine kinase catalyzing the first step in the CDP-ethanolamine pathway (eqn **23**) was demonstrated by the ATP-dependent incorporation of [1,2-^{14}C] ethanolamine into phosphorylethanolamine by a 15 000 *g* supernatant fraction (Marshall and Kates, 1973). The second step of the pathway (eqn **24**), catalyzed by CTP:phosphorylethanolamine cytidyltransferase, could not be demonstrated *in vitro* with the 15 000 *g* supernatant fraction under a variety of experimental conditions, as was also reported by Macher and Mudd (1974). The possibility that any CDP-[1,2-^{14}C] ethanolamine formed was hydrolyzed to phosphoryl-[1,2-^{14}C]ethanolamine and CMP by a pyrophosphatase was ruled out in a separate experiment in which CDP-[1,2-^{14}C]ethanolamine was incubated in the assay system without any formation of phosphoryl-[^{14}C]ethanolamine (Marshall and Kates, 1937). The reason for these negative results is unclear. Cytidyltransferase activity if present would be expected in the 15 000 *g* supernatant by analogy with the localization of the corresponding enzyme in animal tissue (Kennedy and Weiss, 1956; Schneider, 1969). However, the presence in spinach leaves of the other two enzymes of the CDP-enthanolamine pathway, namely ethanolamine kinase and the phosphorylethanolaminetransferase, suggests that the cytidyltransferase probably functions *in vivo* but was inactive under the conditions of assay used *in vitro*. The possibility must also be considered that the cytidyltransferase does not exist, in which case the observed *in vitro* incorporation of [^{14}C]ethanolamine in leaves would be accounted for by exchange reactions (see below). However, this does not seem reasonable in view of the low level of ethanolamine exchange (Table VII) and the relatively high activity of the CDP-ethanolamine-diglyceride transferase (Table XII) observed *in vitro*.

3. *Ethanolamine exchange pathway*

Another pathway observed for ethanolamine incorporation into PE in isolated liver and brain-particulate fractions is by a Ca^{2+}-stimulated exchange reaction in which ethanolamine exchanges with the serine moiety of PS (Borkenhagen *et al.*, 1961: Kanfer, 1972; Porcellati *et al.*, 1971) or with the choline moiety of PC:

$$\text{phosphatidylserine} + \text{ethanolamine} \xrightleftharpoons{Ca^{2+}} \text{phosphatidylethanolamine} + \text{serine} \quad \textbf{(26)}$$

$$\text{phosphatidylcholine} + \text{ethanolamine} \xrightleftharpoons{Ca^{2+}} \text{phosphatidylethanolamine} + \text{choline} \quad \textbf{(27)}$$

In an experiment designed to test the presence of the exchange pathway in spinach leaves we observed a Ca^{2+}-stimulated incorporation of [1,2-^{14}C]-ethanolamine into PE by a 1200–100 000 *g* particulate fraction in the absence of cytidine co-enzymes (Table VII). This result suggests the *in vitro* operation of an ethanolamine exchange reaction in spinach cell-free systems. However, it should be noted that the observed activity was very low, only 0·18% of the [1,2-^{14}C]ethanolamine being incorporated into PE. Furthermore, in view of the well known ability of phospholipase D to catalyze exchange reactions *in vitro* the question arises as to whether the observed ethanolamine exchange was catalyzed by this enzyme. It may be noted, however, that a Ca^{2+}-stimulated ethanolamine exchange reaction has been reported in pea seedlings (Vandor and Richardson, 1968) but it was considered not to be catalyzed by phospholipase D. Clearly, further studies are required to resolve the question of the role of the exchange reactions *in vivo*, in relation to the *de novo* synthesis of phosphatidylethanolamine.

H. PHOSPHATIDYLCHOLINE (PC)

Two major pathways are known for the *de novo* biosynthesis of PC, namely the nucleotide pathway utilizing CDP-choline (eqns **28–30**), which has been found only in animals and plants, and the methylation pathway from PE (eqns **31–33**), which has been found in animals and some bacteria.

1. *The nucleotide pathway*

$$\text{choline} + \text{ATP} \rightarrow \text{phosphorylcholine} + \text{ADP} \quad \textbf{(28)}$$

$$\text{phosphorylcholine} + \text{CTP} \rightarrow \text{CDP-choline} + \text{PP}_i \quad \textbf{(29)}$$

$$\text{CDP-choline} + sn\text{-1,2-diglyceride} \rightarrow \text{phosphatidylcholine} + \text{CMP} \quad \textbf{(30)}$$

Choline kinase was first described by Wittenberg and Kornberg (1953) who partially purified the enzyme from brewers' yeast. Choline kinase activity has since been detected in a variety of animal tissues (McCaman, 1962) and in Ehrlich ascites cells (Sung and Johnstone, 1967).

The involvement of phosphorylcholine in PC biosynthesis in liver was first shown by Kornberg and Pricer (1952), but the intermediate reactions were not known until the classical work of Kennedy and co-workers (see Kennedy, 1961, 1962), who demonstrated that the synthesis of PC in rat liver cell-free extracts involved the intermediate formation of CDP-choline from phosphorylcholine and CTP (eqn **29**) followed by transfer of phosphorylcholine to *sn*-1,2-diglyceride (eqn **30**).

CDP-choline:diglyceride phosphorylcholinetransferase has also been demonstrated in brain (Strickland *et al.*, 1963; Chojnacki, 1964; Ansell and Chojnacki, 1966; Miller and Dawson, 1972), lung (Vereyken *et al.*, 1972) and other animal tissues, in the yeast *Saccharomyces cerevisiae* (Waechter *et al.*, 1969), and in the protozoan *Tetrahymena pyriformis* (Smith and Law, 1970).

In regard to the specificity of the individual enzymes in the nucleotide pathway in animal tissues, the cytidyl transferase catalyzing the second step (eqn **29**) is inactive with non-cytosine nucleotides (Borkenhagen and Kennedy, 1957), but dCTP is as effective as CTP (Kennedy *et al.*, 1959). The phosphorylcholinetransferase catalyzing the third step (eqn **30**) is also inactive with non-cytosine containing nucleoside diphosphate choline esters (Kennedy and Weiss, 1956), and while Kennedy *et al.* (1959) have claimed that dCDP-choline is as effective as CDP-choline, the more detailed results of Schneider and Behki (1963) strongly suggest that CDP-choline is the preferred substrate.

Chojnacki (1964) and Ansell and Chojnacki (1966) found that in brain homogenates, ethanolamine and its mono and dimethyl derivatives and choline were incorporated to the same extent from their CDP esters into the corresponding phospholipid. However, the formation of mono- and dimethyl-PE almost certainly has no significance *in vivo* since these phospholipids have not been detected in brain, neither can their methylation to PC take place in this tissue (see below). It therefore appears that the phosphorylbase-transferase activity in brain may be nonspecific towards the structure of the base and that the synthesis of PE and PC from CDP-ethanolamine and CDP-choline, respectively, may be catalyzed by a single enzyme. In the original studies of Kennedy and Weiss (1956), the two phosphorylbase-transferases in rat liver were both present in the 18 000 *g* pellet fraction and no attempt was made to separate them. The only evidence supporting two separate enzymes was the fact that the phosphorylcholinetransferase activity was retained by the phosphorylethanolaminetransferase activity was lost upon lyophilization of the 18 000 *g* pellet. However, in a recent publication, Dennis and Kennedy (1972) suggested that inactivation of interfering enzymes may have been responsible for these observations. Clearly, the question of whether separate enzymes are involved in the two phosphorylbase-transferase reactions (eqns **25** and **30**) in brain and liver can only be answered by purification and separation of the two enzyme activities.

There is considerable evidence that the nucleotide pathway is present in photosynthetic and non-photosynthetic plant tissue. Choline kinase has been found in rapeseed (Ramasara and Wetter, 1957) and in leaves from a wide variety of species including spinach (Tanaka *et al.*, 1966). Morré *et al.* (1970) have demonstrated the presence of all three enzymes in cell-free fractions from onion stem.

The presence of the phosphorylcholinetransferase has been conclusively demonstrated in spinach leaves by Devor and Mudd (1971). Their evidence was based on the incorporation of ^{14}C into PC both from CDP-[^{14}C]choline and *sn*-1,2-[^{14}C]diglyceride by a microsomal fraction; the incorporation from CDP-[^{14}C] choline was stimulated by added diglyceride, and that from [^{14}C] diglyceride by added CDP-choline.

In our studies with spinach leaves (Marshall and Kates, 1974) high incorporation of ^{14}C from [1,2-^{14}C]choline into the GPC moiety of PC by leaf slices was observed (Tables V and VIII). There are two possible pathways that would give rise to this incorporation, namely the CDP-choline pathway (eqns **28**–**30**) and the exchange pathway (eqn **27**; see below), but the operation of the nucleotide pathway was clearly demonstrated by the finding of incorporation of ^{14}C from CDP-[1,2-^{14}C]choline into PC by subcellular fractions in the presence of *sn*-1,2-diglyceride and Mn^{2+} (Table XIII). The fraction with the highest specific activity and proportion of the total activity was the 100 000 *g* microsomal pellet, which was virtually free of mitochondria. These results are essentially in agreement with the findings of Devor and Mudd (1971) and together establish the presence in spinach leaves of CDP-choline: *sn*-1,2-diglyceride phosphorylcholinetransferase catalyzing the reaction shown in eqn **30**. The question whether in spinach leaves this enzyme and CDP-ethanolamine:diglyceride phosphorylethanolaminetransferase are separate enzymes has been raised by Macher and Mudd (1974), but can only be answered conclusively by purification and separation of the two enzyme activities.

TABLE XIII

Subcellular distribution of phosphatidylcholine synthesis from CDP-[1,2-^{14}C]choline

Cell fraction[a]	Phosphatidylcholine synthesis[b]		
	Activity (nmol/30 min)	%	Specific activity
Cell-free homogenate	76·5	100	0·10
1200 *g* (chloroplast) pellet	8·3	11	0·06
15 000 *g* (mitochondrial) pellet	19·2	25	0·15
100 000 *g* (microsomal) pellet	36·0	47	0·72
100 000 *g* supernatant	12·4	16	0·02
% recovery		99	

[a] Cell fractions were those used for assay of PE synthesis described in Table XII after storage at pH 7·0 for 24 h at −15°C.

[b] Assay system (vol, 1 ml): 52 μM CDP-[1,2-^{14}C]choline ($1{\cdot}1 \times 10^6$ dpm), 100 mM Tris-HCl (pH 8·0), 2 mM $MnCl_2$, *sn*-1,2-diglyceride (300 μg), and an aliquot of cell-free homogenate (0·05 ml) or cell fraction (0·1 ml) suspended in homogenizing medium. The diglyceride was prepared and dispersed in the assay mixture as described in Table XII. Incubation, 30 min at 30°C. PC synthesis was calculated from the incorporation of ^{14}C into total lipids.

2. *The methylation pathway*

This pathway consists of the step-wise methylation of PE to PC by *S*-adenosylmethionine (SAM), with the intermediate formation of phosphatidyl monomethylethanolamine (PE-Me) and phosphatidyl dimethylethanolamine (PE-diMe) (eqns **31**–**33**):

$$\text{PE} \xrightarrow{\text{SAM}} \text{PE-Me} \qquad (31)$$

$$\text{PE-Me} \xrightarrow{\text{SAM}} \text{PE-diMe} \qquad (32)$$

$$\text{PE-diMe} \xrightarrow{\text{SAM}} \text{PC} \qquad (33)$$

The operation of the methylation pathway for phosphatidylcholine synthesis has been established in rat liver (Bremer and Greenberg, 1961; Gibson *et al.*, 1961; Artom, 1964), in which it is localized in the microsomal fraction. However, it should be noted that direct evidence for the existence of the first step, namely the methylation of PE to PE-Me (eqn **31**), is lacking in rat liver.

The methylation pathway also functions in the mold *Neurospora crassa* (Scarborough and Nyc, 1967), the yeast *Saccharomyces cerevisiae* (Waechter *et al.*, 1969; Steiner and Lester, 1969), and the phytoflagellate *Ochromonas malhamenis* and the protozoan *Tetrahymena pyriformis* (Lust and Daniel, 1966; Smith and Law, 1970). The pathway is not generally present in bacteria but is found in those species that contain phosphatidylcholine, e.g. three species of *Agrobacterium* (Kaneshiro and Law, 1964; Goldfine and Ellis, 1964), *Hyphomicrobium vulgare* (Goldfine and Hagen, 1968), and the photosynthetic bacterium *Rhodopseudomonas spheroides* (Gorchein *et al.*, 1968). The bacteria *Clostridium butyricum* and *Proteus vulgaris*, which do not contain PC, are of novel interest since PE-Me is a major phospholipid component and can be labelled with ^{14}C when the cells are grown in [Me-^{14}C]methionine (Goldfine, 1962; Goldfine and Ellis, 1964). Kaneshiro and Law (1964) in their studies with *Agrobacterium tumefaciens*, showed that ^{14}C incorporation into PE-Me, PE-diMe and PC from [^{14}C]SAM by microsomal and cytoplasmic fractions was almost totally dependent upon exogenous PE. They also isolated and partially purified an enzyme from the cytoplasm with PE:SAM methyltransferase activity, thus demonstrating conclusively the first step (eqn **31**) in the methylation pathway. In their studies with *Neurospora*, Scarborough and Nyc (1967) found that microsomes from a mutant of this organism were deficient in PE:SAM methyltransferase activity, while microsomes from a second mutant, where the lesion was at a single gene locus, could not catalyze the methylation of either PE-Me or PE-diMe. This finding was important in showing that in *Neurospora* a single enzyme catalyzes the first step, while a second enzyme catalyzes both the second and third steps. The methylation pathway in other organisms most likely also involves separate enzymes for the first and subsequent methylations. The finding of PE-Me, but not PC, in *C. butyricum* and *P. vulgaris* is consistent with this idea.

The evidence available at present, although indirect, suggests that the methylation pathway also exists in photosynthetic and non-photosynthetic plant tissues. Thus, Sastry and Kates (1965) showed that in *Chlorella vulgaris* grown in [^{32}P]glycerophosphate, ^{32}P-incorporation into and specific activity of PE was greater than that of PC in the early stages of growth. Also, Willemot and Boll (1967) found that ^{14}C was incorporated by excised tomato root from [^{14}C]ethanolamine and [U-^{14}C] serine not only into PE, but also into PC. Tipton and Swords (1966) claimed to have provided evidence for the presence of the methylation pathway in *Euglena gracilis*, but the results were equivocal due to the lack of proper controls.

In our studies with whole spinach leaves and leaf slices (Marshall and Kates, 1974) ^{14}C was found to be incorporated into the GPC moiety of PC from [1-^{14}C]serine, [1,2-^{14}C]ethanolamine and [Me-^{14}C]methionine (Fig. 4; Tables V and VIII), and into the deacylated moieties of PE and PE-Me from [1,2-^{14}C]ethanolamine by whole leaves (Fig. 5). The proportion of total lipid ^{14}C in PC labelled either from ^{14}C-labelled ethanolamine, methionine or serine was considerably less in leaf slices than in whole leaves (Table V).

In vitro studies with [^{14}C]SAM showed that only a small amount of ^{14}C was incorporated into PC by a microsomal fraction in the absence of added PE, PE-Me or PE-diMe (Fig. 6; Table XIV); 13–32% of this ^{14}C was localized in the H_2O-soluble GPC moiety, the remaining ^{14}C being in the chloroform-soluble moieties (fatty acids). However, in the presence of exogenous PE-Me or PE-diMe, the incorporation of ^{14}C into the GPC moiety of PC was stimulated 20-fold and 200-fold, respectively; furthermore, in the presence of PE-Me, ^{14}C was also incorporated into PE-diMe in which 83% was localized in the GPE-diMe moiety (Fig. 6; Table XIV).

The low ^{14}C-incorporation into PC in the absence of added PE or methylated PE was not stimulated when a sonicated preparation of spinach PE was added, and no significant amounts of ^{14}C were detected in the PE (+PE-Me) or PE-diMe fractions (Table XIV, expt 1; Fig. 6). When the added PE was dispersed by dialysis in the presence of deoxycholate, an increase in the ^{14}C-incorporation into PC was observed but this occurred almost entirely in the

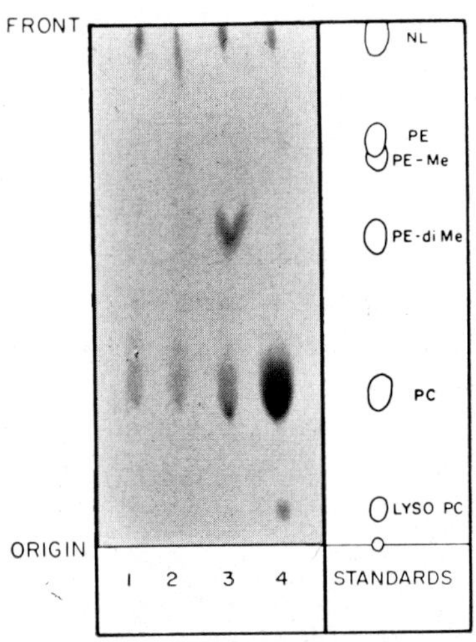

FIG. 6. Autoradiogram of a chromatogram of lipids labelled with ^{14}C from [^{14}C]*S*-adenosyl-methionine by a 100 000 *g* fraction:

1, complete system; 2, complete system + PE; 3, complete system + PE-Me; 4, complete system + PE-diMe. Assay conditions are as described in Table XIV. Chromatography was on a silica gel H t.l.c. plate developed in chloroform–methanol–acetic acid–water (65:35:8:4). (Data from Marshall and Kates, 1974.)

TABLE XIV

The effect of phosphatidylethanolamine, phosphatidyl monomethylethanolamine and phosphatidyl dimethylethanolamine on the biosynthesis of phosphatidyl choline by the methylation pathway in spinach microsomes

Assay system	^{14}C incorporation[a], dpm × 10^{-3}		
	PC	PE (+PE-Me)[b]	PE-diMe
Expt 1[c]			
Complete	1·39 (13)	0·37	0
Complete + PE (0·2 mg)	1·23	0·49	0
Complete + PE-Me (0·4 mg)	4·85 (70)	0·43	3·36 (83)
Complete + PE-diMe (0·4 mg)	32·73 (94)	0·38	0
Complete + 100 000 *g* supernatant (3 mg protein)	0·90	0·49	0
Expt 2[d]			
Complete	3·14 (32)	0·60	0
Complete + PE (0·3 mg)	9·12 (17)	1·89	0

[a] Lipid products are those shown in Fig. 6. Values in parentheses are percentage of counts in the deacylated moiety.

[b] PE and PE-Me are not resolved in solvent used (see Fig. 6) and were counted as one fraction. In a separate experiment more than 82% of the ^{14}C in the PE + PE-Me fraction was found in the fatty acid moieties.

[c] Complete assay system (1 ml): 8 μM [^{14}C]SAM (9·5 × 10^5 dpm), 50 mM Tris HCl (pH 8·5) and 0·47 ml of 100 000 *g* pellet suspension (1·7 mg protein). Incubation, 30 min at 30°C. Spinach PE and PE-Me and PE-diMe were dispersed by sonication and added where indicated.

[d] Complete assay system as in expt. 1, except that 16 μM [^{14}C]SAM (1·9 × 10^6 dpm) and 3·5 mg of 100 000 *g* protein were used, and spinach PE was dispersed prior to assay by the dialysis procedure of Fleischer and Klouwen (1961). (Data from Marshall and Kates, 1974.)

fatty acid moieties (Table XIV; expt 2). Furthermore, in an experiment in which [^{14}C]PE labelled *in situ* from CDP-[^{14}C]ethanolamine (Table XII) was incubated with the 100 000 *g* fraction in the presence of unlabelled SAM, no [^{14}C]PC was formed. Finally, the low methylation of PE could not be stimulated by varying pH or SAM concentration, or by adding metal ions (Na^+, K^+, Mg^{2+}, Mn^{2+}); also, this reaction could not be detected in any cell fractions prepared from leaves homogenized in the presence of β-mercapto-ethanol.

The possibility that PE-Me may be formed by methylation of PE by an as yet uncharacterized methyl donor was also considered, but 5-[Me-^{14}C] methyltetrahydrofolate was found not to act as methyl donor for PE in the *in vitro* system; other methyl donors have not yet been tested.

With regard to the subcellular localization of the methylation pathway, the cell fraction with the highest proportion of total activity and the highest specific activity of the PE-diMe:SAM methyltransferase was the 100 000 *g*

TABLE XV

Subcellular distribution of phosphatidyl dimethylethanolamine: S-adenosylmethionine methyltransferase

	Phosphatidylcholine synthesis[a]		
	Activity (nmol/15 min)	%	Specific activity
Cell-free homogenate	40·6	100	0·03
1200 *g* (chloroplast) pellet	2·1	5	0·02
15 000 *g* (mitochondrial) pellet	17·2	42	0·09
100 000 *g* (microsomal) pellet	37·4	92	0·47
100 000 *g* supernatant	6·9	17	0·01
% recovery		156	

[a] Assay system (vol, 0·5 ml: 2·7 μM SAM (3·1 × 10^5 dpm), 100 mM Tris-HCl (pH 8·0), PE-diMe (0·4 mg, dispersed prior to addition to assay mixture by sonication) and 0·1 ml of cell-free homogenate or cell fraction dispersed in homogenizing medium. Incubation, 15 min at 30°C. ^{14}C-incorporation into total lipids was measured. Control assays were carried out for each fraction in the above assay mixture omitting PE-diMe and the results were corrected for these blank values. (Data from Marshall and Kates, 1974.)

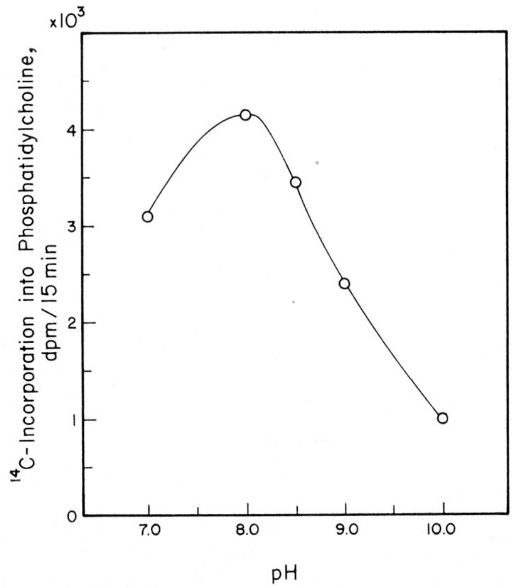

FIG. 7. The effect of pH on the incorporation of ^{14}C into phosphatidylcholine synthesized by methylation of phosphatidyldimethylethanolamine by [^{14}C]*S*-adenosylmethionine.

Assay system as described in Table XIV except that pH of Tris-HCl buffer was varied as indicated. ^{14}C incorporation into PC calculated as the difference between ^{14}C incorporation into total lipid from assays containing PE-diMe and control assays in which PE-diMe was absent.

microsomal pellet (Table XV), which was essentially free of mitochondria and chloroplasts. This methyltransferase was found to have a pH optimum in Tris HCl buffer of 8·0 (Fig. 7) and was apparently specific for SAM since no ^{14}C was found to be incorporated into lipid when [^{14}C]SAM was replaced by 5-[Me-^{14}C]methyltetrahydrofolate in an assay system identical to that described in Table XV.

3. *Choline exchange pathway*

Evidence for the choline exchange pathway (eqn **27**) in spinach leaves was the finding of a Ca^{2+}-stimulated incorporation of [^{14}C]choline into PC by a particulate (1200–100 000 *g*) fraction in the absence of cytidine nucleotides (Table VII). In this case the activity was even lower than that with ethanolamine, only 0·12% of [^{14}C]choline being incorporated into PC. However, these results suggest the operation at least *in vitro* of a choline exchange reaction, but, as discussed above in the case of the ethanolamine exchange, the *in vivo* significance of these reactions still remains to be demonstrated.

V. Conclusions

Pathways for biosynthesis of phospholipids in spinach leaves appear to combine those in bacteria and in yeasts or fungi (Fig. 8). Thus, PG and PS are biosynthesized by the same pathways as those in bacteria (eqns **10–11**, and **21**, respectively). Synthesis of PS by a phosphatidyl transferase reaction between CDP-diglyceride and L-serine (eqn **21**) appears to be the only pathway for PS synthesis in spinach leaves, since the presence of the serine exchange pathway could not be demonstrated (Table VII). The serine:PE exchange pathway (eqn **26**), the only pathway utilized for PS synthesis in animal tissues (Borkenhagen *et al.*, 1961) and protozoa (Dennis and Kennedy, 1970), apparently does not occur in spinach leaves.

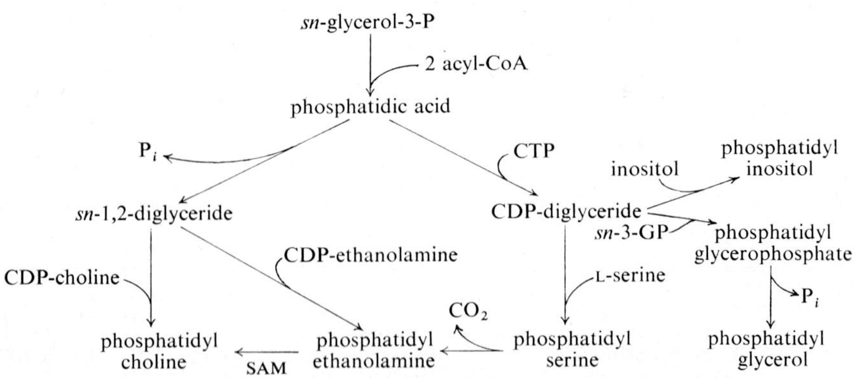

FIG. 8. Biosynthetic pathways for phospholipids in spinach leaves.

PE is clearly biosynthesized in spinach leaves by the decarboxylation of PS (eqn **22**) and the PS decarboxylase resembles the bacterial enzyme in utilizing PS biosynthesized from CDP-diglyceride and serine; in contrast, the decarboxylase in protozoa, liver and pea seedlings presumably utilizes PS synthesized by the exchange pathway. PE is also synthesized *in vivo* from free ethanolamine but the exact pathways utilized have still not been determined unambiguously. It is most probable that the nucleotide pathway (eqns **23–25**) is the predominant route utilized, rather than the ethanolamine exchange pathway (eqns **26–27**). However, the presence of the CTP:phosphorylethanolamine cytidyltransferase remains to be demonstrated, and the question whether a separate CDP-ethanolamine:diglyceride phosphoethanolaminetransferase exists must be resolved.

In regard to PC biosynthesis, it is clear that both the nucleotide pathway (eqns **28–30**) and the methylation pathway (eqns **31–33**) are utilized in spinach leaves, as they are in yeast, fungi, protozoa and animals. However, the failure to demonstrate methylation of PE by SAM in spinach microsomal preparations raises the question whether an unknown methyl-donor might be involved; further investigations will be necessary to resolve this point. Biosynthesis of PI in spinach leaves appears to follow the same pathway (eqns **7–9**) utilized in yeast and animals.

Concerning the subcellular localization of the phospholipid biosynthesizing enzymes in spinach leaves, the microsomal fraction appears to be the primary site for the biosynthesis of PG, PI, and PC. Biosynthesis of PD by the nucleotide pathway may be localized in the microsomal fraction, but synthesis of PE by PS decarboxylase was found to be largely associated with the 100 000 *g* supernatant fraction, although this activity may have arisen by "solubilization" of the enzyme from the mitochondrial fraction. The PS synthetase is a particulate-bound enzyme but its subcellular localization has not yet been determined.

PC and PE are the major phospholipids of plant mitochondria and PG is the major phospholipid in chloroplast lamellae (Table I), but it would appear that these phospholipids are not synthesized *in situ* by mitochondria or chloroplasts, respectively. They may perhaps be derived, however, from "microsomal" synthesized lipid by direct transport or exchange between "microsomal" membranes and promitochondria, or proplastids, respectively. Such a mechanism has been demonstrated previously in non-photosynthetic plant tissues (Abdelkader and Mazliak, 1970) and in animal tissues (Blok *et al.*, 1971; McMurray and Dawson, 1969). However, spinach mitochondria may have some capacity to synthesize PE *in situ* via the PS decarboxylase possibly present in this organelle. Further, more definitive studies on the localization of phospholipid synthesizing enzymes and on the exchange or transport of phospholipids between various organelle membranes would be extremely important for the resolution of the questions raised in the above discussion.

REFERENCES

Abdelkader, A. B. and Mazliak, P. (1970). *Eur. J. Biochem.* **15**, 250–262.
Ailhaud, G. and Vagelos, P. R. (1966). *J. biol. Chem.* **241**, 3866–3869.
Allen, C. F., Good, P., Davis, H. F., Chisum, P. and Fowler, S. D. (1966). *J. Am. Oil. Chem. Soc.* **43**, 223–231.
Aneja, R., Chadha, J. S. and Knaggs, J. A. (1969). *Biochem. biophys. Res. Commun.* **36**, 401–406.
Ansell, G. B. and Chojnacki, T. (1966). *Biochem. J.* **98**, 303–310.
Artom, C. (1964). *Biochem. biophys. Res. Commun.* **15**, 201–206.
Bahl, J., Guillot-Solomon, T. and Douce, R. (1970). *Physiol. Veg.* **8**, 55–74.
Bell, R. M., Mavis, R. D., Osborn, M. J. and Vagelos, P. R. (1971). *Biochim. biophys. Acta* **249**, 628–635.
Benjamins, J. A. and Agranoff, B. W. (1969). *J. Neurochem.* **16**, 513–527.
Benson, A. A. and Maruo, B. (1958). *Biochim. biophys. Acta* **27**, 189–195.
Benson, A. A. and Strickland, E. H. (1960). *Biochim. biophys. Acta* **41**, 328–333.
Benson, A. A. and Miyano, M. (1961). *Biochem. J.* **31**, 31P.
Blok, M. C., Wirtz, K. W. A. and Scherphof, G. L. (1971). *Biochim. biophys. Acta* **233**, 61–75.
Borkenhagen, L. F. and Kennedy, E. P. (1957). *J. biol. Chem.* **227**, 951–962.
Borkenhagen, L. F., Kennedy, E. P. and Fielding, L. (1961). *J. biol. Chem.* **236**, PC28–29.
Bradbeer, D. and Stumpf, P. K. (1960). *J. Lipid Res.* **1**, 214–220.
Branton, D. (1969). *A. Rev. Pl. Physiol.* **20**, 209–238.
Bremer, J. and Greenberg, D. M. (1961). *Biochim. biophys. Acta* **46**, 205–216.
Carter, J. R. and Kennedy, E. P. (1966). *J. Lipid Res.* **7**, 678–683.
Carter, J. R. (1968). *J. Lipid Res.* **9**, 748–754.
Carter, H. E. and Koob, J. L. (1969). *J. Lipid Res.* **10**, 363–369.
Chang, Y.-Y. and Kennedy, E. P. (1967a). *J. Lipid Res.* **8**, 447–455.
Chang, Y.-Y. and Kennedy, E. P. (1967b). *J. Lipid Res.* **8**, 456–462.
Cheniae, G. M. (1965). *Pl. Physiol., Lancaster* **40**, 235–243.
Chojnacki, T. (1964). *Acta biochim. polon.* **11**, 11–23.
Cronan, J. E. Jr. and Vagelos, P. R. (1972). *Biochim. biophys. Acta* **265**, 25–60.
Davidson, J. B. and Stanacev, N. Z. (1970). *Can. J. Biochem.* **48**, 633–642.
Davidson, J. B. and Stanacev, N. Z. (1971). *Can. J. Biochem.* **49**, 1117–1124.
Dawson, R. M. C. (1967). *Biochem. J.* **102**, 205–210.
Dawson, R. M. C., Clarke, N. and Quarles, R. H. (1969). *Biochem. J.* **114**, 265–270.
Debuch, H. and Rotsch, E. (1966). *Z. physiol. Chem.* **346**, 79–86.
Dennis, E. A. and Kennedy, E. P. (1970). *J. Lipid Res.* **11**, 394–403.
Dennis, E. A. and Kennedy, E. P. (1972). *J. Lipid Res.* **13**, 263–267.
DeSiervo, A. J. and Salton, M. R. J. (1971). *Biochim. biophys. Acta* **239**, 280–292.
Devor, K. A. and Mudd, J. B. (1971). *J. Lipid Res.* **12**, 403–411.
Dixon, M. and Webb, E. C. (1964). "Enzymes", p. 72. Longmans, London.
Domazet, Z., Stuhne-Sekalec, L., Davidson, J. B. and Stanacev, N. Z. (1973). *Can. J. Biochem.* **51**, 274–285.
Douce, R. (1968). *C. r. hebd. Seance Acad. Sci., Paris* **267**, 534–536.
Douce, R., Guillot-Salomon, T., Lance, C. and Signol, M. (1968). *Bull. Soc. fr. Physiol. veg.* **14**, 351–373.
Douce, R. and Dupont, J. (1969). *C. r. hebd. Seanc. Acad. Sci., Paris* **268**, 1657–1660.
Eberhardt, F. M. and Kates, M. (1957). *Can. J. Bot.* **35**, 907–921.
Eichenberger, W. and Menke, W. (1966). *Z. Naturforsch.* **21b**, 859–867.

Fleischer, S. and Klouwen, H. (1961). *Biochem. biophys. Res. Commun.* **5**, 378–383.

Fleischer, S. and Rouser, G. (1965). *J. Am. Oil Chem. Soc.* **42**, 588–607.

Gibson, K. D., Wilson, J. D. and Udenfriend, S. (1961). *J. biol. Chem.* **236**, 673–679.

Goldfine, H. (1962). *Biochim. biophys. Acta* **59**, 504–506.

Goldfine, H. (1972). *Adv. Microbial Physiol.* **8**, 1–58.

Goldfine, H. and Ellis, M. E. (1964). *J. Bact.* **87**, 8–15.

Goldfine, H. and Hagen, P.-O. (1968). *J. Bact.* **95**, 367–375.

Gorchein, A., Neuberger, A. and Tait, G. H. (1968). *Proc. R. Soc.* B. **170**, 299–310.

Gray, G. M. (1964). *Biochim. biophys. Acta* **84**, 35–40.

Hajra, A. K. (1968). *J. biol. Chem.* **243**, 3458–3465.

Hajra, A. K. and Agranoff, B. W. (1968). *J. biol. Chem.* **243**, 3542–3543.

Harwood, J. L. and Stumpf, P. K. (1972). *Lipids* **7**, 8–19.

Haverkate, F. and van Deenen, L. L. M. (1965a). *Biochim. biophys. Acta* **106**, 78–92.

Haverkate, F. and van Deenen, L. L. M. (1965b). *Proc. Kon. Ned. Akad. Wetensch.*, Ser. B **68**, 141–153.

Hechemy, K. and Goldfine, H. (1971). *Biochem. biophys. Res. Commun.* **42**, 245–251.

Hirschberg, C. B. and Kennedy, E. P. (1972). *Proc. natn. Acad. Sci. U.S.A.* **69**, 648–651.

Hitchcock, C. and Nichols, B. W. (1971). "Plant Lipid Biochemistry". Academic Press, London and New York.

Hokin, M. R. and Hokin, L. E. (1959). *J. biol. Chem.* **234**, 1381–1386.

Hostetler, K. Y., van den Bosch, H. and van Deenen, L. L. M. (1971). *Biochim. biophys. Acta.* **239**, 113–119.

Hostetler, K. Y., van den Bosch, H. and van Deenen, L. L. M. (1972). *Biochim. biophys. Acta* **260**, 507–513.

Hutchison, H. T. and Cronan, J. E. Jr. (1968). *Biochim. biophys. Acta* **164**, 606–608.

Hutt, H. H., Malkin, T., Poole, A. G. and Watt, P. R. (1950). *Nature, Lond.* **165**, 314–315.

Kaneshiro, T. and Law, J. H. (1964). *J. biol. Chem.* **239**, 1705–1713.

Kanfer, J. and Kennedy, E. P. (1964). *J. biol. Chem.* **239**, 1720–1726.

Kanfer, J. (1972). *J. Lipid Res.* **13**, 468–476.

Kates, M. and Eberhardt, F. M. (1957). *Can. J. Bot.* **35**, 895–905.

Kates, M. and Volcani, B. E. (1966). *Biochim. biophys. Acta* **116**, 264–278.

Kates, M. (1960). *Biochim. biophys. Acta* **41**, 315–328.

Kates, M. (1970). *Adv. Lipid Res.* **8**, 225–265.

Kennedy, E. P. and Weiss, S. B. (1956). *J. biol. Chem.* **222**, 193–214.

Kennedy, E. P., Borkenhagen, L. F. and Smith, S. W. (1959). *J. biol. Chem.* **234**, 1998–2000.

Kennedy, E. P. (1961). *Fed. Proc. Fedn Am. Socs exp. Biol.* **20**, 934–940.

Kennedy, E. P. (1962). The Harvey Lectures, Series 57, pp. 143–171. Academic Press, New York and London.

Kiribuchi, T., Mizunaga, T. and Funahashi, S. (1966). *Agric. biol. Chem.* **30**, 770–778.

Kiribuchi, T., Yasumatsu, N. and Funahashi, S. (1967). *Agric. biol. Chem.* **31**, 1244–1247.

Kiyasu, J. Y., Pieringer, R. A., Paulus, H. and Kennedy, E. P. (1963). *J. biol. Chem.* **238**, 2293–2298.

Korn, E. D. (1969). *A. Rev. Biochem.* **38**, 263–288.

Kornberg, A. and Pricer, W. E. Jr. (1952). *Federation Proc.* **11**, 242.

Lands, W. E. M. and Hart, P. (1964). *J. Lipid Res.* **5**, 81–87.

Lepage, M. (1964). *J. Lipid Res.* **5**, 587–592.
Lust, G. and Daniel, L. J. (1966). *Archs Biochem. Biophys.* **113**, 603–608.
Macher, B. A. and Mudd, J. B. (1974). *Pl. Physiol., Lancaster* **53**, 171–175.
MacKender, R. O. and Leech, R. M. (1974). *Pl. Physiol., Lancaster* **54**, 496–502.
Mangnall, D. and Getz, G. S. (1973). *In* "Lipids and Biomembranes of Eukaryotic Microorganisms" (J. A. Erwin, ed.), pp. 145–195. Academic Press, New York and London.
Marshall, M. O. and Kates, M. (1972). *Biochim. biophys. Acta.* **260**, 558–570.
Marshall, M. O. and Kates, M. (1973). *FEBS Lett.* **31**, 199–203.
Marshall, M. O. and Kates, M. (1974). *Can. J. Biochem.* **52**, 469–482.
McCaman, R. E. (1962). *J. biol. Chem.* **237**, 672–676.
McCaman, R. E. and Finnerty, W. R. (1968). *J. biol. Chem.* **243**, 5074–5080.
McMurray, W. C. and Dawson, R. M. C. (1969). *Biochem. J.* **112**, 91–108.
Miller, E. K. and Dawson, R. M. C. (1972). *Biochem. J.* **126**, 805–821.
Morré, D. J., Nyquist, S. and Rivera, E. (1970). *Pl. Physiol., Lancaster* **45**, 800–804.
Ongun, A. and Mudd, J. B. (1970). *Pl. Physiol., Lancaster* **45**, 255–262.
Ongun, A., Thomson, W. W. and Mudd, J. B. (1968). *J. Lipid Res.* **9**, 409–415.
Patterson, P. H. and Lennarz, W. J. (1971). *J. biol. Chem.* **246**, 1062–1072.
Paulus, H. and Kennedy, E. P. (1960). *J. biol. Chem.* **235**, 1303–1311.
Petzold, C. L. and Agranoff, B. W. (1967). *J. biol. Chem.* **242**, 1187–1191.
Pieringer, R. A. and Hokin, L. E. (1962a). *J. biol. Chem.* **237**, 653–658.
Pieringer, R. A. and Hokin, L. E. (1962b). *J. biol. Chem.* **237**, 659–664.
Pieringer, R. A. and Kunnes, R. S. (1965). *J. biol. Chem.* **240**, 2833–2838.
Poincelot, R. P. (1973). *Archs Biochem. Biophys.* **159**, 134–142.
Porcellati, G., Arienti, G., Pirotta, M. and Giorgini, D. (1971). *J. Neurochem.* **18**, 1395–1417.
Possmayer, F., Balakrishnan, G. and Strickland, K. P. (1968). *Biochim. biophys. Acta* **164**, 79–87.
Prottery, C. and Hawthorne, J. N. (1967). *Biochem. J.* **105**, 379–392.
Puleo, L. E., Rao, G. A. and Reiser, R. (1970). *Lipids* **5**, 770–775.
Raetz, C. R. H. and Kennedy, E. P. (1972). *J. biol. Chem.* **247**, 208–214.
Raetz, C. R. H. and Kennedy, E. P. (1973). *J. biol. Chem.* **248**, 1098–1105.
Ramasarma, T. and Wetter, L. R. (1957). *Can. J. Biochem. Physiol.* **35**, 853–863.
Rampini, C., Barbu, E. and Polonovsky, J. (1971). *C. r. hebd. Seanc. Acad. Sci., Paris* **270**, 882–884.
Ray, T. K., Cronan, J. E. Jr., Mavis, R. D. and Vagelos, P. R. (1970). *J. biol. Chem.* **245**, 6442–6448.
Renkonen, O. and Bloch, K. (1969). *J. biol. Chem.* **244**, 4899–4903.
Roughan, P. G. and Batt, R. D. (1969). *Phytochemistry* **8**, 363–369.
Sastry, P. S. and Kates, M. (1964). *Biochemistry* **3**, 1271–1280.
Sastry, P. S. and Kates, M. (1965). *Can. J. Biochem.* **43**, 1445–1453.
Sastry, P. S. and Kates, M. (1966). *Can. J. Biochem.* **44**, 459–467.
Scarborough, G. A. and Nyc, J. F. (1967). *J. biol. Chem.* **242**, 238–242.
Schantz, R., Douce, R. and Duranton, H. M. (1972). *FEBS Lett.* **20**, 157–161.
Schneider, W. C. and Behki, R. M. (1963). *J. biol. Chem.* **238**, 3565–3571.
Schneider, W. C. (1969). *In* "Methods in Enzymology" (J. M. Lowenstein, ed.), Vol. 14, pp. 684–690. Academic Press, New York and London.
Schwertner, H. A. and Biale, J. B. (1973). *J. Lipid Res.* **14**, 235–242.
Smith, J. D. and Law, J. H. (1970). *Biochim. biophys. Acta* **202**, 141–152.
Stanacev, N. Z., Chang, Y.-Y. and Kennedy, E. P. (1967). *J. biol. Chem.* **242**, 3018–3019.
Stanacev, N. Z., Izaac, D. C. and Brookes, K. B. (1968). *Biochim. biophys. Acta* **152**, 806–807.

Stanacev, N. Z., Stuhne-Sekalec, L., Brookes, K. B. and Davidson, J. B. (1969). *Biochim. biophys. Acta* **176**, 650–653.

Stanacev, N. Z. and Stuhne-Sekalec, L. (1970). *Biochim. biophys. Acta* **210**, 350–353.

Stanacev, N. Z., Stuhne-Sekalec, L. and Anderson, K. M. (1970). *Endocrinology* **86**, 1205–1211.

Stanacev, N. Z., Davidson, J. B., Stuhne-Sekalec, L. and Domazet, Z. (1973). *Can. J. Biochem.* **51**, 286–304.

Steiner, M. R. and Lester, R. L. (1969). *Biochemistry* **9**, 63–69.

Steiner, M. R. and Lester, R. L. (1972). *Biochim. biophys. Acta* **260**, 222–243.

Stoeckenius, W. and Engelman, D. (1969). *J. Cell Biol.* **42**, 613–646.

Strickland, K. P., Subrahmanyam, D., Pritchard, E. T., Thompson, W. and Rossiter, R. J. (1963). *Biochem. J.* **87**, 128–136.

Stuhne-Sekalec, L. and Stanacev, N. Z. (1970). *Can. J. Biochem.* **48**, 1214–1221.

Sumida, S. and Mudd, J. B. (1970a). *Pl. Physiol., Lancaster* **45**, 719–722.

Sumida, S. and Mudd, J. B. (1970b). *Pl. Physiol., Lancaster* **45**, 712–718.

Sung, C. and Johnstone, R. M. (1967). *Biochem. J.* **105**, 497–503.

Tanaka, K., Tolbert, N. E. and Gohlke, A. F. (1966). *Pl. Physiol., Lancaster* **41**, 307–312.

Ter Schegget, J., van den Bosch, H., van Baak, M. A., Hostetler, K. Y. and Borst, P. (1971). *Biochim. biophys. Acta* **239**, 234–242.

Thompson, J. E. (1969). *Can. J. Biochem.* **47**, 685–689.

Tipton, C. L. and Swords, M. D. (1966). *J. Protozool.* **13**, 469–472.

Vandor, S. L. and Richardson, K. E. (1968). *Can. J. Biochem.* **46**, 1309–1315.

Vereyken, J. M., Montfoort, A. and van Golde, L. M. G. (1972). *Biochim. biophys. Acta* **260**, 70–81.

Waechter, C. J., Steiner, M. R. and Lester, R. L. (1969). *J. biol. Chem.* **244**, 3419–3422.

Weinhold, P. A. and Rethy, V. B. (1972). *Biochim. biophys. Acta* **276**, 143–154.

Wells, M. A. and Dittmer, J. C. (1966). *Biochemistry* **5**, 3405–3418.

Wickner, W. T. and Kennedy, E. P. (1971). *Fed. Proc. Fedn Am. Socs exp. Biol.* **30**, 1119.

Wilgram, G. F. and Kennedy, E. P. (1963). *J. biol. Chem.* **238**, 2615–2619.

Willemot, C. and Boll, W. G. (1967). *Can. J. Bot.* **45**, 1863–1876.

Wintermans, J. F. G. M. (1960). *Biochim. biophys. Acta* **44**, 49–54.

Wittenberg, J. and Kornberg, A. (1953). *J. biol. Chem.* **202**, 431–444.

Wood, B. J. B., Nichols, B. W. and James, A. T. (1965). *Biochim. biophys. Acta* **106**, 261–273.

Yang, S. F., Freer, S. and Benson, A. A. (1967). *J. biol. Chem.* **242**, 477–484.

CHAPTER 6

Biosynthesis of Glycolipids

J. B. MUDD and R. E. GARCIA

Department of Biochemistry

and

Air Pollution Research Center,
University of California,
Riverside, California, U.S.A.

Abbreviations

ACP, acyl carrier protein; ASG, acylated steryl glucoside; DGDG, digalactosyl diglyceride; DPG, diphosphatidyl glycerol; MGDG, monogalactosyl diglyceride; PA, phosphatidic acid; PC, phosphatidylcholine; PE, phosphatidylethanolamine; PG, phosphatidyl glycerol; PI, phosphatidylinositol; PS, phosphatidylserine; SG, steryl glucoside; SL sulfoquinovosyl diglyceride

I. Introduction

The glycolipids of plants have not received much attention from biochemists and physiologists interested in biosynthesis. There is a large body of analytical data which includes the glycolipids, and these data give an indication of the quantitative importance of glycolipids. Koenig's (1971) analysis of the lamella system of maize chloroplasts shows 47% of the lipid to be galactosyldiglycerides and 4·5% to be the sulfonolipid (Table I).

Even in non-photosynthetic tissue the glycolipids make an important

TABLE I

Lipid composition of chloroplasts of *Zea mays*

	g per 100 g lipid
chlorophyll *a*	19·8
chlorophyll *b*	3·8
β-carotene	1·18
lutein	1·39
violaxanthin	0·71
neoxanthin	0·38
plastoquinone 45	0·56
tocopherol quinone	0·067
tocopherol	0·202
naphthoquinone	0·20
MGDG	26·9
DGDG	20·5
SL	4·5
PG	4·22
PI	0·67
PE	0·64

(Data from Koenig, 1971)

TABLE II

Glycolipid composition of kernels of *Zea mays*

Days after pollination	μmoles sugar per 100 kernels				
	10	15	20	45	60
MGDG	0·3	1·5	5·6	5·3	4·1
DGDG	2·3	6·8	19·8	29·2	6·0
SG	0·3	0·5	0·6	0·8	1·1
ASG	0·2	0·2	0·4	1·5	2·0
SL[a]	0·1	0·6	3·6	13·5	5·4
cerebroside[a]	0·9	1·9	2·1	2·2	2·4

(Data from Weber, 1970)

[a]May contain other glycolipids.

TABLE III

Lipid composition of potato tuber and apple fruit

	% by weight of total lipid			
	Potato tuber		Apple fruit (post climacteric)	
MGDG	5·7	32·0	1·3	19·8
DGDG	16·2		5·3	
SL	1·3		1·0	
Cerebroside	2·4		5·2	
SG	2·0		5·4	
ASG	6·4		1·6	
Sterol	0·6		20·1	
Sterol ester	0·1		2·1	
PC	25·8		22·7	
PE	12·6		10·7	
PI	6·3		6·2	
PG	0·9		0·8	
DPG	0·6		0·5	
PS	0·6		0·4	
PA	0·3		0·7	

(Data from Galliard, 1968a, b)

contribution. Weber (1970) has shown considerable amounts of various glycolipids in maturing seeds of *Zea mays*, with cerebroside being a noticeable component (Table II). Although phytoglycolipid has been isolated from seeds of *Zea mays*, its quantitative contribution is not known. As far as the glycolipids derived from sterols are concerned, the analyses of Galliard (1968a, b) show them to be significant components of potato tuber and apple fruit lipids (Table III).

In this article four classes of glycolipids will be considered: (a) galactosyldiglycerides, (b) sterol glycosides, (c) sulfoquinovosyldiglycerides and (d) glycosphingolipids (structures: Fig. 1). The functions of these glycolipids are grounds for speculation and experiment. No specific function can be ascribed to any of the plant glycolipids as the consequence of experimental observations. For this reason the present article concentrates on the analysis and biosynthesis of the glycolipids.

II. GALACTOSYLDIGLYCERIDES

A. STRUCTURE

1. *The sugar moiety*

In a series of papers from Carter's laboratory (1956, 1961) the structures of two galactolipids isolated from wheat flour were elucidated. The assignments

Ia

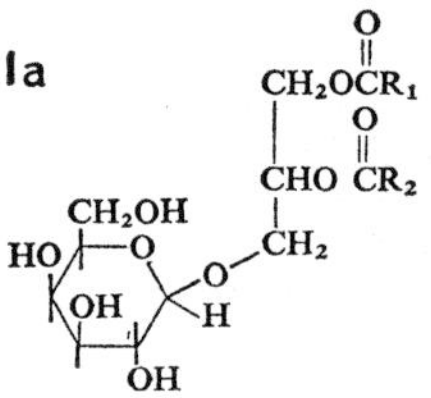

monogalactosyl diglyceride (MGDG)

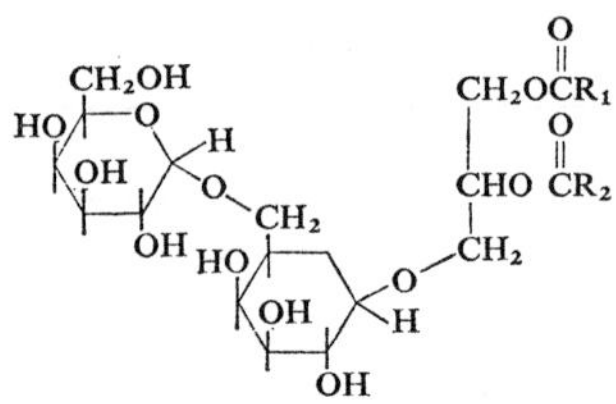

digalactosyl diglyceride (DGDG)

Ib

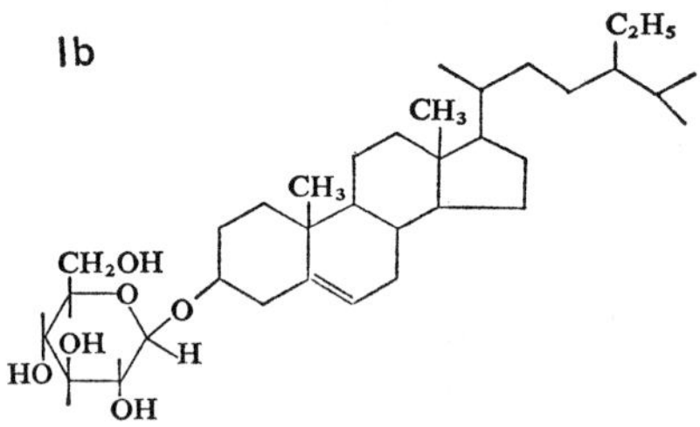

steryl glucoside (SG)

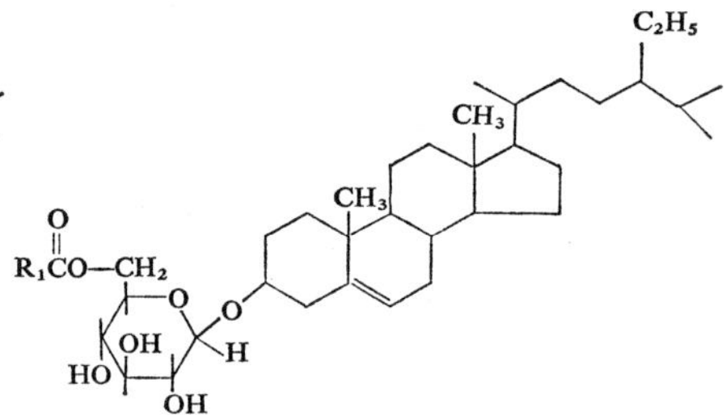

acylated steryl glucoside (ASG)

Ic

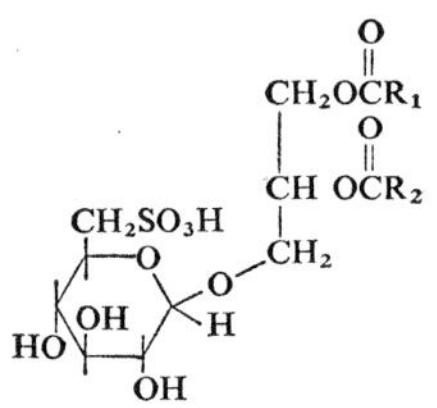

sulfoquinovosyl diglyceride (SL)

Id

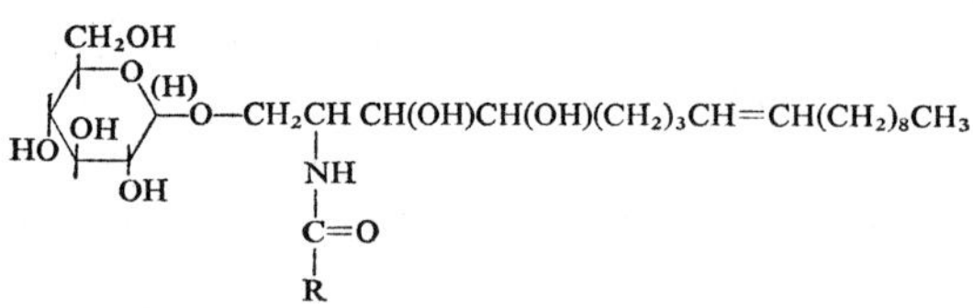

cerebroside

ceramide (1′) phosphate (1) inositol (2←1) mannose

(6 ↑ 1)

glucosamine (1→4) glucuronic acid

phytoglycolipid

FIG. 1. Structures of plant glycolipids: 1a, galactosyl diglycerides; 1b, steryl glucosides; 1c, sulfonolipid; 1d, sphingolipids.

of the two structures were 2,3-diacyl-1-β-D-galactopyranosyl-D-glycerol and 2,3-diacyl-1 (α-D-galactopyranosyl-1,6-β-D-galactopyranosyl)-D-glycerol (Carter *et al.*, 1961). According to the stereochemical numbering system (*sn*) these lipids would be referred to as *sn*-1,2-diacylglycerol derivatives. As far as photosynthetic tissues are concerned, Benson and co-workers (1958, 1959) confirmed that the galactolipids were of the same structure as those found in wheat flour by comparing the deacylated galactolipids from leaves with the authentic galactosylglycerols. Sastry and Kates (1964) also determined the structure of galactolipids from runner bean leaves and found them to be identical with those from wheat flour. The structure of these galactosyldiglycerides appears to be so secure that metabolic studies almost never include thorough characterization of the compounds, depending for identification on chromatographic behavior and color reactions. In material biosynthesized from UDP-galactose by spinach chloroplasts it should be noted that in the deacylation product of the putative DGDG, only 17% of the terminal galactose was released by α-galactosidase (Neufeld and Hall, 1966). In a report on the glycolipids of the Chlorophyceae, Brush and Percival (1972) specifically discussed the susceptibility of the deacylated galactolipids of *Ulva lactuca* to galactosidases, reporting that the digalactosyl glycerol was susceptible to both α- and β-galactosidases.

The existence of higher homologs of the galactosyldiglycerides was recognized by Benson *et al.*, (1958) and reiterated in biosynthetic studies (Neufeld and Hall, 1964; Ongun and Mudd, 1968). The trigalactosyl diglyceride was finally isolated in sufficient quantity for structural studies and it was concluded that the third galactose is linked α(1→6) to the DGDG (Galliard, 1969). The trigalactosyl glycerol obtained from *Ulva lactuca* was also susceptible to α-galactosidase (Brush and Percival, 1972). The possibility exists that even higher homologs exist in plants, but this has not been thoroughly explored so far.

Nichols and co-workers have pointed out that the lipid composition of blue-green algae is relatively simple, comprising only MGDG and DGDG, SL and PG (Nichols *et al.*, 1965; Nichols and Wood, 1968). In some cases photosynthetic bacteria contain glycolipids which have not yet been characterized but clearly cannot be either one of the galactosyl diglycerides common in higher plants (Constantopoulos and Bloch, 1967; Nichols and Wood, 1968). Comparison of the glycosyl diglycerides of bacteria and plants may provide some phylogenetic insights. However, the digalactosyl diglyceride of *Arthrobacter globiformis* has two β-linked galactose moieties and therefore is not the same as the one isolated from wheat flour (Walker and Bastl, 1967). The monogalactosyl diglyceride from the photosynthetic bacterium *Chloropseudomonas ethylicum* was shown to have the β-linkage and is thus the same as the MGDG from higher plants (Constantopoulos and Bloch, 1967). It is apparent that the study of the glycolipids of photosynthetic bacteria, and blue green algae is amenable to further development.

TABLE IV

Fatty acid composition of galactosyl diglycerides

Species	Lipid	% fatty acid composition												Reference
		14:0	16:0	16:1	16:2	16:3	18:0	18:1	18:2	18:3	18:4	20:0	20:5	
Spinacea oleracea	MGDG	tr	0·4	—	0·5	25·0	—	0·6	1·6	71·9	—	—	—	Allen *et al.*
(spinach chloroplast) thylakoids	DGDG	tr	3·4	—	tr	5·4	—	1·8	1·7	86·7	—	—	—	(1966)
Zea mays	MGDG	—	12·9	—	—	—	3·6	20·9	36·4	26·2	—	—	—	Weber (1970)
(maize kernels) 60 DAP	DGDG	—	19·8	—	—	—	3·5	18·7	44·0	14·0	—	—	—	
Pyrus malus	MGDG	—	2·0	—	—	—	0·6	0·6	5·4	91·4	—	—	—	Galliard
(pre-climacteric) apple pulp	DGDG	—	16·9	—	—	—	4·2	2·2	16·9	59·8	—	—	—	(1968b)
Solanum tuberosum	MGDG	—	0·8	—	—	—	—	0·9	58·0	40·3	—	—	—	Galliard
(potato tuber)	DGDG	—	14·1	—	—	—	6·3	7·7	46·5	25·4	—	—	—	(1968a)
Zea mays	MGDG	—	5·0	tr	—	—	2·2	2·3	5·9	84·6	—	—	—	Leech *et al.*
(older section of leaf)	DGDG	—	12·3	—	—	—	3·0	3·0	5·2	76·5	—	—	—	(1973)
Narcissus sp.	MGDG	—	4·4	2·2	0·8	0·8	2·2	5·3	73·8	10·3	—	—	—	Nichols and
(bulbs)	DGDG	—	12·7	2·1	0·7	0·7	2·0	8·1	68·6	4·6	—	—	—	James (1964)
Chlorella vulgaris	MGDG	—	4·5	1·5	18·9	10·0	1·9	2·8	16·8	44·6	—	—	—	Nichols (1965)
(inorganic medium) light	DGDG	2·1	7·6	2·7	5·6	1·4	3·4	2·6	34·9	36·6	—	—	—	
Chlorella vulgaris	MGDG	—	3·3	10·6	28·4	—	tr	17·5	33·7	5·0	—	—	—	Nichols (1965)
(organic medium) dark	DGDG	1·6	9·6	6·8	3·9	—	1·7	16·5	44·9	3·5	—	—	—	
Batrachospermum	MGDG	1·2	18·0	5·0	0·3	2·7	0·7	6·7	2·0	1·9	2·8	—	51·4	Radunz (1968)
moniliforme	DGDG	0·8	38·4	—	—	—	—	9·3	2·0	1·1	—	—	35·2	
Fucus vesiculosus	MGDG	2·7	4·9	0·7	—	6·2[a]	—	8·2	5·3	13·3	30·2	—	24·0	Radunz (1968)
	DGDG	4·0	7·8	0·8	—	0·8[a]	—	13·1	6·6	13·9	23·0	—	25·3	
Spinacea oleracea	MGDG	—	tr	—	—	2·6	—	0·6	1·0	72·0	—	—	—	Douce *et al.*
(thylakoids)	DGDG	—	4·6	—	—	4·0	tr	2·5	0·8	87·0	—	—	—	(1973)
Spinacea oleracea	MGDG	—	tr	—	—	24·6	—	1·0	0·6	73·6	—	—	—	Douce *et al.*
(chloroplast envelope)	DGDG	—	18·8	—	—	4·0	tr	5·1	3·2	68·6	—	—	—	(1973)

	MGDG	[illegible]	[illegible]	[illegible]	0·9	33·4	—	—	1·8	61·9	—	—	—	Auling *et al.*
	DGDG	—	12·2	—	—	3·4	1·8	—	5·3	77·3	—	—	—	(1971)
Brassica oleracea	MGDG	—	—	—	1·6	32·1	—	—	1·8	64·6	—	—	—	Auling *et al.*
var. *acephala*	DGDG	—	12·0	—	—	1·2	0·8	tr	10·8	75·2	—	—	—	(1971)
Ginkgo biloba	MGDG	—	2·3	—	—	11·2	—	tr	tr	86·1	—	0·5	—	Auling *et al.*
	DGDG	—	26·2	5·5	—	—	—	11·2	10·1	44·4	—	2·6	—	(1971)
Pteridium aquilinum	MGDG	—	1·9	tr	1·1	26·6	—	—	1·7	65·9	—	2·9	—	Auling *et al.*
	DGDG	—	33·3	—	—	0·5	—	3·5	6·8	48·5	—	7·5	—	(1971)
Equisetum arvense	MGDG	—	1·0	—	0·8	23·4	—	—	0·8	73·9	—	tr	—	Auling *et al.*
	DGDG	—	23·7	—	—	1·7	—	tr	4·9	61·0	—	8·6	—	(1971)
Catharinea undulata	MGDG	—	2·7	1·9	13·9	12·6	—	0·7	23·4	34·1	—	10·6	—	Auling *et al.*
	DGDG	—	10·5	tr	3·5	1·9	tr	1·8	29·9	46·4	—	6·0	—	(1971)
Triticum vulgare	MGDG	—	5·4	0·2	—	—	0·5	7·3	82·1	4·5	—	—	—	Arunga and
wheat flour	DGDG	—	14·0	0·2	—	—	1·0	5·8	73·5	5·5	—	—	—	Morrison (1971)
Artemisia princeps	MGDG	—	1·5	0·4	—	—	tr	0·3	3·7	94·1	—	—	—	Noda and
	DGDG	tr	16·3	5·0	—	—	2·5	5·7	14·1	56·4	—	—	—	Fujinera (1967)
Anabaena cylindrica	MGDG	—	24·0	1·0	10·0	3·0	—	3·0	22·0	30·0	—	—	—	Safford and Nichols (1970)
Chlorella vulgaris photoautotrophic	MGDG	—	3·0	2·0	18·0	10·0	—	—	22·0	46·0	—	—	—	Safford and Nichols (1970)
Spinacea oleracea	DGDG	—	5·6	—	—	2·3	—	2·0	2·1	88·0	—	—	—	Tulloch *et al.* (1973)
Sinapis alba	DGDG	—	8·0	—	—	4·5	—	—	4·6	83·0	—	—	—	Tulloch *et al.* (1973)
Brassica oleracea	DGDG	—	7·0	—	—	2·3	—	—	5·1	85·6	—	—	—	Tulloch *et al.* (1973)
Vicia faba	DGDG	—	4·2	—	—	—	—	—	2·2	93·6	—	—	—	Tulloch *et al.* (1973)

[a] These percentages were for 16:4, there were only traces of 16:3.

2. *Fatty Acid Composition*

Fatty acid analyses of the galactosyl diglycerides of higher plants show particularly notable features: (a) the galactolipids are highly unsaturated, (b) MGDG appears to be more unsaturated than DGDG, (c) when present, hexadecatrienoic acid is predominantly in MGDG, (d) palmitic acid is predominantly in DGDG. These generalizations are illustrated in the data collected in Table IV. The differences in fatty acid composition of MGDG and DGDG raise questions concerning the biosynthesis of these compounds, not only the possible conversion of MGDG to DGDG but also concerning acyl exchange and modification of acyl groups while in ester linkage.

3. *Fatty Acid Positional Specificity*

The work on both animal and plant phospholipids has made possible the generalization that unsaturated fatty acids are predominantly esterified at position 2 of the glycerol moiety. Analysis of the fatty acid distribution of MGDG and DGDG might permit a similar generalization. Results for MGDG and DGDG are compiled in Tables V and VI. Safford and Nichols (1970) concluded that in *Chlorella vulgaris* the distribution in MGDG showed preference for C_{16} acids at position 2, i.e. chain length specificity, whereas the specificity in higher plants showed preference for unsaturated fatty acids. The data collected in Table V do not entirely bear out these generalizations: in MGDG, there appears to be a preference for 16:2 and 16:3 at position 2 and a remarkable preference for C_{20} acids, when they are present, at that position. Further generalizations would be premature. The data presented in Tables V and VI should be examined in conjunction with the fatty acid analyses in Table IV. When a given fatty acid constitutes 90% of the total there cannot be a great deal of discrimination between positions 1 and 2, and determination of the distribution of the remaining acids becomes subject to large percentage errors. Fatty acid distribution data can provide information on potential biosynthetic pathways. The subject is taken up again in Section II.c.2.

B. GALACTOLIPID COMPOSITION OF TISSUES

1. *Effect of Physiological Conditions*

Roughan and Batt (1969) have compiled analyses of the galactosyl diglyceride composition of twenty plant species. The amount of MGDG consistently exceeded the amount of DGDG: the ratio MGDG/DGDG ranged from 1 to 2. In a study of the lipids of maize leaves, Leech *et al.* (1973) analyzed sections corresponding to differing physiological ages. In sections of increasing ages A, B, C, D and E the MGDG/DGDG ratios were 1·25, 1·37, 1·71, 1·78 and 1·80 respectively. These changes, as well as the increasing ratio of galactolipid to sulfonolipid, correlate with the development of

TABLE V

Fatty acid distribution in monogalactosyl diglycerides

Species	% of fatty acid found at position 2 of the glycerol moiety									Reference
	16:0	16:1	16:2	16:3	18:0	18:1	18:2	18:3	20	
Anabaena cylindrica	90	65	90	83	—	33	15	15	—	Safford and Nichols (1970)
Chlorella vulgaris	100	40	83	90	—	—	31	33	—	Safford and Nichols (1970)
Spinacia oleracea	100	—	—	92	—	—	37	38	—	Auling *et al.* (1971)
Sinapis alba	—	—	94	100	—	—	0	12	—	Auling *et al.* (1971)
Brassica oleracea	—	—	100	100	—	—	22	17	—	Auling *et al.* (1971)
Ginkgo biloba	20	—	—	88	—	—	—	45	100	Auling *et al.* (1971)
Pteridium aquilinun	34	—	100	89	—	—	38	29	100	Auling *et al.* (1971)
Equisetum arvense	85	—	100	98	—	—	0	33	100	Auling *et al.* (1971)
Catharinea undulata	19	45	100	100	—	0	46	11	100	Auling *et al.* (1971)
Triticum vulgare	2	25	—	—	10	63	51	86	—	Arunga and Morrison (1971)

TABLE VI

Fatty acid distribution in digalactosyl diglycerides

Species	% of fatty acid found at position 2 of the glycerol moiety								Reference
	16:0	16:1	16:2	16:3	18:0	18:1	18:2	18:3	
Artemisia princeps	13	1	—	—	9	18	46	72	Noda and Fujiwara (1967)
Spinacia oleracea	64	—	—	70	—	0	43	50	Tulloch *et al.* (1973)
Sinapis alba	49	—	—	97	—	—	39	48	Tulloch *et al.* (1973)
Brassica oleracea	48	—	—	96	—	—	30	50	Tulloch *et al.* (1973)
Vicia faba	0	—	—	—	—	—	75	52	Tulloch *et al.* (1973)
Triticum vulgare	7	40	—	—	15	63	57	63	Arunga and Morrison (1971)

thylakoid structure. Tevini (1971a) has reported changes in the galactosyldiglyceride composition of *Hordeum* (barley) seedlings during the greening process. The MGDG/DGDG ratio increased during this process. These analytical data may be interpreted as a reflection of the relative contributions of the chloroplast thylakoids and chloroplast envelope to the galactolipid composition. The envelope has a low ratio MGDG/DGDG whereas the ratio is high for thylakoids (Table VII).

2. *Subcellular Distribution*

Early analyses of galactolipid subcellular distribution showed a predominance in the chloroplast fraction (Allen *et al.*, 1966; Ongun *et al.*, 1968). Even though the analyses of fractions separated by simple differential centrifugation showed galactosyl diglycerides in fractions other than the chloroplast it was generally believed that all the MGDG and DGDG (and SL and PG) belonged in the chloroplast. However, Schwertner and Biale (1973) concluded that mitochondria of *Persea americana* (avocado) fruit and *Solanum tuberosum* (potato) tuber contain galactolipids. The mitochondria used for these analyses were purified by density gradient centrifugation. In contrast to this report, McCarty *et al.* (1973) have determined the lipid composition of highly purified mitochondria from *Phaseolus aureus* (mung bean) hypocotyls and potato tubers and concluded that galactosyl diglycerides are absent. More recently still, Mackender and Leech (1974) have analyzed various subcellular fractions and found that the mitochondrial fraction from *Vicia faba* (broad bean) does contain galactosyl diglycerides. None of these papers presented data on enzyme markers for the various subcellular fractions. Schwertner and Biale (1973) demonstrated significant differences in the fatty acid analyses

TABLE VII

Lipid composition of chloroplast fractions

	% of total glycerolipid				
	envelope[a]	envelope[b]	envelope[c]	lamellae[c]	microsomes[d]
MGDG	29·1	32·8	22	51	12
DGDG	32·4	33·6	32	26	8
SL	—	0·8	5·0	7·1	2
PG	0·89	2·9	8·4	9·1	3
PC	29·6	9·3	27	3·2	35
PE	0	1·4	tr	0	30
PI	—	1·1	1·3	1·4	7

Data from:
[a] Makender and Leech, 1974 (*Vicia faba*).
[b] Poincelot, 1974 (*Spinacia oleracea*).
[c] Douce *et al.*, 1974 (*Spinacia oleracea*).
[d] Douce, 1974 (*Spinacia oleracea*).

of the galactosyl diglycerides from the mitochondrial and chloroplast fractions. Mackender and Leech (1974) have also determined striking differences in the fatty acid composition of galactosyl diglycerides from different organelles, making it apparent that the finding of MGDG and DGDG in other organelles cannot be attributed to contamination by chloroplasts.

Some correlations have been made of the galactolipid composition with the structure of chloroplasts. Bishop *et al.* (1971) reported that the galactolipid:chlorophyll ratio was higher in bundle sheath chloroplasts (agranal) than in the mesophyll chloroplasts (granal) of maize and sorghum. The ratio of MGDG to DGDG was slightly higher in the mesophyll chloroplasts. In a separate study Costes *et al.* (1971) reported similar results with maize. Discrimination of subchloroplast particles on the basis of galactolipid composition does not seem to be possible. Wintermans (1971) has prepared chloroplast fragments from spinach with different chlorophyll *a*:*b* ratios indicative of the separation of photosystems I and II. However, the galactolipid ratio of the preparations was essentially the same. Wintermans *et al.* (1969) have also demonstrated that the activities of photosystems I and II are maintained in aging chloroplast preparations even though the change in galactolipid composition is quite rapid. These observations show that the intact galactolipids do not have a crucial role in these reactions. Photophosphorylation is labile under the same aging conditions. Such inhibition may be attributable to the release of fatty acids rather than the destruction of the galactolipid (McCarty and Jagendorf, 1965). The galactolipid transformations observed by Wintermans *et al.* (1969) to a large extent fell into the category of transacylations studied by Heinz (1973) in which the galactosyldiglyceride is further acylated at the 6 position of galactose.

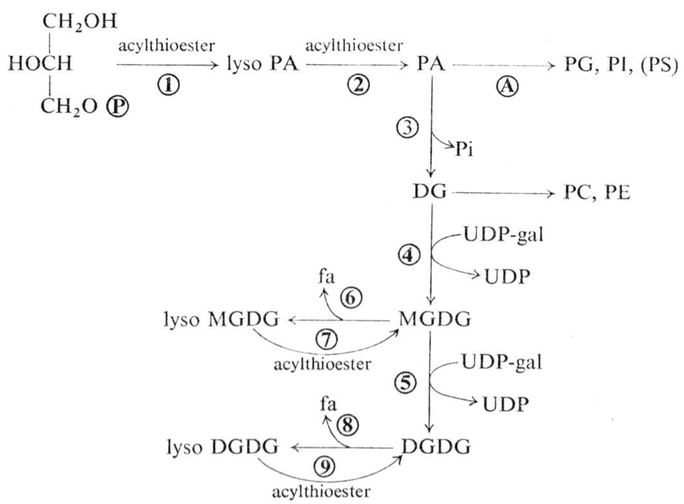

FIG. 2. Schematic representation of galactolipid biosynthesis.

C. BIOSYNTHESIS

Pathways of biosynthesis are represented in Fig. 2.

1. *The Galactose Moiety*

The experiments of Benson *et al.*, (1958) showed that metabolism of $^{14}CO_2$ by *Chlorella* rapidly labeled the galactose moiety of the galactolipids. After five minutes of photosynthesis monogalactosylglycerol (derived from MGDG) contained 39% of the label in lipid and digalactosyl glycerol (derived from DGDG) contained 10%. Benson *et al.* (1958) pointed out that the most reasonable glycosyl donor was UDP-galactose. Presumably the CO_2 was converted first to hexose phosphate and subsequently to nucleotide sugar. In a later paper, Ferrari and Benson (1961) pointed out that the galactosylglycerol (from MGDG) was labeled first and the decrease in its radioactivity was matched by an increase in the digalactosylglycerol (from DGDG), consistent with the supposition that MGDG is the precursor of DGDG. Neufeld and Hall (1964) reported the first *in vitro* studies of galactolipid biosynthesis by spinach chloroplast preparations and they found that UDP-galactose was a very satisfactory substrate for lipid synthesis. UDP-glucose was about half as good as UDP-galactose, but this activity may have been due to (a) conversion to UDP-galactose, and (b) synthesis of sterol glucosides. The experiments of Ongun and Mudd (1968) demonstrated that spinach chloroplast preparations introduced label from UDP-galactose first into MGDG and subsequently into DGDG. This result has been confirmed (Chang and Kulkani, 1970) but is not easily explained. In the experiments referred to, the presence of endogenous lipids in the chloroplast preparations was depended on and, one would have thought, the availability of MGDG (for DGDG synthesis) would exceed that of DG (for MGDG synthesis). One could rationalize that the abundant supplies of MGDG were not juxtaposed with the enzyme or else the MGDG orientation in a membrane structure was not suitable for further galactosylation. In experiments with intact cells of *Chlorella pyrenoidosa*, Sumida (1967) found that [^{14}C]acetate labeled MGDG before DGDG as in the *in vitro* experiments, but that with [^{14}C]glucose DGDG was labeled before MGDG. In the latter instance it was argued that the glucose was converted to UDP-galactose which could indeed react with MGDG (Figs. 3 and 4). In the acetate case apparently *de novo* synthesis of fatty acid and of the galactolipid took place, thus labeling MGDG first. In a time course experiment one would expect the second galactose of DGDG to be labeled first and have a higher specific activity. As time progresses the specific activity in the first (inner) galactose should rise as MGDG is converted to DGDG.

2. *The Diacylglycerol Moiety*

Although chloroplast preparations have sufficient endogenous acceptor (or

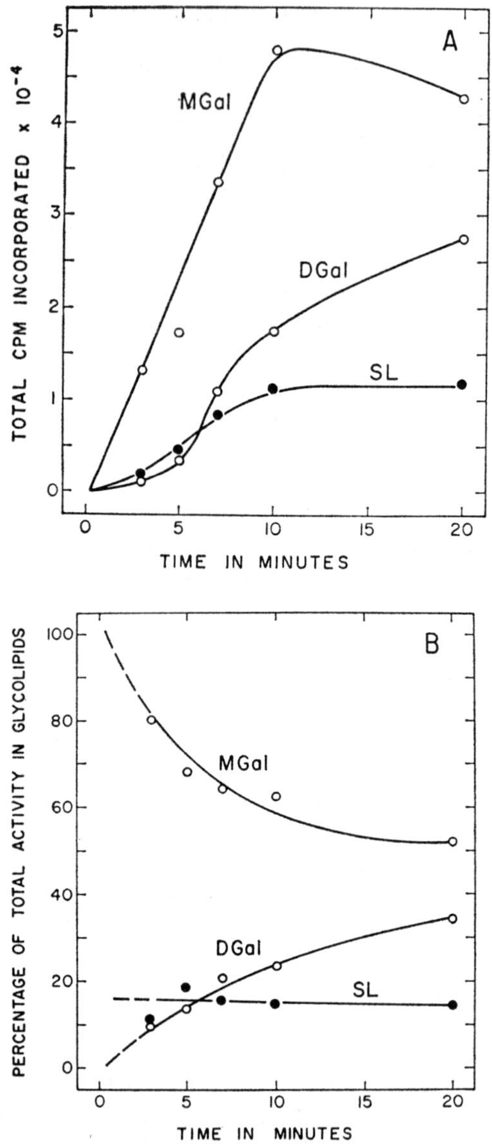

FIG. 3. Incorporation of [^{14}C]acetate into glycolipids by *Chlorella vulgaris*: (a) radioactivity incorporated; (b) percentage distribution of incorporated activity. (From Sumida, 1967.)

the capability to make it from endogenous lipids), acetone washing can create the dependency on added diglyceride (Mudd *et al.*, 1969; Eccleshall and Hawke, 1971). One may conclude that biosynthesis of diglyceride takes place by well established pathways. Indeed, Douce and Guillot-Salomon

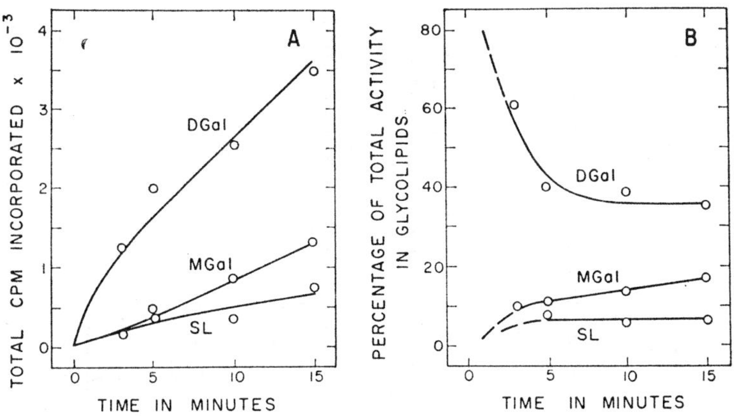

FIG. 4. Incorporation of [^{14}C]glucose into glycolipids by *Chlorella vulgaris*: (a) radioactivity incorporated; (b) percentage distribution of incorporated activity. (From Sumida, 1967.)

(1970) have shown very nicely that etioplasts of maize use [^{14}C]-*sn*-glycerol-3-phosphate in the synthesis of PA, lyso-PA and DG in the absence of UDP-gal, but that in its presence MGDG is synthesized and DG utilized. In these experiments ATP and CoASH were present in the reaction mixture, implying the requirement for activated fatty acids. This requirement was tested directly by Renkonen and Bloch (1969), using cell-free extracts of *Euglena gracilis*. They found that both CoA esters and ACP esters could be used in the synthesis of MGDG. However, the ACP esters did not serve as substrates for phospholipid biosynthesis, but the CoA esters did. The incorporation of fatty acyl groups from ACP esters into MGDG was stimulated by *sn*-glycerol-3-phosphate, indicating total synthesis of MGDG.

There are several possible explanations for the highly unsaturated fatty acids of MGDG: (a) the galactosylation reaction is specific for unsaturated diglycerides; (b) desaturation reactions work well on fatty acid substituents of the galactolipids; (c) acyl exchange reactions adjust the fatty acid substituents in favor of unsaturated fatty acids; (d) at the site of biosynthesis, only unsaturated diglycerides are available. These possibilities are not equally amenable to experimental approach. The specificity of the first galactosylation reaction has been reported in two laboratories; unfortunately, opposite results were obtained. Mudd *et al.* (1969) reported that the biosynthesis of MGDG by spinach chloroplast acetone powder was highly selective of unsaturated diglycerides. Eccleshall and Hawke (1971) with essentially the same enzyme preparation found all diglycerides were used equally well. Table VIII shows some of the characteristics of enzyme systems used in the biosynthesis of glycosyl diglycerides and the results obtained. In comparing the results from spinach chloroplast preparations with all other results it is

TABLE VIII

Characteristics of glycosyldiglyceride-synthesizing enzymes

Product	Enzyme source	Diglyceride source	Dispersal method	Nucleotide sugar concn	Specificity	Reference
monogalactosyl diglyceride	spinach chloroplast acetone powder	synthetic mixture of 1,2- and 2,3-diglycerides	absorbed on acetone powder from acetone solution	12 μM	none	Eccleshall and Hawke (1971)
monogalactosyl diglyceride	spinach chloroplast acetone powder	1,2-diglycerides derived from spinach PL by phospholipase C action	suspended in 0·03 % Tween 20	0·1 μM	specific; unsat. fatty acids preferred	Mudd *et al.* (1969)
mannosyl diglyceride	acetone powder of *Micrococcus lysodeikticus*	natural and synthetic	suspended in detergent, e.g. sodium lauryl sulfate	0·5 mM	specific; branched chain fatty acids preferred	Lennarz and Talamo (1961)
glucosyl diglyceride	lyophilized particulate preparation from *Streptococcus faecalis*	synthetic mixture of 1,2- and 2,3-diglycerides	absorbed in acetone powder from benzene solution	2 mM	specific; unsat. fatty acids preferred	Pieringer (1968)
monogalactosyl diglyceride	lyophilized microsomal preparation from rat brain	synthetic mixture of 1,2- and 2,3-diglycerides	absorbed on acetone powder from benzene solution	1 mM	specific; dipalmitin preferred	Wenger *et al.* (1968)
glucosyl diglyceride	membrane fraction *Mycoplasma laidlawii*	1,2-diglycerides from egg lipids, *M. laidlawii* and synthesis	suspended in sodium lauryl sulfate	1mM	specific; *M. laidlawii* lipid preferred	Smith (1969)
glucuronosyl diglyceride	halotolerant bacterium 140 000 *g* particles	phospholipase C digestion of phospholipids from various sources	(a) from benzene; (b) 0·1 % detergent	0·1 mM	specific; DG containing cyclopropane fatty acids preferred	Stern and Tietz (1973)

not possible to identify the reason for the difference in the two spinach results. In the other results specificity is shown with acetone powders, lyophilized preparations and membrane fractions; both synthetic and natural diglycerides were active, specificity being shown when they were absorbed on the enzyme and when added in detergent. Perhaps the concentration used by Mudd *et al.* (1969) was particularly low; however, specificity was observed in the other systems at relatively high nucleotide sugar concentrations, higher than those used by Eccleshall and Hawke (1971). It should be noted that when specificity was observed, the diglyceride preferred was that which yielded a glycosyl diglyceride of fatty acid composition closest to that naturally occurring in the organism from which the enzyme was obtained.

Evidence for the modification of fatty acid residues after formation of MGDG is presented by Nichols and Moorhouse (1969). Their analyses of the molecular species of MGDG from *Chlorella vulgaris* showed a remarkable relationship between the two fatty acid substituents. Their data are shown in Table IX. The remarkable features are that no "two double-bond" species contains linoleate and no "four double-bond" species contains linolenate. In other words, the two fatty acid substituents do not differ by more than one double bond. Nichols and co-workers point out that this result is consistent with the occurrence of desaturation while the acyl groups are in ester linkage; this point is, however, the subject of controversy (see Chapter 4, p. 109). There is some question as to whether the results obtained with *Chlorella vulgaris* are applicable to dicotyledenous plants. Judging from measurements of specific activity in the glycerolipids of *Cucurbita pepo* (pumpkin) leaves, MGDG does not turn over rapidly and a more likely site for fatty acid changes is PC (Roughan, 1970).

3. *Acyl Exchange*

The final fatty acid composition of MGDG could be determined by an acyl exchange mechanism. This would depend on the removal of a single acyl group from MGDG or DGDG (Enzymes 6 and 8, Fig. 2 and replacement by Enzymes 7 and 9, Fig. 2). Specificity in the action of the latter enzymes could

TABLE IX

Fatty acid composition of molecular species of monogalactosyl diglyceride from *Chlorella vulgaris* (Nichols and Moorhouse, 1969)

Double bonds	16:0	16:1	16:2	18:1	18:2	18:3
1	28	6	0	50	3	0
2	2	39	tr	52	7	0
3	tr	23	18	26	34	0
4	tr	2	41	3	53	0
5	2	2	35	3	14	40

account for the association of more unsaturated fatty acids with MGDG and the virtual exclusion of 16:3 from DGDG. Safford *et al.* (1971) have shown that lyso-MGDG (prepared by action of pancreatic lipase on MGDG) is converted to MGDG in the presence of CoA, ATP and spinach leaf homogenates. Bajwa and Sastry (1973) have refined this experiment slightly by using a spinach chloroplast acetone powder and specific fatty acids. The formation of MGDG with the following fatty acids, 18:3, 18:1, 16:0 was in the ratio 6·9:5·7:1. This approach deserves further development, depending on the availability of lysogalactosyl diglycerides of known structure. The specificity of the hydrolytic enzyme from *Rhizopus delemar* should be particularly useful in this respect.

4. *Subcellular Distribution of the Two Galactosylation Reactions*

Although it is acknowledged that the site of galactolipid synthesis is the chloroplast, the site within the chloroplast and discrimination between the two galactosylation reactions have not been actively studied. Fragmentation of chloroplasts in hypotonic buffer and separation of pellet and supernatant fractions gave enzyme preparations with greatly different capabilities with respect to ratio of MGDG and DGDG synthesized (Ongun and Mudd, 1968). The two galactosylation steps can also be distinguished on the basis of response to temperature and pH (Mudd *et al.*, 1969) and sensitivity to inhibitors (Mudd *et al.*, 1971). The use of acetone powders has permitted the demonstration of direct conversion of MGDG to DGDG (Ongun and Mudd, 1968; Bowden and Williams, 1973). A most significant result has been obtained by Douce (1974) in studying the localization of galactolipid biosynthesis. He found that the chloroplast envelope is the major site of synthesis. The same result has been obtained by von Hummel (1974), but it is not clear at the moment whether both MGDG and DGDG are synthesized by the envelope. This site of galactolipid synthesis is consistent with the role of the envelope in thylakoid synthesis. However the possibility of the conversion of MGDG into DGDG in the thylakoid is a problem yet to be tackled. Is there, for instance, exchange from the thylakoid to the envelope in order to conduct the second galactosylation?

5. *Abnormal Physiological Conditions*

Tevini (1971b) has examined the lipid composition of *Impatiens balsamina* and *Hordeum vulgare* under conditions of phosphate deficiency. In both cases there was a marked decrease in all phospholipids. However, the chlorophyll content was not changed and the amounts of MGDG, DGDG and SL were increased somewhat. This result showed that steps (**1**), (**2**), and (**3**) (Fig. 2) of the phospholipid-synthesizing pathway were functioning normally. A contrasting result was obtained when various herbicides were used to inhibit [^{14}C]acetate metabolism by *Chlorella ellipsoidea* (Sumida and Ueda, 1973). Those herbicides which inhibited the conversion of acetate into lipid under photosynthetic conditions markedly inhibited MGDG and DGDG synthesis. This inhibition was balanced by relatively greater synthesis of PG especially, and also SL and PE. Effective inhibitors were simazine, (2-chloro-4,6-bis (ethylamino)-1,3,5 triazine); DCPA, (N-(3,4-dichlorophenyl)propionamide); DCMU, (3-(3′ 4′-dichlorophenyl)-1,1-dimethylurea); bromacil, (5-bromo-6-methyl-3(1-methyl-*n*-propyl)uracil); and swep, (methyl-*N*-(3,4-dichlorophenyl carbamate). The results are to some extent consistent with the inhibition of the Hill reaction affecting lipid biosynthesis in the chloroplast, but the increase of percentage label in PG, SL and PE is then difficult to explain.

There have been studies of the effects of cycloheximide and chloramphenicol on lipid biosynthesis, both in higher plants (Vichanska *et al.*, 1973) and in *Euglena gracilis* (Bishop and Smillie, 1970). In the former study, with leaves of *Pisum sativum*, both compounds inhibited galactolipid biosynthesis (particularly DGDG) and stimulated phospholipid biosynthesis (particularly PE and PC). The experiments with *Euglena gracilis* used lower concentrations of the two inhibitors and showed that phospholipid biosynthesis was unaffected while galactolipid biosynthesis was inhibited. Cycloheximide raised the MGDG/DGDG ratio as it did in the studies with *Pisum sativum*. Neither study is conclusive regarding the relative importance of chloroplastic and cytoplasmic systems of synthesis of the galactolipid metabolizing enzymes, but they do agree with the other studies mentioned above which show that galactolipid biosynthesis is to some extent independent of phospholipid biosynthesis. These two synthetic pathways (glycolipid and phospholipid) may be roughly viewed as divergent pathways (Fig. 2), but thoughts of end-product or feedback inhibition seem to be profitless.

III. Biosynthesis of Steryl Glucosides and Acylated Steryl Glucosides

A. General

In 1966 Kiribuchi *et al.* (1966) showed that phytosterols of ground soybeans exist in four different forms: free sterols, steryl esters, steryl glucosides, and acylated steryl glucosides. Of these, two are unique to plants, steryl glucosides and acylated steryl glucosides. Steryl glucosides and acylated steryl glucosides are both glycolipids, containing D-glucose attached via a β-glycosidic linkage to the 3-position of several types of sterols (Lepage, 1964; Kiribuchi *et al.*, 1966; Kiribuchi *et al.*, 1967; Grunwald, 1970; Kates, 1970; Bush *et al.*, 1971; Duperon *et al.*, 1972; Wojciechowski, 1972), such as sitosterol, stigmasterol, campesterol and cholesterol. In acylated steryl glucosides the glucose is acylated at the 6-position with several types of fatty acids (Lepage, 1964; Kiribuchi *et al.*, 1972; Eichenberger and Grob, 1972; Holmer *et al.*, 1973), such as palmitic, stearic, oleic, linoleic, and linolenic. Steryl glucodises have been known to exist in plants for some time (Power and Salway, 1913), whereas acylated steryl glucosides were only recently identified in potato tubers and soybean lecithin by Lepage (1964).

The four phytosterols listed above are now known to be of frequent occurrence in plants (Eichenberger and Menke, 1966; Lepage, 1967; Lepage, 1968; Eichenberger and Grob, 1970a; Grunwald, 1970; Kates, 1970; Eichenberger and Grob, 1972; Duperon *et al.*, 1972). Galliard (1968a) found that 9·1 % of the lipids in potato tubers were phytosterols (6·4 % as acylated steryl glucosides, 2·0 % as steryl glucosides, 0·6 % as free sterols, and 0·1 % as steryl esters). In apples Galliard (1968b) found that 29·2 % of the lipids were phytosterols (1·6 % as acylated steryl glucosides, 5·4 % as steryl glucosides, 20·1 % as freesterols, and 2·1 % as steryl esters).

B. PHYTOSTEROL SUBCELLULAR DISTRIBUTION

There are several reports indicating that free sterols (Kemp and Mercer, 1968; Grunwald, 1970; Brandt and Benveniste, 1972; Duperon *et al.*, 1972), steryl glucosides (Grunwald, 1970; Duperon *et al.*, 1972), acylated steryl glucosides (Duperon *et al.*, 1972), and sterol esters (Kemp and Mercer, 1968; Grunwald, 1970; Brandt and Benveniste, 1972; Duperon *et al.*, 1972) are present in all subcellular organelles of various plants and plant tissues. In all cases the highest concentration of phytosterols occurred in the microsomal fraction. The second highest concentration occurred in the mitochondrial fraction. The nuclear and the chloroplast fractions were the poorest phytosterol-containing subcellular organelles. The highest concentration of acylated steryl glucoside (Duperon *et al.*, 1972) and steryl glucoside (Eichenberger and Grob, 1970a; Grunwald, 1970; Duperon *et al.*, 1972) occurred in the microsomal fraction. The presence of these four phytosterols in all plant subcellular organelles implies that they may have a physiological role associated with the structure and function of membranes.

When Eichenberger and Menke (1966) analyzed the phytosterols in spinach leaves, they found that 57–69 % of the sterols were free sterols, 6–9 % were steryl esters, and 25–34 % were steryl glucosides. The spinach chloroplast, in contrast, contained only traces of steryl glucosides and a proportionally greater amount of steryl esters. Using lettuce leaf homogenates, Eichenberger and Grob (1969a) noted that a high-speed sedimentation fraction (22 000 *g*) with little chlorophyll contained most of the steryl glucosides, while the chloroplast fraction with most of the chlorophyll contained very little steryl glucoside. Allen *et al.* (1966) showed that spinach chloroplast lamellar membranes contain no steryl glucosides. Poincelot (1973) reported that spinach chloroplast envelope membranes contain steryl glucosides, acylated steryl glucosides, steryl esters, and free sterols and that these phytosterols were absent in spinach chloroplast lamellar membranes. Of the lipids in the chloroplast envelope membranes, 5–6 % were phytosterols (1·1 % as steryl glucosides, 0·8 % as acylated steryl glucosides, 1·9 % as free sterols, and 1·8 % as steryl esters). The phytosterol composition of spinach chloroplast envelope membranes is, therefore, substantially different from that of the lamellar membranes and these differences are undoubtedly responsible, at least partially, for the different functional properties of these membranes.

C. STERYL GLUCOSIDE BIOSYNTHESIS

1. *UDP-Glucose Requirement*

Particulate enzyme preparations from dark-grown mung bean shoots (Kauss, 1968), immature soybean seeds (Hou *et al.*, 1967, 1968), spinach leaves (Eichenberger and Grob, 1968), lettuce leaves (Eichenberger and Newman, 1968; Eichenberger and Grob, 1968, 1969a, 1969b), starch grains from sweet corn (Lavintman and Cardini, 1970), avocado mitochondria (Ongun and Mudd, 1970), cauliflower mitochondria (Ongun and Mudd, 1970), pea root mitochondria (Ongun and Mudd, 1970), spinach leaf chloroplasts (Ongun and Mudd, 1970), lettuce leaf chloroplasts (Eichenberger and Grob, 1969a), wheat root microsomes (Péaud-Lenoël and Axelos, 1971; Péaud-Lenoël and Axelos, 1972), marigold flowers, leaves, stems, and roots (Wojciechowski, 1972), and developing cotton seeds (Forsee *et al.*, 1972) have all been shown to incorporate [^{14}C] glucose from UDP-[^{14}C] glucose into steryl glucosides and acylated steryl glucosides. Hou *et al.* (1968), Lavintman and Cardini (1970), and Forsee *et al.* (1972) have shown that the enzyme involved in steryl glucoside biosynthesis has a high specificity for UDP-glucose. Hou *et al.* (1968) reported that TDP-glucose is 40% as effective as UDP-glucose in immature soybean seeds, whereas Forsee *et al.* (1972) indicated that it is only 18% as effective in developing cotton seeds. ADP-glucose (Hou *et al.*, 1968; Lavintman and Cardini, 1970; Forsee *et al.*, 1972), CDP-glucose (Hou *et al.*, 1968; Forsee *et al.*, 1972), GDP-glucose (Hou *et al.*, 1968; Lavintman and Cardini, 1970; Forsee *et al.*, 1972), IDP-glucose (Hou *et al.*, 1968), GDP-mannose (Lavintman and Cardini, 1970), UDP-galactose (Lavintman and Cardini, 1970; Forsee *et al.*, 1972), UDP-mannose (Forsee *et al.*, 1972), and UDP-glucuronic acid (Hou *et al.*, 1968) have all been shown to be ineffective as glycosyl donors in steryl glucoside biosynthesis.

If spinach leaf discs were incubated with either UDP-[^{14}C]glucose or UDP-[^{14}C]galactose, Eichenberger and Newman (1968) noted that the labeled hexose in the steryl glucoside fraction was always glucose. Similarly, Ongun and Mudd (1970) always found [^{14}C]glucose in steryl glucosides and acylated steryl glucosides formed by a pea root mitochondrial preparation regardless of whether UDP-[^{14}C]glucose or UDP-[^{14}C]galactose were supplied. Obviously, spinach leaves and pea roots contain an epimerase that converts UDP-galactose to UDP-glucose which is the nucleotide sugar required for steryl glucoside biosynthesis.

An optimal reaction conditions for each plant, the apparent K_m of the transglucosylase for UDP-glucose is 0·22 mM with a particulate preparation from immature soybean seeds (Hou *et al.*, 1968), 0·028 mM with a microsomal preparation from wheat roots (Péaud-Lenoël and Axelos, 1971), 0·1 mM with a particulate preparation from the starch grains from the immature endosperm of sweet corn (Lavintman and Cardini, 1970), and 0·035 mM with a solubilized and partially purified transglucosylase from developing cotton seeds (Forsee *et al.*, 1972).

2. *Sterol Requirement*

Numerous researchers (Eichenberger and Grob, 1968, 1969a; Hou *et al.*, 1968; Lavintman and Cardini, 1970; Ongun and Mudd, 1970; Wojciechowski, 1972; Forsee *et al.*, 1972; Bush and Grunwald, 1974) have established a free sterol requirement for steryl glucoside biosynthesis. Many different techniques have been employed in this endeavor, but the most successful ones have either purified the transglucosylase (Forsee *et al.*, 1972) or removed endogenous sterols by acetone extraction (Ongun and Mudd, 1970; Péaud-Lenoël and Axelos, 1971, 1972; Wojciechowski, 1972) or by petroleum ether extraction (Eichenberger and Grob, 1968, 1969a); before addition of exogenous sterol.

Eichenberger and Grob (1968, 1969a) demonstrated the free sterol requirement for steryl glucoside biosynthesis by lyophilizing a lettuce leaf homogenate and then extracting the endogenous lipids with petroleum ether. When this sterol-free preparation was incubated with UDP-[^{14}C]glucose, little [^{14}C]glucose incorporation into steryl glucoside occurred, unless exogenous free sterols or the petroleum ether extract were added to the mixture. Ongun and Mudd (1970) obtained a 2-fold stimulation in steryl glucoside and acylated glucoside biosynthesis when exogenous free sterols in acetone were added to a pea root mitochondrial acetone powder or when the acetone lipid extract was added back to the acetone powder. Similarly, Wojciechowski (1972) obtained a significant stimulation (20–30%) in steryl glucoside biosynthesis when an emulsion of sitosterol or stigmasterol in a 0·5% aqueous solution of Tween 20 was added to a marigold leaf acetone powder preparation. This acetone powder, unlike the pea root mitochondrial acetone powder, exhibited no biosynthesis of acylated steryl glucoside.

Péaud-Lenoël and Axelos (1971) obtained by differential centrifugation and sucrose density gradient centrifugation a purified particulate fraction from wheat roots that was capable of synthesizing steryl glucosides and acylated steryl glucosides when UDP-[^{14}C]glucose was added. They were able to separate completely the transglucosylase and the transacylase from their endogenous substrates by extraction of the particulate preparation with acetone (Table X). An active steryl glucoside-synthesizing system was reconstituted by the addition of UDP-[^{14}C]glucose and the acetone lipid extract or free sterol (Table X). An active acylated steryl glucoside synthesizing-system was reconstituted by the addition of steryl-[^{14}C]glucoside and the acetone lipid extract (Table X). The lipid substrates were solubilized by Triton X-100.

TABLE X

Synthesis of acylated steryl glucoside by acetone powder of microsomes from wheat

Protein (μg)	Substrates	dpm	
		Added	Incorpd into ASG
35	SG-^{14}C	16 800	150
35	SG-^{14}C + lipid extract of wheat microsomes	16 800	9 040
43	SG-^{14}C + crude soya lecithin	27 400	12 000
59	SG-^{14}C + wheat PE	26 000	4 160
42	SG-^{14}C + soya PE	27 500	14 100
42	SG-^{14}C + spinach PE	29 300	10 100
66	SG-^{14}C + *E. coli* PE	23 400	7 350
42	SG-^{14}C + dipalmitoyl GPE	29 300	1 630
140	SG-^{14}C + PE from *E. coli*	$2{\cdot}2 \times 10^6$	$7{\cdot}8 \times 10^4$

(Data from Axelos and Péaud-Lenoël, 1971)

These reconstitution studies have certainly established a free sterol requirement for steryl glucoside biosynthesis, and this free sterol requirement has been substantiated by Bush and Grunwald (1974), who incubated [4-^{14}C] cholesterol, dissolved in ethanol, with a particulate enzyme preparation from tobacco seeds and obtained radioactively labeled cholesteryl glucoside. Acylated steryl glucoside was not formed during their standard assay conditions. Therefore, the experimental results obtained with exogenous UDP-glucose and exogenous free sterol with *in vitro* systems clearly indicate that plants contain an enzyme capable of synthesizing steryl glucosides according to the following scheme:

$$\text{sterol} + \text{UDP-glucose} \xrightarrow{\text{transglucosylase}} \text{steryl glucoside} + \text{UDP}$$

3. *Sterol Substrate Specificity*

Bush and Grunwald (1974) quantitatively followed the disappearance of free sterols in their *in vitro* reaction mixtures by gas–liquid chromatography and found the percentage disappearance of cholesterol, consisting of endogenous and exogenous cholesterol, was of the same magnitude as the disappearance of endogenous campesterol, stigmasterol and sitosterol (Table XI). This suggests that both exogenous and endogenous free sterols are readily available for steryl glucoside biosynthesis and that steryl glucoside biosynthesis, at least in a particulate enzyme preparation from tobacco seeds, does not exhibit a sterol substrate specificity. When Bush and Grunwald (1974) incubated intact 6-day-old tobacco seedlings with [4-^{14}C] cholesterol or [4-^{14}C]sitosterol for 24 h, 27% of the [4-^{14}C]cholesterol and 32% of the [4-^{14}C]sitosterol obtained from the intact tobacco seedlings were recovered in the steryl glucoside fraction. Although the total uptake of cholesterol by

TABLE XI

Biosynthesis of steryl glycosides as determined by disappearance of free sterol

Reaction time min	Cholesterol		Campesterol		Stigmasterol		Sitosterol	
	μg	%	μg	%	μg	%	μg	%
0	26·5	100·0	21·3	100·0	65·6	100·0	43·1	100·0
10	9·0	34·0	7·4	34·7	21·8	33·2	13·9	32·3
60	5·4	20·4	5·5	25·8	23·2	35·4	12·5	29·0

(Data from Bush and Grunwald, 1974)

the tissue was greater than that of sitosterol, Bush and Grunwald (1974) concluded that this *in vivo* study also showed an absence of a sterol substrate specificity for steryl glucoside biosynthesis.

A comparison of the sterol composition of steryl glucosides and free sterols in potato tubers (Duperon *et al.*, 1972), tobacco leaves (Grunwald, 1970), green and etiolated barley shoots (Bush *et al.*, 1971), and soybeans (Kiribuchi, 1966) also suggests the absence of sterol substrate specificity for the transglucosylase since the sterol composition of steryl glucosides and the sterol composition of free sterols in these plants are nearly identical. In contrast, Péaud-Lenoël and Axelos (1971) showed that sitosterol is better incorporated than stigmasterol and that stigmasterol is better incorporated than cholesterol into steryl glucosides when the Triton X-100 solubilized transglucosylase from a wheat root acetone powder is incubated with UDP-[^{14}C]glucose (Table X); however, they claim these differences are due to the instability of the transglucosylase when isolated in this manner. The solubilized and partially purified transglucosylase obtained by Forsee *et al.* (1972) from developing cotton seeds preferred sitosterol as the sterol substrate. Cholesterol and stigmasterol were incorporated into steryl glucoside only 60% as effectively as sitosterol. A more thorough kinetic study on the sterol substrate specificity of the transglucosylase is apparently needed before the question of whether there is or is not a sterol substrate specificity can be adequately answered.

4. *Transglucosylase Purification*

When Péaud-Lenoël and Axelos (1971) treated their purified particulate fraction with Triton X-100 (0·7 mM) and centrifuged at 130 000 *g* for 2 h, the steryl glucoside and the acylated steryl glucoside-synthesizing activities were found in the supernatant (Table X). It was possible, therefore, to solubilize these two enzymes with a nonionic detergent. When their acetone powder preparation was treated in a similar manner, the steryl glucoside-synthesizing activity was solubilized but the acylated steryl glucoside-synthesizing activity was lost (Table X). Since the transacylase was not recovered in either the supernatant or the centrifuged pellet and since a mixture of the two was also inactive, Axelos and Péaud-Lenoël (1971) concluded that the solubilized transacylase was unstable.

Forsee *et al.* (1972) devised a scheme by which it is possible to solubilize and partially purify the transglucosylase involved in cotton seed steryl glucoside biosynthesis (Table XII). The enzyme was solubilized by treating a 20 000 *g* pellet fraction with 0·25% Triton X-100 and recentrifuging at 30 000 *g*. The resulting supernatant was retained and more Triton X-100 was added to give a final concentration of 2·4%. The solubilized enzyme was then partially purified by DEAE-cellulose (Cl^-) column (3 × 10 cm) chromatography followed by a 55% ammonium sulfate (pH 8·0) precipitation. Their solubilized and partially purified transglucosylase was free of acylated steryl glucoside-synthesizing activity.

The specific activity of the various steps during the course of the purification are given in Table XII, where activity is measured by determining the amount of incorporation of [^{14}C]glucose from UDP-[^{14}C]glucose into steryl glucoside and one unit of enzyme is defined as the amount of enzyme catalyzing the formation of 1 μmol of steryl glucoside per hour. The final purification step resulted in a 23-fold increase in specific activity with a 13% recovery of total enzyme units. The addition of Triton X-100 followed by DEAE-cellulose chromatography yielded an enzyme preparation that was absolutely dependent upon exogenous sitosterol (Table XII). Sitosterol was added as a water-clear micellar dispersion of sitosterol, lecithin, and Triton X-100.

Forsee *et al.*, (1972) obtained more than a 3-fold stimulation in steryl glucoside biosynthesis when lecithin was included with the Triton X-100: sitosterol micelles. However, the apparent K_m of the transglucosylase for sitosterol and UDP-glucose varied depending on the amount of lecithin in the reaction mixture. When the Triton X-100: lecithin: sitosterol ratio was 24:1:1, the apparent K_m for sitosterol was 0·04 mM, and, when the ratio was 24:8:1, the apparent K_m was 0·1 mM. The latter ratio gave maximal reaction

TABLE XII

Purification of UDP-glucose: sitosterol glycosyltransferase

Fraction	Total protein (mg)	Activity in absence of added sterol (endogenous)	Total units	Specific activity
20 000 *g* pellet	680	3·1	15	0·022
0·25% Triton supernatant	89	2·1	11	0·12
2·4% Triton	88	0·11	4·8	0·054
DEAE-cellulose	6·5	0·00	2·6	0·40
Ammonium sulfate	3·9	0·00	2·0	0·51

(Data from Forsee *et al.*, 1972)

rates. The apparent K_m values for UDP-glucose were 0·022 mM for a ratio of 24:1:1 and 0·035 mM for a ratio of 24:8:1. Using a microsomal preparation from wheat roots as the enzyme source, Péaud-Lenoël and Axelos (1971) obtained an apparent K_m of 0·3 mM for sitosterol. They also solubilized the sitosterol in Triton X-100 but lecithin was omitted.

D. ACYLATED STERYL GLUCOSIDE BIOSYNTHESIS

1. *Precursor–Product Relationship*

Two biosynthetic pathways have been proposed for the biosynthesis of steryl glucosides and acylated steryl glucosides (Hou *et al.*, 1968; Eichenberger and Grob, 1969b; Péaud-Lenoël and Axelos, 1971, 1972) and they are as follows:

(1) UDP-glucose —(sterol)→ SG —(acyl donor)→ ASG

(2) UDP-glucose —(acyl donor)→ UDP-glucose-acyl —(sterol)→ ASG —(→ fatty acid)→ SG

In Fig. 5 Péaud-Lenoël and Axelos (1971) show, via an *in vitro* kinetic study on the rate of synthesis of steryl glucoside and acylated steryl glucoside, that steryl glucoside is the precursor of acylated steryl glucoside. The time course for steryl glucoside biosynthesis from UDP-[^{14}C]glucose by a microsomal preparation from wheat roots has no lag period, whereas the time

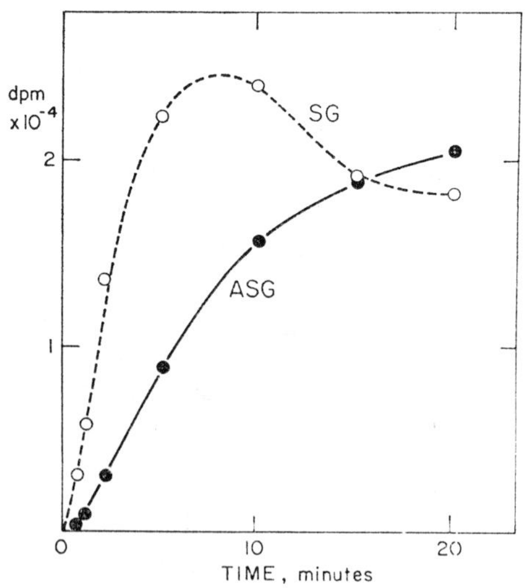

FIG. 5. Biosynthesis of steryl glucoside and acylated steryl glucoside by a microsomal preparation from wheat roots. (Adapted from Péaud-Lenoël and Axelos, 1971.)

course for acylated steryl glucoside biosynthesis has a definite lag period during the intitial stages of synthesis. Steryl glucoside biosynthesis eventually reaches a plateau and declines rapidly thereafter. This rate of decline is proportional to the rate of increase in acylated steryl glucoside. These kinetic results are consistent with the suggestion that steryl glucoside is the precursor for acylated steryl glucoside. Ongun and Mudd (1970) and Wojciechowski (1972) were also able to establish steryl glucoside as the precursor for acylated steryl glucoside from *in vitro* kinetic studies.

Another method used to establish the precursor–product relationship between steryl glucoside and acylated steryl glucoside has been to add labeled steryl glucoside to a particulate enzyme preparation and see if it becomes acylated. One of the first studies of this sort was conducted by Eichenberger and Grob (1969b). They dissolved steryl-[^{14}C]glucoside in dimethyl sulfoxide and added it to a particulate fraction from lettuce leaves. The steryl-[^{14}C] glucoside was converted by fatty acid esterification to acylated steryl glucoside. This established steryl glucoside as the precursor for acylated steryl glucoside. Similar experiments were done by Péaud-Lenoël and Axelos (1971, 1972) and Axelos and Péaud-Lenoël (1971) with a microsomal preparation from wheat roots (Table X) and by Wojciechowski (1972) with a leaf homogenate from marigolds, and they too obtained acylation of steryl-[^{14}C] glucoside.

These experiments, as well as those showing free sterol as the precursor for steryl glucoside biosynthesis, establish steryl glucoside as the precursor of acylated steryl glucoside and reaction 1 as the biosynthetic pathway for steryl glucoside and acylated steryl glucoside biosynthesis. These experiments, however, were done with *in vitro* systems. Bush and Grunwald (1974), using an *in vivo* system, have evidence suggesting that acylated steryl glucoside may be the precursor for steryl glucoside in intact tobacco seedlings. When they followed the incorporation of [4-^{14}C]cholesterol into steryl glucoside and acylated steryl glucoside as a function of time, acylated steryl glucoside increased at a faster rate and obtained a higher final level of radioactivity than did steryl glucoside. These results suggest, at least for one *in vivo* system, that acylated steryl glucoside is formed through the transfer of an acyl–glucosyl unit to the steryl moiety and that steryl glucoside is then formed by deacylation of acylated steryl glucoside. However, steryl glucoside biosynthesis may be the rate limiting step in acylated steryl glucoside biosynthesis during *in vivo* conditions and, thereby, account for the observed kinetics. Although Eichenberger and Grob (1969b) point out that acylated glucose has been isolated, Péaud-Lenoël and Axelos (1971) did not observe [^{14}C]acyl-glucose in their studies.

When authentic [4-^{14}C]cholesteryl glucoside was incubated with intact tobacco seedlings for 12 h, Bush and Grunwald (1974) recovered very little of the radioactivity in acylated steryl glucoside (4–11 %), whereas a considerable amount of radioactivity was recovered in free cholesterol (21–23 %).

When the same experiment was conducted with [4-^{14}C]cholesterol, instead of [4-^{14}C]cholesteryl glucoside, more radioactivity was incorporated into acylated steryl glucoside (13–26%) than into steryl glucoside (3–6%). From these experiments Bush and Grunwald (1974) concluded that steryl glucoside is not acylated very rapidly in *in vivo* systems, thereby substantiating their argument that acylated steryl glucoside is formed prior to steryl glucoside in intact tobacco seedlings. The slow acylation of [4-^{14}C]cholesteryl glucoside observed by Bush and Grunwald (1974), however, may be due to membrane barriers which prevent substrate from coming into contact with the transacylase. Another possibility is that [4-^{14}C]cholesteryl glucoside has to be bound to the membrane in close association with the transacylase before transacylation can occur. Smith (1971) has shown with *Mycoplasma gallinarum* that only membrane-bound cholesterol participates in steryl glucoside biosynthesis. This may also be the case for steryl glucoside in the biosynthesis of ASG in plants. Additional experimentation is needed with *in vivo* systems in order to decide whether reaction (**2**) is or is not a possible biosynthetic pathway within the plant.

2. *Acyl Donor Identification*

Eichenberger and Grob (1970b) found that a soluble enzyme preparation from carrots could catalyze the formation of acylated steryl glucoside from steryl-[^{14}C]glucoside when either monogalactosyl or digalactosyl diglyceride was supplied as the acyl donor. Digalactosyl diglyceride was the more effective acyl donor of the two. Although stimulation of acylated steryl glucoside biosynthesis by unlabeled galactolipids is a necessary condition in proving galactolipids as the acyl donor, it alone is not a sufficient condition. In order to establish digalactosyl diglyceride as the acyl donor, radioactivity from [^{14}C]acyl-labeled digalactosyl diglyceride must be incorporated into acylated steryl glucoside. Without a study of this type, an accurate interpretation of the galactolipid stimulation is dubious. Galactolipids may be activating the enzyme or the substrate or they may be supplying precursors via degradation.

Axelos and Péaud-Lenoël (1971) were able to restore acylated steryl glucoside-synthesizing activity to an acetone-extracted microsomal preparation from wheat roots by the addition of steryl-[^{14}C]glucoside and phosphatidylethanolamine from soya, wheat, spinach, or *E. coli* (Table XIII). Phosphatidylcholine and phosphatidylserine were also tested but could not restore the acylated steryl glucoside-synthesizing activity. A certain percentage of unsaturated fatty acids in phosphatidylethanolamine appeared to be necessary for activating acylated steryl glucoside biosynthesis since dipalmitoylphosphatidylethanolamine had little activity (Table XIII). This, however, may have been a solubility problem rather than an inability to restore activity fully.

When Axelos and Péaud-Lenoël (1971) incubated [^{14}C]acyl-labeled phosphatidylethanolamine and unlabeled steryl glucoside with a microsomal

TABLE XIII

Formation of steryl glucoside and acylated steryl glucoside from UDP-[^{14}C]glucose

Enzyme	Additions	Incubation time (min)	dpm	
			SG	ASG
1. Particles	UDP-G-^{14}C	5	10 800	6 300
	SG-^{14}C	30	—	6 450
2. Acetone powder, suspended in Triton X-100+ mercaptoethanol	UDP-G-^{14}C	5	0	0
	UDP+lipid extract	5	15 150	760
	SG-^{14}C	30	—	185
	SG-^{14}C+lipid extract	30	—	11 300
3. Fraction of acetone powder soluble in Triton X-100	UDPG-^{14}C	5	0	0
	UDPG-^{14}C+lipid extract	5	4 950	0
	UDPG-^{14}C+sitosterol	5	6 300	0
	UDPG-^{14}C+stigmasterol	5	4 800	0
	UDPG-^{14}C+cholesterol	5	3 950	0

(Adapted from Péaud-Lenoël and Axelos, 1971)

acetone powder preparation from wheat roots, they obtained [^{14}C]acyl-labeled steryl glucoside (Table XIII). They concluded, therefore, that phosphatidylethanolamine is the acyl donor involved in acylated steryl glucoside biosynthesis and that the biosynthesis of steryl glucoside and acylated steryl glucoside in wheat roots involves the following sequence of reactions:

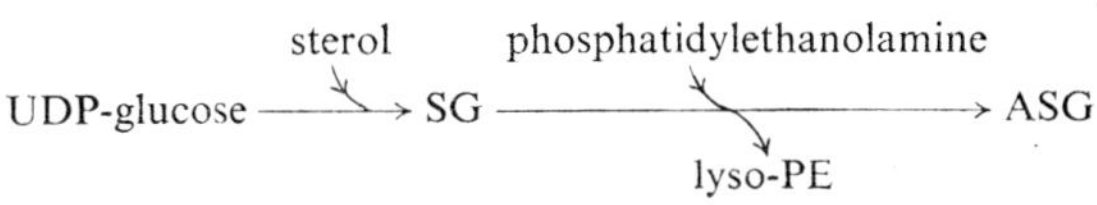

The optimal concentration of phosphatidylethanolamine for restoration of the transacylase activity was 13 mM (Axelos and Péaud-Lenoël, 1971). This is an unusually high concentration and it immediately suggests the possibility of substrate solubility problems.

Péaud-Lenoël and Axelos (1972) indicated via thin-layer chromatography and X-ray autoradiography that acylated steryl glucoside is not the only lipid that becomes labeled when [^{14}C]acyl-labeled phosphatidylethanolamine is added along with steryl glucoside to their reaction mixture. Two unknown neutral lipids are also labeled. One of these labeled lipids, therefore, may be the actual acyl donor rather than phosphatidylethanolamine. Before phosphatidylethanolamine can be positively identified as the acyl donor, these labeled compounds will have to be identified and tested as possible acyl donors. If phosphatidylethanolamine is the acyl donor, then lysophosphatidylethanolamine and acylated steryl glucoside would be the products. Although Péaud-Lenoël and Axelos (1972) did not establish the presence or the absence of labeled lysophosphatidylethanolamine in their reaction mixtures, its absence would not necessarily disprove phosphatidylethanolamine as the acyl donor.

The ability of free fatty acids and diglycerides to serve as acyl donors in acylated steryl glucoside biosynthesis should also be tested. A possibility Péaud-Lenoël and Axelos (1972) may have overlooked is the presence of phospholipase D and phosphatidic acid phosphatase in their microsomal enzyme preparation from wheat roots. Phospholipase D catalyzes the conversion of phosphatidylethanolamine to phosphatidic acid, and phosphatidic acid phosphatase catalyzes the conversion of phosphatidic acid to diglyceride. Diglyceride, a degradation product of phosphatidylethanolamine, may, therefore be the acyl donor involved in acylated steryl glucoside biosynthesis. Quarles and Dawson (1969) have shown that the 13 000 *g* supernatant from oat roots contains a substantial amount of phospholipase D activity and that phospholipase D is synthesized in plant cells undergoing active growth, such as in the rapidly growing tip of a root. Moore *et al.* (1973) found phosphatidic acid phosphatase predominantly, if not exclusively, in the microsomal fraction from castor bean endosperm.

At present, there is obviously a conflict as to the identity of the acyl donor involved in acylated steryl glucoside biosynthesis, it being either digalactosyl diglyceride in carrots (Eichenberger and Grob, 1970b) or phosphatidylethanolamine in wheat roots (Axelos and Péaud-Lenoël, 1971; Péaud-Lenoël and Axelos, 1972). These differences may be due to the different types of plants and/or the different subcellular fractions employed. In either case, the absolute identity of the acyl donor has not been satisfactorily resolved. Not only must the work by Axelos and Péaud-Lenoël (1971), Péaud-Lenoël and Axelos (1972), and Eichenberger and Grob (1970b) be verified, but other possible substrates as well as other mechanistic pathways must be tested before the acyl donor has been positively identified.

Several research groups working with either potato tubers (Lepage, 1964; Galliard, 1968a; Lepage, 1968), *Hosta undulata* (Eichenberger and Grob, 1972), *Hordeum vulgare* (Eichenberger and Grob, 1972), *Allium porrum* (Eichenberger and Grob, 1972), or soybeans (Kiribuchi *et al.*, 1966) report that palmitic acid is the predominant fatty acid esterified to acylated steryl glucoside in these plants. There are also reports indicating that linoleic acid is the predominantly acylated glucoside fatty acid in celery (Eichenberger and Grob, 1972), carrots (Eichenberg and Grob, 1972), apples (Galliard, 1968b), and 6-day-old tobacco seedlings (Bush and Grunwald, 1972) as well as in the coleoptile and coleorhiza of 5-day-old barley seedlings (Holmer *et al.*, 1973). Only via kinetic studies will it be possible to establish unequivocally a specific fatty acid substrate for a particular transacylase. A positive identification of the acyl donor involved in acylated steryl glucoside biosynthesis, however, is an absolute prerequisite before meaningful work on the fatty acid substrate specificities can begin.

E. ENZYME LOCALIZATION

Although the biosynthesis of steryl glucoside and acylated steryl glucoside has been shown to be associated with a high-speed particulate fraction (10 000–100 000 *g*) from plant cell-free homogenates (Hou *et al.*, 1968; Lavintman and Cardini, 1970; Forsee *et al.*, 1972; Wojciechowski, 1972; Bush and Grunwald, 1974), the subcellular organelle or organelles involved in their biosynthesis have not been identified with certainty. An extensive investigation is needed, therefore, to determine not only the subcellular localization of these enzymes but the overall ability of each organelle to synthesize these glycolipids. To date, steryl glucoside and acylated steryl glucoside biosynthesis has been associated with the chloroplast fraction (Eichenberger and Grob, 1969a; Ongun and Mudd, 1970), the mitochondrial fraction (Ongun and Mudd, 1970) and the plasmalemma plus endoplasmic reticulum fraction (Péaud-Lenoël and Axelos, 1971). None of these investigations, however, utilized highly purified subcellular organelles that had been checked by electron microscopy and/or marker enzymes. Moreover all of them depended upon endogenous free sterol for steryl glucoside biosynthesis and/or endogenous acyl donor for acylated steryl glucoside biosynthesis. Since the highest subcellular concentration of steryl glucoside and acylated steryl glucoside occurs in the microsomal fraction (Eichenberger and Grob, 1970a; Grunwald, 1970; Duperon *et al.*, 1972), it will be of interest to see if the endoplasmic reticulum is the site of their biosynthesis, as the high-speed particulate fractions suggest it might be.

F. ACTIVATORS AND INHIBITORS

Hou *et al.*, (1968) did not observe an inhibition of steryl glucoside biosynthesis when an excess of UDP-glucose was added to a particulate preparation from immature soybean seeds. They did observe an inhibition, however, when a small amount of UDP was added to their incubation mixtures; 50% inhibition was obtained with 0·083 mM UDP. Lavintman and Cardini (1970) inhibited the Mg^{2+} activation of glycolipid biosynthesis (steryl glucoside plus acylated steryl glucoside) in a particulate preparation from the starch grains from sweet corn to the extent of 76% with 2 mM UDP and 41% with 2 mM UTP. ADP, on the other hand, had no effect on the Mg^{2+} activation thereby, eliminating the possibility that the nucleotides were chelating the Mg^{2+}.

Kauss (1968) observed no divalent cation requirement for steryl glucoside biosynthesis in a particulate enzyme preparation from dark-grown mung bean shoots. Lavintman and Cardini (1970), however, were able to stimulate glycolipid biosynthesis (steryl glucoside plus acylated steryl glucoside) several-fold by adding either 10 mM Mg^{2+} or 10 mM Mn^{2+}, the latter being the more effective activator of the two. Axelos and Péaud-Lenoël (1971) found that 5 mM Mg^{2+} was a strong inhibitor of acylated steryl glucoside biosynthesis in an acetone powder preparation from wheat roots but had no effect on steryl glucoside biosynthesis. Using a solubilized and partially purified transglucosylase from developing cotton seeds, Forsee *et al.* (1972) obtained a slight stimulation in steryl glucoside biosynthesis with Ca^{2+} and Mg^{2+} and a slight inhibition with Zn^{2+} and Mn^{2+}. To date, only Smith (1971), who studied the biosynthesis of cholesteryl glucoside by *Mycoplasma gallinarum*, has definitely established a metal ion cofactor requirement for steryl glucoside biosynthesis. He was able to inhibit steryl glucoside biosynthesis by extensive dialysis of his enzyme preparation as well as by chelation of endogenous divalent cations with EDTA. The inhibition was alleviated in both cases by exogenous Mg^{2+}. High concentrations of Mn^{2+} and Ca^{2+}, however, inhibited cholesteryl glucoside biosynthesis in a particulate preparation from *Mycoplasma gallinarum* (Smith, 1971).

Hou *et al.* (1967, 1968) noted that ATP was required for optimal steryl glucoside biosynthesis in a particulate enzyme preparation from immature soybean seeds. Similarly, Eichenberger and Grob (1968, 1969a) were able to obtain an ATP stimulation of steryl glucoside and acylated steryl glucoside biosynthesis in a particulate fraction from lettuce leaves. Although Axelos and Péaud-Lenoël (1971) showed that it was possible to activate acylated steryl glucoside biosynthesis 20–30% with 1 mM ATP in an acetone powder preparation from wheat roots, CoA had no effect either in the presence or absence of ATP. The inability of CoA to further activate the ATP stimulation of acylated steryl glucoside biosynthesis negates, but does not eliminate, the possibility of fatty acyl CoA participation in acylated steryl glucoside biosynthesis. Wojciechowski (1972) observed that ATP stimulated the rate of steryl glucoside and acylated steryl glucoside biosynthesis by about 15% in a particulate fraction from marigold leaves. Bush and Grunwald (1974) noted that steryl glucoside biosynthesis required ATP for optimal activity in a particulate enzyme preparation from tobacco seeds. The only explanation for this ATP stimulation of steryl glucoside biosynthesis is given by Hou *et al.* (1968). They postulate that ATP competes with UDP-glucose in a phosphatase reaction and, thereby, prevents the phosphatase from hydrolyzing UDP-glucose. However, when Bush and Grunwald (1974) added UDP-glucose in the presence or absence of ATP, they obtained decreased steryl glucoside biosynthesis, suggesting that UDP-glucose protection from phosphatase may not be the only function of ATP.

Using UDP-[^{14}C]galactose as the radioactive precursor and spinach chloroplasts as the enzyme source, Mudd *et al.* (1971) showed that low concentrations of *N*-ethylmaleimide and $CdCl_2$ stimulated steryl glucoside and acylated steryl glucoside biosynthesis slightly, whereas high concentration of these sulfhydryl reagents inhibited their biosynthesis. Similar results were obtained when *p*-hydroxymercuribenzoate was used, whereas iodoacetate and iodoacetamide had no effect (Mudd *et al.*, 1971). This inhibition of steryl glucoside and acylated steryl glucoside biosynthesis by *N*-ethylmaleimide, $CdCl_2$, and *p*-hydroximercuribenzoate may not be due to sulfhydryl groups in the transglucosylase and transacylase but to sulfhydryl groups in the epimerase, which is required for converting UDP-galactose to UDP-glucose. UDP-glucose is the actual substrate for steryl glucoside biosynthesis not UDP-galactose. More work is required, therefore, before the transglucosylase and/or the transacylase can be identified as sulfhydryl requiring enzymes.

IV. Sulfonolipid

The sulfonolipid found in *Chlorella* by Benson *et al.* (1959) has since been found in a wide range of plants. Literature on the sulfonolipid, particularly on the chemical aspects, has been reviewed by Benson (1961) and this article will concentrate rather on the metabolism.

A. Analytical Studies

In photosynthetic tissues there is no doubt that the sulfonolipid is localized in the chloroplast (Davies *et al.*, 1965; Ongun *et al.*, 1968). Localization within the chloroplast appears to be in the lamellae. Analysis of the chloroplast membrane by Poincelot (1973) shows only traces of sulfonolipid, whereas the analysis of Douce *et al.*, (1973) indicates a lesser amount of sulfonolipid in the chloroplast membrane fraction rather than the lamellae (Table VII). Sulfonolipid is present in non-photosynthetic tissues such as potato tuber and apple fruit (Galliard 1968a, 1968b) but only in small amounts (*ca.* 1% of total lipid). In lipids of etiolated tissues, sulfonolipid is well represented. Tevini (1971a) studied the changes in glycerolipids after illumination of etiolated barley seedlings and found that on a fresh weight basis the sulfonolipid actually decreased after 48 h in the light. Thus there is no absolute requirement for light in the synthesis of sulfonolipid. A somewhat comparable study was made by Leech *et al.*, (1973) who analyzed the lipid composition of maize leaves of different ages. The youngest sections contained proplastids but no chloroplasts. Nevertheless, the youngest sections contained approximately half as much sulfonolipid as the fully developed sections which contained both mesophyll and bundle sheath chloroplasts.

The amount of sulfonolipid found in the leaves of higher plants is invariably less than that of the galactolipids. The ratio of galactosyl diglycerides to sulfonolipid varies somewhat from species to species: 8·8 in *Antirrhinum majus* leaves (Radunz, 1969), 4·3–6·2 in maize leaf sections (Leech *et al.*, 1973), 23 in *Impatiens balsamina* chloroplasts (Tevini, 1971b). In a survey of 20 plant species the ratio varied from 8·0 (*Medicago sativa*) to 45·5 (*Lactuca sativa*), (Roughan and Batt, 1969). Radunz (1969) has also measured this ratio in *Dryopteris filix-mas* (6·5), *Batrachospermum monoliforme* (1·6) and *Fucus vesiculosus* (0·8).

B. PHYSIOLOGICAL STUDIES

Shibuya and Mase (1965) have studied the formation and degradation of the sulfonolipid during bleaching and greening of cells of *Chlorella protothecoides*. Cells were grown in the presence of $^{35}SO_4^{2-}$ either under conditions which produced green cells or under conditions which produced white cells. Sulfoquinovosyl glycerol was equally well labeled in both types of cells, but the white cells had only 22% of the sulfonolipid labeled as compared with green cells and only 9% of the glutathione labeled. During the greening of white cells in the presence of $^{35}SO_4^{2-}$ the sulfonolipid increased 8-fold while the sulfoquinovosyl glycerol was unchanged. In the companion experiment, the bleaching of green cells, the labeling of the sulfonolipid declined while that of sulfoquinovosyl glycerol rose. The changes in the ^{35}S-labeled compounds during bleaching in the absence of $^{35}SO_4^{2-}$ revealed that the increase in sulfoquinovosyl glycerol was greater than the decrease in the sulfonolipid. Shibuya and Mase (1965) concluded from these experiments that the sulfonolipid is rapidly turned over and that during the bleaching process degradation predominates over synthesis. Sulfoquinovosyl glycerol was therefore viewed as a product of sulfonolipid degradation rather than a biosynthetic precursor. This conclusion was endorsed by Miyachi and Miyachi (1966) who found that the transfer of ^{35}S-labeled cells of *Chlorella ellipsoidea* to a system devoid of CO_2 caused a decrease in the sulfonolipid and an increase in sulfoquinovosyl glycerol.

The position of sulfoquinovosyl glycerol in the degradative pathway is also evident in the results of Yagi and Benson (1962). Extracts of *Scenedesmus obliquus* were capable of converting the sulfonolipid to lyso-sulfonolipid and to sulfoquinovosyl glycerol. Further metabolism of sulfoquinovosyl glycerol is different in different plants. Lee and Benson (1972) found that the product in leaves of *Medicago sativa* produced first sulfoquinovose and later sulfolactaldehyde and sulfolactic acid, with the latter accumulating. These results were presented in support of a degradation scheme analogous to the glycolytic sequence for degradation of glucose.

C. BIOSYNTHESIS

There are more suggestions about the biosynthesis of the sulfonolipid than there are data.

1. *Aldolase Pathways for Sulfoquinovose Synthesis*

Davies *et al.*, (1966) incubated various ^{35}S-labeled precursors with *Euglena gracilis*. Cysteic acid was an excellent precursor of the sulfonolipid. When $^{35}SO_4^{2-}$ was used as precursor of the sulfonolipid, cysteic acid did not affect

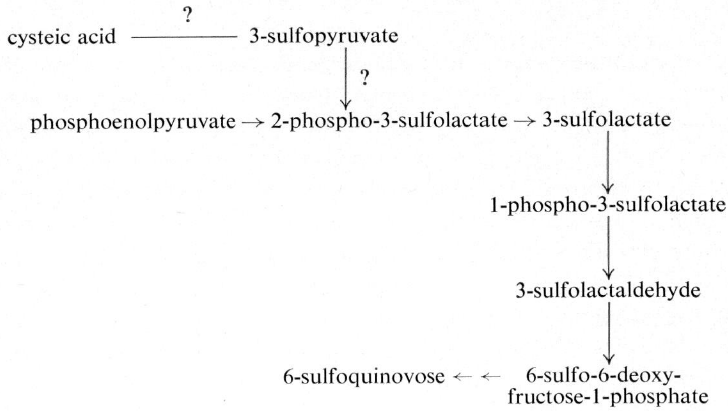

FIG. 6. Scheme for sulfoquinovose synthesis. (Adapted from Davies *et al.*, 1966.)

sulfate uptake, but diluted incorporation into the sulfonolipid. In contrast, cysteine lowered sulfate uptake but did not change the relative conversion to the sulfonolipid. Sulfate was only 10% as effective as cysteic acid in producing the sulfonolipid, and its utilization was inhibited by molybdate. Evidence was presented which indicated that sulfate incorporation into the sulfonolipid was by way of adenosine 3′-phosphate 5′-phosphosulfate. Davies *et al.* (1966) incorporated these results into a hypothetical pathway for sulfoquinovose synthesis (Fig. 6).

This scheme is similar to that suggested by Benson (1963) for the conversion of sulfopyruvate to sulfoquinovose.

2. *Displacement of 0-acyl Groups*

Hodson *et al.* (1971) identified homocysteic acid as a product of sulfate metabolism in a mutant of *Chlorella pyrenoidosa* and suggested it may originate by displacement of an *0*-acetyl or *0*-succinyl group from homoserine. An analogous reaction was suggested for the formation of sulfoquinovose. The donor of the sulfonyl group was not specified. This proposal is related to the suggestion of Zill and Cheniae (1962) that adenosine 3′-phosphate 5′-phosphosulfate is the donor of the sulfonyl group to a lipid or nucleotide precursor.

3. *Addition of Sulfite to Double Bonds*

Lehmann and Benson (1964) have observed that sulfite adds to methyl-α-D-glucoseenide to form 6-sulfoquinovoside. It is possible that such a reaction may give rise to the sulfonyl moiety in the biosynthetic pathway either in sulfoquinovose or in one of the precursors.

4. *Participation of Nucleotide Sugar*

By analogy with the formation of galactosyl diglycerides, one might expect sulfoquinovosyl diglyceride to be formed from 1,2-di-*0*-acyl-3-*0*-*sn*-glycerol and nucleoside diphosphosulfoquinovose. Shibuya *et al.* (1963) have reported the latter compound in extracts of *Chlorella*, but Benson (1963) mentions that it has not been possible to obtain reproducible amounts in other preparations. If such a compound is the precursor of the sulfonolipid, there are further complications. In Fig. 6 the formation of 6-sulfoquinovose is analogous to the formation of glucose 6-phosphate, but conversion to an analog of glucose 1-phosphate is then impossible. It would be necessary to invoke an enzyme which would phosphorylate 6-sulfoquinovose and an enzyme to catalyze the reaction of 6-sulfoquinovose-1-phosphate with a nucleoside triphosphate:

6-sulfoquinovose + ATP → 6-sulfoquinovose-1-phosphate + ADP

6-sulfoquinovose-1-phosphate + NTP → NDP-6-sulfoquinovose + PP

NDP-6-sulfoquinovose + diglyceride → sulfonolipid + NDP.

None of these reactions has been demonstrated.

At the present, the most sophisticated precursor used in the biosynthesis of the sulfonolipid is cysteic acid. Further elucidation of the pathway of biosynthesis should include a study of the early products of cysteic acid metabolism and the utilization of potential precursors such as sulfoquinovose itself. The chemical synthesis of suitable precursors appears to be the greatest barrier at present.

5. *Acyl Exchange*

Wolfberger and Pieringer (1974) have studied the acylation of sulfoquinovosyl monoglyceride, using acyl CoA as the donor. The sulfoquinovosyl monoglyceride was prepared by the action of pancreatic lipase on sulfoquinovosyl diglyceride and the product was presumed to be the 2-acyl isomer. The specificity with respect to acyl donor was not listed, although both palmityl CoA and linoleyl CoA were satisfactory substrates. The studies of Tulloch *et al.* (1974) on the positional distribution of fatty acids in sulfoquinovosyl diglyceride have revealed that the 2 position is preferentially acylated with a saturated fatty acid. This finding raises interesting problems concerning biosynthesis and acyl exchange. The lipase from *Rhizopus delemar* which specifically cleaves fatty acids from position 1 will be valuable in reacylation studies (Tulloch *et al.*, 1974). The reacylation studies of Wolfsberger and Pieringer (1974) demonstrated that sulfoquinovosyl glycerol was not a substrate in the acylation reaction, which supports previous conclusions that sulfoquinovosyl glycerol is not a precursor of the sulfonolipid.

V. Sphingolipids

There are two types of sphingolipids in plants; cerebrosides and phytoglycolipid (Fig. 1). Both contain carbohydrate in addition to the long chain bases and therefore may be called glycolipids as well as sphingolipids.

A. CEREBROSIDES

Cerebrosides were first reported in higher plants by Sastry and Kates (1964). The cerebroside fraction was isolated from runner bean leaves. Characterization of the fraction showed that glucose was the only sugar present, that the fatty acid moieties were all α-hydroxy acids and that the main long chain base was C_{18}-dehydrosphingosine.

Since the report of Sastry and Kates (1964), cerebrosides have often been identified on thin layer chromatograms of plant lipid extracts and occasionally have been analyzed quantitatively. For example, Galliard (1968a, b) found that 2·4% of the lipid extracted from potato tubers and 5·2% of the lipid extracted from the pulp of post-climacteric apples was glucocerebroside. Galliard confirmed the observations of Sastry and Kates that the fatty acid substituents were mainly α-hydroxy acids.

In the plant kingdom, most of the work done on cerebrosides has been with fungi (Wagner and Zofcsik, 1966; Wagner and Fiegert, 1969). Weiss and Stiller (1972) have found both ceramides and cerebrosides in a number of fungi where they amounted to as much as 5% of the dry weight.

The biosynthesis of cerebrosides in higher plants has not been studied. Cerebrosides have not shown significant labeling in experiments using radioactive precursors which work well for the labeling of other compounds, e.g., acetate, carbon dioxide, UDP-glucose, serine. It seems likely that a systematic study of cerebroside synthesis in higher plants should yield results particularly in view of the success of studies on the biosynthesis of the long chain bases in *Hansenula cifferi* (Di Mari *et al.*, 1971). The *N*-acylation and glycosylation reactions are also appropriate areas for study.

B. PHYTOGLYCOLIPIDS

Carter and his collaborators have studied a complex glycolipid found in several plant seeds (Carter *et al.*, 1962, 1964). They called this lipid phytoglycolipid and eventually published the structure of the phytoglycolipid tetrasaccharide (Carter *et al.*, 1969) shown in Fig. 1. Wagner *et al.* (1969) also determined the structure of the phytoglycolipid of peanut, *Arachis hypogea* L., and this structure differed in some of the sugar linkages proposed by Carter *et al.* (1969) for the phytoglycolipid from maize seeds. It seems likely that a variety of oligosaccharides will be found in these glycosphingolipids.

Wagner and Zofcsik (1966) have reported glycolipids in *Candida utilis* and *Saccharomyces cerevisiae* which included compounds of the phytoglycolipid type. They contained C_{18} and C_{20} phytosphingosines, α-hydroxy fatty acids, phosphate, inositol and mannose. These compounds therefore differ from the phytoglycolipids of higher plants in having a somewhat less complicated oligosaccharide moiety. The studies of Wagner *et al.* (1969) have also revealed a phytoglycolipid in *Scenedesmus obliquus.* This compound contained the sphingosine bases, fatty acids, phosphate and carbohydrates. In contrast to the phytoglycolipids from the seeds of higher plants, this glycolipid did not contain inositol or glucuronic acid.

The phytoglycolipids have not received much attention in the quantitative analysis of plant lipids. In developing maize seeds (one of the original sources of phytoglycolipid) analyses do not include phytoglycolipid at all (Weber, 1970), possibly because they may have co-chromatographed with other compounds. As far as the biosynthesis of phytoglycolipids is concerned, there is no work whatsoever. Experiments with developing seeds using potential inositol donors may provide a start to the study of this problem.

REFERENCES

Allen, C. F., Good, P., Davis, H. E., Chisum, P. and Fowler, S. D. (1966). *J. Am. Oil Chem. Soc.* **43**, 223–231.

Arunga, R. O. and Morrison, W. R. (1971). *Lipids* **6**, 768–776.

Auling, G., Heinz, E. and Tulloch, A. P. (1971). *Z. Physiol. Chem.* **352**, 905–912.

Axelos, M. and Péaud-Lenoël, C. (1971). *C. r. hebd. Séanc. Acad. Sci. Paris Ser. D* **273**, 1434–1437.

Bajwa, S. S. and Sastry, P. S. (1973). *Indian J. Biochem. Biophys.* **10**, 65–66.

Benson, A. A. (1961). *Adv. Lipid Res.* **1**, 387–394.

Benson, A. A., Daniel, H. and Wiser, R. (1959). *Proc. natn. Acad. Sci. U.S.A.* **45**, 1582–1587.

Benson, A. A., Wintermans, J. F. G. M. and Wiser, R. (1959). *Pl. Physiol., Lancaster* **34**, 315–317.

Benson, A. A., Wiser, R., Ferrari, R. A. and Miller, J. A. (1958). *J. Am. chem. Soc.* **80**, 4740.

Bishop, D. G., Anderson, K. S. and Smillie, R. M. (1971). *Biochim. biophys. Acta* **231**, 412–414.

Bishop, D. G. and Smillie, R. M. (1970). *Archs Biochem. Biophys.* **137**, 179–189.

Bowden, B. N. and Williams, P. M. (1972). *Phytochemistry* **11**, 1059–1064.

Brandt, R. D. and Benveniste, P. (1972). *Biochim. biophys. Acta* **282**, 85–92.

Brush, P. and Percival, E. (1972). *Phytochemistry* **11**, 1847–1849.

Bush, P. B. and Grunwald, C. (1972). *Pl. Physiol., Lancaster* **50**, 69–72.

Bush, P. B. and Grunwald, C. (1974). *Pl. Physiol., Lancaster* **53**, 131–135.

Bush, P. B., Grunwald, C. and Davis, D. L. (1971). *Pl. Physiol., Lancaster* **47**, 745–749.

Carter, H. E., Betts, B. E. and Strobach, D. R. (1964). *Biochemistry* **3**, 1103–1107.

Carter, H. E., Galanos, D. S., Hendrickson, H. S., Jann, B., Nakayama, T., Nakazawa, Y. and Nichols, B. W. (1962). *J. Am. Oil Chem. Soc.* **39**, 107–115.

Carter, H. E., Hendry, R. A. and Stanacev, N. Z. (1961). *J. Lipid Res.* **2**, 223–227.
Carter, H. E. and Koob, J. L. (1969). *J. Lipid Res.* **10**, 363–369.
Carter, H. E., McCluer, R. H. and Slifer, E. D. (1956). *J. Am. chem. Soc.* **78**, 3735–3738.
Carter, H. E., Ohno, K., Nojima, S., Tipton, C. L. and Stanacev, N. Z. (1961). *J. Lipid Res.* **2**, 215–222.
Carter, H. E., Strobach, D. R. and Hawthorne, J. N. (1969). *Biochemistry* **8**, 383–388.
Chang, S. B. and Kulkani, N. D. (1970). *Phytochemistry* **9**, 927–934.
Constantopoulos, G. and Bloch, K. (1967). *J. Bact.* **93**, 1788–1793.
Costes, C., Bazier, R., Brangeon, J. and Bourdu, R. (1971). *C. r. hebd. Séanc. Acad. Sci. Paris Ser. C* **272**, 1597–1600.
Davies, W. H., Mercer, E. I. and Goodwin, T. W. (1965). *Phytochemistry* **4**, 741–749.
Davies, W. H., Mercer, E. I. and Goodwin, T. W. (1966). *Biochem. J.* **98**, 369–373.
DiMari, S. J., Brady, R. N. and Snell, E. E. (1971). *Archs Biochem. Biophys.* **143**, 553–565.
Douce, R. (1974). *Science, N.Y.* **183**, 852–853.
Douce, R. and Guillot-Salomon, T. (1970). *FEBS Letters* **11**, 121–126.
Douce, R., Holz, R. B. and Benson, A. A. (1973). *J. biol. Chem.* **248**, 7215–7222.
Duperon, R., Brillard, M. and Duperon, P. (1972). *C. r. hebd. Séanc. Acad. Sci., Paris Ser. D* **274**, 2321–2324.
Eccleshall, T. R. and Hawke, J. C. (1971). *Phytochemistry* **10**, 3035–3045.
Eichenberger, W. and Grob, E. C. (1968). *Chimia* **22**, 46–48.
Eichenberger, W. and Grob, E. C. (1969a). *In* "Progress in Photosynthesis Research" (H. Metzner, ed.), Vol. I, pp. 338–344. Verlag C. Lichtenstern, Munich.
Eichenberger, W. and Grob, E. C. (1969b). *Chimia* **23**, 368–369.
Eichenberger, W. and Grob, E. C. (1970a). *FEBS Letters* **11**, 177–180.
Eichenberger, W. and Grob, E. C. (1970b). *Chimia* **24**, 394–396.
Eichenberger, W. and Grob, E. C. (1972). *FEBS Letters* **20**, 245–248.
Eichenberger, W. and Menke, W. (1966). *Z. Naturforsch.* **21b**, 859–867.
Eichenberger, W. and Newman, D. W. (1968). *Biochem. biophys. Res. Commun.* **32**, 366–374.
Ferrari, R. A. and Benson, A. A. (1961). *Archs Biochem. Biophys.* **93**, 185–192.
Forsee, W. T., Laine, R. A., Chambers, J. P. and Elbein, A. D. (1972). *In* "Methods in Enzymology" (V. Ginsburg, ed.), Vol. XXVIII, Part B., pp. 478–482. Academic Press, New York and London.
Galliard, T. (1968a). *Phytochemistry* **7**, 1907–1914.
Galliard, T. (1968b). *Phytochemistry* **7**, 1915–1922.
Galliard, T. (1969). *Biochem. J.* **115**, 335–339.
Grunwald, C. (1970). *Pl. Physiol., Lancaster* **45**, 663–666.
Heinz, E. (1973). *Z. Pflanzenphysiol.* **69**, 359–376.
Hodson, R. C., Schiff, J. A. and Mather, J. P. (1971). *Pl. Physiol., Lancaster* **47**, 306–311.
Holmer, G., Ory, R. L. and Hoy, C. (1973). *Lipids* **8**, 277–283.
Hou, C. T., Umemura, Y., Nakamura, M. and Funahashi, S. (1967). *J. Biochem. (Tokyo)* **62**, 389–391.
Hou, C. T., Umemura, Y., Nakamura, M. and Funahashi, S. (1968). *J. Biochem. (Tokyo)* **63**, 351–360.
van Hummel, H. C. (1974). Doctorate Thesis. University of Nijmegen, The Netherlands.
Kates, M. (1970). *In* "Advances in Lipid Research" (R. Paoletti and D. Kritchevsky, eds), Vol. 8, pp. 225–265. Academic Press, New York and London.

Kauss, H. (1968). *Z. Naturforsch.* **23b**, 1522–1526.
Kemp, R. J. and Mercer, E. I. (1968). *Biochem. J.* **110**, 119–125.
Kiribuchi, T., Mizunaga, T. and Funahashi, S. (1966). *Agric. biol. Chem.* **30**, 770–778.
Kiribuchi, T., Yasumatsu, N. and Funahashi, S. (1967). *Agric. biol. Chem.* **31**, 1244–1247.
Koenig, F. (1971). *Z. Naturforsch.* **26b**, 1180–1187.
Lavintman, N. and Cardini, C. E. (1970). *Biochim. biophys. Acta* **201**, 508–510.
Lee, R. F. and Benson, A. A. (1972). *Biochim. biophys. Acta* **261**, 35–37.
Leech, R. M., Rumsby, M. G. and Thomson, W. W. (1973). *Pl. Physiol., Lancaster* **52**, 240–245.
Lehmann, J. and Benson, A. A. (1964). *J. Am. chem. Soc.* **86**, 4469–4472.
Lennarz, W. J. and Talamo, B. (1966). *J. biol. Chem.* **241**, 2707–2719.
Lepage, M. (1964). *J. Lipid Res.* **5**, 587–592.
Lepage, M. (1967). *Lipids* **2**, 244–250.
Lepage, M. (1968). *Lipids* **3**, 477–481.
Mackender, L. R. O. and Leech, R. M. (1974). *Pl. Physiol., Lancaster* **53**, 496–502.
McCarty, R. E., Douce, R. and Benson, A. A. (1973). *Biochim. biophys. Acta* **316**, 266–270.
McCarty, R. E. and Jagendorf, A. T. (1965). *Pl. Physiol., Lancaster* **40**, 725–735.
Miyachi, S. and Miyachi, S. (1966). *Pl. Physiol., Lancaster* **41**, 479–486.
Moore, T. S., Lord, J. M., Kagawa, T. and Beevers, H. (1973). *Pl. Physiol., Lancaster* **52**, 50–53.
Mudd, J. B., McManus, T. T., Ongun, A. and McCullough, T. (1971). *Pl. Physiol., Lancaster* **48**, 335–339.
Mudd, J. B., van Vliet, H. H. D. M. and van Deenen, L. L. M. (1969). *J. Lipid Res.* **10**, 623–630.
Neufeld, E. F. and Hall, C. W. (1964). *Biochem. biophys. Res. Commun.* **14**, 503–508.
Nichols, B. W. (1965). *Biochim. biophys. Acta* **106**, 274–279.
Nichols, B. W. and Moorhouse, R. (1969). *Lipids* **4**, 311–316.
Nichols, B. W. and Wood, B. J. B. (1968). *Nature, Lond.* **217**, 767–768.
Noda, M. and Fujiwara, N. (1967). *Biochim. biophys. Acta* **137**, 199–201.
Ongun, A. and Mudd, J. B. (1968). *J. biol. Chem.* **243**, 1558–1566.
Ongun, A. and Mudd, J. B. (1970). *Pl. Physiol., Lancaster* **45**, 255–262.
Ongun, A., Thomson, W. W. and Mudd, J. B. (1968). *J. Lipid Res.* **9**, 409–415.
Péaud-Lenoël, C. and Axelos, M. (1971). *C. r. hebd. Séanc. Acad. Sci., Paris Ser. D* **273**, 1057–1060.
Péaud-Lenoël, C. and Axelos, M. (1972). *Carbohydrate Res.* **24**, 247–262.
Pieringer, R. A. (1968). *J. biol. Chem.* **243**, 4896–4903.
Poincelot, R. P. (1973). *Archs Biochem. Biophys.* **159**, 134–142.
Power, F. B. and Salway, A. H. (1913). *J. chem. Soc.* **103**, 399–406.
Quarles, R. H. and Dawson, R. M. C. (1969). *Biochem. J.* **112**, 787–794.
Radunz, A. (1968). *Z. Physiol. Chem.* **349**, 1091–1094.
Radunz, A. (1969). *Z. Physiol. Chem.* **350**, 411–417.
Renkonen, O. and Bloch, K. (1969). *J. biol. Chem.* **244**, 4899–4903.
Roughan, P. G. (1970). *Biochem. J.* **117**, 1–8.
Roughan, P. G. and Batt, R. D. (1969). *Phytochemistry* **8**, 363–369.
Safford, R., Appleby, R. S. and Nichols, B. W. (1971). *Biochim. biophys. Acta* **239**, 509–512.
Safford, R. and Nichols, B. W. (1970). *Biochim. biophys. Acta* **210**, 57–64.
Sastry, P. S. and Kates, M. (1964). *Biochemistry* **3**, 1271–1280.
Schwertner, H. A. and Biale, J. B. (1973). *J. Lipid Res.* **14**, 235–242.

Shibaya, I. and Hase, E. (1965). *Pl. Cell Physiol.* **6**, 267–284.
Shibaya, I., Yagi, I. and Benson, A. A. (1963). *In* "Studies on Microalgae and Photosynthetic Bacteria", p. 627. University of Tokyo Press.
Smith, P. F. (1969). *J. Bact.* **99**, 480–486.
Smith, P. F. (1971). *J. Bact.* **108**, 986–991.
Stern, N. and Tietz, A. (1973). *Biochim. biophys. Acta* **296**, 136–144.
Sumida, S. (1967). M.Sc. Thesis. University of California, Riverside.
Sumida, S. and Ueda, M. (1973). *Pl. Cell. Physiol.* **14**, 781–785.
Tevini, M. (1971a). *Z. Pflanzenphysiol.* **65**, 266–272.
Tevini, M. (1971b). *Z. Pflanzenphysiol.* **66**, 64–72.
Tulloch, A. P., Heinz, E. and Fischer, W. (1973). *Z. Physiol. Chem.* **354**, 879–889.
Vichanska, J., Tremolieres, A. and Mazliak, P. (1973). *C. r. hebd. Séanc. Acad. Sci., Paris Ser. D* **277**, 501–504.
Wagner, H. and Fiegert, E. (1969). *Z. Naturforsch.* **24b**, 359.
Wagner, H., Pohl, P. and Munzing, A. (1969). *Z. Naturforsch.* **24b**, 360.
Wagner, H. and Zofcsik, W. (1966a). *Biochem. Z.* **346**, 333–342.
Wagner, H. and Zofcsik, W. (1966b). *Biochem. Z.* **346**, 343–350.
Wagner, H., Zofcsik, W. and Heng, I. (1969). *Z. Naturforsch.* **24b**, 922–927.
Walker, R. W. and Bastl, C. P. (1967). *Carbohydrate Res.* **4**, 49–54.
Weber, E. (1970). *J. Am. Oil Chem. Soc.* **47**, 340–343.
Weiss, B. and Stiller, R. L. (1972). *Biochemistry* **11**, 4552–4557.
Wenger, D. A., Petitpas, J. W. and Pieringer, R. A. (1968). *Biochemistry* **7**, 3700–3707.
Wintermans, J. F. G. M. (1971). *Biochim. biophys. Acta* **248**, 530–535.
Wintermans, J. F. G. M., Helmsing, P. J., Polman, B. J. J., van Gisberger, J. and Collard, J. (1969). *Biochim. biophys. Acta* **189**, 99–105.
Wojciechowski, Z. (1972). *Acta biochim. polon.* **19**, 43–49.
Wolfsberger, M. G. and Pieringer, R. A. (1974). *J. Lipid Res.* **15**, 1–10.
Yagi, T. and Benson, A. A. (1972). *Biochim. biophys. Acta* **57**, 601–603.
Zill, L. P. and Cheniae, G. M. (1962). *A. Rev. Pl. Physiol.* **13**, 225–264.

CHAPTER 7

Biochemistry of Cutin, Suberin and Waxes, the Lipid Barriers on Plants

P. E. KOLATTUKUDY

Department of Agricultural Chemistry and the Program in Biochemistry and Biophysics, Washington State University, Pullman Washington, U.S.A.

I. Introduction

The outer layer of living organisms consists of a polymeric structural component which, in most cases, is water-proofed with a complex mixture of nonpolar lipids. While chitin and proteins constitute the main polymeric materials of the outer layers of animals, the structural component of plant cuticle is a polymer of hydroxy-fatty acids. Thus, virtually the entire plant cuticle is made of lipids. In recent years considerable progress has been made in our understanding of the chemical nature and biosynthesis of the plant cuticle. Since several review articles on this subject have appeared in recent years (Mazliak, 1968; Martin and Juniper, 1970; Kolattukudy and Walton, 1972b), this discussion will concentrate on the more recent developments with no attempt to be comprehensive.

This chapter is dedicated to the memory of Sir Robert Robinson whose collaboration was invaluable in some of the work discussed in this chapter.

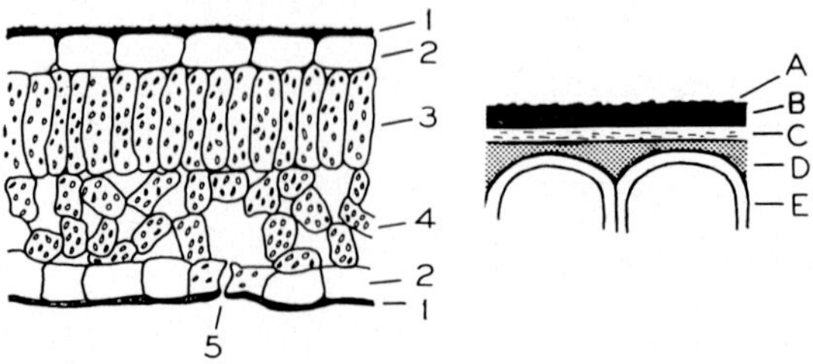

FIG. 1. Left: Cross-section of a leaf. 1, cuticle with surface wax; 2, epidermal cells; 3, palisade cells; 4, spongy tissue; and 5, stomata. Right: Schematic representation of the structure of the leaf cuticle. A, surface wax; B, cutin embedded in wax; C, a mixed layer containing some cutin, wax and carbohydrate polymers with possibly traces of protein; D, pectin; and E, cellulose wall of the epidermal cells.

II. SOLUBLE CUTICULAR LIPIDS

A. ISOLATION

The cuticular structure is fastened to the epidermal layer of cells with a pectinaceous layer (Fig. 1). The cuticle itself consists of a hydroxy-fatty acid polymer, cutin, which is embedded in a complex mixture of very long chain nonpolar lipids collectively called wax. Often wax protrudes through the cuticular structure and gives rise to a crystalline material on the surface (Fig. 2). The soluble cuticular lipids can be easily isolated by dipping the plant tissue in an organic solvent such as chloroform for a brief period (10–30 s). In the case of plant tissues with rather thick cuticular structures, such as apple fruits, more drastic methods must be used to achieve complete extraction of the cuticular lipids buried deep within the cuticle.

B. STRUCTURE

Hydrocarbons constitute one of the most common components of cuticular lipids. Usually the hydrocarbon fraction consists of a mixture of *n*-alkanes with 25–35 carbon atoms with an overwhelming predominance of odd-numbered chains. Often one or a few hydrocarbons predominate and *n*-C_{29} or *n*-C_{31} is usually the dominant alkane in higher plants. For example, *n*-C_{29} alkane constitutes more than 90% of the alkanes of *Brassica oleracea* (Purdy and Truter, 1963; Kolattukudy, 1965), while an equally high proportion of *n*-C_{31} alkane is found in *Pisum sativum* (Macey and Barber, 1970a; Kolattukudy, 1970f) and *Senecio odoris* (Kolattukudy, 1968a). However, there are many exceptions to the above generalization regarding structure and chain-length distribution of hydrocarbons in plants. Substantial amounts of *iso*-(2-methyl) and *anteiso*-(3-methyl) branched alkanes are found in many plants such as tobacco (Carruthers and Johnstone, 1959; Mold *et al.*, 1966; Kaneda, 1967). Usually *iso* alkanes have an odd number of carbon atoms while *anteiso* alkanes have an even number of carbon atoms. In some instances,

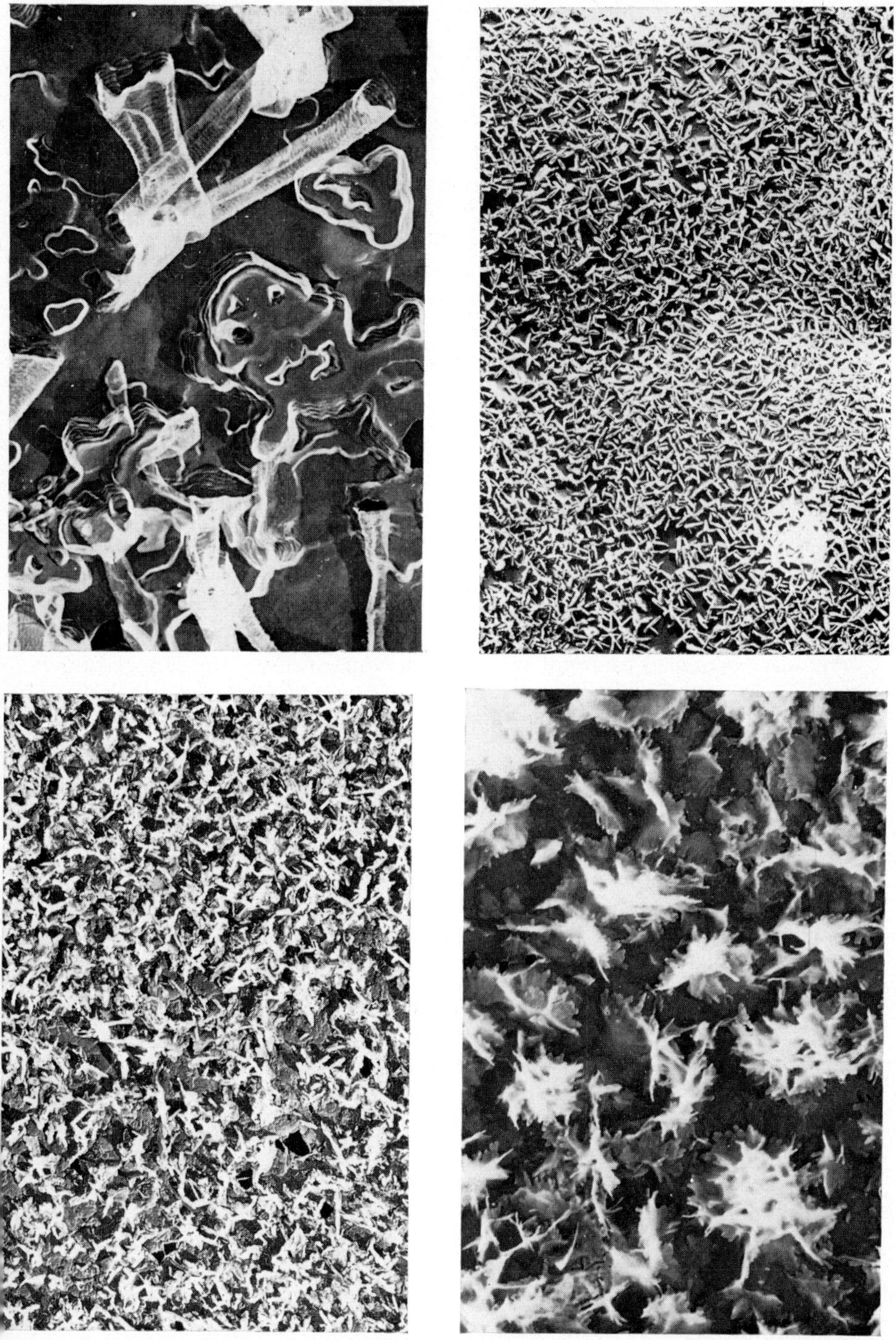

FIG. 2. Electron micrographs of surface lipids of (top left) Cabbage leaf (adaxial surface of the leaf of January King $\times$16 500); (top right) Pea leaves (adaxial surface of the leaf of *Pisum sativum*, var. Alaska $\times$4000); (bottom left) Corn (adaxial surface of the leaf of *Zea maize* $\times$5000); and (bottom right) *Eucalyptus* (adaxial surface of the leaf of *E. cloeziana* $\times$5000). These electron micrographs were prepared by Professor B. E. Juniper of the Oxford University Botany School.

hydrocarbons with internal and/or multiple branches, unsaturation, and cyclic structures (Jarolimek *et al.*, 1964; Stransky *et al.*, 1970; Wollrab *et al.*, 1965, 1967; Mold *et al.*, 1966; Kuksis, 1964) have been found as minor components. In lower plants such as algae n-C_{17} alkane predominates, and internally branched alkanes and unsaturated hydrocarbons are not rare (Youngblood *et al.*, 1971). For example, mono-, di- and tri-unsaturated C_{21} to C_{31} hydrocarbons predominate in *Botryococcus brauni* and *Anacystis montana* (Brown *et al.*, 1969; Gelpi *et al.*, 1968; Knights *et al.*, 1970) and 7- and 8-methylheptadecane occur widely in blue-green algae (Han *et al.*, 1968).

Oxygenated derivatives of alkanes, such as ketones and secondary alcohols with the functional group near the middle of the molecule, occur in the cuticular wax of many plants. The chain-length distribution of these oxygenated compounds usually corresponds to that of the alkanes of the same plant (Channon and Chibnall, 1929; Kolattukudy and Walton, 1972b; Dodova-Anghelova and Ivanov, 1973). For example, nonacosan-15-one and hentriacontan-16-one are found in plants where the corresponding alkanes predominate. In tissues such as rose petals, where a more complex mixture of alkanes are found, mixtures of ketones and secondary alcohols also occur (Wollrab, 1969). In some species such as *Eucalyptus*, β-diketones and hydroxy-β-diketones occur as substantial, if not major, components of the cuticular lipids (Horn *et al.*, 1964; Tulloch and Weenik, 1966). Unlike the monoketones, the β-diketones usually do not correspond in chain length to the major alkanes of the same tissue.

Wax esters are one of the most common constituents of the soluble lipids of the plant cuticle. They usually consist of n-alkanoic acids and n-alkan-1-ols of an even number of carbon atoms in the range of C_{12} to C_{32}. In the acyl portion, carbon chains of C_{20} to C_{24} tend to predominate, while C_{24} to C_{28} are often the major components of the alcohol portion. In most plants free fatty alcohols and free fatty acids occur as significant components of cuticular lipids. The chain-length distribution of the free alcohol is generally similar to that of the esterified alcohols, while free acids tend to have substantially longer chain lengths than those found in wax esters. In recent years fatty aldehydes, with chain lengths similar to those of the alcohols, have been found in several plants (Kranz *et al.*, 1960; Lamberton, 1965; Radler, 1965; Schmid and Bandi, 1969; Kolattukudy, 1970f). Among the less common components of cuticular lipids are alkane diols, ketols, terpenes, flavones, and aromatic hydrocarbons (Kolattukudy and Walton, 1972b; Martin and Juniper, 1970).

C. BIOSYNTHESIS

1. *Biosynthesis of Very Long-chain Acids*

The major distinction between the internal lipids and the cuticular lipids is that the former consist of C_{14} to C_{20} carbon chains while the latter consist mainly of carbon chains longer than C_{20}. This discussion is limited to the synthesis of these very long-chain components. In many young plant tissues, a variety of exogenous labeled fatty acids (C_2 to C_{18}) are incorporated into very long chain acids, most probably by chain elongation. It was found that particulate systems from germinating pea cotyledons (Macey and Stumpf, 1968) and pea leaf epidermis (Kolattukudy and Buckner, 1972) elongated endogenous acyl moieties to C_{20} to C_{28} acids with malonyl-CoA as the elongating agent and NADPH as the preferred reductant. However, no exogenous acyl derivatives were elongated and, therefore, the nature of the reaction cannot be fully defined at this time. Furthermore, it is not known whether all the elongated chains of the cuticular lipids are synthesized by a single elongating system. The following observations raise the possibility that the various classes of elongated products are made by different elongating systems. Whereas certain thiocarbamates (at μM concentrations) severely inhibited incorporation of [1-^{14}C]acetate into alkanes in pea leaf slices, labeling of C_{26} and C_{28} alcohols was not affected and accumulation of label occurred specifically in C_{22} acid (Kolattukudy and Brown, 1974). Similarly, trichloroacetate, at concentrations which severely inhibited the synthesis of C_{29} or C_{31} alkanes and their oxygenated derivatives, did not affect incorporation of labeled acetate into long-chain alcohols in pea leaves (Liu, 1972) and broccoli leaves (Kolattukudy, 1965). Furthermore, the inhibition of incorporation of [1-^{14}C]acetate into C_{20} to C_{28} acids obtained with higher concentrations of trichloroacetate could be fully reversed by the addition of pantothenic acid, whereas inhibition of alkane synthesis could not be reversed to any substantial degree (Liu, 1972). The proportion of ^{14}C found in the carboxyl carbon of C_{18} to C_{30} fatty acids derived from [1-^{14}C]acetate in *Allium porrum* (Lessire, 1973) also suggested that all of these acids were not synthesized by the same elongating system. However, a final answer must await isolation and characterization of the various elongating enzyme systems.

2. *Reduction of Fatty Acids*

Fatty acids were assumed to be the precursors of fatty alcohols ever since naturally occurring fatty alcohols were identified (Chibnall and Piper, 1934). Fatty alcohols are precursors of ether lipids (Snyder, 1969) as well as wax esters (Kolattukudy, 1967a). *In vivo* studies have indicated that fatty acids are converted into the corresponding alcohols in many living organisms. Enzymatic conversion of fatty acids to alcohols was demonstrated for the first time with a protein fraction isolated from acetone powder extracts of young leaves of *B. oleracea* (Kolattukudy, 1969). Since thioesters of fatty acids are rapidly hydrolyzed by crude leaf extracts and acetone powder extracts, protein fractionation was required before acyl-CoA reduction could be demonstrated. Ammonium sulfate fractionation of an acetone powder extract, followed by gel filtration on Sephadex G-100, provided a protein fraction which catalyzed reduction of palmitoyl-CoA to palmitaldehyde with

NADH as the preferred reductant. Another protein fraction obtained from the same gel filtration step catalyzed reduction of palmitaldehyde to the corresponding alcohol with NADPH as the preferred reductant (Kolattukudy, 1971). This aldehyde reductase has a pH optimum of 5·6 and the apparent K_m was 5×10^{-5}M and 7×10^{-5}M for NADPH and palmitaldehyde respectively. This enzyme is severely inhibited by thiol reagents such as iodoacetamide, *N*-ethylmaleimide and *p*-chloromercuric benzoate suggesting the presence of thiol groups essential for activity. A similar aldehyde reductase has been recently isolated from bovine cardiac muscle (Kawalek and Gilbertson, 1974).

Fatty acid reduction to fatty alcohol was also demonstrated in cell-free preparations from dark-grown *Euglena gracilis* which contains large quantities (50% of total lipids) of wax esters (Kolattukudy, 1970d). This conversion required ATP, CoA and NADH as cofactors. The pH optimum was near 6·5 and apparent K_m values were $1{\cdot}6 \times 10^{-5}$ M and $2{\cdot}4 \times 10^{-4}$ M for myristic acid and NADH, respectively. In this case, the reduction was limited to C_{14} to C_{16} acids, and this specificity is consistent with the chain length of the alcohol moieties found in this organism. More recently a similar enzymatic fatty acid reduction has been demonstrated in extracts of perputial gland tumors (Snyder and Malone, 1970), *Clostridium butyricum* (Day *et al.*, 1970), beef heart muscle (Johnson and Gilbertson, 1972), mycobacterium (Wang *et al.*, 1972) and the europygial glands of ring necked pheasant, white crowned sparrow, and goose (unpublished results). The only differences among these systems are in the pyridine nucleotide specificity, the subcellular localization and the tightness of the coupling between the acyl-CoA reductase and the aldehyde reductase. While NADH is the preferred reductant in *B. oleracea*, *Euglena*, bovine heart muscle and *C. butyricum*, NADPH is the preferred reductant in *Mycobacterium*, mouse perputial gland tumors and the europygial glands of goose (Buckner and Kolattukudy, unpublished results). The fatty acyl-CoA reductase appears to be localized in a particulate (microsomal) fraction in *E. gracilis* (Khan and Kolattukudy, 1973b), perputial gland tumors and mycobacterium, while in *B. oleracea*, *C. butyricum* and bovine heart muscle the enzyme appears to be soluble. The coupling between the acyl-CoA reductase and the aldehyde reductase might be related to the subcellular location of these enzymes. In particulate preparations, such as those obtained from *E. gracilis* and perputial gland tumors, the aldehyde intermediate does not seem to accumulate, although such an intermediate could be trapped in the former case with carbonyl reagents. On the other hand, the soluble preparations such as those obtained from *B. oleracea* and bovine heart muscle, could be easily resolved into acyl-CoA reductase and aldehyde reductase.

The membranous acyl-CoA reductase preparation from *E. gracilis* also catalyzes the synthesis of acyl chains from malonyl-CoA and the esterification of the fatty alcohol to wax esters (Khan and Kolattukudy, 1973a). This membranous fatty acid synthetase appears to be coupled to the reductase in the presence of ATP. Also, in the mycobacterial particles, ATP was required for maximal rates of palmitoyl-CoA reduction (Wang *et al.*, 1972). ATP might play a significant role in controlling the synthesis of fatty alcohols, so that under conditions of nutritional abundance the resulting high energy charge could bring about the channeling of the acyl chains into wax esters which are the major energy reserve in *E. gracilis*. More recently, the membranous fatty acid synthetase and fatty acyl-CoA reductase have been solubilized with detergent and high salt concentration (Khan and Kolattukudy, 1974). The reductase appears to be a fairly hydrophobic protein which aggregates in aqueous media of low ionic strength. The detergent-solubilized reductase requires phospholipids for maximal activity. Thus, the fatty acid synthetase located near the surface of the membrane must be functioning in close conjunction with the reductase which is probably located somewhat deeper in the membrane. Gel filtration of the solubilized preparations indicates that the three enzymes, the synthetase, the reductase and the acyl transferase which synthesizes the wax esters, are part of one multienzyme complex functioning in a tightly coupled fashion.

3. *Biosynthesis of Wax Esters*

Wax esters could be synthesized by three possible mechanisms:

(a) direct esterification between a free acid and a fatty alcohol brought about by the reversal of an esterase type reaction

$$RCOOH + R_1OH \rightleftarrows RCOOR_1 + H_2O;$$

(b) acyl transfer from a CoA ester to the hydroxyl group of a fatty alcohol

$$RCO\text{—S-CoA} + R_1OH \rightleftarrows RCOOR_1 + \text{CoA-SH};$$

(c) acyl transfer from a phospholipid to the hydroxyl group of the fatty alcohol

$$\text{phosphatidylcholine} + R_1OH \rightleftarrows RCOOR_1 + \textit{lyso}\text{-phosphatidylcholine}.$$

All three of these possibilities can be demonstrated in cell-free preparations under appropriate experimental conditions. Cell-free preparations from mammalian and dogfish liver catalyzed synthesis of wax ester from free fatty acids and fatty alcohol (Friedberg and Greene, 1967). At high concentrations of alcohol and acid and at low pH, acetone powder preparations from young broccoli leaves also catalyzed the synthesis of wax esters (Kolattukudy, 1967a). However, under physiological conditions the more energetically favorable acyl transfer mechanism, rather than a reversal of esterase, would be expected to be involved in wax ester synthesis. In fact, with acetone powder preparations, it was shown that labeled acyl groups from phospholipids were more efficiently converted into wax esters than would be expected from the labeled free acids generated from the phospholipids. This result strongly suggested a direct acyl transfer from phospholipids to fatty alcohol.

Recently an unusually high concentration of *lyso*-phosphatidylcholine was found in a marine organism which synthesizes large amounts of wax ester (Patton *et al.*, 1972). It is possible that this *lyso*-phosphatidylcholine is a byproduct of wax ester synthesis by an acyl-transfer from the phospholipid to alcohol. However, no experiments to test this possibility have been reported.

An acyl-CoA–fatty alcohol transacylase should be an efficient catalyst for the synthesis of wax ester. However, attempts to demonstrate such an enzyme in the liver preparations and in crude leaf extracts or acetone powder preparations failed. However, ammonium sulfate fractionation of acetone powder extracts of young broccoli leaves, followed by gel filtration on Sephadex G-100 gave an active preparation of acyl-CoA–fatty alcohol transacylase (Kolattukudy, 1967a). This enzyme showed a broad pH optimum between 6·0 and 9·0, while the optimal pH for the esterification of a free acid with an alcohol was 5·0. Requirement for activation of the fatty acid prior to incorporation into wax ester was also clearly demonstrated with cell-free preparations of *E. gracilis* (Kolattukudy, 1970d). More recently, involvement of acyl CoA in the synthesis of such esters has been shown in perputial gland tumors (Snyder and Malone, 1970), rat liver microsomes (Grigor and Bell, (1973), mycobacteria (Wang *et al.*, 1972) and marine organisms (Nevenzel, 1970; Sargent *et al.*, 1974).

The chain length specificity of the esterifying enzymes has not been studied. However, direct g.l.c. analysis of wax esters of pea leaves indicated that there might be some tendency to form esters shorter than those expected from a completely random esterification process (Kolattukudy, 1970f; Tulloch and Hoffman, 1971).

4. *Biosynthesis of Hydrocarbons*

(*a*) *Head-to-head condensation hypothesis.* As a result of the discovery of *n*-nonacosane, *n*-nonacosan-15-one and *n*-nonacosan-15-ol in *B. oleracea*, Channon and Chibnall (1929) proposed a head-to-head condensation mechanism for the biosynthesis of these very long-chain compounds. According to this hypothesis (Fig. 3) two molecules of C_{15} acid undergo head-to-head condensation with decarboxylation of one of them to give the ketone, which could undergo reduction to the secondary alcohol, which in turn could undergo dehydration and reduction to give *n*-nonacosane. Clenshaw and Smedley-Maclean (1929) also proposed a similar mechanism for the synthesis of *n*-C_{31} alkane, which they discovered in spinach. With the demonstration of the presence of odd-numbered chain-length fatty acids in most organisms and the discovery of active α-oxidation enzymes in young leaves (see Chapter 11, p. 330), the head-to-head condensation pathway appeared plausible (Fig. 3). It became possible to test this hypothesis when it was discovered that rapidly expanding leaves readily incorporated exogenous precursors into alkanes (Kolattukudy, 1965). According to the head-to-head condensation

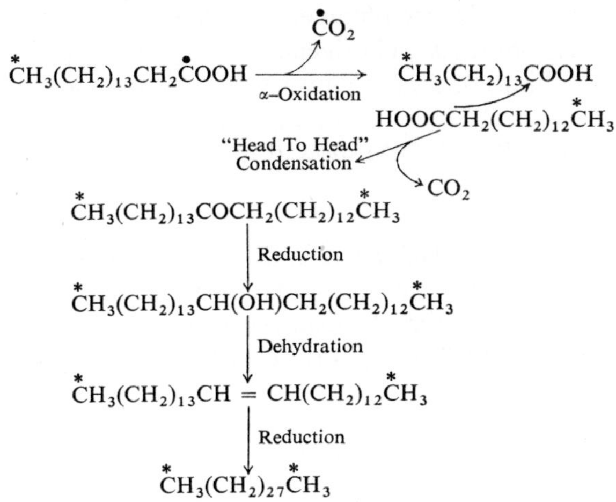

FIG. 3. The head-to-head condensation hypothesis for the biosynthesis of *n*-nonacosane.

hypothesis, the carboxyl carbon of the precursor palmitic acid should be lost during its conversion into C_{15} acid. Thus, incorporation of ^{14}C into C_{29} alkane from [1-^{14}C]palmitic acid should be much smaller (if at all) than that from [U-^{14}C]palmitic acid. However, experimentally it was shown that label from [1-^{14}C]palmitic acid was incorporated into the C_{29} compounds just as efficiently as that from [U-^{14}C]palmitic acid (Kolattukudy, 1966). Double labeling studies showed that the intact carbon chains of C_{12}, C_{14}, C_{16} and C_{18} acids were incorporated into C_{29} alkane in *B. oleracea* and other plants (Kolattukudy, 1968a,b,c). Furthermore, it was shown that the labeled alkane formed in *B. oleracea* was *n*-C_{29}, irrespective of the chain length of the labeled precursor from C_2 to C_{22} (Kolattukudy, 1970e; Kolattukudy and Walton, 1972b). In addition, in time-course experiments the three C_{29} compounds did not show the precursor–product relationships expected from the head-to-head condensation mechanism (Kolattukudy, 1965). At all times the C_{29} alkane had the highest specific activity followed by the ketone and secondary alcohol. According to the Channon-Chibnall hypothesis, the carbonyl group of the ketone is derived from the carboxyl group of the condensing acid. Experimentally, however, it was shown that C-2 of the palmitic acid and C-4 of stearic acid became the carbonyl carbon of nonacosan-15-one during the incorporation of the intact carbon chains of these precursor acids into the ketone (Kolattukudy *et al.*, 1968, 1971a). These experimental results ruled out the head-to-head condensation hypothesis for the synthesis of alkanes in higher plants.

As a possible alternative to the classical head-to-head condensation mechanism, an acceptor–donor condensation hypothesis was proposed, according to which, two biochemically dissimilar fatty acids may condense head-to-head with specific decarboxylation of one of them (Kolattukudy,

1968d). Experimental evidence obtained with *Sarcina lutea*, which synthesizes olefins with *anteiso* branches on both ends, strongly suggested such a mechanism for the biosynthesis of these olefins (Albro and Dittmer, 1970). However, a series of experimental results summarized elsewhere (Kolattukudy, 1970a; Kolattukudy and Walton, 1972b) showed that such a mechanism is probably not operative in higher plants.

(*b*) *Elongation-decarboxylation mechanism.* On the basis of many lines of experimental evidence including those discussed in the previous section, an elongation–decarboxylation mechanism was proposed for the biosynthesis of alkanes (Kolattukudy, 1967b) (Fig. 4). According to this hypothesis, palmitic

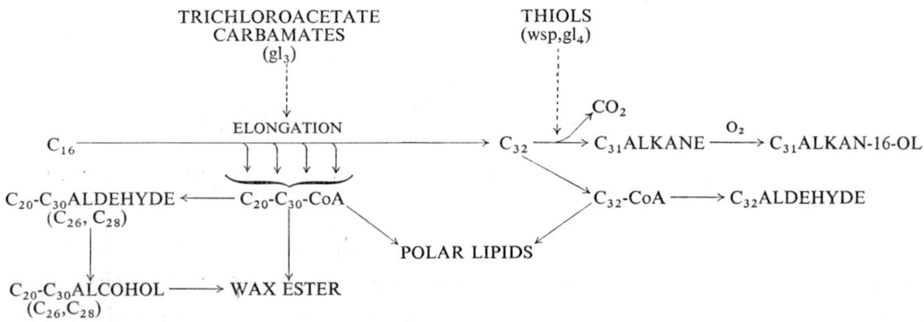

FIG. 4. An elongation–decarboxylation mechanism for the biosynthesis of alkanes and probable biosynthetic relationships among various surface lipids in pea leaves; gl_3 and gl_4 are mutants of *Brassica oleracea* and WSP is a mutant of *Pisum sativum*. The steps probably affected by these mutations and the inhibitors are indicated by the broken arrows (Buckner and Kolattukudy, 1973).

acid, the usual end product of fatty acid synthetase, becomes the substrate for an elongation–decarboxylation system which adds C_2 units from malonyl-CoA until the chain length of the acid reaches C_{30} or C_{32}. The elongated product is then decarboxylated, releasing the hydrocarbon. Involvement of malonyl-CoA in alkane synthesis has been demonstrated with a cell-free preparation from the epidermis of *Allium porrum* (Lessire, 1973). Some of the elongated chains are released from this complex, presumably as CoA esters which, in turn, are incorporated into fatty aldehydes, polar lipids and wax esters. The probable sites of action of various inhibitors as well as the probable sites of genetic blocks observed in certain mutants are also indicated in the pathway shown in Fig. 4. Trichloroacetate (Kolattukudy, 1967b; Kolattukudy *et al.*, 1974) and several thiocarbamates (Kolattukudy and Brown, 1974) as well as the mutation in the gl_3 mutant of *B. oleracea* (Macey and Barber, 1970b) appear to block the elongation process, while certain thiol compounds (Buckner and Kolattukudy, 1973) and the genetic alternation in WSP (*P. sativum*) and gl_4 (*B. oleracea*) mutants (Macey and

Barber, 1970a, b) inhibit the terminal step(s) in alkane synthesis, presumably the decarboxylation step. All the experimental data thus far obtained strongly supports the pathway shown in Fig. 4. Such evidence has been summarized elsewhere (Kolattukudy and Walton, 1972b) and therefore, the following brief outline lists a few major pieces of evidence, particularly those obtained recently, which show that alkanes are probably synthesized in higher plants by an elongation–decarboxylation mechanism.

(a) A variety of experimental evidence summarized elsewhere strongly suggested that alkane synthesis is directly linked to the chain elongation process (Kolattukudy and Walton, 1972b).

(b) Incorporation of intact carbon chains of C_{12}, C_{14}, C_{16} and C_{18} acids into C_{29} alkane was demonstrated with double labeling techniques and all precursors, irrespective of their chain length, were incorporated into C_{29} alkane in *B. oleracea.*

(c) The distribution of ^{14}C incorporated from [U-^{14}C]acetate into fatty acids and alkanes in *Allium porrum* showed the precursor–product relationship expected from the decarboxylation mechanism (Fig. 5). The specific

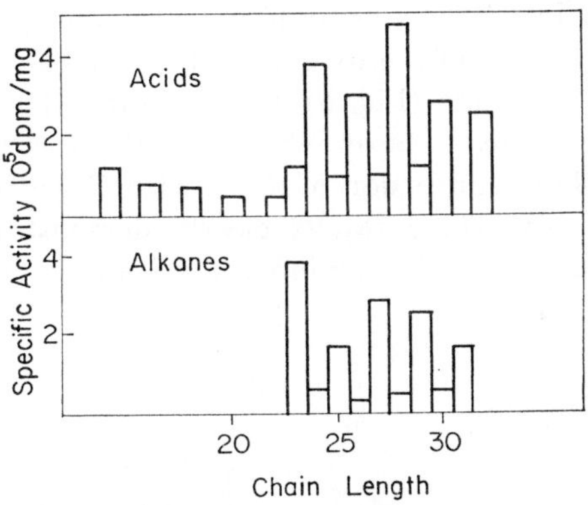

FIG. 5. Specific radioactivity of fatty acids and alkanes isolated from *Allium porrum* leaf slices which metabolized [U-^{14}C]acetate (Cassagne, 1970).

activity of each acid and the corresponding alkane also agreed with such a mechanism (Cassagne, 1970).

(d) In *A. porrum*, exogenous [1-^{14}C]n-C_{24} acid gave rise to C_{25} and longer-chain alkanes but no n-C_{23} alkane could be detected, whereas 10–15% of the label incorporated into alkanes from [1-^{14}C]stearic acid was found in n-C_{23} alkane (Lessire, 1973).

(e) Chemical and genetic blocks at the terminal step of alkane synthesis resulted in accumulation of even chain length acids one carbon longer than the alkane. For example, in the gl_4 mutant of *B. oleracea*, in which C_{29} alkane synthesis was blocked, C_{30} acid accumulated, while in WSP mutant of *P. sativum*, in which C_{31} alkane synthesis was inhibited, C_{32} acid accumulated

(Macey and Barber, 1970a, b). Thiol compounds such as dithioerythritol specifically inhibited incorporation of labeled acetate into C_{31} alkane, with an accumulation of label in elongated chains, particularly in the C_{32} chain (Buckner and Kolattukudy, 1973).

(f) The most direct evidence for the decarboxylation mechanism is the observation that exogenous C_{30} acid was converted into C_{29} alkane, although in low yields, in *B. oleracea* (Kolattukudy *et al.*, 1972). More recently we have obtained even more convincing evidence with the use of synthetic labeled C_{32} acid of high specific activity. This acid was converted into C_{31} alkane in pea leaves and in the flower petals of *Vicia faba* (Kolattukudy *et al.*, 1974). *V. faba* flower petals contain three major alkanes, C_{27}, C_{29} and C_{31}. Exogenous labeled C_{28}, C_{30} and C_{32} acids were converted into alkanes in 0·5–5% yield. Radio–g.l.c. analysis of the labeled alkanes formed showed that exogenous C_{28} acid was converted into C_{29} and C_{31} alkanes in addition to the direct decarboxylation product, namely C_{27} alkane. Exogenous C_{30} acid was incorporated into C_{31} alkane in addition to the direct decarboxylation product, C_{29} alkane, while C_{32} acid gave rise only to C_{31} alkane (Fig. 6). Furthermore, trichloroacetate, which had been postulated to be an elongation inhibitor (Kolattukudy, 1966, 1967b), inhibited the incorporation of [G-^{2}H]-octacosanoic acid specifically into C_{29} and C_{31} alkane but not its direct conversion into C_{27} alkane (Fig. 6). On the other hand, dithioerythritol, which had been postulated to be an inhibitor of the decarboxylation step did, in fact, strongly inhibit conversion of the labeled C_{32} acid to C_{31} alkane by *V. faba* flower petals. These results clearly demonstrate an elongation–decarboxylation mechanism for the biosynthesis of alkanes in higher plants.

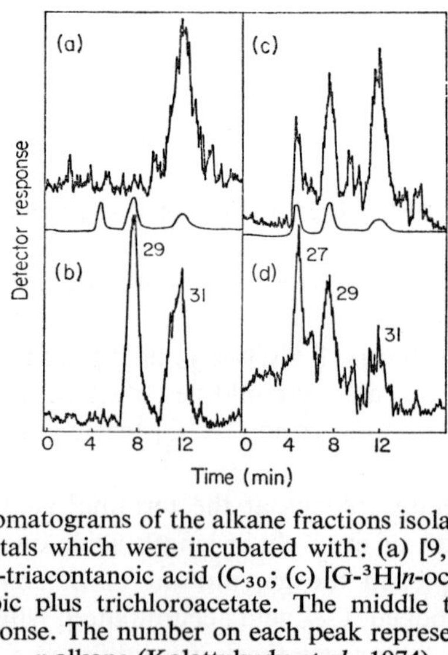

FIG. 6. Radio-g.l.c. chromatograms of the alkane fractions isolated by t.l.c. from the lipids of *Vicia faba* flower petals which were incubated with: (a) [9,10,11-^{3}H]*n*-detriacontanoic acid (C_{32}); (b) [G-^{3}H]*n*-triacontanoic acid (C_{30}; (c) [G-^{3}H]*n*-octacosanoic acid (C_{28}); and (d) [G-^{3}H]*n*-octacosanoic plus trichloroacetate. The middle traces represent the flame ionization detector response. The number on each peak represents the chain-length of the *n*-alkane (Kolattukudy *et al.*, 1974).

In lower plants such as *Nostoc muscorum*, exogenous n-C_{18} acid was converted in low yields into n-C_{17} alkane, the major alkane of this organism. Exogenous n-C_{16} acid gave rise to n-C_{17} alkane in addition to n-C_{15} alkane, which is expected from direct decarboxylation (Han *et al.*, 1969). Thus, in lower plants also, direct decarboxylation of a fatty acid is the most likely mechanism for alkane synthesis.

(g) The direct evidence, discussed in a following section, that alkanes are converted into oxygenated derivatives such as secondary alcohols and ketones is fully consistent with the elongation–decarboxylation mechanism, and negates the major rationale based on which the head-to-head condensation mechanism was originally proposed.

The exact mechanism of the reactions involved in alkane biosynthesis cannot be determined until the enzymes involved are isolated. Recently, a cell-free preparation from the excised epidermis of *A. porrum* was described which catalyzed incorporation of acetate and malonyl-CoA into alkanes (Lessire, 1973). However, the data obtained so far from such experiments are not sufficient to draw any conclusions concerning the mechanism of alkane biosynthesis. We have recently found that homogenates of the excised epidermis of *Senecio odoris* leaves catalyze conversion of synthetic [9,10, 11-^{3}H]n-C_{32} acid of high specific activity to n-C_{31} alkane. However, substantial improvements in the activity of these preparations will be required before the mechanistic questions concerning alkane synthesis can be answered. Chemical considerations would suggest that decarboxylation of a saturated fatty acid is likely to involve oxidation followed by reduction. The observed oxygen requirement for alkane synthesis (Kolattukudy, 1968d) is consistent with this hypothesis. Furthermore, recent experiments in this laboratory indicated that when labeled very long-chain acids (C_n where $n = 28$, 30 or 32) were incubated with petals of developing *V. faba* flowers, detectable amounts of label were found in the even-chain alkanes (C_{n-2} and C_n), even though the major labeled products were those expected from direct decarboxylation and elongation followed by decarboxylation. These even chain alkanes could be a reflection of the possible involvement of an α-oxidative step in alkane biosynthesis. Such possibilities are being explored at this time in our laboratory.

(*c*) *Branched hydrocarbons*. The most abundant branched alkanes found in plant surface lipids contain *iso* or *anteiso* methyl branches. The dominant *iso* alkanes usually contain an odd number of carbon atoms while the *anteiso* alkanes usually have an even number of carbon atoms. This structural pattern may be explained on the basis that the most probable biosynthetic starter moieties for *iso* and *anteiso* alkanes would be *iso*-C_4 and *anteiso*-C_5, respectively. Elongation of these acyl moieties should result in *iso* branched even chain-length acids and *anteiso* branched odd chain-length acids, which on decarboxylation would produce odd chain-length *iso* alkanes and even *anteiso* alkanes (Fig. 7). In strong support of this hypothesis, labeled valine and isobutyric acid fed to tobacco plants (Kaneda, 1967) or excised young tobacco leaves (Kolattukudy, 1968b) and *A. porrum* (Cassagne, 1970) were incorporated preferentially into odd chain-length *iso* alkanes. Furthermore, labeled *iso*-branched C_{16} to C_{28} even chain fatty acids were also detected in tobacco leaves which were fed labeled valine or isobutyrate (Kolattukudy, 1968b). Similarly, labeled isoleucine gave rise specifically to *anteiso* branched, odd chain-length C_{17} to C_{27} acids and *anteiso*, even chain alkanes. In plants such as *B. oleracea* which do not normally synthesize branched alkanes, exogenous labeled branched precursors gave rise to appropriately branched acids shorter than C_{20}, but neither longer branched acids nor branched alkanes were labeled (Kolattukudy, 1968b). Thus, even though branched acids were formed and such acids were incorporated into surface lipids such as wax esters, the branched precursors were excluded from alkane synthesis. Therefore, it is concluded that the specificity of the alkane-synthesizing system, rather than the availability of branched precursors, determines the structural features of the alkane formed in each plant species. In this

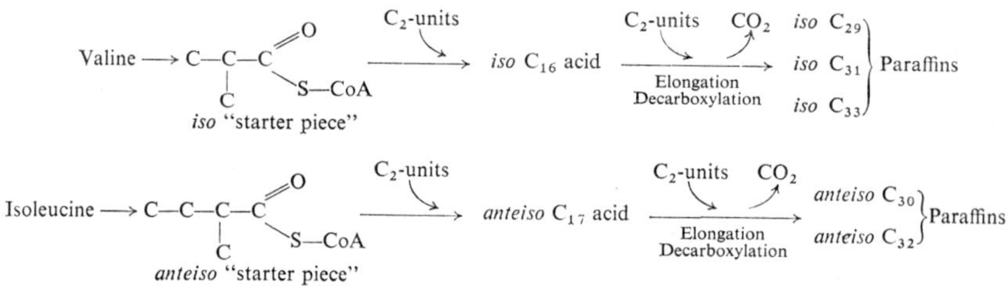

FIG. 7. Proposed biosynthetic pathway for *iso* and *anteiso* acids and alkanes.

connection, it is interesting to note that a mutant of *B. oleracea* has been found which synthesizes *anteiso*-C_{30} alkane (Netting and Macey, 1971).

Even though internally branched alkanes are rare in higher plants, such alkanes have been found as substantial components of the alkane fraction of lower plants, such as algae. Such branches could originate either from methylmalonyl-CoA or by methylation via *S*-adenosylmethionine. Available experimental evidence obtained with *Anabaena variabilis* (Fehler and Light, 1970) and *Nostoc muscorum* (Han *et al.*, 1969) strongly suggests that the methylation mechanism is responsible for the formation of the 50:50 mixture of 7- and 8-methyl C_{17} alkane found in these organisms. Occurrence of 18:1(11*c*) fatty acid in these organisms is consistent with the hypothesis that methylation of the double bond of this acid and decarboxylation is the likely pathway for the formation of the internally branched C_{17} alkane. In fact, a cell-free preparation from *A. variabilis* has been shown to incorporate the methyl group from *S*-adenosyl methionine into 7- and 8-methyl C_{17} alkane, but the nature of the endogenous moiety which was methylated has not been defined (Fehler and Light, 1972). Incorporation of intact CD_3 groups from deutero-methionine into the branched C_{17} alkanes of *A. variabilis* (Fehler and Light, 1970) ruled out the cylopropyl intermediate proposed earlier.

5. *Biosynthesis of Ketones and Secondary Alcohols*

The ketones and secondary alcohols which are found in many plants, were considered to be intermediates of alkane biosynthesis according to the head-to-head condensation hypothesis. However, since all the experimental evidence showed that an elongation–decarboxylation mechanism is responsible for alkane synthesis, the origin of the ketones and secondary alcohols remained unexplained. Specific activity measurements (Kolattukudy, 1965) indicated that the alkanes were precursors of the oxygenated compounds. Experimental results which showed that C-2, C-4 and C-6 of C_{16}, C_{18} and C_{20} acids, respectively, became the carbonyl carbon of the ketone in *B. oleracea*, strongly suggested that the carbonyl oxygen was introduced into a preformed chain of carbon atoms (Kolattukudy *et al.*, 1968, 1971a). It became obvious that the secondary alcohol was not derived from the ketone when it was found that the $^3H:^{14}C$ ratio of the C_{29} secondary alcohol derived from [1-^{14}C-2-^{3}H]C_{16} acid was higher than that in the C_{29} alkane, while the C_{29} ketone contained very little ^3H (Kolattukudy, 1970b). The high $^3H:^{14}C$ ratio of the secondary alcohol was attributed to a selective oxidation of the nontritiated secondary alcohol to the ketone. Thus it was proposed that hydroxylation of the alkane at a specific carbon in the middle of the chain generates the secondary alcohol, which in turn is oxidized to the corresponding ketone. This hypothesis is in good agreement with the observation that the chain-length distribution of the oxygenated derivatives usually corresponds to that of the alkanes. Direct evidence for the above hypothesis was provided by the demonstration that leaves of *B. oleracea* converted exogenous [G-^{3}H]C_{29} alkane into nonacosan-15-one (Kolattukudy and Liu, 1970). Furthermore, exogenous [G-^{3}H]nonacosan-15-ol was converted into the corresponding ketone by broccoli leaves. Considering the technical difficulties involved in getting such very long chain non-polar compounds to the metabolic sites, it is not surprising that low yields of products were obtained. However, the positional specificity of oxidation, demonstrated by the chemical degradation of the oxygenated products, clearly demonstrates that the observed conversions do indeed represent the usual reactions which occur in these leaves. A more detailed examination of the labeled secondary alcohol fraction derived from exogenous labeled nonacosane gave further support to this conclusion. Chemical degradation studies proved that the exogenous nonacosane was converted into nonacosan-15-ol and nonacosan-14-ol in a ratio of 3:2, which is identical to the ratio of the naturally occurring positional isomers in *B. oleracea* (Kolattukudy *et al.*, 1973a). Although the secondary alcohol fraction is a mixture of positional isomers, virtually all the C_{29} ketone of this plant is the 15-oxo isomer. Therefore, it appeared likely that oxidation of the secondary alcohol is highly specific for the position of the hydroxyl group. Such a specificity could be demonstrated by the fact that *B. oleracea* leaves preferentially converted nonacosan-15-ol to the corresponding ketone when presented with a mixture of generally ^{3}H-labeled nonacosan-15-ol and nonacosan-14-ol. On the basis of the experimental results described above, the biosynthetic relationships among alkanes, secondary alcohols and ketones can be represented as shown in Fig. 8.

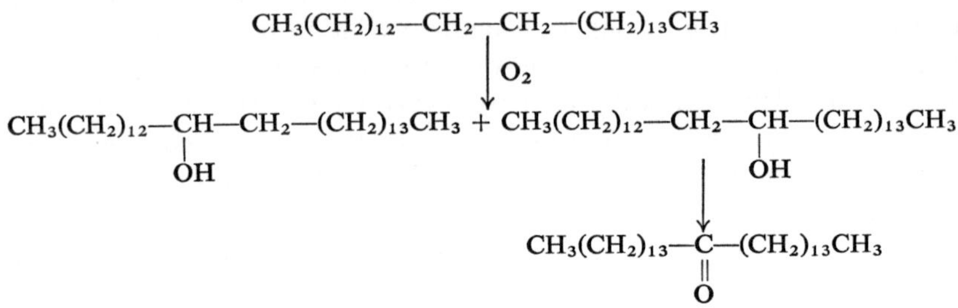

FIG. 8. Biosynthetic relationships among alkanes, secondary alcohols and ketones as postulated from the experimental results obtained with *Brassica oleracea*.

III. CUTIN AND SUBERIN

A. ISOLATION

Cutin can be isolated by disrupting the pectinaceous layer chemically with agents such as oxalates or enzymatically with pectinase preparations (Orgell, 1955). From the cuticular layers thus obtained, the soluble lipids can be removed by a thorough solvent extraction. Usually, repeated treatment with a mixture of cellulase and pectinase helps to remove any remaining carbohydrate materials (Walton and Kolattukudy, 1972b). Small quantities of phenolic components and flavanoids, which might be covalently linked to the cutin, are often found in such cutin preparations.

The polymeric material which covers the underground parts of plants, such as potato tubers, is quite different in composition from the polymer which covers the aerial parts. From the limited amount of data available (Rodriguez-Miguens and Ribas-Marques, 1972; Kolattukudy and Agrawal, 1974) one may tentatively classify this material as suberin. From tissues such as potato tubers, one can easily remove the polymer mechanically after boiling the tubers in water or with ammonium oxalate and oxalic acid. The adhering carbohydrate material can be removed by enzymes such as amylases, pectinase and cellulase. About one-third of the insoluble material remaining after the above treatment consists of aliphatic components while the rest appears to be phenolic in nature. In our experience it is not convenient to apply similar methods of isolation to root-covering polymers. In such cases the entire bark of the root has been subjected to chemical analysis just as crude cork of tree barks has been analyzed (Holloway, 1972a, b).

B. STRUCTURE

1. *Depolymerization*

Since the monomers of cutin are held together mainly by ester linkages, alkaline hydrolysis was used to cleave the polymer (Lee, 1925; Legg and Wheeler, 1929; Matic, 1956). However, a substantial proportion of the polymer was left behind after this treatment (Crisp, 1965), presumably because of the presence of interchain peroxide bridges and ether linkages. Therefore, sequential treatment of the residue with NaI and HI has been used in conjunction with alkaline hydrolysis. More recently, exhaustive hydrogenolysis with $LiAlH_4$ in tetrahydrofuran was introduced as a technique for a more complete depolymerization (Kolattukudy, 1970c; Walton and Kolattukudy, 1972b). This treatment converts carboxyl groups of the monomer into primary alcohol functions, which are indistinguishable from the naturally occurring primary alcohol functions, Furthermore, other $LiAlH_4$-susceptible functions such as aldehydes and epoxides are also converted into alcohols and therefore such structural details are lost. This difficulty is easily overcome with the use of $LiAlD_4$, which introduces deuterium atoms where the reducible groups were originally present. Therefore, with the use of this technique functional groups such as epoxides and aldehydes can be readily detected. More recently, transesterification techniques were employed in which sodium methoxide or BF_3 were used in absolute methanol (Brieskorn and Kabelitz, 1971; Kolattukudy *et al.*, 1973b; Holloway and Deas, 1973; Kolattukudy and Agrawal, 1974; Kolattukudy and Dean, 1974.) Under absolutely anhydrous conditions and low concentrations of sodium methoxide, functional groups such as epoxide were not affected. With higher concentrations of sodium methoxide in anhydrous methanol, quantitative methanolysis of the oxirane ring takes place, and this derivative was used for identification of epoxy acids (Hollaway and Deas, 1973). The commercially available 14% BF_3 in methanol is a good reagent for depolymerization of cutin and suberin by transesterification, and under such acid conditions epoxides are cleaved by methanolysis and aldehydes are converted into acetals. However, some BF_3-catalyzed addition of methanol at the double bonds of monomers occurs during prolonged treatment with this reagent. Since the major components of suberin appear to be ω-hydroxy acids and dicarboxylic acids, this transesterification technique is preferred over the hydrogenolysis technique because the latter converts both of these classes of monomers into indistinguishable diols (Kolattukudy and Agrawal, 1974; Kolattukudy and Dean, 1974). When comparable quantities of saturated and unsaturated ω-hydroxy acids and dicarboxylic acids are present, the deuterium labeling technique is not as convenient as the transesterification technique for mass spectral analysis.

2. *Structure Determination of Monomers*

Irrespective of the depolymerization technique employed, the complex mixture of monomer derivatives obtained is usually subjected to t.l.c. The various fractions thus obtained can be further separated by argentation t.l.c., or they can be subjected directly to g.l.c. as the trimethylsilyl (TMSi) ethers (Kolattukudy and Walton, 1972b). The most convenient and reliable means for structure determination of the monomers is by mass spectrometry (m.s.). Therefore, the availability of the combined g.l.c.–m.s. systems has played a key role in furthering our understanding of the structure of the monomers of cutin and suberin. In order to illustrate this technique, the mass spectra of certain derivatives of a few major components of cutin and suberin are discussed below. In most cases either the TMSi ether of the methyl ester or the TMSi ether of the polyhydroxy alkanes derived from the $LiAlH_4$ or $LiAlD_4$ treatment have been used for structural studies (Eglinton and Hunneman, 1968; Eglinton *et al.*, 1968; Brieskorn and Kabelitz, 1971; Croteau and Fagerson, 1972; Holloway and Deas, 1971; Walton and Kolattukudy, 1972b) and, therefore, the following discussion is based on these studies.

(*a*) *ω-Hydroxyfatty Acids and Dicarboxylic Acids.* In the case of an unsaturated ω-trimethylsiloxy methyl ester, the characteristic ions are M^+, M^+-CH_3, M^+-CH_3O and $M^+-CH_3-CH_3OH$. The relative intensities of these ions vary depending on the degree of unsaturation. A characteristic metastable ion representing the transition $(M^+-CH_3)\rightarrow(M^+-CH_3-CH_3OH)$ is also found in such spectra. The mass spectra of the TMSi ethers of the diols derived from the ω-hydroxy acids contain significant ions at M^+, M^+-CH_3, $M^+-CH_3-CH_4$ and $M^+-CH_3-(CH_3)_3SiOH$. In this case, a characteristic doubly charged ion representing $[M-2CH_3]^{2+}$ can be used for diagnostic purposes. In the case of dicarboxylic acids such as octadecenedioic acid, which is a major component of suberin, the dimethyl esters give major characteristic ions at M^+, M^+-32, M^+-50, M^+-64 and M^+-92.

(*b*) *Dihydroxyacids* The dihydroxyacids of cutin and suberin usually contain a primary alcohol function at the ω-carbon and a secondary alcohol near the middle of the carbon chain. Either the TMSi ethers of the methylesters, or the TMSi ethers of the alkane triols derived from the dihydroxyacids, give characteristic mass spectra which lend themselves to unambiguous interpretation. For example, in the mass spectrum of the methyl dihydroxypalmitate of potato suberin (Kolattukudy and Agrawal, 1974) the major ions in the high mass region are at M^+, M^+-15, M^+-31, M^+-47, $M^+-47-32$, $M^+-15-90$, $M^+-31-90$ and $M^+-47-90$ (Fig. 9). The position of the in-chain hydroxyl group can be determined from the intense ions derived by cleavage on either side of the secondary hydroxyl group. For example, the ion pairs at m/e 273–275, 259–289 and 245–303 showed that the dihydroxypalmitic acid of potato suberin was a mixture of 10,16-, 9,16-, and 8,16-dihydroxy-16:0 acids.

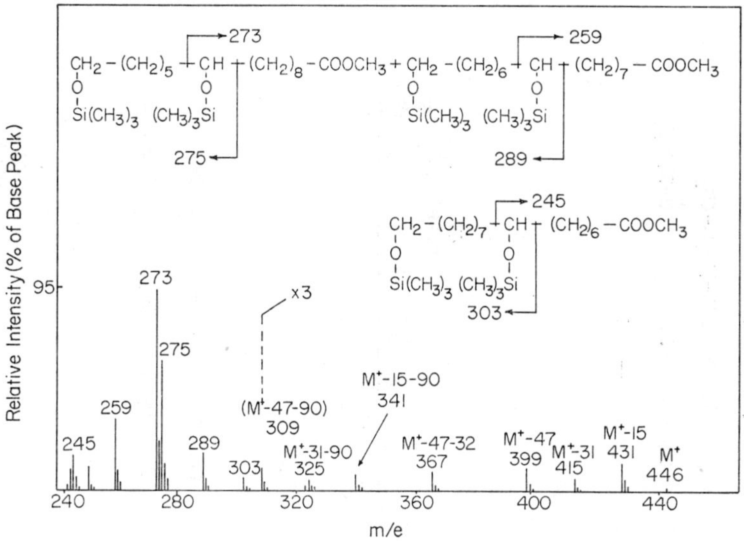

FIG. 9. Partial mass spectrum of the TMSi ether of methyl dihydroxypalmitic acid from potato tuber suberin.

(*c*) *Epoxyacids* 18-Hydroxy-9,10-epoxy-18:0 acid has been identified by the mass spectrum of the methyl ester (Croteau and Fagerson, 1972), the TMSi ether of its deuterolysis product (Kolattukudy *et al.*, 1971b; Walton and Kolattukudy, 1972b) and the TMSi ether of the methanolysis or ethanolysis product (Holloway and Deas, 1973; Kolattukudy and Agrawal, 1974). Since the α-cleavages on either side of the oxirane ring itself are not dominant cleavages, the mass spectrum of the intact epoxyacid is not a convenient means of identifying such acids. At least partly for this reason, epoxyacids went undetected by some investigators. The deuterolysis technique is a very sensitive and unambiguous means of identifying epoxyacids. The one deuterium atom introduced at either of the carbon atoms of the oxirane ring and the two deuterium atoms on the carboxyl carbon result in the incorporation of a total of three deuterium atoms, which can be easily detected by the ions at the high mass region: M^+, M^+-15 and M^+-90. Furthermore, the dominant α-cleavages on either side of the secondary alcohol function (TMSi ether) of this trideuterated alkane triol give rise to a pair of intense ions containing one or three deuterium atoms (Fig. 10). These unusual even number mass ions are quite characteristic of the epoxide function.

More recently, the methanolysis derivative of the oxirane function has been used for identification of the epoxyacids. This technique involves cleavage of the polymer with either sodium methoxide or BF_3 in absolute methanol, which simultaneously cleaves the oxirane function by methanolysis. The TMSi ether of the mixture of the two isomeric α-methoxy alcohols provides characteristic mass spectra (Fig. 11). In the high mass region, the major ions are

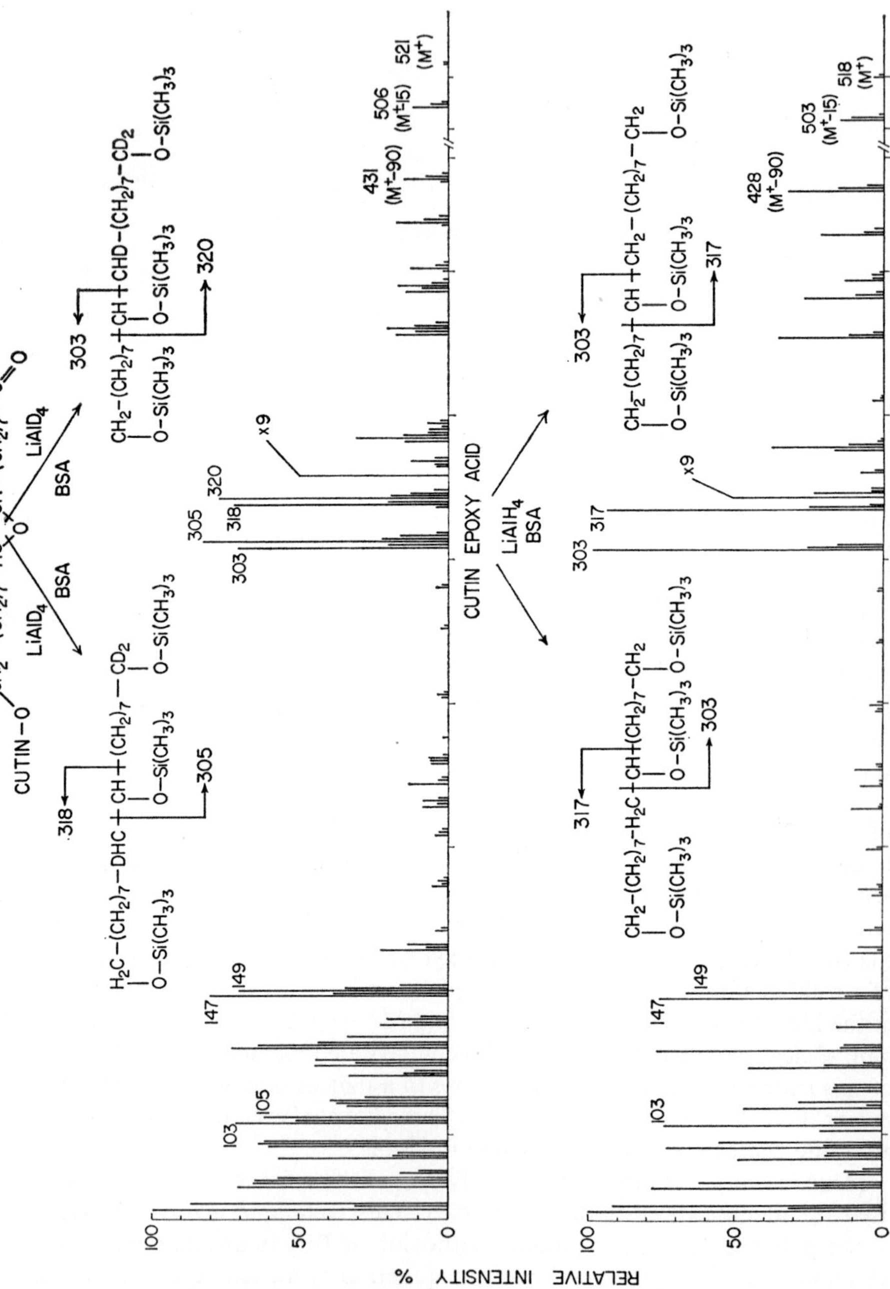
O—CUTIN
CUTIN—O
$CH_2—(CH_2)_7—HC—CH—(CH_2)_7—C=O$
LiAlD$_4$
BSA
318
305
303
320
$H_2C—(CH_2)_7—DHC—CH—(CH_2)_7—CD_2$
O—Si(CH$_3$)$_3$
$CH_2—(CH_2)_7—CH—CHD—(CH_2)_7—CD_2$
CUTIN EPOXY ACID
LiAlH$_4$
317
$CH_2—(CH_2)_7—H_2C—CH—(CH_2)_7—CH_2$
$CH_2—(CH_2)_7—CH—CH_2—(CH_2)_7—CH_2$
x9
103
105
147
149
431 (M$^+$-90)
506 (M$^+$-15)
521 (M$^+$)
428 (M$^+$-90)
503 (M$^+$-15)
518 (M$^+$)
RELATIVE INTENSITY %
m/e

FIG. 10. Partial mass spectrum of the TMSi ether of saturated C_{18} triol obtained by hydrogenolysis (bottom) and deuterolysis (top) of apple cutin.

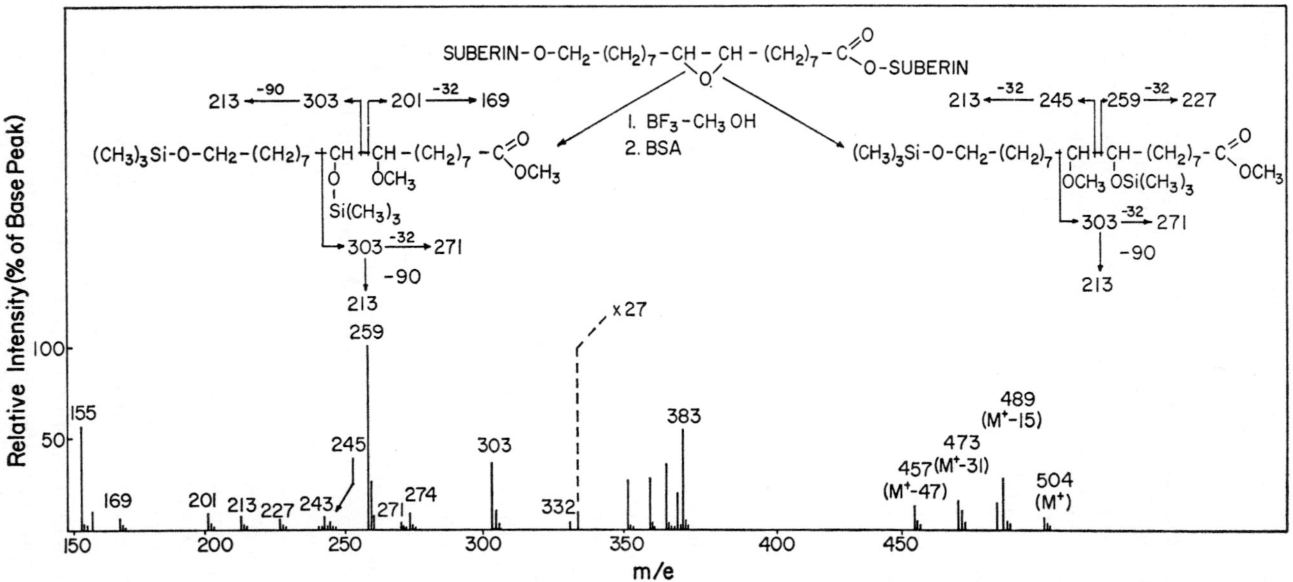

FIG. 11. Partial mass spectrum of the methanolysis product of the 18-hydroxy-9,10-epoxy-stearic acid of potato tuber suberin.

M^+, M^+-15, M^+-31, M^+-47 and $M^+-90-31$. In the α-cleavage region the most intense ions are at m/e 259 and 303. As illustrated in Fig. 11, the other ions are those expected from a 50:50 mixture of methyl 9-methoxy-10, 18-dihydroxy-18:0 and methyl 10-methoxy-9:18-dihydroxy-18:0.

(*d*) *Trihydroxyacids* In this class a *vic*-diol function is found usually at the 9,10-position. The dominant cleavage between the two adjacent substituted carbon atoms provides a fairly simple mass spectrum from either the alkanetetraol derived from the trihydroxyacid by reductive cleavage or the methyl ester produced by the transesterification technique. Therefore, such trihydroxyacids and the corresponding dicarboxylic acids can be readily identified.

(*e*) *Pentahydroxy Acids and Trihydroxy Epoxyacids* Combined g.l.c.–m.s. of the TMSi derivatives of the methyl esters of cutin from *Rosmarimus officinalis* showed that 3·2% of the cutin acids was, 9,10,12,13,18-pentahydroxy-18:0 (Brieskorn and Kabelitz, 1971). We have confirmed the identity of this compound by depolymerization of the crude cutin of *R. officinalis* by treatment with $LiAlH_4$ and $LiAlD_4$ followed by g.l.c.–m.s. (Croteau and Kolattukudy, 1974a). Even though the earlier identification of the pentahydroxy acid was thus confirmed, our analysis showed that the proportion of this component was much less than that reported earlier. However, in our experience, the recovery of such highly hydroxylated materials is much poorer than that of the less hydroxylated monomers. With hydrogenolysis and deuterolysis techniques, in conjunction with g.l.c.–m.s., we have obtained evidence also for the presence of 9,10,18-trihydroxy-12,13-epoxy-18:0, which is the likely precursor of the pentahydroxy acid.

3. *Composition of Cutin*

Recent structural studies discussed above showed that cutin is composed of two families of acids: a C_{16} family and a C_{18} family. The most rapidly growing plants appear to have mainly the C_{16} family of acids while the slower growing plants, with thicker cuticle, contain both the C_{16} family and the C_{18} family of acids. The most common components of the C_{16} family of acids are 16:0, 16-hydroxy-16:0 and dihydroxy-16:0. Usually the secondary hydroxyl group of the dihydroxypalmitic acid is on the C-10, C-9, C-8 or C-7 position, and in a given plant one of these isomers usually predominates with a small quantity ($\simeq 10\%$) of its nearest positional isomer (Holloway and Deas, 1971; Walton and Kolattukudy, 1972b). For example, in *V. faba* the major dihydroxy C_{16} acid is 10,16-dihydroxy acid (90%) with about 10% 9,16-dihydroxy acid (Kolattukudy and Walton, 1972a). More recently 16-oxo-9-hydroxy-16:0 and 16-oxo-10-hydroxy-16:0 were identified as subtantial components of the cutin of very young tissues (Kolattukudy, 1972, 1974).

The most common components of the C_{18} family of acids are oleic acid

(18:1 9), 18-hydroxy-18:1(9), 18-hydroxy-9,10-epoxy-18:0 and 9, 10,18-trihydroxy-18:0, together with analogues related to 18:1(12). In young tissues 12,15-diunsaturated analogues are also found. At least in some plants, such as *R. officinalis*, 9,10,18-trihydroxy-12,13-epoxy-18:0 and 9,10,12,13,18-pentahydroxy-18:0 also occur. Since substantial proportions of these polar compounds appear to be lost during the usual extraction procedures their occurrence in other plants could have gone undetected. More recently 18-oxo-9,10-epoxy-18:0 was detected in the cutin of young apple fruits (Kolattukudy, 1973). 12-Hydroxy-18:1(9) (ricinoleic acid) (Brieskorn and Kabelitz, 1971), 12,18-dihydroxy-18:1(9) and 12,18-dihydroxy-9,10-epoxy-18:0 acid were also detected in the cutin of *R. officinalis* (Croteau and Kolattukudy, 1974a), but these acids are not as common as the others described above.

4. *Composition of Suberin*

The major aliphatic components of suberin are ω-hydroxy acids and dicarboxylic acids (Table I). Among the dicarboxylic acids, octadec-9-ene-1,18-dioic acid is usually dominant, although C_{16}, C_{18}, C_{20}, C_{22} and longer saturated dicarboxylic acids are not rare (Rodriguez-Miguens and Ribas-Marques, 1972; Holloway, 1972a, b; Kolattukudy and Agrawal, 1974). Similarly, among the ω-hydroxyacids, 18-hydroxy-18:1(9) is usually a major component. The proportion of the very long chains ($>C_{20}$) contained in the ω-hydroxyacid fraction is usually much higher than that in the dicarboxylic acids. Among the α,ω-diols, fatty alcohols and fatty acids, which are often found as significant components of suberin, very long chains are common. Small quantities of in-chain hydroxylated C_{16} and C_{18} acids and epoxyacids

TABLE I

The major aliphatic components of cutin and suberin from different sources[a]

Fraction	% composition: Suberin		Cutin	
Source:	*Ribes americanum*	Potato tuber	Potato leaf	Apple fruit
α,ω-diol	10	—	—	—
ω-hydroxy acid	50	30	—	16
dicarboxylic acid	36	44	—	—
dihydroxy-C_{16} acid[b]	—	—	94	23
epoxy- and trihydroxy-C_{18} acid	—	—	5	56

[a] The values were calculated from the literature (Holloway, 1972a, b; Kolattukudy and Agrawal, 1974) and unpublished results from this laboratory. Components which constitute <5% of the aliphatic fractions are not included.

[b] The dihydroxy C_{16} acid of potato leaf was a mixture of 10,16-dihydroxy-16:0 (61%), 9,16-dihydroxy-16:0 (26%) and 8,16-dihydroxy-16:0 (13%).

have been found in suberin (Holloway, 1972a; Kolattukudy and Agrawal, 1974). Among them, the corresponding dicarboxylic acids are also found. The generalizations outlined above also hold good for the suberins of carrot parsnip, rutabaga, turnip and sweet potatoes, although each of these species has a characteristic quantitative monomer distribution (Kolattukudy *et al.*, 1975b).

5. *Suberin of Wound Periderm*

Wounds in plants are known to become covered with a material derived from lipids. In storage organs such as potato tubers, a material which stains like a lipid accumulates at the walls of the cells which are adjacent to the cut surfaces. This process is generally referred to as suberization. However, little was known until recently about the structure of this material and its possible similarity to the naturally occurring suberin of potato tuber. Recent structural studies showed that the major aliphatic components of the wound periderm material are 18-hydroxy-18:1(9) acid and 18:1(9)-1,18-dioic acid (Kolattukudy and Dean, 1974). The same acids were identified to be the major components of the naturally occurring skin polymer of potato tuber. Very long (C_{20} to C_{26}) acids and alcohols characteristic of suberin were also found in this wound periderm. Since the major aliphatic components of the suberin of wound periderm are ω-hydroxy-18:1 acid and the corresponding dicarboxylic acid, the quantity of these acids should give a measure of the suberization process. On this basis, a g.l.c. assay for suberization was developed. This assay involves treatment of lyophylized, powdered tissue with $LiAlH_4$ in tetrahydrofuran, followed by gas chromatographic measurement of octadec-9-ene-1,18-diol, which is derived from both ω-hydroxyoleic acid and the corresponding dicarboxylic acid of suberin.

6. *Distinction Between Cutin and Suberin*

It is clear that the aliphatic components of suberin are structurally similar to those found in cutin, but usually there are dramatic quantitative differences (Table 1). On the basis of these differences the tentative guidelines summarized

TABLE II

A proposed guideline for the classification of hydroxyacid phytopolymers as cutin or suberin

Monomer	Cutin	Suberin
dicarboxylic acids	minor	major
in-chain substituted acids	major	minor
phenolics	low	high
very long chain (C_{20}–C_{26}) acids	rare and minor	common and substantial
very long chain alcohols	rare and minor	common and substantial

in Table II can be used to classify the naturally occurring hydroxyacid polymers of plants. The in-chain hydroxylated monomers, such as the dihydroxy-16:0 and 9,10,18-trihydroxy-18:0 acids are major components of cutin, while in suberin they are only minor components. The major components of suberin appear to be ω-hydroxy acids and dicarboxylic acids while dicarboxylic acids are only minor components of cutin (Rodriguez-Miguens and Ribas-Marques, 1972; Kolattukudy and Agrawal, 1974; Holloway, 1972a, b). Even among the in-chain hydroxylated minor components of suberin, dicarboxylic acids dominate. Although many cutin samples contain some phenolic components, and recent structural studies show that some phenolic components are covalently linked to cutin (Riley and Kolattukudy, unpublished results), such compounds are usually only minor components of cutin, while up to two thirds of suberin may consist of phenolic compounds (Holloway, 1972a, b). Although these appear to be the major differences between cutin and suberin, a firm classification must await reliable structural studies on cutin and suberin from many more species of plants. A recent analysis of the bark of *Betula pendula* (Holloway, 1972b) showed a monomer composition similar to that of cutin rather than suberin as defined above. However, it is not clear whether proper precautions were taken to exclude the outer cutin layer from the bark sample which was analyzed.

7. *Developmental Changes in Monomer Composition of Cutin*

Structural analysis of the cutin from the shoots of embryonic *V. faba* with a deuterium labeling technique revealed that 16-oxo-9-hydroxy-16:0 was a major component of this polymer (Kolattukudy, 1972). The occurrence of the aldehyde function and the position of the secondary hydroxyl group were unexpected because the major component of the cutin of the leaves of *V. faba* was known to be 10,16-dihydroxy-16:0. The above observation suggested that monomer composition of cutin may change with the development of the plant tissue. A detailed examination of the monomer structure of *V. faba* leaves during their development showed that, in the disubstituted C_{16} acid fraction, the major portion of the 9-hydroxy acid contained an aldehyde function at the 16-position, while only a small portion of the 10-hydroxy isomer contained a 16-oxo function (Kolattukudy, 1974). As the leaves developed, the 9-hydroxy isomer decreased from 50% in the youngest tissue to 10% in the mature leaf. Such structural changes during development were also demonstrated by incorporation of [1-^{14}C]palmitic acid into the 9-hydroxy and 10-hydroxy isomers as measured by the distribution of radioactivity among the CrO_3 oxidation products. Treatment of cutin with hydroxylamine, followed by $LiAlH_4$ or $LiAlD_4$ gave rise to aminodiol (Fig. 12) which could be readily identified by g.l.c.–m.s. of its *N*-acetyl, *O*-trimethylsilyl derivative (Fig. 13). In the high mass region of the spectrum of the hydrogenolysis product a weak molecular ion was observed at m/e 459 with a strong ion at m/e 444 ($M^+ - CH_3$). Fairly intense ions were observed at m/e 369 $M^+ - (CH_3)_3SiOH$) and m/e 354 ($M^+ - CH_3 - (CH_3)_3SiOH$). The corresponding ions in the deuterolysis products were displaced by three atomic mass units, showing the expected incorporation of three deuterium atoms. The α-cleavage ions originating from the amino end of the molecule (m/e 244 and m/e 258 from the hydrogenolysis product and m/e 245 and m/e 259 from the deuterolysis product) were much stronger than those originating from the alcohol end (m/e 303 and m/e 317 in the hydrogenolysis product and m/e 305 and m/e 319

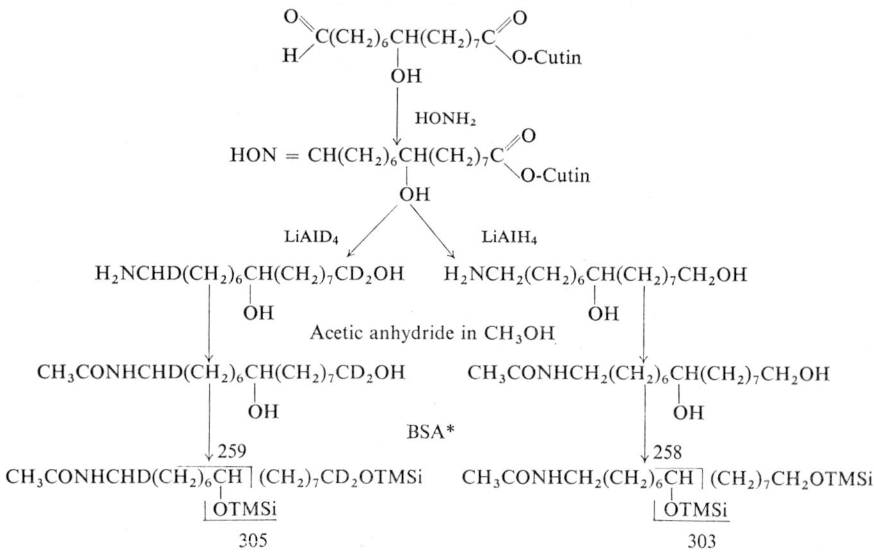

FIG. 12. Reactions used for the isolation and identification of 16-oxo-9-hydroxyhexadecanoic acid from *Vicia faba* (Kolattukudy, 1974). *BSA = *N,O*-bis(trimethylsilyl)-acetamide. m/e values are given for the significant ions formed by α-cleavage of the TMSi ethers.

in the deuterolysis products). This technique allows the isolation of the aldehyde-containing molecules even when they are present only as minor components. For example, with this technique the 16-oxo compound was identified as a minor component in other cutins such as that from apple fruit (Agrawal and Kolattukudy, unpublished results). The monomer composition of cutin of apple fruits also changed with development. For example, ω-hydroxy-18:2(9,12) and 9,10-epoxy-18-hydroxy-18:2(12,15) were found to be significant components of the cutin of very young apple fruits, but they were only minor components of the cutin of mature fruits (Kolattukudy, *et al.*, 1973b). 18-oxo-9,10-epoxy-18:0 was also identified in the cutin of very young fruits (Kolattukudy, 1973). Thus, significant changes in monomer composition occur in the cutin of plant tissues during development.

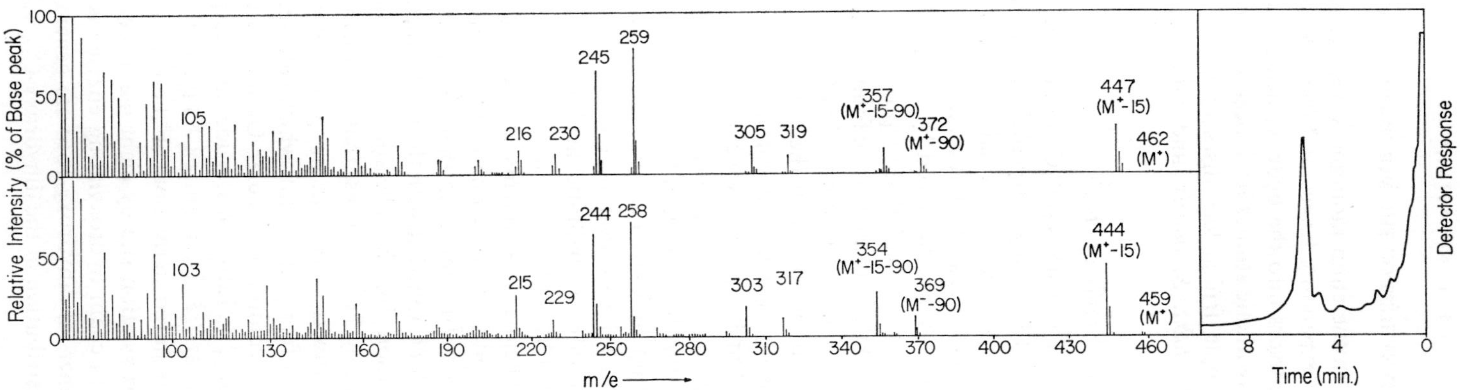

FIG. 13. g.l.c.–m.s. analysis of TMSi ethers of the *N*-acetylated aminoalkane diols obtained from NH_2OH-treated cutin of young *Vicia faba* leaves by treatment with $LiAlD_4$ (upper) or $LiAlH_4$ (lower).

C. BIOSYNTHESIS OF THE MONOMERS OF CUTIN

Biosynthetic studies on the cutin monomers became possible when it was discovered that rapidly expanding leaves of *Vicia faba* readily incorporated exogenous labeled precursors into the hydroxy acids of cutin (Kolattukudy, 1970c). Since then, biosynthetic studies have been performed with skin slices of grape berries and apple fruits, leaf discs of apple and *Senecio odoris*, excised epidermis of *V. faba*, *S. odoris*, and *Pisum sativum*, and cell-free preparations.

1. *Biosynthesis of the* C_{16} *Family of Acids*

(*a*) *Hydroxy* C_{16}*-acids*. Exogenous [1-^{14}C]palmitic acid was incorporated into 16-hydroxy-16:0 and 10,16-dihydroxy-16:0 of the cutin in the leaves of *V. faba*, *S. odoris* and apple, as well as skin slices of young apple fruits (Kolattukudy and Walton, 1972a; Kolattukudy *et al.*, 1973b). Identity of these biosynthetic products was established by t.l.c. and g.l.c. as well as by identification of their chemical degradation products. In all these tissues, virtually all of the hydroxylated products were found in the insoluble polymer with only a very small amount in the soluble lipids.

Hydroxylation appears to occur first at the C-16 position because 9 or 10-hydroxy-16:0 is not usually detected in cutin, while 16-hydroxy-16:0 is a common component. Furthermore, exogenous labeled 16-hydroxy-16:0 (chemically synthesized as well as biosynthetically labeled) was readily incorporated into 10,16-dihydroxy-16:0 of the cutin in *V. faba* leaf discs. The secondary hydroxyl group at C-10 may be introduced by either a direct hydroxylation or by an olefin hydration mechanism. Under conditions which allowed rapid incorporation of palmitic acid into 10,16-dihydroxy-16:0, neither 16:1(9*c*) nor 16:1 (9*t*) acid was incorporated into the dihydroxy acid. Furthermore, double labeling experiments showed that, out of the four hydrogen atoms at C-9 and C-10, only one was lost during the introduction of the hydroxyl group at C-10, indicating that a 9-ene-intermediate was not involved. These observations, together with the requirement for molecular O_2, inhibition of hydroxylation by the chelating agent phenanthroline and reversal of this inhibition by Fe^{2+}, suggest that ω-hydroxy-16:0 is hydroxylated at C-10 by a mixed function oxidase. In agreement with this conclusion was the finding that a cell-free preparation which catalyzed C-10 hydroxylation of 16-hydroxy-16:0 required molecular O_2 and NADPH as cofactors (Walton and Kolattukudy, 1972b). This crude enzyme preparation, which showed a pH optimum near 7·3 also required ATP and CoA, suggesting that hydroxylation at C-10 required activation of the carboxyl group which is separated from the hydroxylation site by eight methylene groups. The positional specificity of the in-chain hydroxylating enzyme is not absolute and, therefore, positional isomers of dihydroxy-16:0 are generally found in cutin. The positional isomer distribution of the dihydroxy-C_{16} acids of cutin appear to be identical to that found in the suberin from the same species suggesting that the same hydroxylating enzyme is involved in the synthesis of both polymers. For example, it was found that the ratio 10,16-dihydroxy-: 9,16-dihydroxy-: 8,16-dihydroxy-C_{16} acids was 67:23:10 in the potato tuber suberin (Kolattukudy and Agrawal, 1974) while the same isomers were present in the potato leaf cutin in the ratio of 61:26:13 (unpublished results).

The ω-hydroxylating enzyme involved in cutin synthesis does not appear to have a stringent specificity. All fatty acids thus far examined were incorporated into cutin as their ω-hydroxy derivatives. Even exogenous 9-hydroxy and 10-hydroxy-[10-^{14}C]palmitic acid were readily incorporated into cutin in *V. faba* leaves. Hydrogenolysis of the labeled cutin and the structural examination of the labeled products by chromatographic as well as by chemical degradation, showed that the exogenous hydroxy acids were incorporated as such, and also after conversion into 9,16- or 10,16-dihydroxyacids (unpublished results). This finding may also be taken to mean that it is possible that the in-chain hydroxylation occurs prior to ω-hydroxylation, however, at the present time this possibility appears unlikely.

(*b*) *Oxidation of the ω-hydroxyl of cutin monomers to an aldehyde.* The recent discovery of 16-oxo-9-hydroxy-16:0 and 16-oxo-10-hydroxy-16:0 in young *V. faba* leaves suggested that, at least in very young tissues, the primary alcohol function was further oxidized. In fact, exogenous [G-^{3}H]10,16-dihydroxy-16:0 was directly converted into the corresponding 16-oxo-compound in very young tissues (Kolattukudy, 1974). On the basis of the following evidence, at least a portion of the aldehyde groups appears to be involved in acetal-type linkages: chromic acid oxidation of the diol fraction isolated from the hydrogenolysis products of labeled cutin derived from [1-^{14}C]palmitic acid in very young tissues gave C_8, C_9, and C_{10} dicarboxylic acids, in addition to the expected C_{16} dicarboxylic acid. Radio-g.l.c. and other evidence ruled out the possibility that the unknown compound, which gave rise to the short dicarboxylic acids, arose from 9- or 10-hydroxy-16:0. Acetylation of the labeled diol fraction allowed the separation of the unknown from the true diol (diacetate) by t.l.c. Chromic acid oxidation of these two separated fractions (after hydrolysis) clearly showed that the true diol gave C_{16} dicarboxylic acid while the unknown gave C_8, C_9 and C_{10}-labeled dicarboxylic acids. The correlation between the quantity of the unknown and the aldehyde content in leaves of different ages suggested that the unknown was derived mainly from 16-oxo-9-hydroxy-16:0 and a small amount from its 10-hydroxy isomer. In support of this conclusion, the distribution of ^{14}C among C_8, C_9 and C_{10} dicarboxylic acids obtained by CrO_3 oxidation of the unknown was similar to that obtained from the 16-oxo compound of the same tissue. This distribution pattern was different from that obtained from the dihydroxy C_{16} acid, in that the 9-hydroxy isomer predominated in the aldehyde compound and the unknown, while the 10-hydroxy isomer was the major one in the dihydroxy acid. Thus, the unknown compound was presumed to arise from the aldehyde and, therefore, an acetal type structure was suspected. The following experimental evidence supports this tentative conclusion: (i) the labeled unknown derived from [1-^{14}C]palmitic acid was treated with H_2SO_4 to cleave the suspected acetal linkage and the lipid products were isolated and reduced with $LiAlH_4$ (Fig. 14); t.l.c. of the products showed labeled hexadecanetriol; (ii) the unknown was treated with hydroxylamine and the products were treated with $LiAlH_4$. The oxime, if formed, should have been converted into aminohexadecane diol (Fig. 14). In fact, about a quarter of the total ^{14}C was recovered in a basic fraction. Thin-layer chromatography of this basic fraction showed that all the ^{14}C was

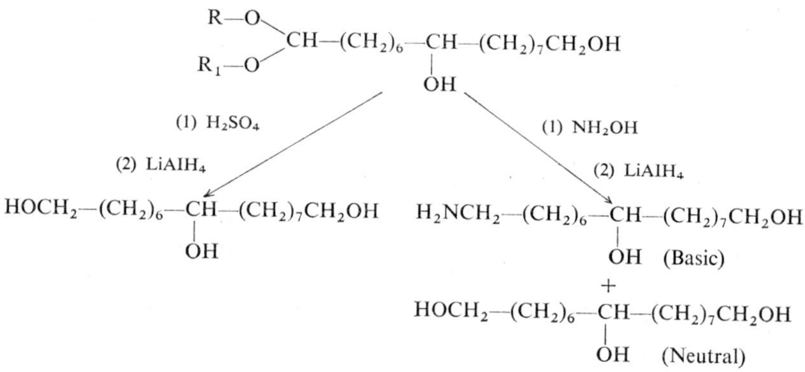

FIG. 14. Reaction sequence used for the tentative identification of acetal type structures in the cutin of young *Vicia faba* leaves.

contained in a component with an R_f identical to that of 16-aminohexadecane-1,10-diol. The relative amount of ^{14}C recovered in the basic fraction was closely correlated to the aldehyde content of the leaf, which had been shown to change dramatically with age. For example, in a young leaf (8 mm) where the free aldehyde content was found to be maximal, more than two-thirds of the total radioactive products of the reaction (excluding unreacted material) could be recovered in the basic material, while in the more developed leaf (20 mm) which contained much less aldehyde, less than one-third of the label in the products was found to be in the basic fraction. Thus, the basic material was probably derived from the 16-oxo compound contained in the unknown. Analysis by t.l.c. of the neutral fraction obtained by the above treatment gave two components, one of which was shown to be the unreacted unknown, while the other component had an R_f identical to that of hexadecanetriol. Thus, if the unknown represents an acetal, at least one of the hydroxyl groups which participated in acetal formation was that of the dihydroxy C_{16} acid.

This acetal-type compound does not appear to be an artifact formed during isolation. For example, isolation in basic buffers, which would minimize acetal formation, did not significantly affect the amount of the labeled acetal-type compound formed from [1-^{14}C]palmitic acid. Another attempt to prevent formation of acetal during the isolation procedure was made in the following manner: leaf slices were frozen in liquid nitrogen immediately after incubation with [1-^{14}C]palmitic acid and the tissue was finely powdered. This powder was immediately treated with $NaBH_4$ to reduce the free 16-oxo function. This treatment also did not significantly affect the amount of the labeled acetal-type unknown formed. From these results it is tentatively concluded that the 16-oxo compound, which is dominant only in rapidly expanding very young tissues, forms acetal-type cross linkages and possibly helps to anchor the starting points of the polymer to the plant. However, further work is required before the type of linkages involved can be defined.

(*c*) *Synthesis of dicarboxylic acids.* Dicarboxylic acids occur in most cutin preparations as minor components (Eglinton and Hunneman, 1968; Croteau and Fagerson, 1972; Walton and Kolattukudy, 1972b). These compounds are probably derived by further oxidation of the primary alcohol group at the ω-carbon. We have found an ω-hydroxy acid dehydrogenase which was located in the 100 000 *g* supernatant of extracts prepared from excised epidermis of *V. faba* leaves (Kolattukudy *et al.*, 1975a). With 16-hydroxy-16:0 as the substrate, the enzyme showed a pH optimum near 8·0 and a high degree of preference for NADP. With this enzyme preparation, the apparent K_m values for 16-hydroxy-16:0 and NADP were $1{\cdot}3 \times 10^{-5}$ M and $3{\cdot}6 \times 10^{-4}$ M, respectively. Even though free aldehyde intermediate did not accumulate, the formation of such an intermediate could be demonstrated by trapping it with dinitrophenylhydrazine. 16-Hydroxy-16:0 was the preferred substrate, but other compounds such as methyl 16-hydroxy-16:0, 10,16-dihydroxy-16:0, and its methyl ester, were also oxidized at rates 20–35% of that observed with 16-hydroxy-16:0. This dehydrogenase was severely inhibited by thiol reagents such as *N*-ethyl maleimide and *p*-chloromercuric benzoate.

(*d*) *The proposed pathway for the biosynthesis of the* C_{16}-*family of acids.* On the basis of all the experimental evidence thus far obtained it appears that palmitoyl-CoA or some other thioester such as palmitoyl-ACP undergoes ω-hydroxylation followed by hydroxylation at C-10 or C-9 (Fig. 15). Both hydroxylation steps are probably catalyzed by mixed function oxidases. The specificity of the direct hydroxylation in the middle of the chain apparently varies from species to species resulting in the different positional isomer patterns observed in different species. As soon as these monomers are formed they are transferred to the growing polymer. The involvement of the thioester postulated here is based on the following experimental observations: (i) the C-10 hydroxylase requires ATP and CoA providing direct evidence for the

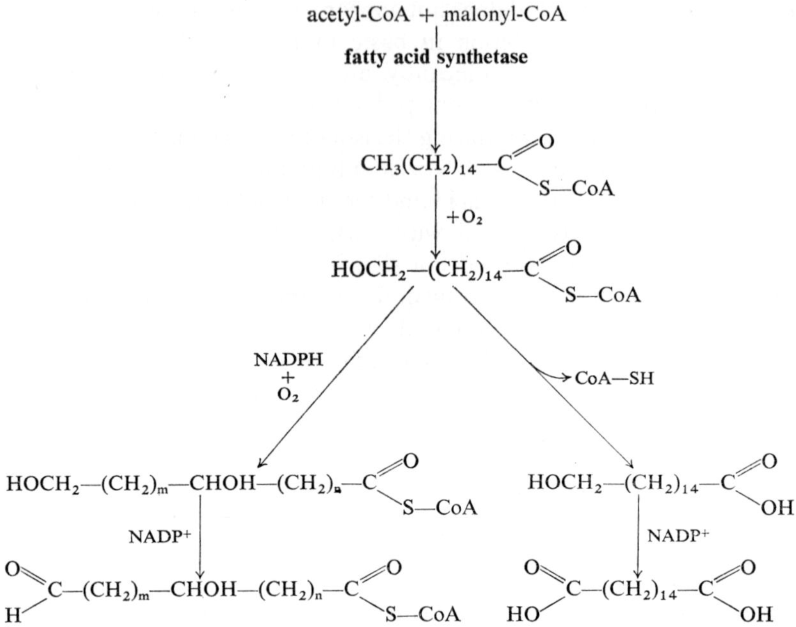

FIG. 15. The proposed pathway for the biosynthesis of the C_{16}-family of cutin acids.

involvement of a thioester in this step; (ii) incorporation of monomers into the polymer catalyzed by cell-free preparations from *V. faba* epidermis required ATP and CoA; (iii) the observation that only minor amounts of free intermediates are found in tissues which rapidly synthesize cutin is consistent with the notion that all monomers are synthesized as thioesters which are efficient acyl donors for the synthesis of the polymer.

2. *Biosynthesis of the* C_{18}*-Family of Acids*

Skin slices of young apple fruits and grape berries, as well as leaf disks of apple and *Senecio odoris* have been used for biosynthetic studies on the C_{18}-family of cutin acids (Kolattukudy *et al.*, 1971b, 1973b; Croteau and Kolattukudy, 1974b). Unlike the synthesis of the C_{16}-family of acids, saturated C_{18} acid (18:0) was not incorporated into any in-chain hydroxylated monomers, which constitute the bulk of the C_{18}-family. However, [1-^{14}C]oleic acid (18:1 9*c*) was readily converted specifically into the C_{18}-family of acids in all of the tissues mentioned above. The labeled products were identified by subjecting their derivatives to t.l.c. and g.l.c. analyses as well as by chemical degradation followed by product identification. Thus, oleic acid was shown to be converted into 18-hydroxy-18:1(9), 18-hydroxy-9,10-epoxy and 9,10,18-trihydroxy-18:0. Exogenous labeled 18-hydroxyoleic acid was similarly shown to be converted into 18-hydroxy-9,10-epoxy-18:0

and 9,10,18-trihydroxy-18:0 (Kolattukudy *et al.*, 1973b; Croteau and Kolattukudy, 1974b). More recently we have synthesized 18-hydroxy-9,10-epoxy-[18-^{3}H]-18:0 of high specific radioactivity and this acid was shown to be directly converted into 9,10,18-trihydroxy-18:0 (Croteau and Kolattukudy, 1974b). The experimental evidence discussed above strongly suggests the pathway shown in Fig. 16. Since we have no direct experimental evidence concerning the nature of the acyl derivative which undergoes the sequential reactions of ω-hydroxylation, epoxidation and epoxide hydration, the carboxyl group is shown free. However, *in vivo*, thioesters are likely to be the actual substrates because only minor amounts of free hydroxylated intermediates are found.

More recently, we have extended this pathway to include the biosynthesis of the more unsaturated members of the C_{18}-family of acids. The 12-ene and 12,15-diene members of the C_{18}-family of acids could be formed from linoleic acid (18:2 9, 12) and linolenic acid (18:3 9,12,15), respectively, by a sequence of reactions analogous to those shown in Fig. 16. Specific epoxidation of the double bond at C-9 followed by hydration of this epoxide could then account for the presence of the unsaturated members of the C_{18}-family of acids. Recently, the incorporation of [1-^{14}C]linoleic acid and [1-^{14}C]linolenic acid specifically into the C_{18}-family of hydroxy acids was demonstrated (Kolattukudy *et al.*, 1973b). Furthermore, each labeled product was shown to contain the number of double bonds predicted from the above hypothesis.

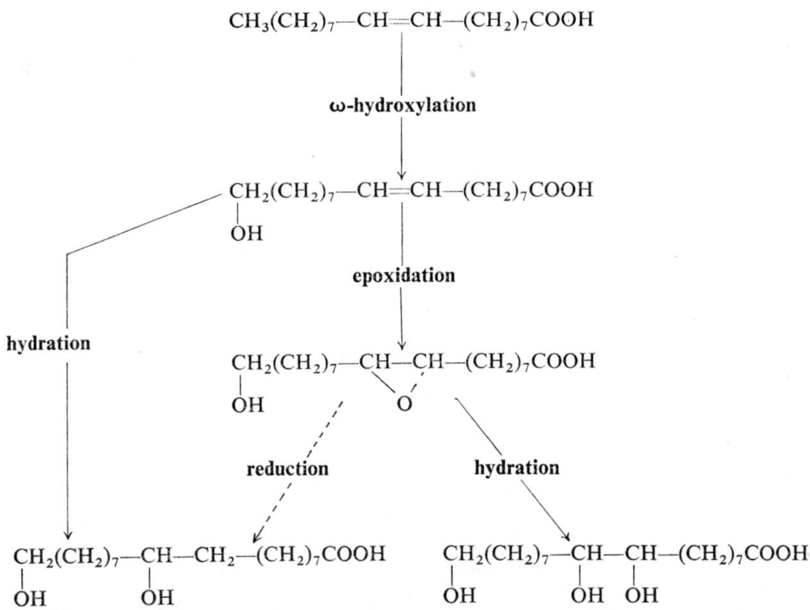

FIG. 16. The proposed pathway for the biosynthesis of the C_{18}-family of cutin acids.

Thus, specific epoxidation of the double bond at C-9 of linoleic acid and linolenic acid, followed by epoxide hydration, was shown to be the mechanism of formation of the more unsaturated cutin acids of the C_{18}-family.

The most probable mechanism for the biosynthesis of 9,10,12,13,18-pentahydroxy-18:0 was suggested to be ω-hydroxylation followed by a step-wise epoxidation and hydration of the 9- and 12-double bonds of linoleic acid (Kolattukudy and Walton, 1972b). Recently, all the intermediates expected from such a hypothesis (Fig. 17) were identified in the cutin of *Rosmarinus officinalis* (Croteau and Kolattukudy, 1974b). Furthermore, labeled linoleic acid was shown to be incorporated specifically into the pentahydroxy acid and all the expected intermediates. All the labeled products were identified by t.l.c. and g.l.c. analyses of their derivatives, as well as by chemical degradation followed by product analysis. For example, the hexaol obtained by reduction of the pentahydroxy acid derived from [U-^{14}C]linoleic acid gave rise to the expected C_6 and C_9 bifunctional molecules and malonic acid upon treatment with $NaIO_4$ and $KMnO_4$. Thus, the pathway (Fig. 16) for the biosynthesis of pentahydroxy-18:0 was established.

Recently an enzyme which catalyzes the epoxide hydration step involved in the biosynthesis of 9,10,18-trihydroxy-18:0 acid has been demonstrated in cell-free preparations of the skin of young apple fruits (Croteau and Kolattukudy, unpublished). This enzyme is localized exclusively in a cutin-containing 3 000 *g* particulate fraction and it shows a high degree of preference for 18-hydroxy-9,10-epoxy-18:0 acid. Its optimal pH is 6·5 and it shows an apparent K_m of $1{\cdot}6 \times 10^{-4}$M. This enzyme is inhibited by thiol reagents such as

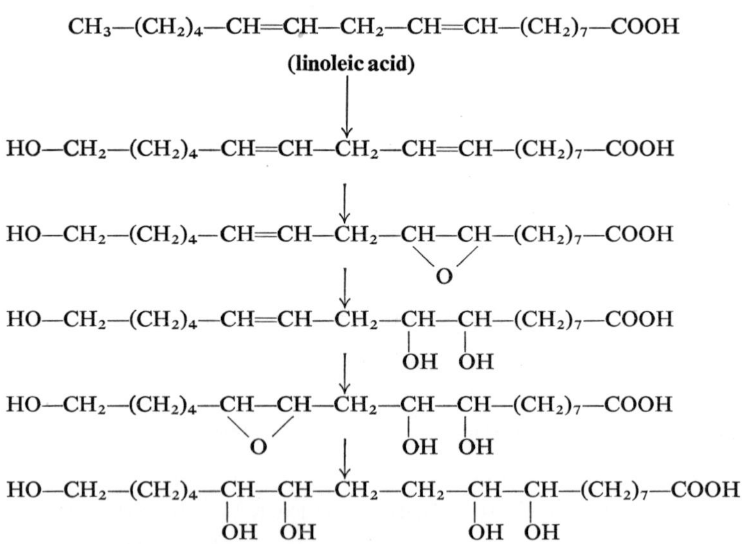

FIG. 17. The proposed pathway for the biosynthesis of the pentahydroxyoctadecanoic acid of cutin in *Rosmarinus officinalis*.

N-ethylmaleimide and *p*-chloromercuribenzoate and high concentrations (5 mM) of trichloropropyleneoxide. A much less active epoxide hydrase has been found in 100 000 *g* supernatant fractions of the epidermal extracts of *S. odoris* leaves which contain much less 9,10,18-trihydroxy-18:0 acid than the apple fruits. We have also obtained cell-free preparations which catalyze epoxidation of 18-hydroxyoleic acid, from young spinach leaves which contain 18-hydroxy-9,10-epoxy stearic acid as the major component of their cutin. However, this reaction has yet to be biochemically defined. The enzyme which catalyzes the ω-hydroxylation step remains to be demonstrated in higher plants.

D. BIOSYNTHESIS OF CUTIN FROM THE MONOMERS

In our attempts to demonstrate conversion of 16-hydroxy-16:0 into 10,16-dihydroxy-16:0 in homogenates of excised *V. faba* epidermis, it was found that a major portion of the hydroxylated product was incorporated into an insoluble polymer (Kolattukudy, unpublished results). Recently, we have examined this system, which catalyzes the synthesis of the polymer, in greater detail. A particulate preparation obtained from the above homogenate catalyzed the incorporation of 16:0, 16-hydroxy-16:0 and 10,16-dihydroxy-16:0 into an insoluble material with ATP and CoA as the required cofactors (Croteau and Kolattukudy, 1973). Sequential treatment of the insoluble residue with hydrolytic enzymes, and chemical depolymerization studies, demonstrated that these acids were esterified to cutin. The apparent K_m values for 16:0, 16-hydroxy-16:0 and 10,16-dihydroxy-16:0 were $2{\cdot}0 \times 10^{-5}$M, $6{\cdot}7 \times 10^{-5}$M and $1{\cdot}1 \times 10^{-4}$M, respectively (Croteau and Kolattukudy, 1974c). Fatty acids from C_{10} to C_{18} and C_{18}-hydroxy acids could function as substrates, but the C_{16}-family of acids was preferred, consistent with the chain length of the acyl moieties found in *V. faba* cutin. Methyl esters and fatty alcohols were not significantly incorporated into cutin, but acetates of hydroxy acids were incorporated with 60–70% efficiency of the parent acid, suggesting that the carboxyl end of the incoming monomer was transferred to free hydroxyls of the polymer. With free acids as substrates, the system had two pH optima, near 7·0 and near 8·5, while with palmitoyl-CoA a single optimal pH (near 8·5) was observed, indicating that activation of the carboxyl group and transacylation occur optimally at pH 7·0 and pH 8·5, respectively. Sulfhydryl inhibitors did not decrease the incorporation of palmitoyl-CoA into cutin, but diisopropylfluorophosphate (10^{-4}M) inhibited incorporation by 50%.

Mild sonication of the particulate preparation provided an enzyme system that required exogenous purified cutin as a primer. The transacylase activity was proportional to the amount of *V. faba* cutin added up to a level of 3 mg/ml. This primer-dependent system required the same cofactors and utilized the same substrates as the particulate preparation. A variety of chemically distinct cutins from several plant species could function as primers, but *V. faba* cutin was strongly preferred. Cellulose, glycerol, hexadecanol, cholesterol and polyethylene glycol were ineffective as acyl acceptors. Methylation of the carboxyl groups of *V. faba* cutin had no effect on its efficiency as a primer, indicating that free carboxyl groups of cutin were not involved in the esterification process. Acetylation of *V. faba* cutin decreased primer efficiency by 70%, confirming the requirement of cutin hydroxyl groups as acyl acceptors. Cutin prepared from young *V. faba* leaves was a more efficient primer than that prepared from mature, fully expanded leaves, suggesting that the enzyme prefers the less developed polymer. Treatment of *V. faba* cutin with cutinase doubled the efficiency of the primer, but extended treatment decreased the priming efficiency. Chemical treatments of cutin which increased the number of hydroxyl groups or opened the polymer matrix also increased priming efficiency. Enzyme preparations capable of esterifying acyl moieties to cutin were also obtained from *V. faba* flowers and the epidermis of *S. odoris* leaves. Regarding the biogenesis of other types of bonds, such as interchain peroxide bridges and ether bonds, no experimental results have been published since this subject was last reviewed.

E. SUBERIZATION

From the structural studies discussed in an earlier section, it appears that the major aliphatic monomers of suberin are ω-hydroxy acids, dicarboxylic acids, very long ($> C_{20}$) acids and alcohols with smaller quantities of in-chain hydroxylated and epoxy acids. The synthesis of the very long acids and alcohols probably occurs as discussed in earlier sections and the in-chain epoxy and hydroxy acids of suberin are probably formed in a manner similar to that discussed under cutin. The key reactions involved in suberin synthesis are ω-hydroxylation and oxidation of the ω-hydroxy acid to the dicarboxylic acid. Young, intact potato tubers were used in the preliminary studies on suberin biosynthesis in this laboratory. Exogenous [1-C^{14}]oleic acid administered to the tubers was incorporated into suberin. Depolymerization with $LiAlH_4$, followed by t.l.c. and radio-g.l.c. analysis of the labeled products, showed the presence of octadec-9-ene-1,18-diol, which was derived from 18-hydroxy-18:1(9) and/or the corresponding dicarboxylic acid. Since the suberin formed at the wound surface of potato slices was shown to be identical to that of the natural skin of the tuber (Kolattukudy and Dean, 1974), at least with respect to the aliphatic components, we used a tissue slice system for biochemical studies on suberization.

With the use of a g.l.c. assay for suberin, it was found that, during the first several days after the potato slices were excised, little deposition of suberin occurred. The major part of the suberin formed in these tissue slices was synthesized during the period between 5 and 7 days, after which little increase in suberin content could be detected (Fig. 18). Incorporation of [1-^{14}C]oleic acid into suberin by these tissue slices was also consistent with this time-course of suberization.

Since oxidation of ω-hydroxy acid to the corresponding dicarboxylic acid is the key reaction which distinguishes suberization from cutin synthesis, we examined the extracts of suberizing potato slices for such an ω-hydroxy acid dehydrogenase activity with 16-hydroxy-16:0 as the substrate (Agrawal and Kolattukudy, 1974, unpublished results). A fairly active enzyme was found to be located in the 100 000 *g* supernatant. This enzyme required

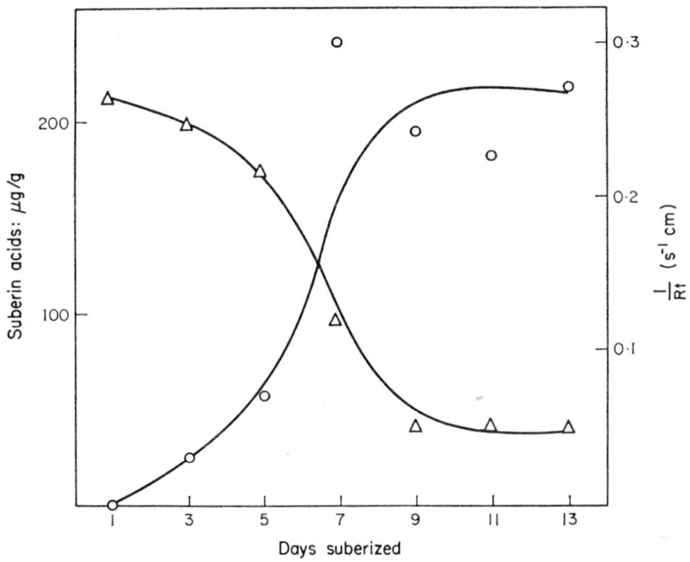

FIG. 18. Time course of development of suberin as measured by g.l.c. and the development of resistance to diffusion of water vapor. The suberin expressed as μg/g represents only the octadecene diol derived from ω-hydroxy-octadec-9-enoic acid and octadec-9-ene dioic acid. The reciprocal of diffusion resistance is plotted to show the inverse relationship between water loss from the tissue and formation of the major aliphatic components of suberin.

NADP for the conversion of 16-hydroxy-16:0 to the corresponding dicarboxylic acid. Even though the corresponding aldehyde did not accumulate, preliminary experimental results show that the aldehyde intermediate can be trapped with carbonyl reagents. Furthermore, the same enzyme preparation catalyzed conversion of 16-oxo-16:0 to the corresponding dicarboxylic acids much faster than the rate of conversion of 16-hydroxy-16:0 to the dicarboxylic acid, suggesting that, under normal conditions, conversion of the primary alcohol to the aldehyde is rate limiting. The conversion of the aldehyde intermediate into the dicarboxylic acid also showed a clear preference for NADP. The overall conversion of 16-hydroxy-16:0 acid to the dicarboxylic acid showed a pH optimum near 9·5, significantly higher than that observed with the ω-hydroxy acid dehydrogenase from the epidermis of *V. faba*. The apparent K_m of the potato dehydrogenase for 16-hydroxy-16:0 acid was 5×10^{-5} M, which is fairly close to that observed with the *V. faba* epidermal enzyme. The ω-hydroxy acid dehydrogenase of potato slices appears to be induced as a result of wounding. During the first several days after the preparation of the slices very little dehydrogenase activity could be detected. The dramatic increase in the ω-hydroxy acid dehydrogenase correlated quite well with the rapid deposition of suberin, as determined previously by g.l.c. Further work with the suberizing potato slices should enable us to define the biochemical events that lead to suberization.

F. FUNCTION OF CUTIN AND SUBERIN

Since cutin and suberin are the structural components of the outer layer of plants, it is obvious that they play a key role in the interaction of the plant with its environment. Therefore, the major role often assigned to these polymers is one of protection of the plant from dehydration, microorganisms and other harmful agents. There is a distinct possibility that cuticular components provide more than a physical barrier to pathogens (Martin, 1964). It appears that flavanoids with potential toxicity to pathogens are covalently attached to the cutin of tomato fruits. More recently, we have found phenolic material in apple fruit cutin which had been treated with 1·0 M KOH so extensively that 42% of the polymer had undergone hydrolysis. The major component of this apparently covalently linked phenolic material was identified as *p*-hydroxycinnamic acid by t.l.c. and g.l.c. techniques in conjunction with deuterium labeling and m.s. (Riley and Kolattukudy, 1974, unpublished results). Thus, preliminary evidence suggests that these protective polymers might contain covalently attached, antifungal and antibacterial agents. Even some of the aliphatic components, such as the epoxy acids, may be highly toxic to microorganisms. If the pathogen must hydrolyze the polymer to gain entry into the plant, during this process the covalently attached toxic materials would be released and thus afford protection to the plant. The covalent attachment of toxic materials to the cuticular polymer thus provides protection when needed, but the insoluble storage form is not toxic to the plant itself. For example, the epoxy acid which is a major component of cutin of many plants, including apple, is highly toxic when administered as the free monomer acid (Kolattukudy *et al.*, 1973b). Thus, in order to successfully invade a plant, a pathogen must not only develop the enzymic capability to hydrolyze the polymer, but also develop biochemical means to detoxify the materials released from the polymer during the hydrolytic entry.

Recent biochemical studies on a plant pathogen serve to illustrate the above point of view. *Fusarium solani pisi*, a pea root rotting organism, was found to be capable of growing on the hydroxy acid polymer, cutin, as the sole source of carbon. Under these conditions it produces extracellular hydrolytic enzymes which cleave the polymer (Purdy and Kolattukudy, 1973). From the extracellular fluid two enzymes have been isolated in essentially pure form, a cutinase and a *p*-nitrophenyl ester hydrolase (Purdy and Kolattukudy, 1975). The former (the cutinase) appears to be a serine hydrolase of molecular weight 22 000 while the latter is about twice as large. The cutinase has been shown to release all types of cutin monomers and it is routinely assayed with labeled cutin as the substrate. Its pH optimum is broad, between 8 and 10·5. The *p*-nitrophenyl ester hydrolase, which is routinely assayed with *p*-nitrophenyl palmitate, is also induced as a result of growth on cutin. This enzyme may be specific for the hydrolysis of the phenolic components, such as *p*-hydroxycinnamic acid, which are apparently covalently attached to cutin. In view of the toxicity of the epoxy acids generated by the hydrolytic enzymes, the fungus could not have grown on the polymer without evolving a detoxifying mechanism. In fact, a very active epoxide hydrase has been partially purified from this organism (Kolattukudy and Brown, 1975). This hydrase has a molecular weight of about 35 000, and its pH optimum is at about 8·5. With 18-hydroxy-9,10-epoxystearic acid, it showed an apparent K_m of about 6×10^{-5}. Unlike the animal epoxide hydrase (Oesch *et al.*, 1971; Oesch, 1973), agents such as trichloropropylene oxide inhibited the fungal enzyme only at fairly high concentrations, but thiol reagents severely inhibited the fungal epoxide hydrase. Thus, this pathogen has developed hydrolytic enzymes which might enable it to enter the plant, and enzymes to detoxify toxic agents released during the hydrolytic entry. The point of view expressed here is only a tentative one and much more biochemical work on many plant pathogens is required before we can understand the interaction between the protective polymers and the potential pathogens.

Prevention of pathogen entry and water loss are two functions usually attributed to suberin, and suberization has been considered to be of great economic importance in the potato industry. Recently we have obtained experimental evidence that the time-course of formation of the aliphatic components of the suberin of the wound periderm is identical to the time-course of development of diffusion resistance to moisture (Kolattukudy and Dean, 1974). The time-course of development of resistance to pathogen entry appears to be quite different from that of the formation of the aliphatic components of suberin (unpublished results). It is possible that the phenolic components of suberin provide the resistance against pathogens while the aliphatic constitutents provide the moisture diffusion resistance.

IV. Site of Cuticular Lipid Biosynthesis

The physical and chemical properties of the cuticular lipids would suggest that they are formed very near the cuticle. Experimentally it was shown that the major site of alkane synthesis in the leaves of *Senecio odoris* is the epidermis (Kolattukudy, 1968a). This conclusion was confirmed with *Allium porrum* (Cassagne, 1970). In fact, a fatty acid elongating preparation was obtained from the epidermis of *Pisum sativum* (Kolattukudy and Buckner, 1972). Wax ester synthesis was also found to occur predominantly in the epidermis (Kolattukudy, 1967a). Thus, the soluble cuticular lipids are probably all synthesized in the epidermal layer of cells. Excised epidermis of *Vicia faba*, *S. odoris* and *P. sativum* incorporated labeled acetate into cutin, showing that, in these plants, the epidermal layer of cells contains all the enzymes necessary to synthesize fatty acids and convert these acids into cutin (Kolattukudy and Buckner, unpublished results). Also, in growing apple fruits, incorporation of labeled acetate and long chain acids into cutin took place specifically in the skin, but not in the internal tissue. The enzyme preparation which hydroxylates the C-10 of 16-hydroxy-16:0 was also prepared from the excised epidermis of *V. faba* (Walton and Kolattukudy, 1972a). Hydroxyacyl-CoA-cutin transacylase (Croteau and Kolattukudy, 1973), the enzyme which builds the polymer from monomers, was actually associated with a heavy pellet fraction containing pieces of the cuticle, strongly suggesting that this enzyme is very closely associated with the growing polymer. Thus, it is likely that one of the major functions of the epidermal layer of cells is the biosynthesis of the cuticle. However, it is not known to what extent fatty acids and/or other components formed in the mesophyll cells participate in the biosynthesis of cuticular lipids. Incorporation of [1-^{14}C]acetate into alkanes in broccoli leaf slices was inhibited by DCMU, a photosynthesis inhibitor, while incorporation of palmitic acid and stearic acid into alkanes was not inhibited (Kolattukudy, 1966). These results suggest that the fatty acids synthesized from the labeled acetate within the chloroplasts participate in alkane synthesis in the epidermal cells which contain no chloroplasts. It is also possible that the major contribution of the mesophyll cells to the epidermal cells is an energy source, such as sucrose. But there appear to be few experimental data bearing on this problem.

V. CONCLUSION

It is well known that the cuticle, which is derived entirely from lipids, plays an important role in the interaction between the plant and its environment. In order to understand these interactions a thorough knowledge of the structure of the cuticle and its biochemistry is required. Such an understanding has practical applications in agriculture in that agricultural chemicals must interact with this layer before they enter the plant. Furthermore, knowledge about the ways in which a plant uses its cuticle for protection could allow us to devise ways for efficient but natural protection of crops. With the application of modern analytical and biochemical techniques we are beginning to understand the structure and biochemistry of the protective lipid barriers of plants. However, our knowledge of the biochemistry of these unique lipids is indeed limited. For example, we know little about the structural details of cutin and suberin, and virtually nothing about the three-dimensional structure of these polymers. Many key enzymes involved in the biosynthesis of the cuticle have not been isolated and the nature of several reactions has yet to be defined. We know practically nothing about the regulation of cuticle biosynthesis. The biochemical nature of the interaction between the lipid barriers of plants and pathogens is also poorly understood. Since most of these areas can be investigated with the presently available techniques, they are likely to be fruitful fields for investigation.

ACKNOWLEDGEMENTS

The work from my laboratory, discussed in this paper was supported in part by U.S.P.H.S. Grant GM 18278 and N.S.F. Grants GB–23081 and GB–43076X. Scientific paper No. 4405, Project 2001, College of Agriculture Research Center, Washington State University, Pullman, Washington 99163.

REFERENCES

Albro, P. W. and Dittmer, J. C. (1970). *Lipids* **5**, 320–325.

Brieskorn, C. H. and Kabelitz, L. (1971). *Phytochemistry* **10**, 3195–3204.

Brown, A. C., Knights, B. A. and Conway, E. (1969). *Phytochemistry* **8**, 543–547.

Buckner, J. S. and Kolattukudy, P. E. (1973). *Archs Biochem. Biophys.* **156**, 34–45.

Carruthers, W. and Johnstone, R. A. (1959). *Nature, Lond.* **184**, 1131–1132.

Cassagne, C. (1970). "Les Hydrocarbures Végétaux: Biosynthèse et Localisation Cellulaire". Ph.D. Thesis, University of Bordeaux, France.

Channon, H. J. and Chibnall, A. C. (1929). *Biochem. J.* **23**, 168–175.

Chibnall, A. C. and Piper, S. H. (1934). *Biochem. J.* **28**, 2209–2219.

Clenshaw, E. and Smedley-Maclean, I. (1929). *Biochem. J.* **23**, 107–109.

Crisp, C. E. (1965). "The Biopolymer Cutin". Ph.D. Thesis, University of California, Davis.

Croteau, R. and Fagerson, I. S. (1972). *Phytochemistry* **11**, 353–363.

Croteau, R. and Kolattukudy, P. E. (1973). *Biochem. biophys. Res. Commun.* **52**, 863–869.

Croteau, R. and Kolattukudy, P. E. (1974a). *Archs Biochem. Biophys.* **162**, 458–470.

Croteau, R. and Kolattukudy, P. E. (1974b). *Archs Biochem. Biophys.* **162**, 471–480.

Croteau, R. and Kolattukudy, P. E. (1974c). *Biochemistry* **13**, 3193–3202.

Day, J. I. E., Goldfine, H. and Hagen, P. O. (1970). *Biochim. biophys. Acta* **218**, 179–182.

Dodova-Anghelova, M. S. and Ivanov, C. P. (1973). *Phytochemistry* **12**, 2239–2242.

Eglinton, G. and Hunneman, D. H. (1968). *Phytochemistry*, **7**, 313–322.

Eglinton, G., Hunneman, D. H. and McCormick, A. (1968). *Organic Mass Spectrometry* **1**, 593–611.

Fehler, S. W. G. and Light, R. J. (1970). *Biochemistry* **9**, 418–422.

Fehler, S. W. G. and Light, R. J. (1972). *Biochemistry* **11**, 2411–2416.

Friedberg, S. J. and Greene, R. C. (1967). *J. biol. Chem.* **242**, 234–237.
Gelpi, E., Oro, J., Schneider, H. J. and Bennett, E. O. (1968). *Science, N.Y.* **161**, 700–702.
Grigor, M. R. and Bell, I. C. (1973). *Biochim. biophys. Acta* **306**, 26–30.
Han, J., Chan, H. W.-S. and Calvin, M. (1969). *J. Am. chem. Soc.* **91**, 5156–5159.
Han, J., McCarthy, E. D. and Calvin, M. (1968). *J. chem. Soc.* (*C*) 2785–2791.
Holloway, P. J. (1972a). *Chem. Phys. Lipids* **9**, 158–170.
Holloway, P. J. (1972b). *Chem. Phys. Lipids* **9**, 171–179.
Holloway, P. J. and Deas, A. H. B. (1971). *Phytochemistry* **10**, 2781–2787.
Holloway, P. J. and Deas, A. H. B. (1973). *Phytochemistry* **12**, 1721–1735.
Horn, D. H. S., Kranz, Z. H. and Lamberton, J. A. (1964). *Aust. J. Chem.* **17**, 464–476.
Jarolimek, P., Wollrab, V., Streibl, M. and Sorm, F. (1964). *Chemy Ind.* 237–238.
Johnson, R. C. and Gilbertson, J. R. (1972). *J. biol. Chem.* **247**, 6991–6998.
Kaneda, T. (1967). *Biochemistry* **6**, 2023–2032.
Kawalek, J. C. and Gilbertson, J. R. (1974). *Fedn Proc. Fedn Am. Socs exp. Biol.* **33**, 1378.
Khan, A. A. and Kolattukudy, P. E. (1973a). *Archs Biochem. Biophys.* **158**, 411–420.
Khan, A. A. and Kolattukudy, P. E. (1973b). *Biochemistry* **12**, 1939–1948.
Khan, A. A. and Kolattukudy, P. E. (1974). Manuscript in preparation.
Knights, B. A., Brown, A. C., Conway, E. and Middleditch, B. S. (1970). *Phytochemistry* **9**, 1317–1324.
Kolattukudy, P. E. (1965). *Biochemistry* **4**, 1844–1855.
Kolattukudy, P. E. (1966). *Biochemistry* **5**, 2265–2275.
Kolattukudy, P. E. (1967a). *Biochemistry* **6**, 2705–2715.
Kolattukudy, P. E. (1967b). *Phytochemistry* **6**, 963–975.
Kolattukudy, P. E. (1968a). *Pl. Physiol., Lancaster* **43**, 375–383.
Kolattukudy, P. E. (1968b). *Pl. Physiol., Lancaster* **43**, 1423–1429.
Kolattukudy, P. E. (1968c). *Pl. Physiol., Lancaster* **43**, 1466–1470.
Kolattukudy, P. E. (1968d). *Science, N.Y.* **159**, 498–505.
Kolattukudy, P. E. (April, 1969). Am. Oil Chem. Soc. meeting, San Francisco.
Kolattukudy, P. E. (1970a). *A. Rev. Pl. Physiol.* **21**, 163–192.
Kolattukudy, P. E. (1970b). *Archs Biochem. Biophys.* **141**, 381–383.
Kolattukudy, P. E. (1970c). *Biochem. biophys. Res. Commun.* **41**, 299–305.
Kolattukudy, P. E. (1970d). *Biochemistry* **9**, 1095–1102.
Kolattukudy, P. E. (1970e). *Lipids* **5**, 259–275.
Kolattukudy, P. E. (1970f). *Lipids* **5**, 398–402.
Kolattukudy, P. E. (1971). *Archs Biochem. Biophys.* **142**, 701–709.
Kolattukudy, P. E. (1972). *Biochem. biophys. Res. Commun.* **49**, 1040–1046.
Kolattukudy, P. E. (1973). *Lipids* **8**, 90–92.
Kolattukudy, P. E. (1974). *Biochemistry* **13**, 1354–1363.
Kolattukudy, P. E. and Agrawal, V. P. (1974). *Lipids* **9**, 682–691.
Kolattukudy, P. E. and Brown, L. (1974). *Pl. Physiol., Lancaster* **53**, 903–906.
Kolattukudy, P. E. and Brown, L. (1975). *Archs Biochem. Biophys.* **166**, 599–607.
Kolattukudy, P. E. and Buckner, J. S. (1972). *Biochem. biophys. Res. Commun.* **46**, 801–807.
Kolattukudy, P. E. and Dean, B. B. (1974). *Pl. Physiol., Lancaster* **54**, 116–121.
Kolattukudy, P. E. and Liu, T. J. (1970). *Biochem. biophys. Res. Commun.* **41**, 1369–1374.
Kolattukudy, P. E. and Walton, T. J. (1972a). *Biochemistry* **11**, 1897–1907.
Kolattukudy, P. E. and Walton, T. J. (1972b). *Progr. Chem. Fats Lipids* **13**, 119–175.

Kolattukudy, P. E., Jaeger, R. H. and Robinson, R. (1968). *Nature, Lond.* **219**, 1038–1040.

Kolattukudy, P. E., Jaeger, R. H. and Robinson, R. (1971a). *Phytochemistry* **10**, 3047–3051.

Kolattukudy, P. E., Walton, T. J. and Kushwaha, R. P. S. (1971b). *Biochem. biophys. Res. Commun.* **42**, 739–745.

Kolattukudy, P. E., Buckner, J. S. and Brown, L. (1972). *Biochem. biophys. Res. Commun.* **47**, 1306–1313.

Kolattukudy, P. E., Buckner, J. S. and Liu, T. J. (1973a). *Archs Biochem. Biophys.* **156**, 613–620.

Kolattukudy, P. E., Walton, T. J. and Kushawaha, R. P. S. (1973b). *Biochemistry* **12**, 4488–4498.

Kolattukudy, P. E., Croteau, R. and Brown, L. (1974). *Pl. Physiol., Lancaster* **54**, 670–677.

Kolattukudy, P. E., Croteau, R. and Walton, T. J. (1975). *Pl. Physiol., Lancaster* (in press).

Kolattukudy, P. E., Kronman, K. and Poulose, A. J. (1975b). *Pl. Physiol., Lancaster* (in press).

Kranz, Z. H., Lamberton, J. A., Murray, K. E. and Redcliffe, A. H. (1960). *Aust. J. Chem.* **13**, 498–505.

Kuksis, A. (1964). *Biochemistry*, **3**, 1086–1093.

Lamberton, J. A. (1965). *Aust. J. Chem.* **18**, 911–913.

Lee, B. (1925). *Ann. Bot.* **39**, 755–768.

Legg, V. H. and Wheeler, R. V. (1929). *J. chem. Soc.* **131**, 2444–2449.

Lessire, R. (1973). "Étude de la biosynthèse des acides gras à très longue chaîne et leurs relations avec les alcanes dans la cellule d'épiderme d'*Allium porrum*." Ph.D. Thesis, University of Bordeaux, France.

Liu, T. J. (1972). "The Biosynthetic Relationship Among Long Chain Fatty Acids, Alkanes, Secondary Alcohols, and Ketones in Plants." M.S. Thesis, Washington State University.

Macey, M. J. K. and Barber, H. N. (1970a). *Phytochemistry* **9**, 5–12.

Macey, M. J. K. and Barber, H. N. (1970b). *Phytochemistry* **9**, 13–23.

Macey, M. J. K. and Stumpf, P. K. (1968). *Pl. Physiol., Lancaster* **43**, 1637–1647.

Martin, J. T. (1964). *A. Rev. Phytopathology* **2**, 81–100.

Martin, J. T. and Juniper, B. E. (1970). "The Cuticles of Plants". St. Martins Press, New York.

Matic, M. (1956). *Biochem. J.* **63**, 168–178.

Mazliak, P. (1968). *In* "Progress in Phytochemistry" (L. Reinhold and Y. Liwschitz, eds), Vol. 1, pp. 49–111. Interscience, New York.

Mold, J. D., Means, R. E. and Ruth, J. M. (1966). *Phytochemistry* **5**, 59–66.

Netting, A. G. and Macey, M. J. K. (1971). Personal communication.

Nevenzel, J. C. (1970). *Lipids* **5**, 308–319.

Oesch, F. (1973). *Xenobiotica* **3**, 305–340.

Oesch, F., Kaubisch, N., Jerina, D. M. and Daly, J. W. (1971). *Biochemistry* **10**, 4858–4866.

Orgell, W. H. (1955). *Pl. Physiol., Lancaster* **30**, 78–80.

Patton, S., Lee, R. F. and Benson, A. A. (1972). *Biochim. biophys. Acta* **270**, 479–488.

Purdy, R. and Kolattukudy, P. E. (1973). *Archs Biochem. Biophys.* **159**, 61–69.

Purdy, R. and Kolattukudy, P. E. (1975). *Biochemistry* (in press).

Purdy, S. J. and Truter, E. V. (1963). *Proc. R. Soc. Lond.* B **158**, 553–565.

Radler, F. (1965). *Aust. J. biol. Sci.* **18**, 1045–1056.

Rodriguez-Miguens, B. and Ribas-Marques, I. (1972). *Ann. R. Soc. Esp. Fis. Quim.* **68**, 303–308.

Sargent, J. R., Gatten, R. R. and McIntosh, R. (1974). *Comp. Biochem. Physiol.* **47B**, 217–227.

Sargent, J. R., Gatten, R. R. and McIntosh, R. (1974). *Comp. Biochem. Physiol.* **47B**, 217–227.

Schmid, H. H. O. and Bandi, P. C. (1969). *Hoppe-Seyler's Z. physiol. Chem.* **350S**, 462–466.

Snyder, F. (1969). *Progr. Chem. Fats Lipids* **10**, 287–335.

Snyder, F. and Malone, B. (1970). *Biochem. Biophys. Res. Commun.* **41**, 1382–1387.

Stransky, K., Streibl, M. and Kubelka, V. (1970). *Coll. Czech. Chem. Commun.* **35**, 882–891.

Tulloch, A. P. and Hoffman, L. L. (1971). *Phytochemistry* **10**, 871–876.

Tulloch, A. P. and Weenik, R. O. (1966). *Chem. Commun.* **8**, 225–226.

Walton, T. J. and Kolattukudy, P. E. (1972a). *Biochem. biophys. Res. Commun.* **46**, 16–21.

Walton, T. J. and Kolattukudy, P. E. (1972b). *Biochemistry, Eastern* **11**, 1885–1897.

Wang, L., Takayama, K., Goldman, D. S. and Schroes, H. K. (1972). *Biochim. biophys. Acta* **260**, 41–48.

Wollrab, V. (1969). *Phytochemistry* **8**, 623–627.

Wollrab, V., Streibl, M. and Sorm, F. (1965). *Coll. Czech. Chem. Commun.* **30**, 1654–1669.

Wollrab, V., Streibl, M. and Sorm, F. (1967). *Chemy Ind.*, 1872–1873.

Youngblood, W. W., Blumer, M., Guillard, R. L. and Fiore, F. (1971). *Marine Biol.* **8**, 190–201.

CHAPTER 8

Biochemical and Structural Aspects of Storage and Membrane Lipids in Developing Oil Seeds

L-Å. APPELQVIST

Department of Food Hygiene, Royal Veterinary College, Stockholm, Sweden

I. Introduction

The "storage lipids" of oil-rich seeds, viz. the triacylglycerols of the vegetable oils, are a major food source. As a consequence of their importance in the food industry, much has been published on the lipid composition of the dry oil-rich seeds (see e.g. Hilditch and Williams, 1964; Hitchcock and Nichols, 1971; Gunstone, 1967 with an appendix listing many books in this area). In marked contrast to the large amount of data accumulated on the lipid patterns of mature seeds, the biosynthesis of "storage lipids" in developing oil seeds has been studied to a very minor extent by lipid biochemists. An important reason for this discrepancy probably resides in the technical problems associated with the acquisition of enough starting material of active tissue of well defined status in the biochemist's laboratory. As is well known among those familiar with the literature in this area, favourite tissues for biochemists studying higher plants are either mature leaf tissues, e.g. spinach or lettuce leaves which are easily available at a food store, or tissues which are rapidly prepared in a laboratory under reproducible conditions, namely germinating seeds and small seedlings.

Those oil seeds which during ontogeny have a translucent seed coat, are rich in chloroplasts at the early stages of seed development. The lipids of chloroplasts, mitochondria and other membranous cell constituents are discussed in this paper under the common name "membrane lipids". Generally the discussion will be restricted to fatty acid derivatives and thus exclude polyisoprenoids. As will be seen in the following pages, the lipid biosynthesis both in the chloroplast and in the cytosol of the developing oil-rich tissue must be considered in any attempt to acquire a better understanding of the anabolic metabolism of such seed storage tissue. Whereas most reviews on lipid biochemistry of higher plants focus attention on *pathways* and *enzymes* regardless of tissue type (see e.g. Mazliak, 1973 and loc. cit.), this review will emphasize the *oil seed tissue* and its stage of *development*. It will not present all the literature in the area which has been published during a certain time period. Instead selected papers, some very recent, some rather old, are used in order to present glimpses of the front line in this area of plant lipid biochemistry.

It might also be appropriate to mention that the literature relevant to this presentation is scattered in publications within the various branches of science. This fact might sometimes be overlooked in starting new projects. Valuable information in this area is obtained, for example, from journals within the fields of agriculture and lipid chemistry as well as in biochemical and physiological journals.

II. Lipid Patterns in Mature Seeds and During Seed Development

A. MATURE SEEDS

Leaf tissues from higher plants regardless of species are generally rather similar in fatty acid composition, with palmitic, linoleic and linolenic acids as the dominant fatty acids. On the other hand seed lipids display tremendous variability in fatty acid composition (Hilditch and Williams, 1964; Wolff, 1966; Gunstone, 1967; Hitchcock and Nichols, 1971; Smith, 1970), often overlooked in presentations on plant lipids in biochemical texts. The economically most important species however, have predominantly the common fatty acids in their seed oils. Groundnuts, sunflower, safflower and cottonseed have oleic and linoleic acids as major fatty acids; soyabeans, "zero-erucic acid" rapeseed and flax have these plus linolenic acid.

With few exceptions the data on lipid accumulation in developing oil-rich seeds have been acquired on the oleic, linoleic and linolenic acid-rich species. Important exceptions are *Ricinus communis* (castor bean), rich in ricinoleic acid, *Brassica napus* (rapeseed), *Crambe abyssinica* and *Sinapis alba* rich in erucic acid, *Vernonica anthelmintica*, rich in epoxyoleic acid and *Crepis rubra*, rich in crepenynic acid (for references, see p. 251).

Before discussing seed development in various species, it is interesting to recall some fundamental anatomical differences in oil seeds as presented by Thies (1971). It is seen from Fig. 1 that seeds which accumulate substantial amounts of linolenic acids in their storage lipids are generally those which have cotyledons as major storage organs and which have a seed coat that is at

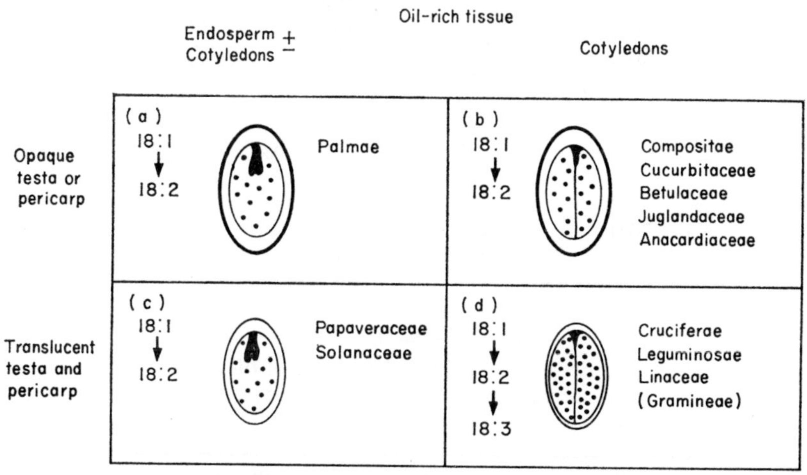

FIG. 1. The four main types of fat-storing angiosperm seeds. Generally, only the plant families named in group (d) synthesize substantial amounts of linolenic acid (18:3) in addition to oleic (18:1) and linoleic (18:2) acids. (After Thies, 1971.)

least partially translucent. (However, there appear to be exceptions to the rule: the lipids from walnuts (family Juglandaceae), which during their development appear to have an opaque seed coat, contain linolenic acid, *ca.*14% in a sample analysed at our laboratory.) Incidentally, data on the extent of light permeability of seed coats of developing oil-rich seeds seem not to be easily obtainable. The presence of chloroplasts with normal grana and starch grains in the immature cotyledons (see Fig. 10), indicates that biosynthetic activities in the chloroplasts may have significance for the amount and possibly also composition of the lipids which accumulate in the oil bodies of the cytosol. This will be discussed further below.

The lipid composition of seed parts (e.g. cotyledons, endosperm, hypocotyl, seed coat) differ in several respects as demonstrated for rapeseed (*Brassica napus*) in Table I and castor bean (*Ricinus communis*) in Table II. The data

TABLE I

Fatty acid composition of cotyledons, hypocotyl and seed coat of rapeseed (*Brassica napus* cv. Regina) (From Appelqvist, 1969)

Fatty acid	Cotyledons	Hypocotyl	Seed coat
	Fatty acid composition, (%)		
16:0	3·3	6·6	5·2
16:1	0·3	0·6	2·0
18:0	1·2	1·6	1·2
18:1	12·4	15·3	18·7
18:2	15·2	21·3	17·0
18:3	9·0	9·8	6·1
20:0	0·8	0·8	0·6
20:1	9·5	10·3	14·4
22:0	0·5	0·5	0·7
22:1	44·6	29·9	30·0
	Isomer proportions ($\omega 9/\omega 7$)		
18:1	12	8	0·9
20:1	8	8	0·3
22:1	25	16	0·3

TABLE II

Percentage fatty acid composition of oil from germinating seeds of castor bean (From Canvin, 1963)

Fatty acid	Embryo[a]	Endosperm[a]
palmitic	5·9	1·2
oleic + linoleic	31·2	9·9
linolenic	7·6	0·0
ricinoleic	55·1	88·4

[a] Germinated in vermiculite for 4 days at 30°C.

in Table II refer to 4-day-old seedlings but mirror the conditions in the dry seed. Although the data refer to total lipids, they largely represent triacylglycerols. It is important to note that the monoenes of the seed coat of rapeseed are very rich in the $\omega 7$*-isomers which are present in minor amounts in the cotyledons and the hypocotyl. These parts predominantly have the more common $\omega 9$-isomers. This difference in lipid composition in the various anatomical parts of a mature seed or bean should be considered when seed tissue is used for biosynthetic studies. A striking example occurs in studies of

* See footnote on page 6.

the biosynthesis of erucic acid, 22:1(ω9), in developing rapeseed where the precursor supplied under some incubation conditions could be used preferentially by the seed coat (unpublished observations) giving rise to mixtures of monoenoic acids which yield complex patterns in degradation studies.

B. DEVELOPING SEEDS

Studies on the accumulation of total lipids (increase in the oil content) of developing seeds have been reported in numerous publications, not listed in this review. There are many papers which also report changes in fatty acid composition. Among those which report gas-chromatographic data for fatty acid composition the following may be mentioned: studies on soyabeans by Fehr *et al.* (1971), Rubel *et al.* (1972) and by Privett *et al.* (1973); on *Crambe abyssinica* by McKillican (1966) and Gurr *et al.* (1972); on rapeseed by Sims (1964), McKillican (1966), Zeman and Kratochvíl (1967) and Appelqvist (unpublished results); on white mustard by Dasgupta and Friend (1973); on sunflower by Hopkins and Chisholm (1961); on flax by Dybing and Zimmerman (1966); on castor bean by Canvin (1963); on *Crepis rubra* by Haigh *et al.* (1968) and on *Vernonica anthelmintica* by Miwa *et al.* (1963). This list is certainly not complete but could aid the interested reader in finding the way into pertinent literature.

The accumulation of organic material in developing oil seeds from petal fall to full maturity can generally be divided into three periods. The first period which, depending upon the species and the environmental conditions, ranges from 10 to 30 days is characterized by a very slow accumulation of lipids and, in chloroplast-containing cotyledonary tissue, but not in castor bean endosperm or sunflower cotyledons (cf. Fig. 1), rather substantial accumulation of carbohydrates. During this period the seed has a very high moisture content. The second period stretching from *ca.* 2 to 5 weeks after the first period in different oil seeds, is characterized by rapid accumulation of lipids. Figure 2 illustrates this for *Crambe abyssinica* (Gurr *et al.*, 1972).

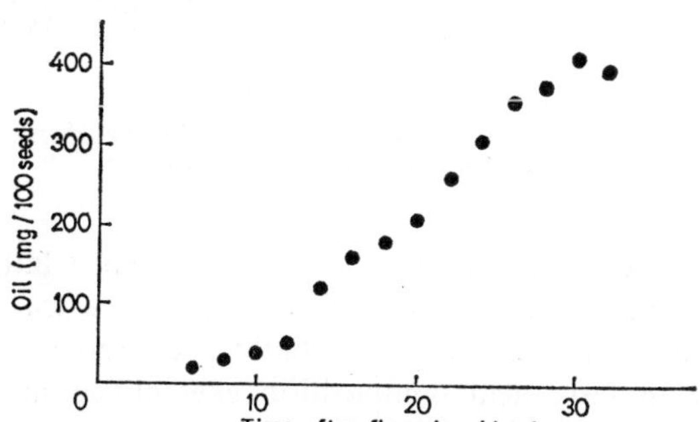

FIG. 2. The change in oil content of maturing seeds of *Crambe abyssinica*, expressed as weight of oil per seed. (From Gurr *et al.*, 1972.)

The general pattern of Fig. 2 is typical for most of the oil seeds reported in the literature. In the third stage, ending at seed maturity, generally only minor amounts of lipids accumulate. If starch was present in the early period it has disappeared at maturity. Besides a loss in moisture, there is often a continuous accumulation of protein during the third stage.

Among recent papers in this field one should also mention the report by Rest and Vaughan (1972) on the development of seed in white mustard as studied by light microscopy and electron microscopy. However, no concerted electron microscopic and lipid analytical studies on the seed development period have been reported in the literature besides a preliminary report by Norton and Harris (1975) from such research on developing seeds of rapeseed.

Data from two very recent series of studies, those of Gurr *et al.* (1972, 1974) on *Crambe abyssinica* and those of Privett *et al.* (1973 and loc. cit.) on soyabeans, together with data on rapeseed from the author's laboratory, will be used as examples of developmental changes in this review. As demonstrated in Fig. 3, the total fatty acid composition changes dramatically during seed

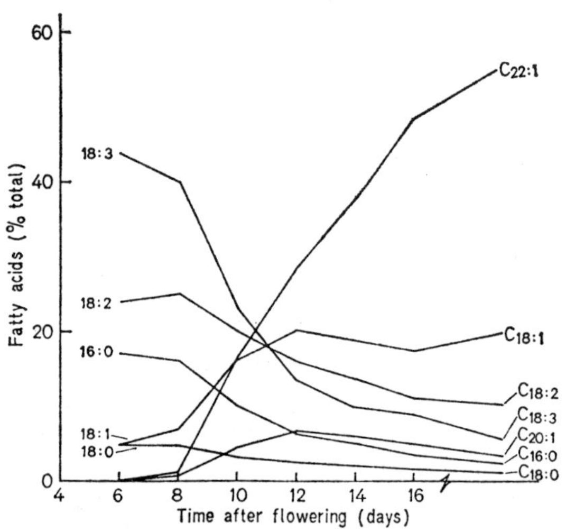

FIG. 3. Changes in fatty acid composition of seeds of *Crambe abyssinica* during seed development. (From Gurr *et al.*, 1972.)

formation in *Crambe abyssinica.* Erucic acid, the major fatty acid of the triacylglycerols of the mature seed, is found only as a trace at the early stages of seed development. This is partially a consequence of the lipid moiety-specificity in fatty acid composition exemplified in Table III for another erucic acid-rich species, namely rapeseed. At the early stages of seed development a major portion of the total lipids is comprised of phospho- and galactolipids (cf. Table V), very low in erucic acid. The almost complete

TABLE III

Fatty acid composition of lipids or lipid classes of siliquae and seeds from rapeseed (*Brassica napus*, L. cv. Gulle). The plants were grown outdoors and were in a stage of late maturation (Appelqvist, unpublished)

Lipid class[a] (Major constituents)	Fatty acid composition (%)										
	16:0	16:3	18:0	18:1	18:2	18:3	20:0	20:1	22:0	22:1	Other acids
Siliquae											
PC+PI+PE	16·3	0·3	6·8	6·4	33·2	35·2	tr	—	—	—	1·8
PG+SQDG	23·6	0·4	6·9	3·2	29·5	32·1	tr	—	—	—	4·3
DGDG	16·7	1·6	4·8	2·3	19·0	52·2	—	—	—	—	3·4
MGDG	10·5	10·2	5·8	2·2	16·0	51·6	1·6	—	—	—	2·1
Neutral lipids	12·6	2·9	4·1	15·5	19·8	21·3	3·7	5·1	0·5	10·2	34·3
Seeds											
PC+PI+PE	16·9	1·0	5·4	9·4	39·1	21·7	3·3	0·9	—	—	2·3
PG+SQDG	26·2	0·2	4·7	9·4	38·7	18·8	0·2	0·8	—	—	1·0
DGDG	16·0	0·8	3·0	12·6	28·9	35·9	—	1·0	—	—	1·8
MGDG	8·8	7·2	2·0	14·5	29·6	28·6	0·1	3·0	4·0	0·8	1·4
Neutral lipids	4·9	0·1	1·5	18·4	17·7	9·9	0·8	12·0	tr	34·3	0·4

[a] SQDG = sulphoquinovosyldiglyceride; for other abbreviations, see footnote to Table VI.

exclusion of erucic acid from membrane lipids in rapeseed is an example of the general exclusion of unusual fatty acids from membrane lipids. However, the triacylglycerols are also low in erucic acid at early stages of development of cruciferous seeds. Actually erucic acid was not detectable before 6 days after petal fall in *Crambe abyssinica* (Gurr *et al.*, 1972). Similarly, ricinoleic acid was not found in the very young castor bean embryos (Canvin, 1963) and crepenynic acid was not detected in 4 day old embryos of *Crepis rubra* (Haigh *et al.*, 1968). The data in Fig. 3 and Table III, besides illustrating tissue specificity *and* lipid moiety specificity in fatty acid patterns, demonstrate that data on the composition of total fatty acids *vs* stages of development can only be a guide for selecting the proper stage of a tissue for biosynthetic work or detailed compositional analysis in which the individual lipid classes are analysed separately for fatty acid composition.

In the past, gross changes in fatty acid composition of oil seed tissue have been interpreted in terms of metabolic interconversions. The danger of such interpretations is very obvious to readers with a biochemical background. Nevertheless it appears appropriate to present some figures which demonstrate how misleading percentage data can be. Values for fatty acids for developing zero-erucic acid rapeseed are presented in Table IV first on a percentage basis and then on an absolute basis, i.e. mg fatty acids per 100 seeds. Thereby it is easily seen that on a "per seed" basis there is no loss of any fatty acid but a difference in the rate of accumulation for the various fatty acids with stage of development. Consequently in any study of this kind the percentage distribution of total fatty acids gives little aid in efforts to explain the events occurring in such tissues.

Reference should next be made to the studies by Privett and collaborators on developing soyabeans (Roehm and Privett, 1970, Singh and Privett 1970a, b, Privett *et al.*, 1973). In the latest study figures for total lipids, neutral lipids, phospholipids and glycolipids on a per seed basis at various stages after flowering were presented. From Fig. 4 it is easily seen that there is an active accumulation period from about 20 days after flowering to about 60 days after flowering (under the field conditions used) in which the neutral lipids, mainly triacylglycerols, increased from *ca.* 0·5 mg per bean to *ca.* 50 mg per bean. At the same time the phospho- and galactolipids increased. At the end of the neutral lipid accumulation period, there was a drop in the amount of both the phospho- and the galactolipids, except for the very last sampling date at which there was a total increase in all lipid components. (The atypical figure at 97 days after flowering could be related to some sampling problem or an analytical artifact. Studies on zero-erucic acid rapeseed in the author's laboratory have demonstrated a loss in phospho- and galactolipids in the last stage of seed development; see further page 259 and Table VII.) However, it has often been observed with rapeseed that the seeds start to germinate in their siliquae under certain climatic conditions. If such seeds are subjected to analysis, one expects to find more membranous lipids.

TABLE IV

Changes in fatty acid patterns and amounts of fatty acids per seed with stage of development of a "zero-erucic acid" genotype of rapeseed, planted in a growth chamber (Appelqvist, unpublished)

Moisture content (%)	Seed weight (mg dry material)	Oil content (% dry matter)	Oil content (mg/seed)	12:0	14:0	16:0	16:1	16:3	18:0	18:1	18:2	18:3	20:0	20:1
				Fatty acid composition (%)										
79	0·7	3·76	0·03	0·85	0·69	21·71	0·85	0·57	6·26	27·15	29·83	11·83	0·29	—
76	1·3	9·15	0·11	0·27	0·40	10·99	0·33	0·14	3·05	54·75	19·85	8·87	0·66	0·73
75	1·5	41·6	0·60	0·10	0·10	6·11	0·35	0·15	1·89	61·09	19·99	8·61	0·57	1·06
61	3·6	34·2	1·24	0·07	0·07	5·92	0·47	0·25	1·62	59·29	20·87	9·94	0·45	1·06
				Content of individual fatty acids (mg/100 seeds)										
79	0·7	3·76	0·03	0·02	0·02	0·6	0·02	0·02	0·2	0·7	0·8	0·3	0·01	—
76	1·3	9·15	0·11	0·03	0·05	1·3	0·04	0·02	0·4	6·2	2·3	1·0	0·08	0·08
75	1·5	41·6	0·60	0·06	0·06	3·7	0·21	0·09	1·1	36·9	12·1	5·2	0·34	0·64
61	3·6	34·2	1·24	0·08	0·09	7·3	0·58	0·31	2·0	73·5	25·9	12·3	0·56	1·31

The proportions of neutral lipids, glycolipids and phospholipids among the total lipids at the various stages of development of the soyabean are presented in Table V and the distribution of individual lipid classes in Table VI. From these tables it is easily seen that the proportion of neutral lipids gradually increases and that of glycolipids and phospholipids decreases. Furthermore, in the neutral lipid fraction the triacylglycerols rise from a trace at 9 days to 97–98 % at maturity. One should also note the fact that only traces of free fatty acids were detected after 40 days and no monoacylglycerols. Considerable amounts of free fatty acids and partial glycerides were previously reported from studies of maturing white mustard (Dasgupta and Friend, 1973) rapeseed (McKillican, 1966) and soyabeans (Hirayama and Hujii, 1965). Probably, these earlier findings are related to degradation during sampling and/or extraction. The glycolipid fraction progressively decreased as per-

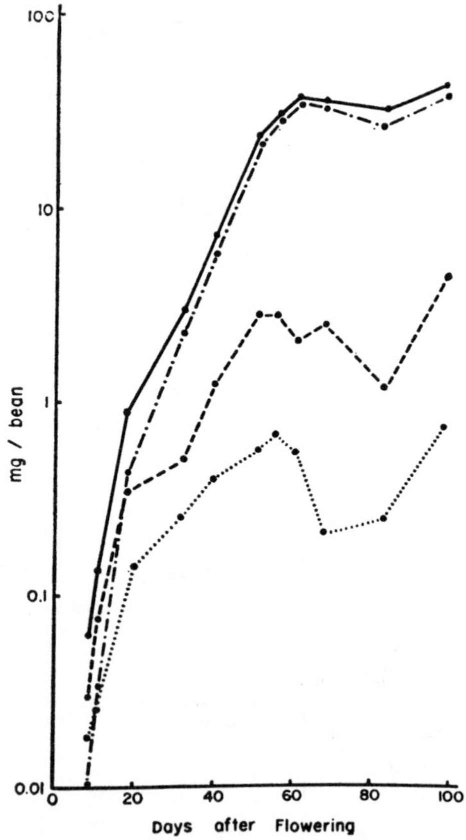

FIG. 4. Changes in major lipid groups, expressed on a logarithmic scale as weight per bean, in developing soyabeans. ——— total oil, — · — · — · neutral lipids, - - - - - phospholipid, and · · · · · glycolipids. (From Privett *et al.*, 1973.)

TABLE V

Changes in proportions of lipid classes in developing soya beans (From Privett *et al.*, 1973)

Days after flowering[a]	Neutral lipid (% wt)	Glycolipid (% wt)	Phospholipid (% wt)	Highly polar lipid (% wt)	Recovery (% wt)
9	17·0	29·2	49·0	4·8	95·2
11	24·3	21·5	54·2	0·0	96·8
18	46·4	14·7	36·5	2·4	95·8
32	75·0	8·3	16·6	0·0	99·6
40	78·1	5·3	16·5	0·1	103·9
50	86·2	2·2	11·0	0·6	100·2
55	88·6	2·1	8·6	0·7	105·6
60	90·0	2·3	5·5	1·2	101·9
67	92·2	0·6	6·9	0·3	104·5
80	92·4	1·6	3·8	2·2	102·6
97	88·1	1·6	9·8	0·5	96·3

[a] Based upon date of picking and average weight of beans.

centage of total lipids from about 30% at 9 days after flowering to approximately 1% in the mature seeds. An unknown glycophospholipid was a major component in immature soyabeans. In the phospholipid fraction, which at 9 days after flowering represents *ca.* 50% of the total lipids, compared to 8–9% in the mature beans, there is a dominance of phosphatidylcholine, phosphatidylethanolamine and phosphatidylinositol in the mature beans compared to phosphatidic acid and unidentified phospholipids in the immature beans. The composition of these complex fractions and labelling experiments are reported by Singh and Privett (1970a, b).

Roehm and Privett (1970) in a detailed study on the composition of soyabean triacylglycerols reported that triacylglycerol species, rich in polyunsaturated acids, pass through a peak at a rather early stage in development and then can not be detected. The authors point out that this finding cannot be interpreted as due to "dilution" but is indicative of triacylglycerol turnover by lipolysis or acyl exchange. It is possible that the triacylglycerols are metabolically active at a very early stage of triacylglycerol accumulation but at later stages it appears that this is not the case (see p. 276 and table XI). The most likely explanation of this appears to be that once the triacylglycerols have been synthesized they are metabolically inert, deposited in the oil droplets until the onset of germination.

An entirely different situation is obvious for the galacto- and phospholipids (of the chloroplasts and mitochondria). Assuming that all the chlorophyll in any of these oil seeds which have chloroplasts in their developing embryos is located in the chloroplast grana, Fig. 5 indicates that there is a distinct peak in the amount of chloroplast grana per seed of zero-erucic acid rapeseed

TABLE VI

Changes in the proportions of individual lipids of the neutral lipid class, the glycolipid class and the phospholipid class during soyabean development (From Privett *et al.*, 1973)

Lipid class	Days after flowering					
	9	18	40	55	67	97
Neutral lipid[a]						
TG	tr	25·9	57·0	82·4	97·0	98·0
FFA	3·7	4·6	tr	tr	tr	tr
ST	8·5	3·3	0·4	0·3	0·7	0·5
UN	87·6	66·2	42·6	17·3	2·3	1·4
Glycolipid[a]						
ESG	9·8	6·6	5·6	6·1	9·4	33·3
MGDG	1·5	1·4	9·1	6·6	6·8	4·6
SG	3·0	2·9	5·3	3·9	7·3	17·3
CE	11·0	21·3	11·6	16·9	10·4	20·2
DGDG	16·7	20·4	7·5	16·6	15·1	16·8
GPL	26·3	18·7	7·5	10·0	10·8	14·4
UN	30·9	28·8	35·5	39·4	40·4	6·8
Phospholipid[a]						
PG+DPG	3·3	3·1	3·5	2·7	6·7	3·3
PE	3·5	3·1	5·0	1·4	3·5	26·3
PC	2·0	1·2	4·4	3·6	10·2	45·0
PI	6·3	8·9	19·9	15·3	12·5	14·1
PA	64·2	75·9	51·0	51·8	39·2	5·0
UN	20·6	7·8	25·4	25·4	27·9	6·3

[a] TG = triglycerides, FFA = free fatty acid, ST = sterols, ESG = esterified sterol glucoside, MGDG = monogalactosyldiglyceride, SG = sterol glucoside, CE = cerebroside, DGDG = digalactosyldiglyceride, GPL = glycophospholipid, PG = phosphatidylglycerol, DPG = diphosphatidylglycerol, PE = phosphatidylethanolamine, PC = phosphatidylcholine, PI = phosphatidylinositol, PA = phosphatidic acid, UN = unidentified material.

when plotted *vs.* time after fertilization. This peak approximately coincides with the peak in dry matter production in the developing seed. The parallel between the peak in chlorophyll content and the peak in rate of dry matter production could indicate that the photosynthetic activities of the cotyledonary chloroplasts play a major role in the accumulation of organic material in these seeds.

It has been reported that both oil content and percentage of linolenic acid in the oil of one genotype of rapeseed were reduced when the siliquae were covered by a light-impermeable material during seed formation whereas another was essentially independent of such a treatment (Rakow and McGregor, 1974). However the genotype of soyabeans studied by Fehr *et al.* (1971) yielded essentially the same oil content and fatty acid composition regardless of whether the pods were illuminated or covered by a foil during seed formation. It should also be mentioned that a high-linolenic acid mutant of zero-erucic acid rapeseed at most of the stages of development had more chlorophyll per seed than a low-linolenic acid mutant (Thies, 1970).

Much further work seems to be necessary in order to establish the contributions of assimilate produced in the pods (siliquae) and the developing seeds to the dry matter accumulation in the seeds. It is also noteworthy that the drop in chlorophyll content in zero-erucic acid rapeseed at the later stage of seed development is paralleled by a drop in amount of phospho- and galactolipids on a per seed basis (Table VII). In the decrease in amount of phospho- and galactolipids there is a preferential loss in those membrane lipids rich in polyunsaturated fatty acids. It would be interesting to know the metabolic fate of the fatty acids lost from the phospho- and galactolipids during the latter part of seed development. It appears possible that they are fully degraded, primarily to acetyl CoA, or they could be re-acylated to diacylglycerol whereby they could raise the level of polyunsaturated fatty acids in the triacylglycerols.

In summary, sections A and B have provided examples of compositional differences with tissues type and stage of development of oil seed tissue.

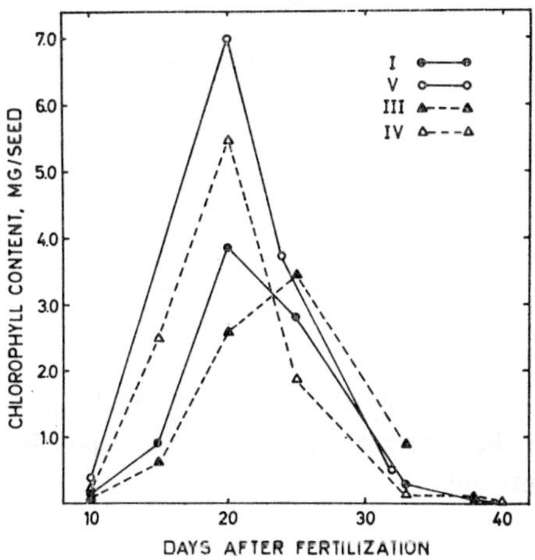

FIG. 5. Changes in chlorophyll content during seed development in rapeseed, planted in growth chambers.

Roman numerals indicate different chamber temperatures and photoperiods applied from flowering to seed maturity.

Climate	Day temp. °C	Night temp. °C	Photoperiod hrs
I	25	25	16
II	25	25	24
III	25	12	16
IV	25	12	21

(Appelqvist, unpublished data.)

TABLE VII

Changes in fatty acid content of phospholipids and glycolipids during seed development in a "zero-erucic" acid genotype of rapeseed (Appelqvist, unpublished)

Days after flowering	Content of individual fatty acids (μg/100 seeds)					Total fatty acids (mg/100 seeds)
	16:0	16:3	18:1	18:2	18:3	
			Phospholipids + SQDG[a]			
10	86	—	16	220	36	0·40
17	171	—	253	329	90	0·89
25	227	—	401	416	138	1·24
33	133	—	230	206	58	0·66
42	149	—	437	233	34	0·87
			Digalactosyldiglycerides			
10	51	—	16	135	67	0·30
17	253	—	1085	698	400	2·50
15	371	—	1564	1193	564	3·79
33	194	—	863	587	148	1·85
42	106	—	594	303	44	1·06
			Monogalactosyldiglycerides			
10	10	6	4	17	4	0·09
17	45	110	93	172	308	0·81
25	54	111	142	187	266	0·85
33	50	23	114	85	55	0·39
42	55	7	87	40	15	0·29

[a] SQDG = sulphoquinovosyldiglyceride.

These compositional facts supply the backbone for biosynthetic experiments attempting to answer the many questions on how the observed changes are accomplished and controlled.

C. ASPECTS OF THE CONTROL OF LIPID PATTERNS IN SEEDS

From the data presented in the previous sections of this review, it is apparent that very strict mechanisms are operating in the plant tissues causing well-defined differences in fatty acid patterns with acyl lipid class, subcellular location, anatomical part and stage of development. Some data on regulatory mechanisms at the enzyme level are presented under section III. Substantial information has been accumulated on the genetic control of fatty acid patterns in mature seeds which reflect mainly the integral of the biosynthetic capacity of the oil bodies.

1. *Genetic control of fatty acid patterns*

By crossing genetic lines of certain species which when selfed give rise to seeds with different fatty acid patterns, and studying the offspring in the F_1 and F_2 generations, information on the mode of control of various fatty acids in soyabeans, safflower, flax and rapeseed has accumulated. In such studies only the cotyledons should be used for analysis (cf. Table I) since the seed coat is of maternal tissue and the endosperm has twice as much genetic material from the maternal as from the paternal plant.

The content of erucic acid (22:1 ω9) in seeds of rape (*Brassica napus*) and turnip rape (*B. campestris*) is determined by the genetic constitution of the embryo (nuclear gene control) with little or no contribution from the genotype of the maternal plant (Table VIII).

For further original data see Lööf and Appelqvist, 1964; Harvey and Downey, 1964; Downey and Craig, 1964; Dorrel and Downey, 1964; Stefansson and Hougen, 1964; Kondra and Stefansson, 1965; Krzymanski and Downey, 1969; Jonsson, 1973.

In the diploid species *B. campestris*, the level of erucic acid appears to be governed mainly by a single, non-dominant gene (= a single elongating enzyme or a single cofactor controlling the elongation from oleic acid (18:1 ω9) via eicosenoic acid (20:1 ω9) to erucic acid (22:1 ω9)). In the allo-tetraploid species *B. napus*, different parental combinations have yielded slightly different results. In many cases it appears, however, that there are two "major" genes acting in an additive manner, which control the level of erucic acid. This pattern indicates that the nuclear genes of both the parental diploid species of *B. napus*, viz. *B. campestris* and *B. oleracea*, exert quantitative influence in the hybrid species on the capacity to elongate oleic acid to erucic acid. The observations of one-gene control by *B. napus* could mean that the genome from one of the parental diploid species in this hybrid is not influencing the quantities produced. To the best of the knowledge of the reviewer the genetic studies on *B. napus* and *B. campestris* are the only ones dealing with the genetic control of an "unusual" fatty acid of seed tissue (cf. Nelson and Burr, 1973).

The mode of control of the percentages of oleic, linoleic and linolenic acids in oil seeds differs among the species studied. In safflower (*Carthamus tinctorius*), with an opaque testa, the oleic and linoleic acid percentages are entirely under embryonic control (Knowles and Mutwakil, 1963; Knowles and Hill, 1964). There seem to exist at least three alleles at one locus controlling the level of linoleic acid in safflower (Knowles, 1972). In soyabeans with translucent seed coat and chloroplasts in their developing cotyledons, the situation is entirely different: the levels of oleic, linoleic and linolenic acids are essentially controlled by the genotype of the maternal plant with minor contributions from the genotype of the embryo (Brim *et al.*, 1968). Flax is somewhat intermediary: the oleic, linolenic and linoleic acid contents are determined largely by the genotype of the seed but there is some influence

TABLE VIII

The fatty acid patterns of seed coats, cotyledons and hypocotyls of seeds of single plants of the Oro and Gulle cultivars of *Brassica napus* (rapeseed) and F_1 seeds of their reciprocal crossings. (Appelqvist, unpublished)

Genotype	Seed part	Fatty acid composition, (%)									
		16:0	16:1	18:0	18:1	18:2	18:3	20:0	20:1	22:1	Other acids
Oro	Seed coat	7·5	3·1	1·3	54·5	25·7	7·3	0·1	0·4	tr	0·1
	Cotyledons	4·2	0·1	1·1	61·8	22·7	9·1	0·1	0·9	tr	—
	Hypocotyl	11·2	0·5	1·5	47·2	29·6	9·1	0·4	0·5	tr	—
Oro × Gulle[a]	Seed coat	7·7	2·2	1·6	37·2	21·9	8·6	0·5	13·6	6·4	0·3
	Cotyledons	4·2	0·2	1·3	33·3	18·6	12·1	0·4	14·3	15·3	0·3
	Hypocotyl	10·2	0·7	2·1	31·4	28·1	10·4	0·5	9·7	6·7	0·2
Gulle × Oro	Seed coat	7·6	2·7	0·6	29·0	22·2	7·5	0·2	19·2	10·8	0·2
	Cotyledons	3·7	0·2	1·1	28·0	20·5	13·5	0·3	14·3	18·0	0·4
	Hypocotyl	10·7	0·6	0·8	29·0	30·0	10·7	0·3	9·6	8·1	0·2
Oro	Seed coat	5·7	2·4	0·7	20·6	17·8	6·7	0·3	26·0	19·1	0·7
	Cotyledons	3·6	tr	1·1	16·7	15·5	13·7	0·2	12·0	37·0	0·2
	Hypocotyl	10·1	0·5	1·0	20·0	26·0	12·7	0·3	11·2	17·8	0·4

[a] The name of the maternal plant is mentioned first in the designation of a hybrid seed.

from the genotype of the maternal plant (Yermanos and Knowles, 1962). With zero-erucic acid rapeseed different parental combinations have yielded slightly different results (Kondra and Stefanson, 1970; Thomas and Kondra, 1973). The linolenic acid content was, in all combinations, controlled both by the genotype of the embryo and that of the maternal plant, whereas with oleic and linoleic acids, two combinations indicated a definite maternal genotype effect and a third one demonstrated embryonic control.

The variation in the mode of control of oleic, linoleic and linolenic acid levels in oil seeds might be related to the degree of physiological autonomy of the developing cotyledons of chloroplast-rich seeds and the extent of contribution of lipid biosynthesis in the chloroplast to the accumulation of storage lipids, triacylglycerols, in the oil bodies of the cytosol (cf. pp. 279 and 281). A contribution, direct or indirect, by the fully functional chloroplast of developing cotyledons to the accumulation of triacylglycerols and their linolenic acid content has been observed from studies on chloroplast mutants of rapeseed (Thies, 1971; Röbbelen, 1972; Appelqvist unpublished). The seeds of mutants had less oil and lower percentages of linolenic acid in their oils.

2. *Environmental modifications of the genetically controlled patterns*

In general, the seed oil (essentially the triacylglycerols of oil bodies) produced in plants in a warmer climate has lower percentages of polyunsaturated acids than when produced in a colder climate (Hilditch and Williams, 1964; Hitchcock and Nichols, 1971). Growth chamber experiments have, however, demonstrated that the species used can respond very differently (For detailed literature references, see Appelqvist, 1971). It has, for example, been shown that the fatty acid patterns of normal lines of rapeseed, sunflower and flax vary markedly with temperature (Canvin, 1965a) whereas those of castor beans and high-linoleic acid safflower vary only to a small extent with temperature from flowering to maturity (Fig. 6).

Studies at our laboratory have demonstrated an exception to the rule that the mean unsaturation decreases as the growth temperature increases. Two sister lines of zero-erucic acid rapeseed yielded more linoleic and essentially unchanged linolenic acid percentages with increasing day temperature (Appelqvist, 1971). Unpublished results from our laboratory have demonstrated that two different zero-erucic acid lines can respond differently to the same climatic variation (Fig. 7). Whereas the high-linoleic acid line (no. 802) responded like the lines previously used in our experiments, the low-linoleic acid line (no. 314) behaved as expected from the classical observations. Interspecific genetic variability in reaction to environmental temperature has also been observed in rapeseed by van Hal and van Baarsel (1974) and in safflower by Knowles (1972). Knowles noted that the fatty acid patterns of high-linoleic acid (*ca.* 80%) and low-linoleic acid (*ca.* 20%) lines of safflower were affected to a rather small extent by temperature during seed develop-

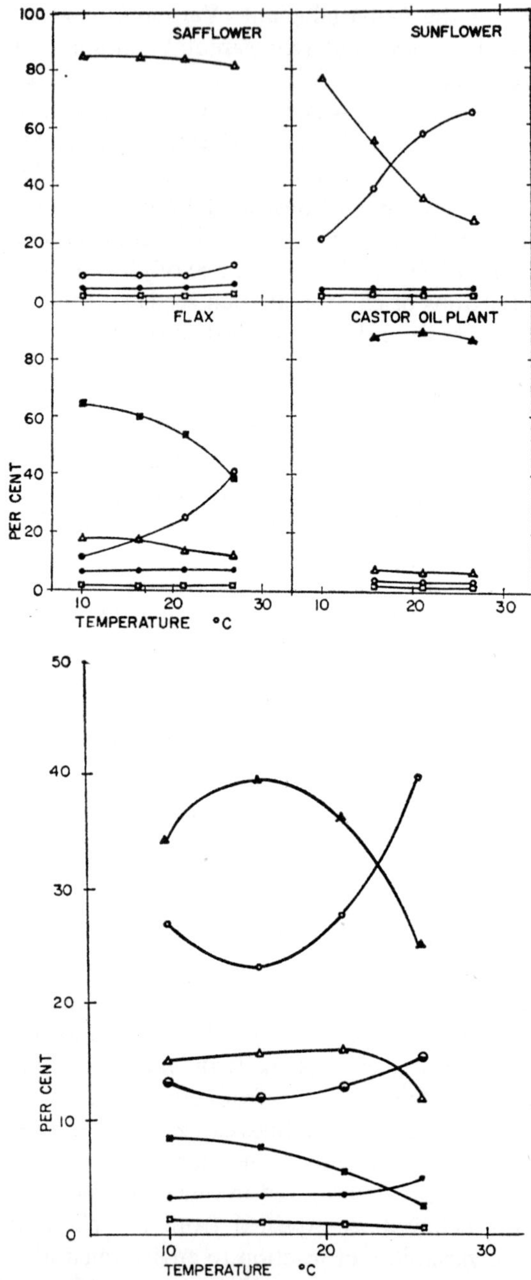

FIG. 6. A. Effect of temperature on the fatty acid composition of the oil from several oil seed crops. ● Palmitic, □ stearic, ○ oleic, △ linoleic, ■ linolenic, ▲ ricinoleic acid. B. Effect of temperature on the fatty acid composition of rapeseed oil. ● Palmitic, □ stearic, ○ oleic, △ linoleic, ■ linolenic, ◒ eicosenoic, ▲ erucic acids. (From Canvin, 1965a.)

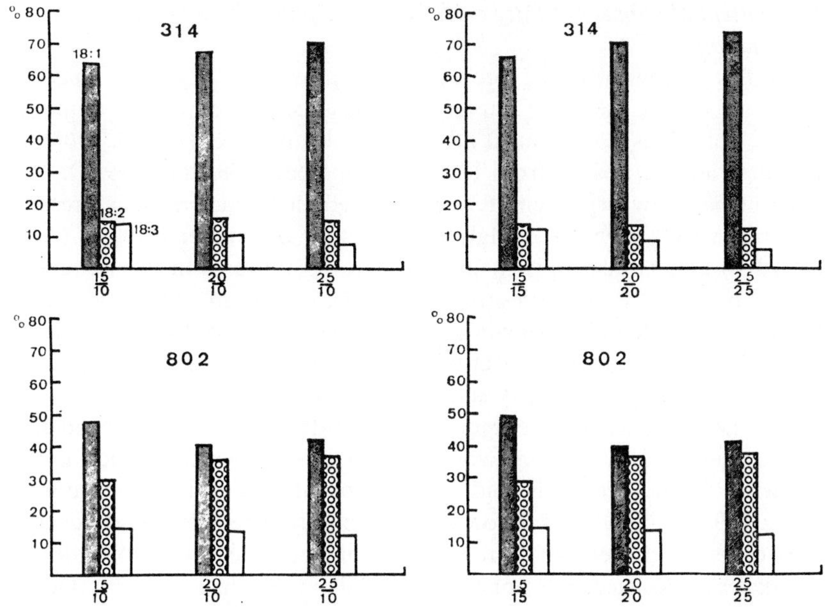

FIG. 7. Effect of environmental temperature during the seed development period on the content of oleic, linoleic and linolenic acids in seeds of two genetic lines of zero-erucic acid rapeseed. The figures under the bars indicate day/night temperature (°C) from flowering to maturity. (Appelqvist, unpublished.)

ment whereas an intermediate-linoleic acid genotype was very temperature-sensitive; at the lowest growth temperature the linoleic acid concentration in the seed oil was 75% and at the highest 39%.

The reduction in formation of linoleic and/or linolenic acids in plant tissues grown at higher temperatures was explained by Harris and James (1969) as being due to reduced tissue concentration of molecular oxygen, a necessary cofactor in the biosynthesis of linoleic and linolenic acids. Experiments using different oxygen pressures during seed formation in sunflower have supported this idea (Dompert and Beringer, 1970). However, the fact that different genotypes of a species can react differently and the fact that molecular oxygen is produced in the chloroplasts of immature cotyledonary tissue of some species seem to warrant a less firm conclusion. Rather than accepting oxygen tension as the general explanation, it appears it can be a key factor in producing the phenotype of some genotypes of certain species.

These inter- and intra-specific variations in both extent and direction of temperature effects on synthesis of linoleic and linolenic acids certainly makes this issue much more complicated than is often assumed. It appears appropriate to mention that Patterson (1970) recently found an optimum in average degree of fatty acid unsaturation of *Chlorella sorokiniana* lipids as the growth temperature was varied in 7 stages from 15 to 35°C.

3. *The content of "unusual" fatty acids in oil seed plants in relation to differentiation*

With few exceptions, the presence of "unusual" fatty acids has been confined to seed tissue (see Chapter 1). It has been stated, for example, that ricinoleic acid does not occur at all in castor bean leaves (James, 1963), and that erucic acid is absent from leaves of rapeseed (Shorland, 1945).

Since it has now been definitely established that erucic acid is present in the quantitatively small, triacylglycerol fraction of rapeseed leaves (Appelqvist, 1970) and is also present in considerable amounts in the siliquae (i.e. maternal tissue) of that species (see Table III), it is of interest to take a fresh look at the possible modes of control operating to achieve the tissue specificities in fatty acid composition. Actually, erucic acid in rapeseed is not the only example of "seed-specific" fatty acids in leaf tissue or other, entirely maternal, tissues. The presence of the cyclopropene fatty acids (malvalic acid and sterculic acid) in leaves from two *Malva* species known to be rich in these fatty acids in the seed oils had been reported earlier by Shenstone and Vickery (1961). Now it is known that cyclopropene fatty acids occur in tissues other than seeds of four plant families of the order Malvales, viz. Sterculiaceae, Malvaceae, Bombaceae and Tiliaceae (Christie, 1970). Detailed studies of some Malvaceae species have demonstrated that the cyclopropene and cyclopropane acids occur not only in leaf and seed tissue but also in stems and roots as well as in callus tissue derived from such plants (Yano *et al.*, 1972a). Generally the highest percentages of the cyclic acids were found in the immature seeds, which is opposite to the situation for erucic acid (cf. Fig. 3). Most of the cyclic acids were generally esterified as neutral lipids, as with other "unusual" fatty acids in higher plants. The opposite situation is found in bacteria and protozoa (see review by Christie, 1970) where they are apparently synthesized from the appropriate monoenoic acid moiety esterified to glycerol in phosphatidylethanolamine. Studies on the occurrence of cyclopropene fatty acids in various plant tissues under different stages of development have also been published by Kaimal and Lakshminarayana (1970) and Sudershan and Lakshminarayana (1972).

Furthermore it is interesting to observe that ximenynic acid (octadec-*trans*-11-ene-9-ynoic acid) is present in the seed fat of *Ximenia* species and several other genera of Santalaceae, whereas the stem and root are rich in the related acids, octadeca-11,13-dien-9-ynoic acid and octadeca-13-ene-9,11-diynoic acid respectively (Hatt *et al.*, 1959, 1960).

Two more examples of occurrence of "unusual" fatty acids in tissues other than seed have recently been reported. Five species of the family Boraginaceae known to contain many species with seed oils rich in 18:3(6,9,12), γ-linolenic acid, and 18:4(6,9,12,15) had considerable amounts of these unusual fatty acids in their leaf lipids (Jamieson and Reid, 1969). It is remarkable that the γ-linolenic acid was preferentially located in the neutral lipids whereas the 18:4 acid was enriched in the monogalactosyldiglyceride fraction.

The ivy plant (*Hedera helix*) which contains very large amounts of petroselinic acid (18:1Δ6) in the mature seeds, also has this unusual fatty acid in the pericarp, which is maternal tissue (Grosbois, 1971).

With all these exceptions to the rule that "unusual fatty acids do not occur in tissues other than seed" at hand, it is tempting to suggest that the presence of unusual fatty acids in leaf tissues has been missed in many studies for one or both of two reasons: like erucic acid in rapeseed leaves (Appelqvist, 1970) they may be present only during the seed-forming period when many leaves are decaying. (Quite naturally, most investigators prefer plants with healthy green leaves). The other fact is the almost exclusive esterification of erucic acid in the leaves to the triacylglycerols which contribute a very small fraction to the total leaf lipids. Thereby analysis of total leaf fatty acids would not yield an observable peak for the "unusual" acid typical for the species being investigated.

The aforementioned data also raise another important question, viz. where are these "unusual" fatty acids being synthesized? It has been repeatedly stated that lipids are not translocated in plants since the seeds and the leaves differ so markedly in fatty acid composition. Although maternal tissue (mainly siliquae) of high-erucic acid *Brassica campestris* of very different stages of development in the author's laboratory have been supplied with ^{14}C-acetate no labelling in erucic acid has been found in the siliquae under conditions which yield rapid labelling of erucic acid in the seeds of these siliquae. Yet saturated long chain acids (20:0, 22:0 and 24:0) were readily labelled (unpublished data). Hence, so far the translocation of minute amounts of erucic acid to maternal tissue cannot be excluded. Data from experiments with ^{14}C-labelled cholesterol applied to stems of tobacco plants (Tso and Cheng, 1971) and with labelled glycerolipids added to root tissues (Kuiper, 1969) have been interpreted as indicating that translocation of some lipids can occur in the vascular system of plants.

On the other hand it is definitely shown that cyclopropene fatty acids are synthesized in leaf tissue of *Malva* spp. (Yano *et al.*, 1972b). Whether maternal tissue such as leaves of rapeseed (*Brassica napus*) can synthesize erucic acid by elongation of oleic acid (although all our experiments so far have been negative) or minute amounts of erucic acid are translocated remains to be investigated.

Two likely explanations of the facts presented on erucic acid in maternal tissue would be: either translocation of a controlling factor, e.g. a hormone (regulating the elongation of oleic acid to erucic acid) produced in the developing cotyledons or translocation of erucic acid or erucic acid-containing lipids. In efforts to discriminate between these possibilities, some grafting experiments of plant parts from zero-erucic acid genotypes onto high-erucic acid plants were undertaken. From the data in Table IX it is not possible to distinguish between the two alternatives. It is, however, apparent that the developing high-erucic acid siliqua-half has produced a gradient of erucic

TABLE IX

The fatty acid patterns of seeds from specific siliqua parts of Oro and Gulle cultivars of *Brassica napus* (rapeseed) bearing intact siliquae and reciprocally grafted siliquae. The grafting was done by cutting siliquae with highly immature seeds diagonally with respect to the length axis of the siliqua at approximately half its length. A special "band aid" and a horticulturist's glue were used to keep the graft tight to the "mother" plant portion (Appelqvist, unpublished)

Seed sample	Fatty acid composition (%)								
	16:0	16:1	18:0	18:1	18:2	18:3[b]	20:1	22:0	22:1
Oro	4·8	0·3	1·4	70·5	15·9	6·2	0·9	tr	—
GULLE GRAFTED ON ORO PLANT									
Oro, distal portion[a]	4·7	0·3	1·0	68·5	16·4	7·9	1·2	tr	—
Oro, proximal portion	4·9	0·2	0·9	68·9	16·6	7·3	1·1	0·1	—
Gulle, proximal portion	6·0	0·4	1·2	22·6	22·2	9·4	12·7	—	25·5
Gulle, distal portion	6·4	0·5	1·5	21·1	22·8	10·1	12·0	—	25·6
Gulle	4·4	0·4	0·9	24·9	15·8	9·4	13·9	—	30·3
ORO GRAFTED ON GULLE PLANT									
Gulle, distal portion	4·4	0·4	0·7	25·4	15·4	9·6	13·8	—	30·3
Gulle, proximal portion	4·5	0·4	0·6	24·7	15·8	8·7	14·1	—	31·2
Oro, proximal portion	8·0	0·9	2·1	43·7	29·4	10·0	2·5	tr	3·4
Oro, distal portion	8·0	1·0	2·1	43·4	31·7	11·6	1·4	tr	0·8

[a] Proximal and distal are related to the cut surface of the graft.
[b] Including *ca.* 0·5% 20:0.

acid in the, otherwise, erucic acid-free graft. Incubation of the siliquae with ^{14}C-acetate failed to reveal any synthesis of erucic acid in the grafted portion thus leaving translocation as the slightly more likely alternative.

Growth of callus tissue from species rich in some "unusual" fatty acid is also expected to yield interesting data on the control mechanisms. Whereas Yano *et al.* (1972a) observed high percentages of cyclopropene fatty acids in callus established from stem tissue of a mature plant of *Malva parviflora* as well as from germinating seeds of *M. sylvestris*, only a trace of erucic acid was found in callus from germinating seeds of *Brassica campestris* and *B. napus* although the seeds were rich in erucic acid (Staba *et al.*, 1971). This is expected, since it has been observed that germinating seeds of *B. napus* do not synthesize erucic acid (Appelqvist, unpublished). Obviously the mode of control of synthesis of various "unusual" fatty acids in plant tissue can vary with the type of biosynthetic reaction. Whereas some modifications of "usual" fatty acids, like elongation of oleic acid to erucic acid in rapeseed and hydroxylation of oleic acid (oleyl CoA) to ricinoleic acid in castor bean, primarily (or solely) occur in the embryonic tissues of the developing seeds, other modifications obviously occur in all tissues of the species involved.

III. Lipid Biosynthesis in Developing Oil Seed Tissues

The state-of-the-art as regards fatty acid biosynthesis in plant tissues is presented in Chapters 3 and 4 and is also excellently discussed in other recent reviews (Stumpf, 1970; Mazliak 1973; Gurr 1974). The discussson will therefore be focused on some problems of special interest in oilseed tissue.

As has been pointed out previously and also discussed at some length (Stumpf, 1970) there are problems associated with use of tissue slices as well as isolated subcellular fractions in the studies of lipid biosynthesis of higher plants:

(a) Administration of a suitable radioactive precursor for fatty acids, often ^{14}C-labelled sodium acetate, to a tissue slice preparation (or the tissue *in vivo*) is a useful technique to demonstrate that a certain lipid component in the tissue is being synthesized under the incubation conditions but is of limited value in efforts to elucidate pathways or mechanisms.

(b) Preparation of cell organelle fractions or a crude soluble enzyme before the administration of the precursor involves other problems. The capacity to synthesize a component, known to be formed in the tissue under *in vivo* or tissue slice conditions is often lost entirely. Furthermore, the acquisition of really clean preparations of organelles from oil seed tissue, which might also contain chloroplasts, is very laborious (Harwood and Stumpf, 1972). Leakage of chloroplastic enzymes through a ruptured chloroplast envelope is another problem. Keeping these difficulties in mind, we shall now briefly discuss some literature data in this area.

A. THE BIOSYNTHESIS OF OLEIC, LINOLEIC AND LINOLENIC ACIDS

The rapid synthesis of oleic acid from exogenous acetate and the inability of stearic acid added to a tissue slice or subcellular fraction to become desaturated to oleic acid have been puzzling facts for many years. However, subcellular preparations from developing soyabeans (both microsomal and 105 000 *g* soluble fractions) have been shown to desaturate added lauric, myristic, palmitic and stearic acid (Inkpen and Quackenbush, 1969). Oleic acid was further desaturated by a similar preparation.

The very recent confirmation of the previous reports by Nagai and Bloch (1968) that a chloroplast preparation from spinach leaves can desaturate stearyl-ACP to oleyl-ACP and the finding that this reaction also occurs in non-photosynthetic tissue, i.e. developing safflower seeds (Jaworski and Stumpf, 1974), is of key interest. A separate plant pathway for oleic acid biosynthesis as suggested previously is still possible as an alternative to explain the inability of exogenous stearic acid to become desaturated (see pp. 63 and 106). However, the suggestion that the proper activating enzyme is missing appears more likely (see Gurr, 1974 for further details).

The true substrate for oleic acid desaturation to form linoleic acid in plant tissues has been debated over recent years. From labelling data in experiments with *Chlorella vulgaris*, Gurr *et al.* (1969) suggested, as one alternative explanation of the results, that the desaturation of oleic acid occurred after it was esterified to glycerol in phosphatidylcholine. Several recent reports implicate phosphatidylcholine as the major recipient of newly formed linoleic acid if not the primary product of desaturation. Such observations have been made on flax seeds by Dybing and Craig (1970), on soyabean cotyledons by Stearns and Morton (1973) and on "aged" potato tuber slices by Abdelkader *et al.* (1973).

On the other hand, Vijay and Stumpf (1971) presented well documented evidence that oleyl CoA is the substrate and linoleyl CoA is the primary product in a microsomal preparation from developing safflower seed (reaction 1 in Fig. 8). This system also contained very active enzymes which formed phosphatidylcholine, species labelled both with the substrate oleic acid and the product, linoleic acid (reactions 2 and 3 in Fig. 8). The existence of at least two oleic acid desaturases in developing castor beans has been postulated from studies by Howling *et al.* (1972). Their conclusions, presented in Table X, could reconcile most of the, at least superficially contradictory, data which have been previously reported on the true substrate. It appears plausible that

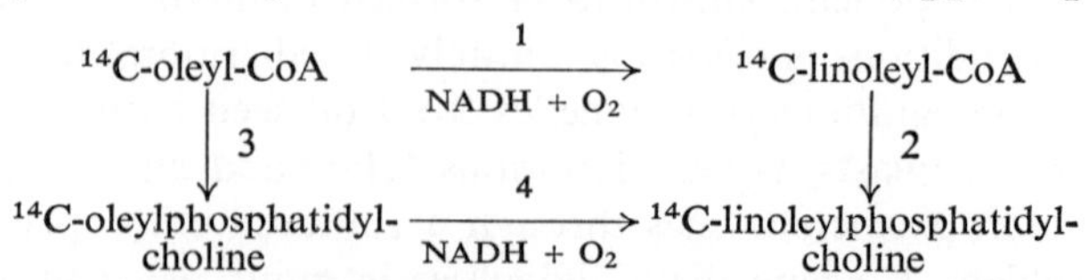

FIG. 8. Pathways of linoleic acid biosynthesis investigated in maturing safflower seeds. (From Vijay and Stumpf, 1971.)

TABLE X

Substrate recognition patterns of different desaturases
(From Howling *et al.*, 1972)

Substrate recognition pattern	Suggested substrate	Reaction catalysed	Enzyme nomenclature
$HOOC(CH_2)_7\overline{\underline{CH_2—CH_2(CH_2)_n}}CH_3$	acyl-S-CoA or acyl-S-acyl carrier protein	saturated C_{14}–C_{18} → Δ9-monoene	Desaturase IΔ
$\overline{\underline{HOOC(CH_2)_7CH{=}CH}}—(CH_2)_nCH_3$	acyl-S-CoA or acyl-S-acyl carrier protein	oleic → linoleic	Desaturase IIΔ
$HOOC(CH_2)_n\overline{\underline{CH{=}CH(CH_2)_7CH_3}}$	acyl-O-phospholipid	oleic → linoleic	Desaturase IIω

the desaturase enzyme recognizing the distance between the methyl end and the existing double bond of fatty acids—the desaturase IIω—acts on the *O*-esters (oleyl phosphatidylcholines), whereas the desaturase IIΔ uses an *S*-ester (CoA or ACP-derivative) as its substrate. The singularity of pathways and subcellular sites in any one lipid metabolic reaction very often searched for apparently has to yield to the more likely situation of different mechanisms in different subcellular locations for several lipid metabolizing enzymes. Since Howling *et al.* used "three embryos and part of endosperm" from developing castor beans in their enzyme source, it cannot be excluded that there is a tissue specificity in content of desaturases IIΔ and IIω.

For many years, the biosynthesis of linolenic acid in plant tissue has been assumed to take place by desaturation of linoleic acid. The sequential appearance of label in oleic/linoleic/linolenic acids, e.g. from $^{14}CO_2$ supplied to whole plants of soyabeans and flax by Dutton and Mounts (1966), from ^{14}C-acetate supplied to developing soyabeans by Rinne and Canvin (1971a) or from oleic acid supplied to green leaves by Harris and James (1965), has been taken to mean that there is a simple precursor–product relationship between the oleic, linoleic and linolenic acids. However, the very slow appearance of label in linolenic acid in photosynthetic tissues known to be very rich in polyunsaturated fatty acids remained a mystery and a challenge. In a very recent series of papers from Stumpf's laboratory, the results of which are discussed in Chapter 4, clear evidence has been presented to show that the synthesis of linolenic acid (18:3) in various kinds of photosynthetic tissues can occur by elongation from 16:3, which in turn is probably formed from 12:0 via 12:3. On the other hand experiments by Trémolières and Mazliak (1974), using pea leaf preparations, demonstrated that linoleic acid can be synthesized from oleic acid in the microsomes and then further desaturated to linolenic acid in the chloroplasts.

Considering the data so far available, it appears likely that linolenic acid in higher plants is formed by at least two systems, one involving 16:3 as substrate and the other using 18:2. If one linolenic acid-synthesizing enzyme were located in the cytosol part of the cell, a genetic block could cause the production of triacylglycerols very low in linolenic acid without interfering with the structure and function of the chloroplast. Such a block is being actively searched for in the current efforts to upgrade soyabean and zero-erucic acid oils by genetic engineering (Howell *et al.*, 1972; Hammond *et al.*, 1972; Lööf and Appelqvist, 1972; Röbbelen, 1972).

The further elucidation of the pathway(s) of linolenic acid biosynthesis in developing oil seed tissue is therefore not only of considerable scientific interest but also of great economic significance.

B. BIOSYNTHESIS OF "UNUSUAL" FATTY ACIDS

The text of a fairly recent review on the biosynthesis of unsaturated fatty acids in plants strikingly revealed how much speculation and reference to lower plants, but how little experiment data could be found in the literature up to 1969 on the biosynthesis of "unusual" fatty acids in higher plants (Stearns, 1970). Only a few papers in this area seem to have been added recently (see also Chapter 3).

The hydroxylation of oleyl CoA to ricinoleyl CoA by a mixed function oxygenase of a microsomal fraction of developing castor beans has been reported by Galliard and Stumpf (1966). Other acyl CoA derivatives were hydroxylated to a much smaller extent. Probably the biosynthesis of ricinoleic acid is the reaction best characterized among all in the field of "unusual" fatty acids (see Galliard and Stumpf (1966) for references to earlier work).

The biosynthesis of the cyclopropane and cyclopropene acids of some *Malva* and other species of Malvales have been studied by several groups, more recently by Johnson *et al.* (1967) and Yano *et al.* (1972b) both using tissue slices. The ring methylene group is derived from the methyl group of methionine and the pathway appears to involve initial formation of dihydrosterculic acid from oleic acid with subsequent desaturation to sterculic acid and α-oxidation to malvalic and dihydromalvalic acids.

In the biosynthesis of acetylenic acids, ximenynic acid in *Santalum acuminatum* (Bu'Lock and Smith, 1963) and crepenynic acid in *Crepis rubra* (Haigh *et al.*, 1968), it appears that the introduction of the acetylenic bond occurs on preformed long-chain fatty acids and not during the *de novo* fatty acid biosynthesis. Oleic acid was the nearest precursor found for crepenynic acid but the degrees of conversion were rather small with the best tissue slice preparations used (Haigh *et al.*, 1968).

The biosynthesis of erucic acid in developing seeds of *Brassica napus* and *Crambe abyssinica* has been studied by several authors (Downey and Craig, 1964; Canvin, 1965b; Appelqvist, 1973; Gurr *et al.*, 1974). Downey and Craig demonstrated that erucic acid is synthesized from oleic acid by two successive additions of two C_2 units (from acetate). The most effective substrate for the elongation of oleic acid appears to be malonyl CoA. The newly formed eicosenoic and erucic acids are preferentially esterified to glycerol as triacylglycerols (Appelqvist, 1973; Gurr *et al.*, 1974).

C. THE SUBCELLULAR LOCATIONS OF FATTY ACID BIOSYNTHESIS

The sites of various lipid synthesizing enzymes have already been touched upon. In a detailed study of four different seed tissues, one of which was developing castor beans, Harwood and Stumpf (1972) have separated the seed homogenate into seven fractions and investigated their capacity for fatty acid biosynthesis from acetate and malonyl CoA (see Fig. 3 in Chapter 3, p. 65). It is obvious that a fat fraction, corresponding to the "oil bodies" *in vivo*, (see p. 279) and the 105 000 *g* supernatant are both very active with malonyl CoA as substrate but with acetate as substrate, the supernatant shows very little activity. However, all fractions demonstrated some activity. Repeated washing of the fat fraction demonstrated that the enzymatic activity was not due to unspecific absorption of soluble enzymes. Also, the fatty acids formed under the incubation conditions differed in the two systems. Using immunological techniques, they were able to present evidence for at least two separate types of fatty acid synthetase enzymes in another tissue investigated, viz. avocado mesocarp, one residing in a plastid lamellae fraction and other(s) associated with stromal or cytosolic proteins.

Other studies on developing castor beans have demonstrated that a particle resembling proplastids is a major site of oleic acid biosynthesis in such tissue (Zilkey and Canvin, 1969; Drennan and Canvin, 1969; Zilkey and Canvin, 1971). The fat fraction was not assayed for fatty acid synthesis and the supernatant was assayed with acetyl CoA, which is known to be ineffective in soluble plant systems. Taken together with the figures of Harwood and Stumpf (1972), these data could indicate a third, quantitatively less important, site of *de novo* fatty acid biosynthesis in developing castor beans. It may also be commented that one of these could be specific for either the embryo or the endosperm (cf. comment on p. 272 on the data of Howling *et al.*, 1972).

Studies on developing soyabean cotyledons have yielded results which appear to differ substantially from those on castor beans (Rinne, 1969; Rinne and Canvin, 1971b). Malonyl CoA was a very poor substrate since a very active malonyl CoA decarboxylase was present. Activities with both soluble and particulate systems have been described but further work seems to be necessary on this point.

Further aspects of the subcellular location of fatty acid synthesizing enzymes are treated under IV. Although specific subcellular fractions of plant tissues have been shown to be the major or only site for specific reactions (see Chapters 3 and 4) the work of Harwood and Stumpf (1972) seems to be the only attempt to give a comprehensive picture of this complicated issue in developing oil seeds.

D. BIOSYNTHESIS OF TRIACYLGLYCEROLS

Recent data on flax embryos by Dybing and Craig (1970) and on developing seeds of *Crambe abyssinica* by Gurr *et al.* (1974) have supported earlier findings indicating the operation of the classical Kennedy pathway in the biosynthesis of triacylglycerols in developing seed tissue.

It is further interesting to know whether the newly formed triacylglycerols are metabolically inert or, like the conditions in adipose tissue of animals, subject to turnover. As seen in Table XI, it appears that the neutral lipids, mainly triacylglycerols, of the maturing rapeseed investigated, were not subject to any substantial metabolism whereas the galacto- and phospholipids seem to lose considerable activity during post-incubation in buffer. Based on these data, one can assume that the physiological conditions, such as tempera-

TABLE XI

Incorporation of ^{14}C-acetate into seed brei from developing seeds of a "zero-erucic acid", genotype of rapeseed. One series of samples (A) were kept in flasks containing buffered substrate for various time periods, whereas series (B) was withdrawn from the ^{14}C-acetate and reincubated in buffer in the light for various time periods (Appelqvist, unpublished)

	Time in		Acetate incorporation in lipids	Incorporation into lipid classes[a]				
	^{14}C-acetate (min)	buffer (min)	(%)	PC+PE+PI	PG+SQDG	DGDG	MGDG	Neutral lipids
				(nmol/g f.w.)				
A	15	0	0·76	0·40	0·29	0·17	0·52	0·92
	60	0	1·28	0·37	0·55	0·33	0·60	2·50
	180	0	1·91	0·54	0·52	0·48	0·57	4·13
	300	0	2·22	0·91	0·88	0·16	0·48	4·18
B	15	0	0·76	0·40	0·29	0·17	0·52	0·92
	15	45	0·48	0·18	0·12	0·17	0·06	1·15
	15	165	0·44	0·17	0·08	0·14	0·07	0·97
	15	285	0·49	0·15	0·08	0·11	0·06	0·97

[a] For abbreviations, see footnote to Table VI.

ture and light intensity, only influence the composition of the triacylglycerols laid down during the period of treatment. However, at the very early stages in triacylglycerol accumulation it is possible that they are subject to catabolism (cf. the data on soyabeans by Roehm and Privett, 1970). The loss in ^{14}C-activity of monogalactosyldiglyceride and also the other polar lipids (Table XI) was reasonable in relationship to the loss in total amounts of polar lipids at later stages of development.

The organelle probably unique in lipid-rich seeds, namely the oil body or spherosome which yields the active fat fraction *in vitro*, will now be discussed in some detail.

IV. Oil Bodies; Ultrastructure, Chemical Composition and Biosynthetic Activity

In mature oil-rich seeds, a predominant portion of the lipids are localized in so-called oil bodies which vary in size and relative abundance in different cells (see Fig. 9). As already mentioned the "fat fraction" of developing seeds has definite biosynthetic capacity. This section deals with these subcellular structures in some detail.

A. Nomenclature and Variability

The subcellular particle under discussion has been named microsome, spherosome, oleosome, oil body and lipid-containing vesicle. Frey-Wyssling *et al.* (1963), tracing the history of the nomenclature of the oil-rich bodies in the cytoplasm of seed tissue, suggest that "spherosomes" from endoplasmic reticulum later develop into "oil bodies" (see Fig. 10). From light-microscopic observations on "high-lipid" tissues (avocado pericarp and coconut endosperm) "intermediate-lipid" tissues (sunflower and rapeseed cotyledons) and a "low-lipid" tissue (the fruit mesocarp of squash, *Cucurbita pepo*, Sorokin (1967) concluded that *spherosomes* which occur in both low- and high-fat tissues, and *oil droplets* occurring only in higher-fat tissues are entirely separate particles. She claimed many differences, e.g. that spherosomes are rather constant in size (0·8–1·0 μm in diameter) in contrast to a wide range in size of the oil droplets and that the former were rich in phospholipids but did not contain neutral lipids in contrast to the neutral-fat abundance in the oil droplets. Her claims were based on various established histochemical techniques. Furthermore, she claimed that the spherosomes were surrounded by a unit membrane which prevented the spherosomes from coalescing, whereas the oil bodies easily coalesced during preparation of microscopic specimens and sometimes even penetrated into the interstitial areas.

Yatsu *et al.* (1971), in an attempt to solve this controversy, isolated "spherosomes" from onion and cabbage as well as "oil droplets" from cottonseed. They found that the particles isolated from low-fat tissues and high-fat tissue were morphologically very similar, even at the ultrastructural level, and also chemically very similar, rich in lipids with very little phospholipids and little protein. In contrast to Sorokin (1967), they found that the size of cabbage sperosomes ranged from 0·4 μm to slightly over 2 μm in diameter. The spherosomes of cottonseed were 2–3 times larger in diameter. They concluded that spherosomes are lipid droplets bound by a single fine-line membrane and that they are the same as the so-called "oil droplets" found in oleaginaceous plant tissues. They therefore coined the name "oleosome" for this particle.

Reports from Mollenhauer and Totten (1971a, b) seem to further complicate the issue. They claim that there are two distinct classes of what they call "lipid-containing vesicles" in pea and bean (including soyabean) cotyledons. The "simple vesicles" by their nomenclature are generally 0·5–3 μm in diameter and contain predominantly linolenic acid-rich triacylglycerols with phospholipids as minor components, with a molar ratio of 40/1. (Allen *et al.*, 1971). These particles would resemble the oil bodies or spherosomes of Yatsu *et al.*, (1971). The "composite vesicles" appear in groups together with membranes (sheets), are smaller in diameter (0·1–1 μm) and contain a molar ratio of phospholipids to triacylglycerols of about 1/10. The "composite vesicles" are associated with plastids or plasma membranes during dormancy and do not seem to occur generally among plant species. Related to this it should be mentioned that chloroplasts often contain so-called osmiophilic droplets (see Fig. 10). According to measurements at our laboratory (Appelqvist and von Wettstein, to be published) these intraplastidic lipid droplets of various tissues of rapeseed on the average measure 0·1–0·3 μm in diameter. If a tissue is homogenized under conditions where the chloroplast envelope is ruptured, which is easily done, these intraplastidic lipid droplets float to the surface during centrifugation, just like spherosomes and lipid vesicles. This fact adds another complication to the analysis of lipid rich subcellular particles from oil seed tissue.

B. ONTOGENY

Since the number of oil bodies per cell increases very rapidly over a few weeks in the developing cotyledon, its subcellular origin is naturally of key interest.

Frey-Wyssling *et al.* (1963) suggest an origin from endoplasmic reticulum according to the scheme presented in Fig. 9. According to this hypothesis, the oil synthesis in cotyledons occurs first in a granular lipid–protein matrix and then, as the spherosome grows, from the surface of the globular particle. A problem with this interpretation is the fact that the assumed progenitors of the lipid-rich spherosomes are never seen in electron micrographs obtained

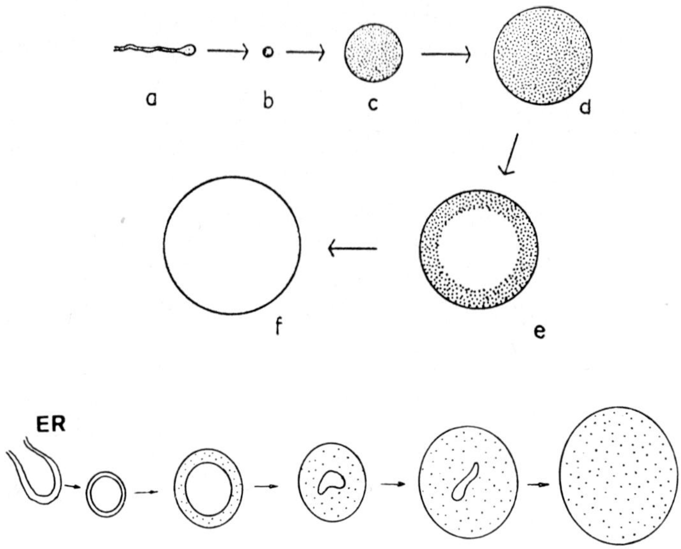

FIG. 9. A. Development of spherosomes and oil droplets in cotyledons: (a) terminal strand of endoplasmic reticulum; (b) detached vesicle; (c) juvenile spherosome; (d) spherosome; (e) transition stage; (f) fat body, oil droplet. (From Frey-Wyssling *et al.*, 1964.) B. Development of spherosomes in castor bean endosperm. The endoplasmic reticulum forms a provacuole which accumulates lipids between the halves of the unit membrane. The border of the spherosome is developed from the outer half-membrane whereas the inner half disappears during the development. (From Schwarzenbach, 1971.)

by glutaraldehyde prefixation followed by osmium tetroxide fixation; see e.g. Fig. 10 and Rest and Vaughan (1972).

On the other hand, Schwarzenbach (1971), working with castor bean endosperm, suggested that the oil bodies, although still originating from the endoplasmic reticulum, grow by lipid biosynthesis between the two membranes of the endoplasmic reticulum (Fig. 9B).

The inclusions of membraneous materials in the oil droplets that he reports are similar to the vacuolar-like inclusions of Harwood *et al.* (1971) from developing castor bean endosperm (see Fig. 11) and isolated fat bodies from such tissue.

A reasonable explanation to this controversy at the moment appears to be that endosperm tissue might differ from embryonic tissue as regards mode of depot fat accumulation. Also Schwarzenbach (1971) claims that the digestion of triacylglycerols in germinating castor bean occurs by different mechanisms in the cotyledons and the endosperm. A problem with the lipid synthesis occurring at the centre of the developing oil body (as in castor bean) would be the transport of precursors through a layer of triacylglycerols, since the inclusions (i.e. centres of biosynthesis) in the pictures so far presented always are surrounded by oil. However, under *in vivo* conditions, part of the inclusion at some point of the oil globule could border the cytosol.

C. BOUNDARY WITH THE CYTOSOL

Some controversy seems to exist on the nature of the boundary between the oil phase and the cytosol, notably in the large bodies or maturing and mature cotyledons (see Fig. 10). If one accepts the theory of formation for oil bodies in castor bean, it is then reasonable to assume that the oil bodies of

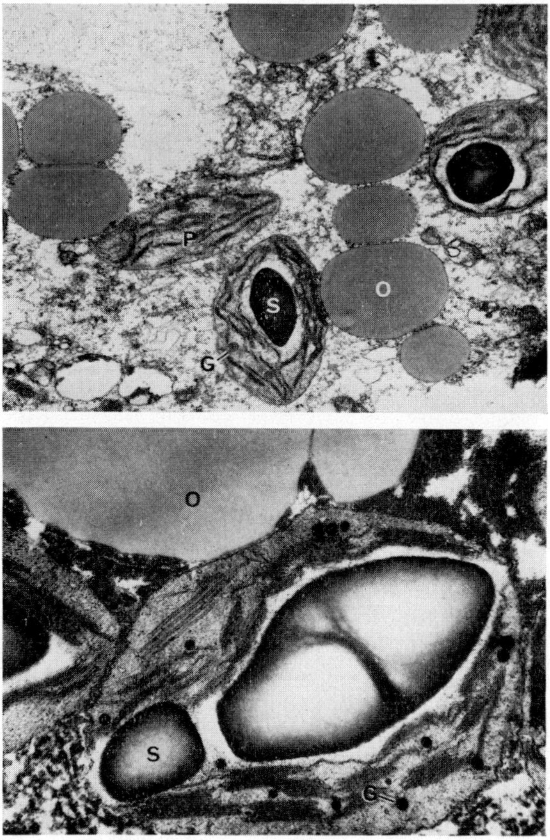

FIG. 10. Portions of cotyledon cells of *Brassica napus* during seed development. O = oil body, P = plastid, S = starch grain, G = intraplastid lipid globule. (Appelqvist and von Wettstein, unpublished.)

the endosperm are bound by a half-unit membrane. Studies by Yatsu and Jacks (1972), on peanut "spherosomes" demonstrated that the boundary had the character of a single-line membrane, 2–3·5 nm wide. Apposed internally, two spherosome membranes touch to form a structure resembling a unit-

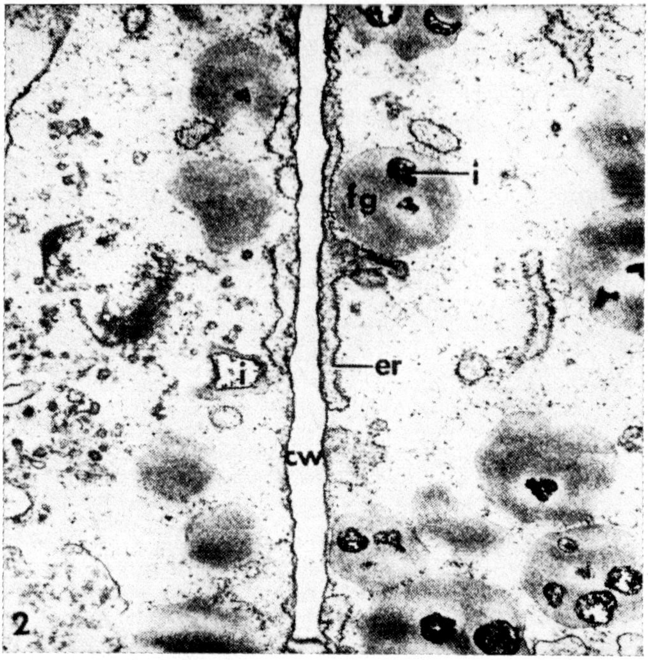

FIG. 11. Portions of two endosperm cells of *Ricinus communis*, showing profiles of developing fat globules with vacuole-like inclusions containing particulates: fg = fat globule, i = inclusion, er = endoplasmic reticulum, cw = cell wall. (From Harwood *et al.*, 1972.)

membrane (often 6–8·5 nm wide). The amino acid composition of this membrane protein had a striking resemblance to typical membrane protein of other tissues and differed greatly from that of the peanut storage protein (of the protein bodies). Frey-Wyssling *et al.* (1963) also state that the conspicuous unit membrane of the fat particles, which prevents any confluence of the oil droplets "must be an organized membrane". Bancher *et al.* (1972) from studies on peanut and walnut spherosomes report that the phospholipid content of the spherosome fraction "points to a spherosomal membrane". Others, however, are of a different opinion. Rest and Vaughan (1972) from detailed studies of maturing seeds of white mustard, conclude that no unit membrane is found around the oil droplets. Also Harwood *et al.* (1971) in studies of maturing endosperm of castor bean report that "the fat droplets lack a peripheral membrane". Comparisons of amino acid patterns of the protein of thoroughly washed preparations of oil droplets to those of the storage proteins of many species of oil seeds might solve this controversy.

D. CHEMICAL COMPOSITION

Slight variations in chemical composition of oil bodies isolated from oil-rich seeds have been reported.

Jacks *et al.* (1967) found that spherosomes isolated 11 days after the onset of germination of peanuts contained 98·1% total lipids (including *ca.* 0·8% phospholipid) and 1·3% protein. Later, Yatsu *et al.* (1971) reported 99·55% total lipid, 0·09% phospholipid and 0·22% protein in repeatedly washed preparations of peanut spherosomes. Preparations from cottonseed, washed five times, had 98·8% fat and 0·63% protein whereas cabbage (leaf) spherosomes washed only twice (due to shortage of material?) had 81% fat and 4% protein.

Gurr *et al.* (1974) have demonstrated considerable interspecific variability in chemical composition of an isolated fat fraction representing oil bodies *in vivo* (Table XII). These fat fractions were isolated from mature seeds by centrifugation and repeated washing. The data for peanuts are close to those previously discussed. Therefore the large amounts of protein in the preparations from *Crambe abyssinica* is noteworthy and could merit further examination of the fraction obtained.

TABLE XII

The chemical composition of a fat fraction, obtained as a floating material after differential centrifugation of seed tissues
(From Gurr *et al.*, 1974)

Component	Crambé	Peanut	Sunflower	Safflower
		% by weight		
Lipid	75·2 ± 3·2	98·3	89·3	85·5
Protein	19·1	1·5	4·3	10·9
Other	5·7	0·2	6·4	3·6

E. BIOSYNTHETIC ACTIVITY

Studies on immature endosperm of castor beans have shown that a floating fat layer, "oil bodies" obtained after differential centrifugation, contains (on a protein content basis) most of the tissue's enzymatic activity transferring malonyl CoA into fatty acids and further into complex lipids (Harwood and Stumpf, 1972; p. 65, Fig. 3). These activities (*ca.* 80% of total) were not due to contamination with other subcellular elements. The fat fraction was also able to synthesize complex lipids from oleyl CoA.

Data presented by Gurr *et al.* (1974) from studies on maturing seeds of *Crambe abyssinica* have shown that the fat fraction readily incorporates malonyl CoA into fatty acids, as well as palmityl CoA and glycerol-3-phosphate into complex lipids, mainly triacylglycerols. The chain length of the fatty acids synthesized by the fat fraction, predominantly C_{22}, was different from that of the other fractions. These data fit nicely to compositional data and support the view that the spherosomes or oil bodies are unique organelles capable of specific lipid biosynthetic capacity. The fact that particles sedimenting at 500 *g* and 24 000 *g* were also capable of incorporating malonyl CoA into C_{20} and C_{22} fatty acids, characteristic of the storage lipids of *Crambe abyssinica* and rapeseed, could be taken to mean that, *in vivo*, there are low-lipid structures in the cytoplasm like the supposed progenitors of spherosomes of Frey-Wyssling *et al.* (Fig. 9A) which are capable of activities similar to those of the larger oil bodies found floating after centrifugation.

In castor bean, the synthesis of ricinoleic acid has been located in a microsomal fraction (Galliard and Stumpf, 1966) but both *de novo* synthesis of fatty acids and their incorporation into triacylglycerols in castor beans are primarily associated with a "fat fraction" (Harwood *et al.*, 1971). It is thus obvious that triacylglycerols are rapidly synthesized at or in the oil bodies in periods of rapid net accumulation of triacylglycerols of oil seeds.

In conclusion, it appears that the fundamental questions of how the oil bodies originate and where the oil is synthesized are still very much open.

V. Concluding Remarks

By presenting data from genetic studies and physiological studies as well as compositional and biosynthetic investigations on oil rich seeds, the author has attempted to induce some cross-fertilization of ideas, relevant to attacks on the scientifically interesting and economically very important problems of the regulation of triacylglycerol production and composition in oil seeds. Most lipid biochemists are reluctant to choose developing oil seed tissue as working material. Problems with greenhouse facilities, careful timing of the experiments, time-consuming preparation of starting material and a complex mixture of cell organelles (at least in the chloroplast-rich cotyledonary tissues) all contribute to this reluctance. There are, however, several interesting problems which have to be solved when using such tissues e.g.: the degree of autonomy of the chloroplast *vs.* the oil-accumulating cytosol to provide background information for genetic engineering of improved oil seed crops. Furthermore, nothing seems to be known about the control mechanisms which guide the rather abrupt initiation of triacylglycerol accumulation in developing seeds and the later decrease in this capacity. The control of the rates of synthesis of the various fatty acids which change during the oil accumulation period is another interesting problem. Studies on developing castor beans (James *et al.*, 1965), safflower seeds (McMahon and Stumpf, 1966) and *Crambe abyssinica* seeds (Gurr *et al.*, 1974) have demonstrated that the capacity to synthesize the major "storage" fatty acids (ricinoleic, linoleic and erucic acid respectively) varies during seed development and has a peak at the most rapid accumulation of triacylglycerols. Perhaps most intriguing to the author is the mechanism for the initiation and turn-off of synthesis of some but not all (cf. p. 266), of the "unusual" fatty acids (like ricinoleic acid in castor beans and erucic acid in rapeseed). These fatty acids are apparently not synthesized in the vegetative plant before seed-setting. Formation starts in the embryo when this has reached a certain age and ceases when the seed has reached maturity; mature and germinating seeds have lost this fatty acid-forming capacity. Of similarly great interest is the mechanism behind the presence of small amounts of erucic acid in tissue entirely of maternal origin, i.e. is this translocation or biosynthetic capacity obtained as a consequence of the capacity in the cotyledons? Hopefully this review will stimulate further attempts to answer some of these questions.

ACKNOWLEDGEMENTS

The investigations in the author's own laboratory were given financial support by The Swedish Council for Forestry and Agricultural Research, the Swedish Natural Sciences Research Council and the Bank of Sweden Tricentennial Fund. Travelling funds were provided by The Royal Society, London, and the Bank of Sweden Tricentennial Fund.

REFERENCES

Abdelkader, A. B., Abdelkader, C., Demandre, C. and Mazliak, P. (1973). *Eur. J. Biochem.* **32**, 155–165.

Allen, F. C., Good, P., Mollenhauer, H. H. and Totten, C. (1971). *J. Cell Biol.* **48**, 542–546.

Appelqvist, L-Å. (1969). *Hereditas* **61**, 9–44.

Appelqvist, L-Å. (1970). Paper no. 303, 44th Fall Meeting of the Amer. Oil Chem. Soc., Chicago, Sept.–Oct. 1970. *J. Am. Oil Chem. Soc.* **47** (7), 1970.

Appelqvist, L-Å. (1971). *Physiologia Pl.* **25**, 493–502.

Appelqvist, L-Å. (1973). Paper no. 130, 64th Spring Meeting of the Amer. Oil Chem. Soc., New Orleans, Apr.–May 1973. *J. Am. Oil Chem. Soc.*, **50** (2), 1973.

Bancher, E., Washüttle, J. and Goller, H. J. (1972). *Z. Pflanzenphysiol.* **67**, 399–403.

Brim, C. A., Schutz, W. M. and Collins, F. I. (1968). *Crop Sci.* **8**, 517–518.

Bu-Lock, L. D. and Smith, G. N. (1963). *Phytochemistry* **2**, 289–296.

Canvin, D. T. (1963). *Can. J. Biochem. Physiol.* **41**, 1879–1885.

Canvin, D. T. (1965a). *Can. J. Bot.* **43**, 63–69.

Canvin, D. T. (1965b). *Can. J. Bot.* **43**, 71–74.

Christie, W. W. (1970). *In* "Topics in Lipid Chemistry" (F. D. Gunstone, ed.), vol. 1, pp. 1–49. Logos Press, London.

Dasgupta, S. K. and Friend, J. (1973). *J. Sci. Fd Agric.* **24**, 463–470.

Dompert, W. and Beringer, H. (1970). *Naturwissenschaften* **57**, 40.

Dorrel, D. C. and Downey, R. K. (1964). *Can. J. Pl. Sci. Sci.* **44**, 499–504.

Downey, R. K. and Craig, B. M. (1964). *J. Am. Oil Chem. Soc.* **41**, 475–478.
Drennan, C. H. and Canvin, D. T. (1969). *Biochim. biophys. Acta* **187**, 193–200.
Dutton, H. J. and Mounts, T. L. (1966). *J. Lipid Res.* **7**, 221–225.
Dybing, C. D. and Zimmerman, D. C. (1966). *Pl. Physiol., Lancaster* **41**, 1465–1470.
Dybing, C. D. and Craig, B. M. (1970). *Lipids* **5**, 422–429.
Fehr, W. R., Thorne, J. C. and Hammond, E. G. (1971). *Crop Sci.* **11**, 211–213.
Frey-Wyssling, A., Grieshaber, E. and Mühletahler, K. (1963). *J. Ultrastruct. Res.* **8**, 506–516.
Galliard, T. and Stumpf, P. K. (1966). *J. biol. Chem.* **241**, 5806–5812.
Grosbois, M. (1971). *Phytochemistry* **10**, 1261–1273.
Gunstone, F. D. (1967). "An Introduction to the Chemistry and Biochemistry of Fatty Acids and their Glycerides", 2nd edition. Chapman and Hall, London.
Gurr, M. I. (1974). *In* "Biochemistry Series One", Vol. 4, "Biochemistry of Lipids", pp. 181–235 (T. W. Goodwin, ed.). Butterworths, London.
Gurr, M. I., Robinson, M. P. and James, A. T. (1969). *Eur. J. Biochem.* **9**, 70–78.
Gurr, M. I., Blades, J. and Appleby, R. S. (1972). *Eur. J. Biochem.* **29**, 362–368.
Gurr, M. I., Blades, J., Appleby, R. S., Smith, C. G., Robinson, M. P. and Nichols, B. W. (1974). *Eur. J. Biochem.* **43**, 281–290.
Haigh, W. G., Morris, L. J. and James, A. T. (1968). *Lipids* **3**, 307–312.
Hammond, E. G., Fehr, W. R. and Snyder, H. E. (1972). *J. Am. Oil Chem. Soc.* **49**, 33–35.
Harris, P. and James, A. T. (1969). *Biochim. biophys. Acta* **187**, 13–18.
Harris, R. V. and James, A. T. (1965). *Biochim. biophys. Acta* **106**, 456–464.
Harvey, B. L. and Downey, R. K. (1964). *Can. J. Pl. Sci.* **44**, 104–111.
Harwood, J. L. and Stumpf, P. K. (1972). *Lipids* **7**, 8–19.
Harwood, J. L., Sodja, A., Stumpf, P. K. and Spurr, A. R. (1971). *Lipids* **6**, 851–854.
Hatt, H. H., Triffett, A. C. K. and Wailes, P. C. (1959). *Aust. J. Chem.* **12**, 190–195.
Hatt, H. H., Triffett, A. C. K. and Wailes, P. C. (1960). *Aust. J. Chem.* **13**, 488–497.
Hilditch, T. P. and Williams, P. N. (1964). "The Chemical Constitution of Natural Fats", 4th edition. Chapman and Hall, London.
Hirayama, O. and Hujii, K. (1965). *Agric. biol. Chem.* **29**, 1–6.
Hitchcock, C. and Nichols, B. W. (1971). "Plant Lipid Biochemistry". Academic Press, London and New York.
Hopkins, C. Y. and Chisholm, M. J. (1961). *Can. J. Biochem. Physiol.* **39**, 1481–1487.
Howell, R. W., Brim, C. A. and Rinne, R. W. (1972). *J. Am. Oil Chem. Soc.* **49**, 30–32.
Howling, D., Morris, L. J., Gurr, M. I. and James, A. T. (1972). *Biochim. biophys. Acta* **260**, 10–19.
Inkpen, J. A. and Quackenbush, F. W. (1969). *Lipids* **4**, 539–543.
Jacks, T. J., Yatsu, L. Y. and Altschul, A. M. (1967). *Pl. Physiol., Lancaster* **42**, 585–597.
James, A. T. (1963). *Biochim. biophys. Acta* **70**, 9–19.
James, A. T., Hadaway, H. C. and Webb, J. P. W. (1965). *Biochem. J.* **95**, 448–452.
Jamieson, G. R. and Reid, E. H. (1969). *Phytochemistry* **8**, 1489–1494.
Jaworski, J. G. and Stumpf, P. K. (1974). *Archs Biochem. Biophys.* **162**, 158–173.
Johnson, A. R., Pearson, J. A., Shenstone, F. S., Fogerty, A. C. and Giovanelli, J. (1967). *Lipids* **2**, 308–315.
Jönsson, R. (1973). *Z. Pflanzenzücht.* **69**, 1–18.
Kaimal, T. N. B. and Lakshminarayana, G. (1970). *Phytochemistry* **9**, 2225–2229.
Knowles, P. F. (1972). *J. Am. Oil Chem. Soc.* **49**, 27–29.
Knowles, P. F. and Hill, A. B. (1964). *Crop Sci.* **4**, 406–409.
Knowles, P. F. and Mutwakil, A. (1963). *Econ. Bot.* **17**, 139–145.

Kondra, Z. P. and Stefansson, B. R. (1965). *Can. J. Genet. Cytol.* **7**, 500–510.
Kondra, Z. P. and Stefansson, B. R. (1970). *Can. J. Pl. Sci.* **50**, 345–346.
Krzymanski, J. and Downey, R. K. (1969). *Can. J. Pl. Sci.* **49**, 313–319.
Kuiper, P. J. C. (1969). *Pl. Physiol., Lancaster* **44**, 968–972.
Lööf, B. and Appelqvist, L-Å. (1964). *Z. Pflanzenzücht.* **52**, 113–126.
Lööf, B. and Appelqvist, L-Å. (1972). *In* "Rapeseed" (L-Å. Appelqvist and R. Ohlson, eds), pp. 101–122. Elsevier, Amsterdam.
Mazliak, P. (1973). *A. Rev. Pl. Physiol.* **24**, 287–310.
McKillican, M. E. (1966). *J. Am. Oil Chem. Soc.* **43**, 461–465.
McMahon, V. and Stumpf, P. K. (1966). *Pl. Physiol., Lancaster* **41**, 148–156.
Miwa, T. K., Earle, F. R., Miwa, G. C. and Wolff, I. A. (1963). *J. Am. Oil Chem. Soc.* **40**, 225–229.
Mollenhauer, H. H. and Totten, C. (1971a). *J. Cell. Biol.* **48**, 387–394.
Mollenhauer, H. H. and Totten, C. (1971b). *J. Cell Biol.* **48**, 395–405.
Nagai, J. and Bloch, K. (1968). *J. biol. Chem.* **243**, 4626–4633.
Nelson, O. E. Jr. and Burr, B. (1973). *A. Rev. Pl. Physiol.* **24**, 493–518.
Norton, G. and Harris, J. F. (1974). Personal communication.
Patterson, G. W. (1970). *Lipids* **5**, 597–600.
Privett, O. S., Dougherty, K. A., Erdahl, W. L. and Stolyhwo, A. (1973). *J. Am. Oil Chem. Soc.* **50**, 516–520.
Rakow, G. and McGregor, D. I. (1974). Abstr. no. 27, 4th International Rapeseed Conference, Giessen, 1974 in *Fette-Seifen-Anstrichmittel*, **76**, 636.
Rest, J. A. and Vaughan, J. G. (1972). *Planta* **105**, 245–262.
Rinne, R. W. (1969). *Pl. Physiol., Lancaster* **44**, 89–94.
Rinne, R. W. and Canvin, D. T. (1971a). *Pl. Cell Physiol.* **12**, 387–393.
Rinne, R. W. and Canvin, D. T. (1971b). *Pl. Cell. Physiol.* **12**, 395–403.
Roehm, J. N. and Privett, O. S. (1970). *Lipids* **5**, 353–358.
Rubel, A., Rinne, R. W. and Canvin, D. T. (1972). *Crop Sci.* **12**, 739–741.
Röbbelen, G. (1972). *In* "The Way Ahead in Plant Breeding". (Plant Breeding Institute, Cambridge 1972), pp. 207–214. Proc. 6 Congr. Eucarpia, Cambridge, 1971.
Schwarzenbach, A. M. (1971). *Ber. Schweiz. Bot. Ges.* **81**, 70–91.
Shenstone, F. S. and Vickery, J. R. (1961). *Nature, Lond.* **190**, 168–169.
Shorland, F. B. (1945). *Nature, Lond.* **156**, 269–270.
Shorland, F. B. (1963). *In* "Chemical Plant Taxonomy" (T. Swain, ed.), pp. 253–303. Academic Press, London and New York.
Sims, R. P. A. (1964). *Can. J. Pl. Sci.* **44**, 217–218.
Singh, H. and Privett, O. S. (1970a). *Biochim. biophys. Acta* **202**, 200–202.
Singh, H. and Privett, O. S. (1970b). *Lipids* **5**, 692–697.
Smith, C. R. Jr. (1970). *In* "Progress in the Chemistry of Fats and other Lipids", Vol. XI, pp. 139–177. Pergamon Press, Oxford.
Sorokin, H. P. (1967). *Am. J. Bot.* **54**, 1008–1016.
Staba, E. J., Shik Shin, B. and Mangold, H. K. (1971). *Chem. Phys. Lipids* **6**, 291–295.
Stearns, E. M. Jr. (1970). *In* "Progress in the Chemistry of Fats and Other Lipids", vol. IX, "Polyunsaturated Acids" (R. T. Holman, ed.), pp. 455–516. Pergamon Press, Oxford.
Stearns, E. M. Jr. and Morton, W. T. (1973). *Lipids* **8**, 668–674.
Stefansson, B. R. and Hougen, F. W. (1964). *Can. J. Pl. Sci.* **44**, 359–364.
Stumpf, P. K. (1970). *In* "Lipid Metabolism" (S. J. Wakil, ed.), pp. 79–106. Academic Press, New York and London.
Sudershan, P. and Lakshminarayana, G. (1972). *Oléagineux* **27**, 495–498.
Thies, W. (1970). *In* "Proc. Intern. Conference on Sci., Technol., Marketing of

Rapeseed". St. Adèle, Canada, pp. 348–356. Publ. by Rapeseed Assoc. of Canada.

Thies, W. (1971). *Fette-Seifen-Anstrichmittel* **73**, 710–715.

Thomas, P. M. and Kondra, Z. P. (1973). *Can. J. Pl. Sci.* **53**, 221–225.

Trémolières, A. and Mazliak, P. (1974). *Pl. Sci. Lett.* **2**, 193–201.

Tso, T. C. and Cheng, A. L. S. (1971). *Phytochemistry* **10**, 2133–2137.

van Hal, J. G. and van Baarsel, H. M. (1974). Abstr. no. 26, 4th International Rapeseed Conference, Giessen, 1974 in *Fette-Seifen-Anstrichmittel* **76**, 362–363.

Vijay, I. K. and Stumpf, P. K. (1971). *J. biol. Chem.* **246**, 2910–2917.

Vijaylakshmi, B. and Rao, S. V. (1972). *Chem. Phys. Lipids* **9**, 82–86.

Wolff, I. A. (1966). *Science* **154**, 1140–1149.

Yano, I., Nichols, B. W., Morris, L. J. and James, A. T. (1972a). *Lipids* **7**, 30–34.

Yano, I., Morris, L. J., Nichols, B. W. and James, A. T. (1972b). *Lipids* **7**, 35–45.

Yatsu, L. Y. and Jacks, T. J. (1972). *Pl. Physiol., Lancaster* **49**, 937–943.

Yatsu, L. Y., Jacks, T. J. and Hensarling, T. P. (1971). *P. Physiol., Lancaster* **48**, 675–682.

Yermanos, D. M. and Knowles, P. F. (1962). *Crop Sci.* **2**, 109–111.

Zeman, I. and Kratochvíl, V. (1967). *Biol. Plant.* **9**, 1–14.

Zilkey, B. and Canvin, D. T. (1969). *Biochem. Biophys. Res. Commun.* **34**, 646–653.

Zilkey, B. and Canvin, D. T. (1971). *Canad. J. Bot.* **50**, 323–326.

CHAPTER 9

Organelles from Castor Bean Seedlings: Biochemical Roles in Gluconeogenesis and Phospholipid Biosynthesis

HARRY BEEVERS

Division of Natural Sciences, University of California Santa Cruz, California, U.S.A.

"Castor Oil: Mild purgative for most animals but unreliable in adult horses." (Merck Index, 8th Edition, 1968).

I. Introduction

The castor bean seed is rich in oil reserves. Some 60% of its dry weight is triglyceride, stored in the endosperm, and when seeds are germinated at 30°C this tissue is the site of intense metabolic activity for a period of 7 days. The predominant metabolic event is the conversion of fat to sucrose and this is accomplished with remarkable efficiency; the fat is completely utilized and some 70% of the carbon is recovered in sucrose (Beevers, 1961). Qualitatively similar changes occur in other fatty seedlings and, as in the castor bean, the sucrose is transferred to the embryonic axis where it is utilized during its early growth (Kriedmann and Beevers, 1967).

The endosperm tissue in castor bean is not merely an inert package of reserve material that is broken down by enzymes produced in adjacent tissues, as it is in germinating cereal seeds. The whole complex mechanism for converting fat to sucrose is produced within the endosperm cells themselves and its development is independent even of the presence of the embryo and its cotyledons (Huang and Beevers, 1974). No cell division occurs in the endosperm tissue during germination but there is an active synthesis of many enzymes and the production of an array of intracellular organelles (including ribosomes, endoplasmic reticulum, mitochondria, plastids, and glyoxysomes) at the expense of protein present in the ungerminated seed. Some of the enzymes and organelles are not found elsewhere in the plant. They are formed only in the endosperm and only during the first few days of germination and they disappear when fat utilization is completed (Gerhardt and Beevers, 1970). This restriction in time and space poses interesting developmental problems at the level of genetic programming. In the present context we will be interested more in the functional and organizational aspects of the metabolic machinery.

II. Pathway of Sucrose Formation from Fat

The biochemical pathway whereby fat is converted to sucrose is now firmly established (Fig. 1). The requisite enzymes are present in adequate amounts (Beevers, 1961; Cooper and Beevers, 1969a) and a variety of labelling experiments, particularly those with [^{14}C]acetate, confirm that the sequence actually operates *in vivo* (Canvin and Beevers, 1961). The pathway shown in summary form in Fig. 1 satisfies the observed stoichiometry of the conversion of fat to sucrose and the very low r.q. (0·3–0·4) of the endosperm during this conversion. Three very important aspects of the process in castor bean endosperm are:

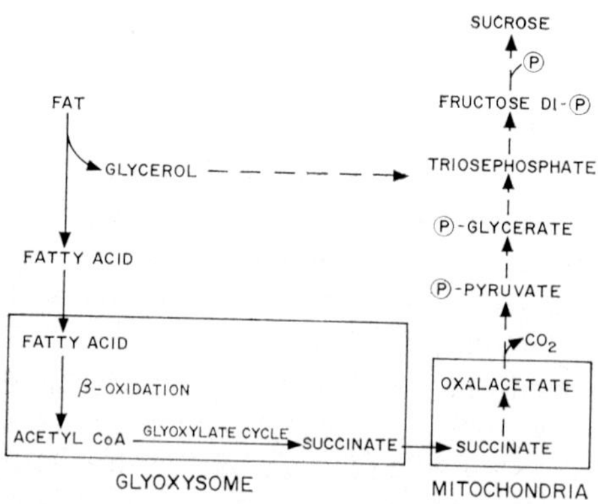

FIG. 1. Reaction pathway in the conversion of fat to sucrose. Note that NAD is reduced during β-oxidation, in the glyoxylate cycle, and in the conversion of succinate to oxaloacetate. Part of this $NADH_2$ is then used in the reduction of phosphoglycerate to triosephosphate and the rest is oxidized, by the electron transport system in the mitochondria, with accompanying ATP formation. ATP is consumed in the activation of fatty acids and of glycerol, and in the reactions producing phosphopyruvate from oxaloacetate and triosephosphate from phosphoglycerate.

1. The oxidation of fatty acid to acetyl CoA by itself provides more than sufficient $NADH_2$ to provide the reductant and (from coupled oxidation in the mitochondria) ATP to drive the reactions from oxaloacetate to sucrose.

2. It can be concluded from a variety of considerations (Canvin and Beevers, 1961) that virtually none of the acetyl CoA derived from fat is oxidized to CO_2 and H_2O in the tricarboxylic acid cycle; it is channeled exclusively through the glyoxylate cycle.

3. The conversion of fat to sucrose with its associated oxidations and decarboxylations constitute the respiration of this tissue. The process is not one superimposed over a background of conventional respiratory glucose utilization.

The most distinctive feature of the pathway is the glyoxylate cycle. That is to say, conversion of fatty acids to acetyl units follows the general pattern established in other tissues and the conversion of phosphoenolpyruvate to hexose by reversed glycolysis occurs by reactions which occur elsewhere in plants.

III. Intracellular Locations of Enzymes in the Pathway

A. The Glyoxylate Cycle

In our early work establishing the role of the glyoxylate cycle in gluconeogenesis from acetate we showed that the key enzymes, isocitrate lyase and malate synthetase, were most active at day 5, when fat utilization was at its most rapid. The enzymes are barely detectable in ungerminated seed and their activities decline as the fat is depleted (Beevers, 1961). Inhibitor evidence suggested that the enzymes were produced *de novo* and not merely activated during germination (Hock and Beevers, 1966) and later work by Longo (1968) proved that this was so. Until 1967 there was uncertainty about the intracellular location of the enzymes of the glyoxylate cycle; most of the activity was recovered on the soluble fraction after standard procedures for preparation of mitochondria, but some 30% of the activity was recovered in the 10 000 *g* pellet (Marcus and Velasco, 1960; Tanner and Beevers, 1965).

Using gentle grinding procedures and a modified extraction medium Breidenbach and Beevers (1967) were able to recover > 60% of the enzyme activity in the crude particulate pellets. When these were resolved on sucrose gradients three distinct protein bands were obtained. The uppermost, at equilibrium density 1·19 g/cc, contained the mitochondria, as seen from electronmicrographs and the distinctive distribution of the mitochondrial markers cytochrome oxidase and fumarase. Isocitrate lyase and malate synthetase were absent from this fraction and were recovered only in the lowermost band at density 1·25 g/cc. (See Fig. 5 for distribution of marker enzymes in a sucrose gradient.) Examination of this material by electron microscopy (Breidenbach *et al.*, 1968) showed that it contained distinctive

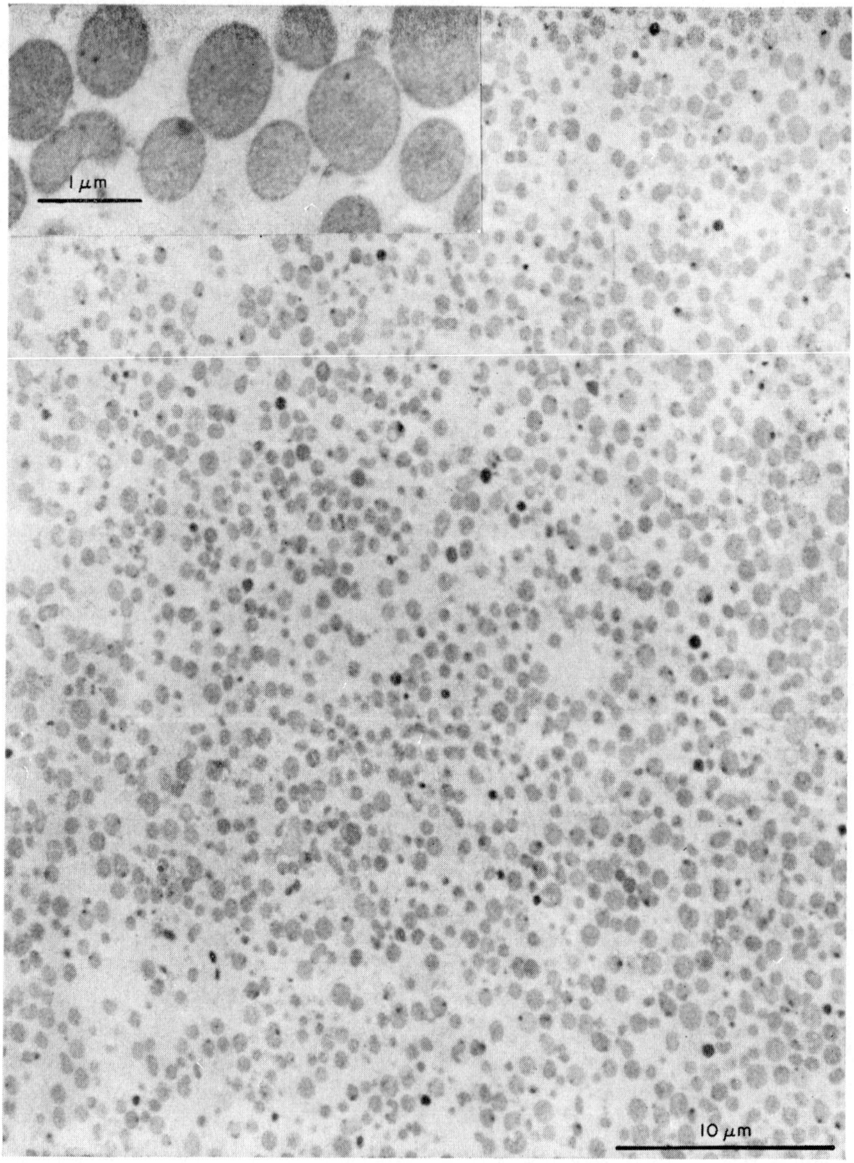

FIG. 2. Electron micrographs of glyoxysomes isolated from castor bean endosperm. The main part of the figure, at lower magnification, shows that the preparation is quite homogeneous. The inset shows a few glyoxysomes at high magnification (Huang and Beevers, 1973).

organelles about 1 μm in diameter with a single bounding membrane and amorphous interior (Fig. 2). Each of the five enzymes of the glyoxylate cycle, citrate synthetase, aconitase, isocitrate lyase, malate synthetase and malate dehydrogenase, was found to be present in these organelles (Breidenbach and Beevers, 1967; Cooper and Beevers, 1969) which were therefore called glyoxysomes.

Until that time the glyoxylate cycle had usually been considered as an appendage or by-pass of the tricarboxylic acid cycle and its operation had been thought to depend on the citrate synthetase, aconitase and malate dehydrogenase in the mitochondria. The discovery of the glyoxysome showed that, at least for the castor bean endosperm, this view was no longer tenable. These three enzymes are present both in mitochondria and glyoxysomes and thus the tricarboxylic acid cycle and the glyoxylate cycle operate independently in separate cell compartments. Glyoxysomes have since been isolated from

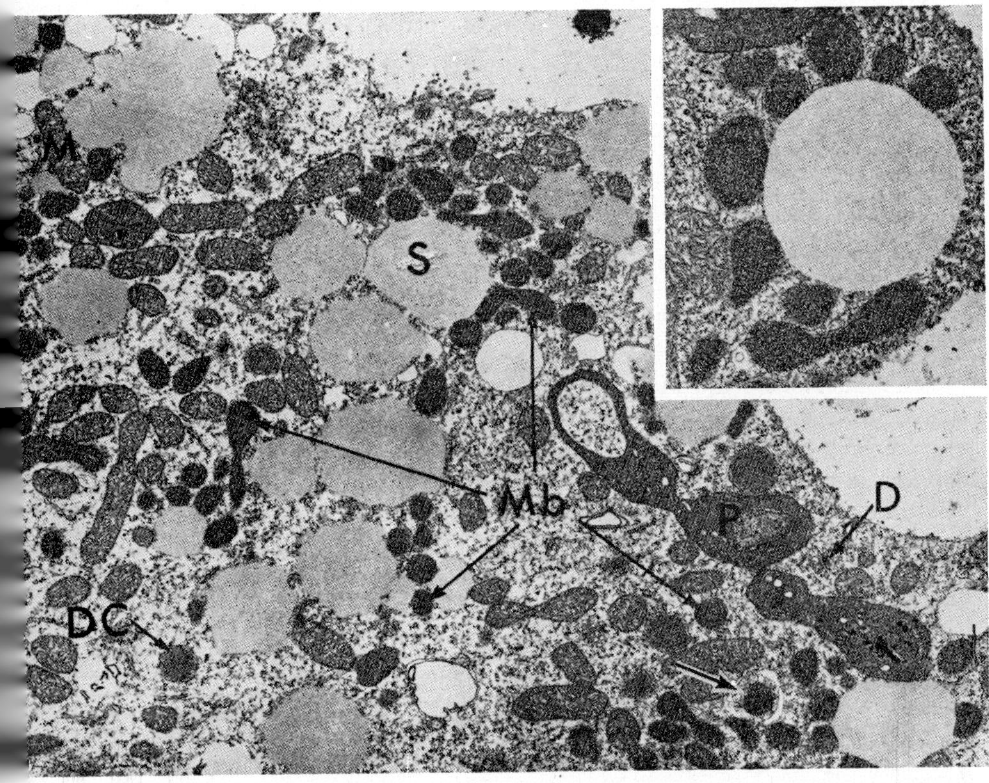

FIG. 3. Electron micrograph of castor bean endosperm tissue after 6 days of germination (Vigil, 1970). Most of the fat has disappeared by this time, but some spherosomes (S) remain. Plastids (P), Mitochondria (M), and numerous glyoxysomes (Mb) are present. Magnification, ×7400. In the inset several glyoxysomes are seen in intimate association with a spherosome (magnification, ×15 700). Vigil's paper should be consulted for further details of the fine structure of the castor bean endosperm.

a variety of fatty seedlings and are seen in abundance in electron micrographs of the appropriate tissue (Fig. 3).

It was recognized that in their morphology and sedimentation behavior on sucrose gradients the glyoxysomes are similar to the microbodies first described in liver and isolated and intensively studied as peroxisomes by de Duve and his colleagues (see de Duve and Baudhuin, 1966, for a review). Indeed, the glyoxysomes were subsequently found to contain catalase, glycollate oxidase (Breidenbach *et al.*, 1968), and uricase (Theimer and Beevers, 1971), the distinctive enzymes of animal peroxisomes. It thus became clear that the glyoxysome is a highly specialized kind of microbody. In relatively small numbers such organelles had been observed in electronmicrographs from a variety of plant cells (Mollenhauer *et al.*, 1966). We have since isolated microbodies from a range of plant tissues. From most of these the microbodies have a limited enzyme complement (catalase, uricase, and glycollate oxidase) and they are regarded as non-specialized (Huang and Beevers, 1971). Leaves contain a specialized type of microbody, the leaf peroxisome, which is responsible for the metabolism of glycollate generated as a by-product of photosynthesis (Tolbert, 1971).

B. CONVERSION OF SUCCINATE TO SUCROSE

The result of acetate utilization within the glyoxysome is the production of succinate and $NADH_2$, neither of which can be oxidized by this organelle (Cooper and Beevers, 1969). Enzymes in the mitochondria must be responsible for converting succinate to oxalacetate and for reoxidation of part of the $NADH_2$ generated in the glyoxysome (Fig. 1). $NADH_2$ is also consumed as the reductant for the triosephosphate dehydrogenase reaction (Fig. 1). ATP the other component required for conversion of oxalacetate to phosphoenolpyruvate (PEP) and for reversed glycolysis (Benedict and Beevers, 1961, 1962) is generated in the electron transport system of the mitochondria where further reducing equivalents are generated in converting succinate to oxaloacetate. At present it appears that the reaction sequence from PEP to sucrose occurs in the soluble phase of the cytoplasm. However, small amounts of the enzymes of the EMP pathway (possibly distinct isozymes) are recovered in the plastids and it is conceivable that gluconeogenesis from PEP occurs *in vivo* in these organelles (Kobr and Beevers, 1968). Improved methods for plastid isolation (Miflin and Beevers, 1974) will allow an investigation of this possibility.

C. FATTY ACID OXIDATION AND LIPOLYSIS

Until 1969 it had been assumed that oxidation of the long-chain fatty acids to acetyl CoA occurred in the mitochondria of the endosperm tissue. In that year Cooper and Beevers (1969) and independently, Hutton and Stumpf

(1969) showed, surprisingly, that this was not so. Instead, individual enzymes of the β-oxidation pathway and the overall oxidation of long-chain fatty acyl CoA derivatives to acetyl CoA were shown to occur within the glyoxysomes. The acyl-CoA dehydrogenase reaction apparently reduces a flavin which is directly reoxidized with O_2 as the acceptor. Catalase in the glyoxysome decomposes the resulting H_2O_2. $NADH_2$ produced in the hydroxyacyl-CoA dehydrogenase reaction cannot be reoxidized within the glyoxysome. Thus the conversion of fatty acyl CoA to acetyl CoA in this organelle is accomplished with concomitant O_2 uptake and $NADH_2$ accumulation (Cooper and Beevers, 1969).

It thus became clear that the role of the glyoxysome in the conversion of fat to sucrose is not restricted to the crucial steps of the glyoxylate cycle. The acetyl CoA consumed in this cycle is apparently generated from fatty acids within the organelle itself. An important consequence of this arrangement whereby acetyl units are channeled directly into the glyoxylate cycle is that they escape oxidation in the tricarboxylic acid cycle. Other controls are not of course excluded.

It was subsequently shown that the activation of long chain fatty acids also occurs in the glyoxysome (Cooper, 1971) and thus that the entire reaction sequence from fatty acids to succinate can be accomplished in this single organelle (Fig. 1).

The starting point in the overall pathway (Fig. 1) is the storage triglyceride which is found in the spherosomes (Fig. 3). In ungerminated castor bean these are seen as discrete packages of lipid contained within a membrane, and in electronmicrographs they appear to occupy most of the intracellular space (Vigil, 1970). Ory and his colleagues (1960, 1968) have shown that the spherosomes (which comprise the fatty layer that floats to the surface when extracts are centrifuged; Fig. 4) contain in their membranes an active acid lipase which is probably responsible for the hydrolysis of fat contained within them. We have recently shown that the endosperm contains an additional lipase which increases in activity up to day 5 of germination (Muto and Beevers, 1974). This lipase, with an alkaline pH optimum, is specifically associated with the glyoxysome and is firmly bound to the membrane. However, the enzyme appears to hydrolyze only monoglycerides, whereas the lipase on the spherosomes can bring about complete hydrolysis of triglycerides.

We are thus approaching a fairly complete accounting of the intracellular localization of each of the steps in the conversion of fat to sucrose (Table I). Clearly the overall process requires the sequential operation of enzymes in spherosomes, glyoxysomes, mitochondria and the soluble phase (and possibly plastids). Specific intermediates in the pathway must move between organelles and there is, in addition, a shuttling of $NAD(H_2)$ (or of cognate reductants) and adenylates. In the castor bean endosperm the whole metabolism seems to be focused on gluconeogenesis from fat (and from amino acids related to the glyoxylate cycle (Stewart and Beevers, 1967) and as the fat is

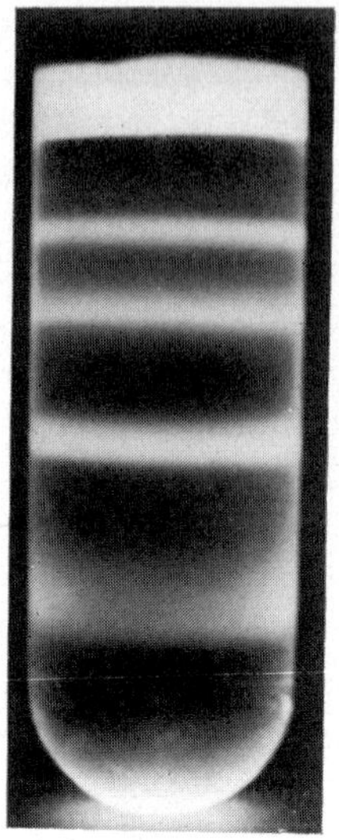

FIG. 4. Appearance of the sucrose gradient after centrifuging the total homogenate from 4 day castor bean endosperm. Band B is made up of vesicles of the endoplasmic reticulum, and is the intracellular site of various enzymes of phospholid metabolism (Lord *et al.*, 1972).

depleted the endosperm tissue is completely consumed. In other fatty tissues such as the cotyledons of cucurbits, the situation is more complex since the storage organ subsequently becomes photosynthetic.

However, that is not to say that no other important synthetic events occur in the endosperm of the castor bean during its brief existence. As we have indicated above the whole of the enzymatic machinery for the conversion of fat is made *de novo* once germination begins, and assembled into the appro-

TABLE I

Intracellular localization of enzyme sequences participating in conversion of fat to sucrose

Triglyceride hydrolysis	Spherosomes
Monoglyceride hydrolysis	Glyoxysomes
Fatty acid activation	Glyoxysomes
β-oxidation	Glyoxysomes
Glyoxylate cycle	Glyoxysomes
(Acetyl CoA $\rightarrow$ succinate)	Glyoxysomes
$NADH_2$ oxidation, ATP synthesis	Mitochondria
Succinate $\rightarrow$ oxalacetate	Mitochondria
Oxalacetate $\rightarrow$ PEP	Soluble
PEP $\rightarrow$ sucrose	Soluble (plastids?)

priate organelles. During days 2–4 when the amounts of the various enzymes are increasing there are striking increases in the amounts of organelles such as glyoxysomes and mitochondria (Gerhardt and Beevers, 1970). This can be seen by measuring the amounts of these organelles recovered from endosperm tissue when extracts are separated on sucrose gradients at daily intervals during early growth. The pattern of rise and fall in enzyme activities during the 7-day interval is matched by changes in the numbers of glyoxysomes and mitochondria, and electron micrographs graphically illustrate these events (Vigil, 1970).

IV. Organelle Biogenesis: Role of the Endoplasmic Reticulum in Phospholipid Synthesis

Our recent work on the biogenesis of the glyoxysome has shown that yet another component of the cell, the endoplasmic reticulum, is intimately involved, specifically in the production of phospholipids destined for inclusion in the membranes of glyoxysomes and other organelles. Glyoxysomal membranes can be obtained by subjecting isolated organelles to a hypotonic medium and salt treatment (Huang and Beevers, 1973). The phospholipid constituents of these and other membranes from the endosperm have been measured by Donaldson *et al.* (1972).

Phosphatidylcholine (lecithin) is a major component of the membranes and lecithin readily becomes labeled when [^{14}C]choline is supplied to endosperm tissue (Kagawa, *et al.*, 1973). We were able to show that the ^{14}C content of lecithin in the membranes of mitochondria and glyoxysomes increases with time after application of [^{14}C]choline, but that the first particulate cellular component to become labeled is a "light membrane fraction" which we now know (Lord *et al.*, 1973) is comprised of vesicles of the endoplasmic reticulum (ER). Part of the labeled lecithin in this component subsequently appears in the membranes of glyoxysomes and mitochondria.

For these experiments the procedure for isolation of organelles begins with a prolonged chopping of the endosperm with a razor blade which, compared to previous methods, further minimizes breakage of organelles. After a brief period of centrifugation at 280 *g*, the whole homogenate (rather than the crude particulate fraction obtained at 10 000 *g*) is layered onto a sucrose gradient (Lord *et al.*, 1972). Such a gradient after high-speed centrifugation for 4 h is shown in Fig. 4. Five distinct bands are resolved. The uppermost, floating, layer is comprised of spherosomes (fat) and the mitochondria and glyoxysomes are obtained, as before, at their respective equilibrium densities. Band B contains the characteristic cytochromes and reductases of "microsomes" and examination in the electron microscope shows that it is indeed very rich in vesicles of the endoplasmic reticulum (ER) separated from the ribosomes with which it is associated *in vivo* (Vigil, 1970). In such gradients

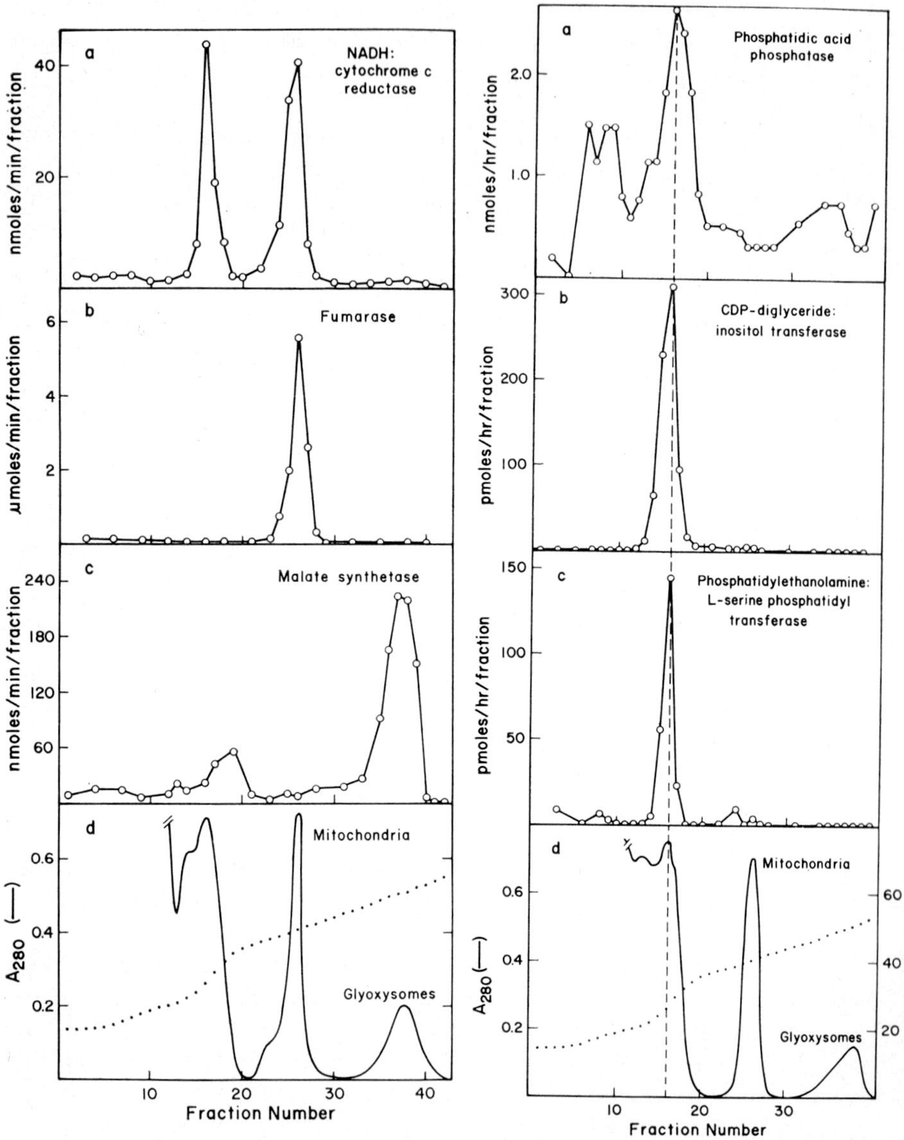

FIG. 5. Distribution of enzymes and protein (A_{280}) in a gradient such as that in Fig. 4 (Moore *et al.*, 1973). Left: The lighter peak of NADH cytochrome *c* reductase marks the endoplasmic reticulum (band B) and the heavier one, coinciding with fumarase, the mitochondria. Malate synthetase is the glyoxysome marker. Right: The major particulate peak of phosphatidic acid phosphatase, and the single peak of CDP-inositol transferase and of phosphatidyl ethanolamine: L-serine transferase all coincide with that of the NADH cytochrome reductase in the endoplasmic reticulum. CDP-diglyceride transferase is also found exclusively in this band (Lord *et al.*, 1972).

truly soluble enzymes and those released from organelles during preparation appear in the upper portions, just below the fatty layer, and do not penetrate into the gradient proper. As seen from the distribution of mitochondrial and glyoxysomal marker enzymes, remarkably little breakage of these organelles occurs (Fig. 5).

Analysis of sequential fractions from the gradient for enzymes concerned with lecithin biosynthesis showed that the phosphorylcholine glyceride transferase, the final enzyme in the sequence, was exclusively confined to the ER (Band B). This finding gives a biochemical explanation for the observation with intact tissue, that [^{14}C]choline appears in glyoxysomes and mitochondria only after it is incorporated into the ER; the confinement of synthesis of lecithin from its precursors in the ER dictates such a relationship. In previous work on fractions obtained by *differential* centrifugation some lecithin synthesis was found to be associated with other organelles. We showed that this is due to contamination of these fractions by ER (Lord *et al.*, 1972).

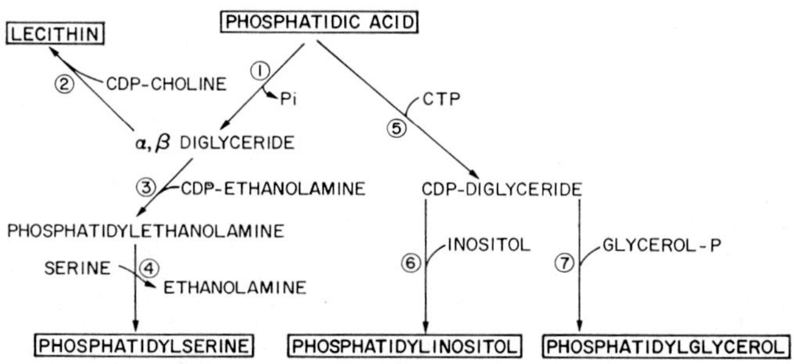

FIG. 6. Reactions of phospholipid metabolism (after Kennedy, 1961). Phosphatidic acid phosphatase (1) as well as the enzymes catalyzing reactions 2, 4, 6 and 7 producing important phospholipids of membranes have all been shown to occur in the membranes of the endoplasmic reticulum (Moore *et al.*, 1973). Reaction 7 occurs additionally in the mitochondria (Moore, 1974).

The ER has since been shown to contain several other enzymes of phospholipid metabolism (Moore *et al.*, 1973). Thus phosphatidic acid phosphatase which produces 1,2-diglyceride precursor for synthesis of lecithin and other phospholipids (Fig. 6) is present in this component. CDP-inositol diglyceride transferase and the enzyme incorporating serine into phosphatidyl serine (phosphatidylethanolamine-serine phosphatidyl transferase) are also confined to this region of the gradient (Fig. 5). By contrast, the production of phosphatidyl glycerol occurs both in the ER and the mitochondria (Moore, 1974).

It is thus clear that the glyoxysome does not contain the enzymes required for the synthesis of the phospholipids present in its membrane. They are synthesized in the ER (Table II).

Attention has been drawn to the close association frequently seen in electron micrographs of whole cells between fragments of the ER and microbodies

TABLE II

Intracellular distribution of enzymes of phospholipid metabolism in castor bean endosperm

Reaction	Product	Location
Choline kinase	Phosphoryl choline	Soluble
P-choline incorporation	CDP-choline	Band A (density 1·08 g/cc)
CDP-choline incorporation	Phosphatidylcholine	Endoplasmic reticulum
Inositol incorporation	Phosphatidylinositol	Endoplasmic reticulum
Serine incorporation	Phosphatidylserine	Endoplasmic reticulum
Phosphatidic acid phosphatase	1,2-diglyceride	Endoplasmic reticulum
Glycerolphosphate incorporation	Phosphatidylglycerol	Endoplasmic reticulum and mitochondria

such as glyoxysomes (e.g. Vigil, 1970). The possibility has been raised that microbodies arise by dilation and vesiculation of the ER tubules (Vigil, 1970; Gruber *et al.*, 1970). Our experimental observations clearly give biochemical support to the possibility that the glyoxysomal membrane is derived fairly directly from the ER. The alternative possibility that phospholipid components made in the ER are individually transported to membranes of developing organelles has been invoked for mitochondria (Wirtz, *et al.*, 1972) which are at least partly autonomous genetically. (Mazliak discusses this process in Chapter 10.) The glyoxysomes themselves contain no DNA (Douglas *et al.*, 1973) or ribosomes and the synthesis of all of their proteins, including enzymes and structural membrane constituents, must occur in cytoplasmic ribosomes. Here again the relationship to the ER is an intriguing one. For if the glyoxysomal membrane is indeed derived from the ER there is a real possibility that proteins made on the ribosomes with which the ER is closely associated *in vivo* are incorporated immediately prior to the vesiculation process. Encouraging to this view is the observation that, at early stages of development, activity of several of the glyoxysomal enzymes can be detected in the ER band. Finally, Dr. Elma Gonzalez, in my laboratory, has shown, using antibodies against soluble glyoxysomal enzymes (Huang *et al.*, 1974) that, at the 2 day stage the ER fraction shows a strong immunological reaction.

ACKNOWLEDGEMENT

Our work in this field has been generously supported by grants from the National Science Foundation and the Atomic Energy Commission.

REFERENCES

Beevers, H. (1961). *Nature, Lond.* **191**, 433–436.

Benedict, C. R. and Beevers, H. (1961). *Pl. Physiol., Lancaster* **36**, 540–544.

Benedict, C. R. and Beevers, H. (1962). *Pl. Physiol., Lancaster* **37**, 176–178.

Breidenbach, R. W. and Beevers, H. (1967). *Biochem. biophys. Res. Comm.* **27**, 462–469.

Breidenbach, R. W., Kahn, A. and Beevers, H. (1968). *Pl. Physiol., Lancaster* **43**, 705–713.

Canvin, S. T. and Beevers, H. (1961). *J. biol. Chem.* **236**, 988–995.

Cooper, T. G. (1971). *J. biol. Chem.* **246**, 3451–3455.

Cooper, T. G. and Beevers, H. (1969a). *J. biol. Chem.* **244**, 3507–3513.

Cooper, T. G. and Beevers, H. (1969b). *J. biol. Chem.* **244**, 3514–3520.

De Duve, C. E. and Baudhuin, P. (1966). *Physiol. Rev.* **46**, 323–357.

Donaldson, R. P., Tolbert, N. E. and Schnarrenberger, C. (1972). *Archs Biochem. Biophys.* **152**, 199–215.

Douglass, S. A., Criddle, R. S. and Breidenbach, R. W. (1973). *Pl. Physiol., Lancaster* **51**, 902–906.

Gerhardt, B. P. and Beevers, H. (1970). *J. Cell Biol.* **44**, 94–102.

Gruber, P. J., Trelease, R. N., Becker, W. M. and Newcomb, E. H. (1970). *Planta* **93**, 269–288.

Hock, B. and Beevers, H. (1966). *Z. Pflanzenphysiol.* **55**, 405–414.

Huang, A. H. C. and Beevers, H. (1971). *Pl. Physiol., Lancaster* **48**, 637–641.

Huang, A. H. C. and Beevers, H. (1973). *J. Cell Biol.* **58**, 379–389.

Huang, A. H. C. and Beevers, H. (1974). *Pl. Physiol., Lancaster* **54**, 277–279.

Huang, A. H. C., Bowman, P. D. and Beevers, H. (1974). *Pl. Physiol., Lancaster* **54**, 364–368.

Hutton, D. and Stumpf, P. K. (1969). *Pl. Physiol., Lancaster* **44**, 508–516.

Kagawa, T., Lord, J. M. and Beevers, H. (1973). *Pl. Physiol., Lancaster* **51**, 61–65.

Kennedy, E. P. (1961). *Fedn Proc. Fedn Am. Socs exp. Biol.* **20**, 934–940.

Kobr, M. J. and Beevers, H. (1968). *Pl. Physiol., Lancaster* **43**, S–17.

Kriedman, P. and Beevers, H. (1967). *Pl. Physiol., Lancaster* **42**, 161–173.

Longo, C. P. (1968). *Pl. Physiol., Lancaster* **43**, 660–664.

Lord, J. M., Kagawa, T. and Beevers, H. (1972). *Proc. Natn. Acad. Sci. U.S.A.* **69**, 2429–2432.

Lord, J. M., Kagawa, T., Moore, T. S. and Beevers, H. (1973). *J. Cell. Biol.* **57**. 659–667.

Marcus, A. and Velasco, J. (1960). *J. biol. Chem.* **235**, 563–567.

Miflin, B. and Beevers, H. (1974). *Pl. Physiol., Lancaster* **53**, 870–874.

Mollenhauer, H. H., Morré, D. J. and Kelley, A. G. (1966). *Protoplasma* **62**, 44–52.

Moore, T. S. (1974). *Pl. Physiol., Lancaster* **54**, 164–168.

Moore, T. S., Lord, J. M., Kagawa, T. and Beevers, H. (1973). *Pl. Physiol., Lancaster* **52**, 50–53.

Muto, S. and Beevers, H. (1974). *Pl. Physiol., Lancaster* **54**, 23–28.

Ory, R. L., St. Angelo, A. J. and Altschul, A. M. (1960). *J. Lipid. Res.* **1**, 208–213.

Ory, R. L., Yatsu, L. Y. and Kircher, H. W. (1968). *Archs Biochem. Biophys.* **264**, 255–264.

Stewart, C. S. and Beevers, H. (1967). *Pl. Physiol., Lancaster* **42**, 1587–1595.

Tanner, W. and Beevers, H. (1965). *Z. Pflanzenphysiol.* **53**, 72–85.

Theimer, R. R. and Beevers, H. (1971). *Pl. Physiol., Lancaster* **47**, 246–251.

Tolbert, N. E. (1971). *A. Rev. Pl. Physiol.* **22**, 45–74.

Vigil, E. L. (1970). *J. Cell Biol.* **46**, 434–454.

Wirtz, K. W. A., Kaup, H. H. and Van Deenen, L. M. M. (1972). *Biochim. biophys. Acta* **274**, 606–617.

CHAPTER 10

Exchange Processes between Organelles Involved in Membrane Lipid Biosynthesis

P. MAZLIAK, D. DOUADY, C. DEMANDRE and J. C. KADER

Université de Paris, Laboratoire de Physiologie cellulaire, Paris, France

I. Introduction

Research in the field of lipids has become considerably more interesting since it is now clearly recognized that lipids are major constitutents of all biological membranes and play a central role in their molecular organization. The membrane model of a lipid bilayer with partially embedded globular proteins is now widely accepted (Singer and Nicolson, 1972).

With this importance of membrane lipids in mind we set out, in our laboratory, to answer two main questions:

(a) Are the membrane lipids identical in the different organelles of the cell?
(b) Where are the membrane lipids synthesized in the plant cell?

These two questions lead the work of our group towards the study of a central phenomenon, that of the *exchange of lipids amongst the different cell compartments.* The greater part of this chapter is devoted to the description of this phenomenon.

II. Plant Membrane Lipids

In this discussion the lipids of the chloroplast membranes which are mainly galactolipids rich in linolenic acid (see Chapters 1 and 6) have been excluded. We shall limit ourselves to the lipids present in non-green tissues such as those of potato tubers and cauliflower florets. The main lipids in the different membranes or organelles of these tissues are phospholipids. Tables I and II show that there is a striking uniformity in the phospholipid composition of different intracellular organelles from the same tissue. This is particularly apparent in the case of phosphatidylcholine and phosphatidylethanolamine which are the most abundant of the membrane phospholipids, together constituting more than 70% of the total. There are, however, some minor variations, for instance the diphosphatidylglycerol is concentrated in the inner membrane of mitochondria. Indeed the two mitochondrial membranes do not have the same phospholipid composition, the outer containing about twice as much phosphatidylcholine as phosphatidylethanolamine whilst the inner has almost equal quantities of these compounds. Nevertheless, in spite of these minor variations, the overwhelming impression is one of great uniformity of lipid composition amongst the different cell membranes.

The same is true of the fatty acid composition of the membrane lipids of different intracellular organelles. Table III, showing the fatty acid compositions of the different membranes of the potato tuber, exemplifies this. There are,

Table I

Percentage composition of the membrane phospholipids of potato tubers

Lipid[a]	Cytoplasmic membranes[1] (plasmalemma)	Mitochondria[2] whole organelles	Mitochondria[2] inner membranes	Mitochondria[2] outer membranes	Peroxisomes[3]	Microsomes[4]
PC	38	43	27	52·6	61	44·8
PE	47[b]	30	29	25	19·6	32·5
PGP	—	8	19·5	12·1	—	1
PA	—	5	—	—	—	3·5
PI	15	7			4·3	16
PS	—	3	24·5	10·3	—	1
PG	—	3			15·1	1

[a] The following abbreviations have been used: PC = phosphatidylcholine, PE = phosphatidylethanolamine, PGP = diphosphatidylglycerol, PA = phosphatidic acid, PI = phosphatidylinositol, PS = phosphatidylserine, PG = phosphatidylglycerol.
[b] Includes PS and PG.

[1] Demandre, C. (1974), unpublished results.
[2] Meunier and Mazliak (1972).
[3] Tchang, F. (1974), unpublished results.
[4] Ben Abdelkader and Mazliak (1970).

TABLE II

Percentage composition of the membrane phospholipids of cauliflower florets

Lipid[a]	Cytoplasmic membranes[1] (plasmalemma)	Mitochondria whole organelles[2]	Mitochondria inner membranes[3]	Mitochondria outer membranes[3]	Peroxisomes[1]	Microsomes[2]
PC	66	43	41	42	48·0	50
PE	21	32	37	24	45·6	36
PGP	—	7	14	3	—	tr
PA	—	1·9	—	—	—	—
PI	13	11	5	21	6·4	6
PS	—	1	—	—	—	—
PG	—	3	3	10	—	8

[a] For abbreviations see footnote to Table I.
[1] Mazliak, P. (1974), unpublished observations.
[2] Ben Abdelkader and Mazliak (1970).
[3] Chauveau, F. and Lance, C. (1973), personal communication.

however, some minor differences; for example, the fatty acids of the outer mitochondrial membrane are more saturated than those of the inner, and the fatty acids of the plasmalemma are more saturated than those of the intracellular organelles. Nevertheless the same fatty acids appear in the different membranes. It has also been found that within the same membrane phosphatidylinositol contains generally more saturated acyl moieties than phosphatidylcholine or phosphatidylethanolamine whilst diphosphatidylglycerol

TABLE III

Percentage composition of the fatty acids of potato tuber membranes

Fatty acid[a]	Cytoplasmic membranes[1] (plasmalemma)	Mitochondria[2] whole organelles	Mitochondria[2] inner membranes	Mitochondria[2] outer membranes	Peroxisomes[3]	Microsomes[4]
14:0	—	—	—	—	2	—
16:0	27·8	14·7	12·5	24	22	31
18:0	8·4	2·7	2·3	3·5	2·1	7
18:1	4·3	1·4	0·8	1·8	4·3	0·5
18:2	40	63·5	67·4	58·0	56	49·5
18:3	8·5	16·9	16·8	12·7	13·3	12·0

[a] The number before the colon represents the number of carbon atoms in the fatty acid whilst the number after the colon indicates the number of double bonds present.
[1] Demandre, C. (1974), unpublished results.
[2] Meunier and Mazliak (1972).
[3] Tchang, F. (1974), unpublished results.
[4] Ben Abdelkader and Mazliak (1970).

TABLE IV

Lipid synthesis by isolated intracellular organelles

Organelle	Tissue	Precursor administered or enzyme assayed	Fatty acids[a] synthesized	Phospholipids[b] synthesized	Authors
Mitochondria	Cauliflower floret; Potato tuber; *Vicia faba* root; *Lupinus albus* root	[1-^{14}C]- and [2-^{14}C]acetate; [1-^{14}C]acetyl-CoA; [2-^{14}C]malonate; [2-^{14}C]malonyl-CoA	14:0, 16:0, 18:0, 18:1		Mazliak *et al.* (1972)
	Peanut cotyledon	AT^{32}P		PA	Bradbeer and Stumpf (1960)
	Cauliflower floret	[^{32}P]glycerophosphate		lyso PA, PA	Bahl *et al.* (1970)
		CT^{32}P		PG	Douce (1968)
		[^{14}C]CTP		PG, DPG, PI	Sumida and Mudd (1970)

		[^{32}P]glycerophosphate		lyso PA, PA	Sastry and Kates (1966)
	Spinach leaves	[^{14}C]glycerophosphate		PA	Cheniae (1965)
		[1,3-^{14}C]glycerol		PG	Marshall and Kates (1972)
Microsomes		[1,2-^{14}C]choline		PC	Devor and Mudd (1971)
	Pea seedlings	[1,2-^{14}C]ethanolamine		PE	Vandor and Richardson (1968)
	Safflower (maturing seeds)	Oleyl-CoA desaturase	18:2		McMahon and Stumpf (1966) Vijay and Stumpf (1971)
	Potato	Oleyl-CoA desaturase	18:2		Ben Abdelkader *et al.* (1973)
	Castor bean endosperm	PA-phosphatase CDP-diglyceride:*myo*-inositol-phosphatidyltransferase PE:L-serine-phosphatidyltransferase		PE, PS, PI	Moore *et al.* (1973)
Glyoxysomes	Castor bean endosperm	Fatty acid β-oxidation enzymes			Cooper and Beevers (1969).
Golgi vesicles	Onion stem	CTP:choline phosphate cytidyltransferase		PC	Morre *et al.* (1970)

[a] For abbreviations see footnote to Table III.
[b] For abbreviations see footnote to Table I.

contains much more unsaturated fatty acids than other lipids (Meunier and Mazliak, 1972).

In conclusion one can say that in all the cell membranes of potato tubers or cauliflower florets phosphatidylcholine and phosphatidylethanolamine are the major phospholipids and that the same polyunsaturated fatty acids, linoleic and linolenic, are dominant.

III. Organelles Involved in Lipid Metabolism

To locate the sites of lipid synthesis in plant cells, one can isolate the intracellular organelles (mitochondria, microsomes, peroxisomes, plasmalemma) and see what fatty acids and phospholipids they are capable of synthesizing from radioactively labelled precursors. It is, however, important to ensure that the isolation procedure has not caused any loss of enzymes from the organelles or, indeed, any adsorption of enzymes from some other part of the cell and this is a rather difficult task. Similarly it is essential that the isolated organelles retain their physiological activity, which means that mitochondria, for instance, must exhibit good respiratory control (Mazliak *et al.*, 1972). Table IV summarizes our present knowledge of the synthetic capabilities of isolated plant organelles. Although the table still contains many blanks several points are clear:

(a) Isolated mitochondria are only capable of synthesizing saturated and monounsaturated fatty acids, despite the fact that polyunsaturated fatty acids are the main fatty acids present in mitochondrial membranes (see p. 303).

(b) Isolated mitochondria are only capable of synthesizing acidic phospholipids such as phosphatidic acid, phosphatidylglycerol, diphosphatidylglycerol and phosphatidylinositol, despite the fact that the main phospholipids of mitochondrial membranes, and those of other intracellular organelles, are phosphatidylcholine and phosphatidylethanolamine (see p. 302).

(c) The microsomes of plant cells contain all the enzymes necessary for the biosynthesis of either polyunsaturated fatty acids (at least linoleic acid) or the major phospholipids, phosphatidylcholine and phosphatidylethanolamine.

Therefore one is forced to the conclusion that there must be a transfer of lipids from microsomes to other intracellular organelles to correlate the synthetic capabilities of the various organelles with the uniformity of their lipid composition. In proposing this inter-membrane lipid exchange (Mazliak *et al.*, 1968) we referred to the cooperation between microsomes and mitochondria which was necessary for the biogenesis of the mitochondrial membranes. Since then we have devoted a great deal of time to a study of the inter-membrane exchange of phospholipids and fatty acids in plant cells, *in vivo* and *in vitro*.

IV. Lipid Exchange *in Vivo*

The phenomenon of inter-membrane lipid exchange has been studied by pulse-chase experiments using "aged" potato tuber slices (Ben Abdelkader, 1969, 1972). The "ageing" process was accomplished by maintaining freshly cut slices in water for 24 h and was shown to be accompanied by the formation of many new membranes. Careful examination of the newly formed membranes led to the conclusion that the "ageing" process induced the appearance of new mitochondrial membranes as well as new endoplasmic reticular membranes. Of great interest was the observation that all these processes could be stopped by antibiotics such as cycloheximide. The appearance of the new membranes in "ageing" potato tuber slices was accompanied by the appearance of newly-formed lipid, whose synthesis could be followed by the incorporation of radioactively labelled percursors. It was found that an initial 2 h pulse of [1-^{14}C]acetate labelled the phospholipids of the microsomal fraction more rapidly and more intensively than those of the mitochondria. During the 40 h chase in a medium containing non-radioactive acetate it was found that the total and specific radioactivity of the microsomal phospholipid decreases continuously whilst that of the mitochondrial phospholipid is still increasing (Fig. 1A). A similar result was also obtained with [^{14}C]choline (Fig. 1B) and leads to the conclusion that the newly synthesized, radioactive lipids are transferred from the microsomes to the mitochondria in the chase period. The efficiency of this transfer appears greater for acetate than for choline. This *in vivo* inter-membrane lipid transfer is also seen with polyunsaturated fatty acids such as linoleic acid.

V. Lipid Exchange *in Vitro*

The lipid exchange process has also been studied *in vitro* (Ben Abdelkader and Mazliak, 1970). Initial experiments demonstrating an *in vitro* exchange between microsomes and mitochondria were carried out. Microsomes and mitochondria, containing radioactively labelled lipids, were isolated from potato tuber or cauliflower floret slices previously incubated with [^{14}C]-acetate, [^{14}C]glycerol, [^{14}C]choline or $Na_2H^{32}PO_4$. Labelled microsomes were then mixed with unlabelled mitochondria and vice versa. The resulting mixtures were suspended in a small volume of cytoplasmic supernatant, incubated for varying lengths of time and then separated, by density-gradient centrifugation, into microsomal and mitochondrial fractions which were then analysed. This experiment yielded the following results:

(a) Radioactive lipids leave the initially labelled (donor) organelle and are transferred to the initially unlabelled (acceptor) organelle. Transfers of up to 40 % have been recorded (Fig. 2).

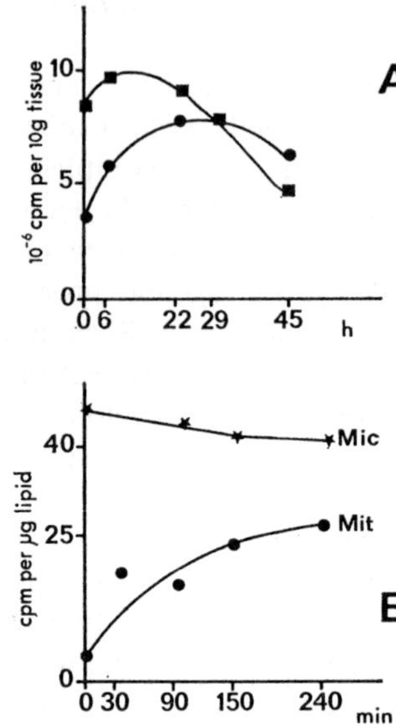

FIG. 1. Lipid exchange between microsomes and mitochondria *in vivo*. A. Potato tuber slices were incubated for 2 h in a solution of [1-^{14}C]acetate. They were then rinsed and transferred to a solution containing unlabelled acetate. Samples were removed after various times of incubation in the unlabelled acetate and the total radioactivity per 10 g of tissue in the microsomal lipid (squares) and mitochondrial lipid (circles) determined. B. The same experiment as A was carried out using a 30 min pulse of [^{14}C]choline instead of a pulse of [1-^{14}C]acetate. The specific radioactivity (cpm/μg) of microsomal lipid (stars) and mitochondrial lipid (circles) were determined after chasing with unlabelled choline for various times.

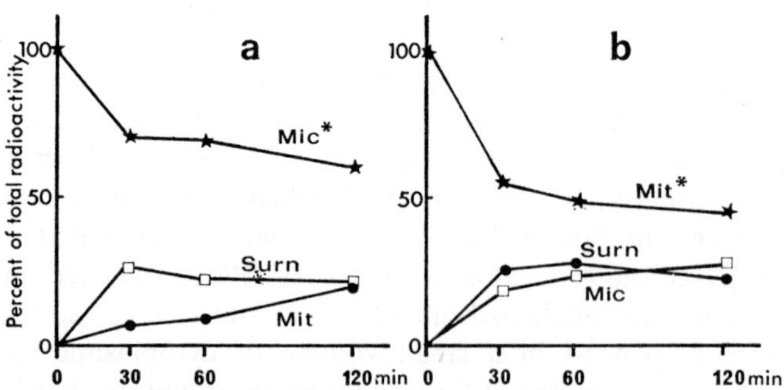

FIG. 2. Lipid exchange between microsomes and mitochondria *in vitro*. a. Incubation of radioactive lipid-labelled microsomes with unlabelled mitochondria, followed by re-isolation of each organelle and determination of content of radioactivity. The latter is expressed as the percentage of the total radioactivity present. b. Incubation of radioactive lipid-labelled mitochondria with unlabelled microsomes (then as for "a").

Mic = microsomes; Mit = mitochondria; Surn = cytoplasmic supernatant; * = initial location of radioactive lipid.

(b) The exchange process operates equally well in either direction, microsomes to mitochondria or vice versa. An examination of the specific radioactivity (Fig. 3) of the phospholipid in the "donor" and "acceptor" fractions shows that it decreases with time in the former fraction and increases with time in the latter. This indicates, not only that labelled phospholipid is continuously transferred from the "donor" organelle to the "acceptor" organelle, but also that the "donor" organelle is continuously receiving unlabelled phospholipid from the "acceptor" organelle.

(c) Calculation of the quantities of lipid exchanged between the organelles shows (i) that it is twice as high when starting with lipids labelled by [^{14}C]-acetate as with lipids labelled by [^{14}C]glycerol or [^{32}P]phosphate, indicating that fatty acids are exchanged more easily than entire phospholipid molecules,

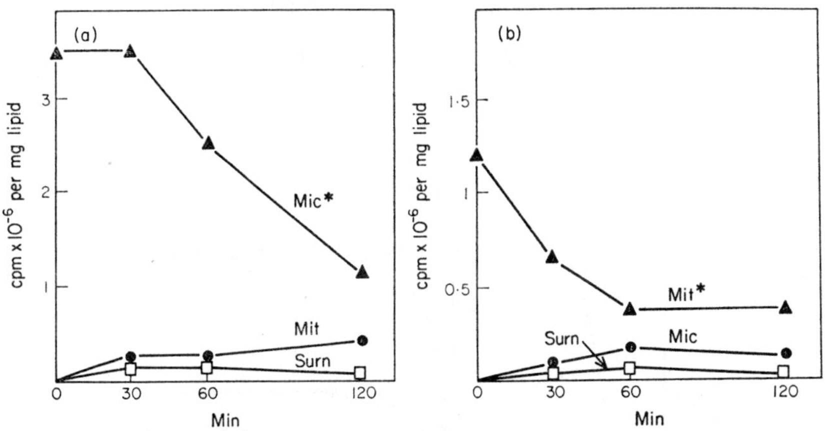

FIG. 3. Lipid exchange between microsomes and mitochondria *in vitro*. a. Incubation of radioactive lipid-labelled microsomes with unlabelled mitochondria, followed by re-isolation of each organelle and determination of the specific radioactivity of its lipid. b. Incubation of radioactive lipid-labelled mitochondria with unlabelled microsomes (then as for "a").

Mic = microsomes; Mit = mitochondria; Surn = cytoplasmic supernatant; * = initial location of radioactive lipid.

and (ii) that phosphatidylcholine and phosphatidylethanolamine are exchanged to a greater extent, regardless of the radioactive precursor used, than any of the other phospholipids, showing that the more abundant a phospholipid is in the membrane the more actively exchanged it will be. This suggests that the exchange process is a physical rather than an enzymic phenomenon. A study of the exchange of labelled fatty acids between microsomes and mitochondria (Mazliak and Ben Abdelkader, 1971) has led to the same conclusion.

VI. Lipid Exchange in a Model System

Since the exchange process appears to be physical rather than enzymic, we have studied the exchange of lipid between radioactively labelled liposomes and unlabelled cauliflower floret mitochondria suspended in an appropriate buffer. As described previously the liposomes and mitochondria were incubated together for different times and under various conditions. In this very simple system we made the following observations:

(a) The quantity of exchanged lipid (i.e. from liposomes to mitochondria) increases with time during the first 15–30 min and then reaches a plateau

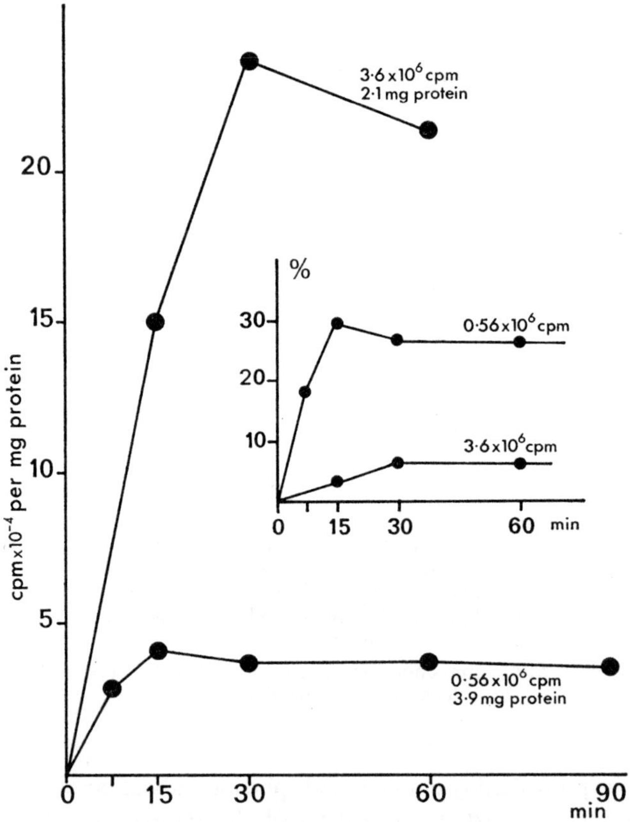

Fig. 4. Transfer of radioactive lipids from liposomes to cauliflower floret mitochondria. Radioactively labelled liposomes were incubated with cauliflower floret mitochondria and after various times the radioactivity in the mitochondria per meg of protein was determined. Two experiments were carried out; in the first (lower curve) 0·56 × 10^6 cpm in the liposomes were offered to 3·9 mg of mitochondrial protein whilst in the second (upper curve) 3·6 × 10^6 cpm in the liposomes were offered to 2·1 mg of mitochondrial protein. The inset gives the plot of the percentage of the initial radioactivity recovered in the mitochondria *vs.* time, in the same two experiments.

(Fig. 4). It is also apparent from Fig. 4 that the quantity of lipid transferred from liposomes to mitochondria per mg of protein increases with the quantity of radioactive lipid made available to the mitochondria in the initial mixture. Also the quantity of lipid transferred per mg of protein increases when the quantity of mitochondrial protein present in the initial mixture is decreased. The percentage of radioactive lipid (Fig. 4, inset) which is transferred to the mitochondria increases with an increasing quantity of mitochondrial protein present in the initial mixture. Moreover, the percentage of transferred radioactivity is increased by diminishing the initial quantity of radioactive lipid. Under optimal exchange conditions of low initial radioactivity and high protein concentration, up to 30% of the radioactivity initially in the liposomes is transferred to the mitochondria in 15 min.

(b) The quantity of exchanged lipid (or the percentage of the initial liposome radioactivity exchanged) depends upon the protein concentration of the "acceptor" (Fig. 5a). It increases up to a protein concentration of 7–8 mg/ml and then reaches a plateau.

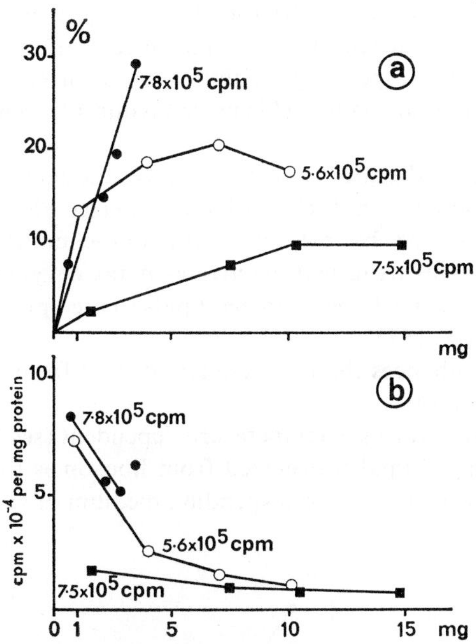

FIG. 5. Transfer of radioactive lipids from liposomes to cauliflower floret mitochondria. a. Radioactively labelled liposomes were incubated with different quantities of cauliflower floret mitochondria (measured as mg protein) for the same period of time. The percentage of the initial radioactivity recovered in the mitochondria was then determined and plotted against the quantity of mitochondrial protein present. Three experiments were carried out with different initial liposome radioactivities, $7{\cdot}8 \times 10^5$ cpm (filled circles), $5{\cdot}6 \times 10^5$ cpm (open circles) and $7{\cdot}5 \times 10^5$ cpm (filled squares). b. The quantity of exchanged lipid per mg of protein in the three experiments of "a" are plotted against the quantity of mitochondria present.

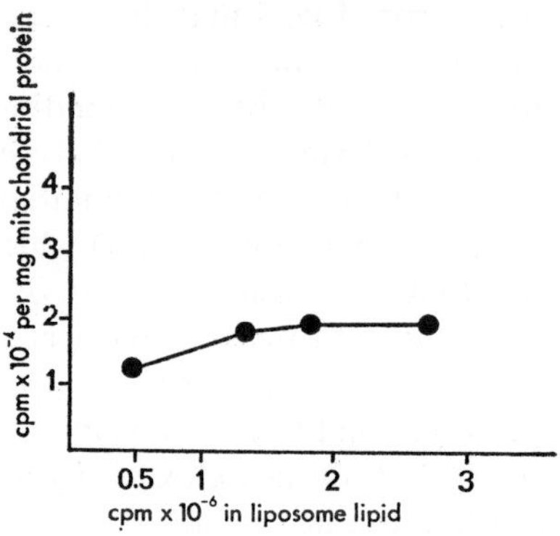

FIG. 6. Effect of liposome concentration on the transfer of radioactive lipids from liposomes to cauliflower floret mitochondria.

(c) The quantity of lipid exchanged per mg of protein is independent of the liposome concentration if the latter is in excess (Fig. 5b). In contrast, at low liposome concentrations, the quantity of lipid exchanged per mg of protein increases with increasing liposome concentration (Fig. 6).

(d) The pH of the suspending medium plays a prominent role in the exchange phenomenon: an acidic pH is more favourable to lipid transfer than a basic pH (Fig. 7).

(e) Delipidation of the mitochondria enhances the extent of the transferance of lipid from liposomes to mitochondrial protein (see Fig. 8). Partial delipidation was accomplished by extracting the mitochondria with 50% (v/v) aqueous acetone. This treatment inactivates many enzymes but they may be reactivated by adding back the extracted lipids to the lipid-depleted organelles (Jolliot and Mazliak, 1973).

(f) Sonication enhances the transferance of lipid from liposomes to mitochondria (see Table V).

(g) The exchange process is temperature dependent (see Fig. 9).

(h) The quantity of lipid transferred from liposomes to mitochondria can be altered by the presence in the suspending medium of various proteins. For

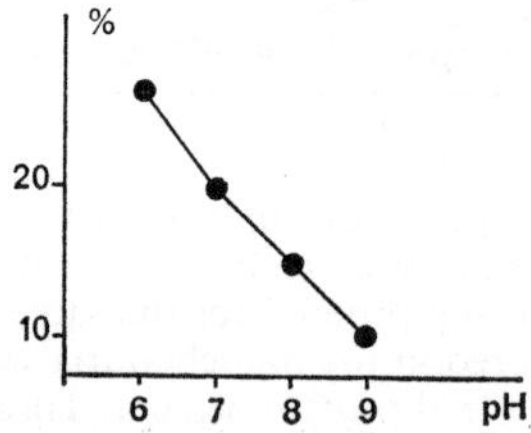

FIG. 7. Effect of pH on the transfer of radioactive lipids from liposomes to cauliflower floret mitochondria. The extent of lipid transfer is expressed as the percentage of the initial radioactivity transferred in 15 min to the cauliflower floret mitochondria.

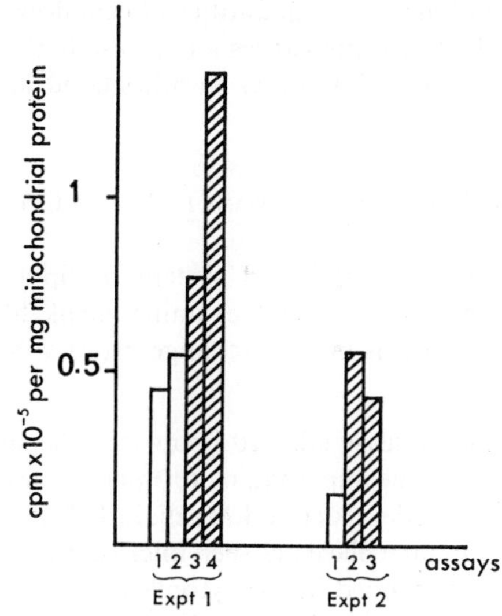

FIG. 8. Effect of delipidation of mitochondria on lipid transfer from liposomes to cauliflower floret mitochondria. The quantity of lipid transferred is expressed as cpm of radioactive lipid per mg of mitochondrial protein. The open bars represent the results with standard mitochondria and the hatched bars those with mitochondria which have been extracted with 50% (v/v) aqueous acetone.

Table V

Effect of sonication of the mitochondria on the transfer of lipid from liposomes to mitochondria

Mitochondria[a]	Radioactivity present expressed as a percentage of that initially in the liposomes
Standard	15·6
Sonicated	29·6

[a] Mitochondria, equivalent to 3 mg of protein, were incubated directly, or after 90 s of sonication under N_2, for 30 min with a 1 ml suspension of radioactive liposomes (6×10^5 cpm).

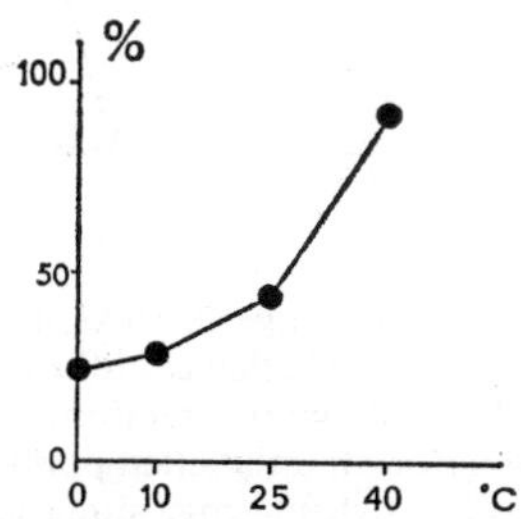

FIG. 9. Effect of temperature on the transfer of radioactive lipid from liposomes to cauliflower floret mitochondria. The extent of lipid transfer is expressed as the percentage of the initial radioactivity transferred to the cauliflower floret mitochondria in 30 min.

instance, the addition of increasing quantities of cauliflower total cytoplasmic protein to the suspending medium causes a decrease in the quantity of radioactive lipids transferred from liposomes to mitochondria.

VII. Exchange Proteins in Plant Cells

Returning to the more complicated system of lipid exchange between microsomes and mitochondria, we shall examine the likelihood of the participation of cytoplasmic proteins in the exchange process. Several facts suggest that this may be the case:

(a) Labelled organelles release labelled lipids into the supernatant (Fig. 2).

(b) Lipid transfer from the supernatant to microsomes and mitochondria has been observed (Ben Abdelkader and Mazliak, 1970).

(c) Lipid exchange can take place between microsomes and mitochondria of different plant species suspended in the cytoplasmic fraction of one or other of the species (Fig. 10). However, it is frequently found that radioactive lipid accumulates in the cytoplasmic fraction to a greater extent than in the "acceptor" organelle; this is shown particularly well with labelled cauliflower microsomes and unlabelled potato mitochondria suspended in potato cytoplasm (see Fig. 10b). Here the potato mitochondria appear to be virtually incapable of accepting the labelled lipids from the cauliflower microsomes; the labelled lipid therefore accumulates in the cytoplasmic fraction.

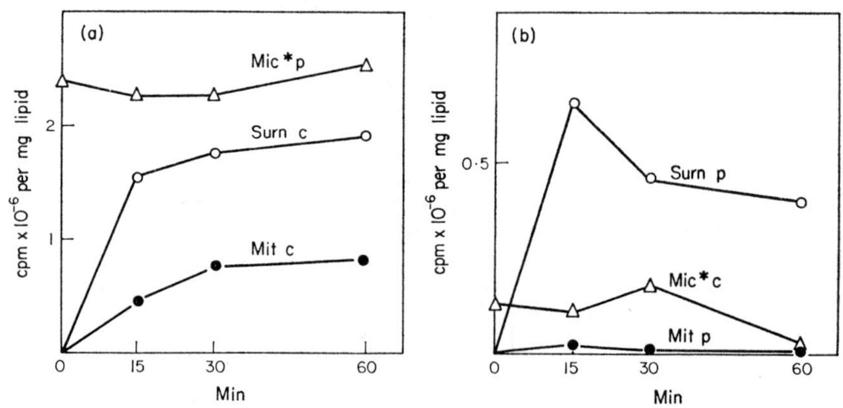

Fig. 10. Effect of cytoplasm on lipid exchange *in vitro*. a. Incubation of radioactive lipid-labelled potato tuber microsomes with unlabelled cauliflower floret mitochondria suspended in the cauliflower floret cytoplasmic supernatant fraction. b. Incubation of radioactive lipid-labelled cauliflower floret microsomes with unlabelled potato tuber mitochondria suspended in the potato tuber cytoplasmic supernatant fraction.

Mic = microsomes; Mit = mitochondria; Surn = cytoplasmic supernatant; * = initial location of radioactive lipid; p = organelle derived from potato tuber; c = organelle derived from cauliflower floret.

These results point to the participation of cytoplasmic proteins in the exchange process. That this is so has been borne out by the isolation of a protein from the cytoplasmic supernatant fraction of potato cells which is involved in this process (Ben Abdelkader, 1973). The post-mitochondrial supernatant was treated according to the technique of Wirtz and Zilversmit (1969). Its pH was lowered to 5·1 with 1 N HCl and it was centrifuged at 15 000 *g* for 15 min. The resulting clear supernatant was adjusted to pH 7·5 with Tris and ammonium sulphate added to a final concentration of 70%. The resulting protein precipitate was collected by centrifugation, freed of ammonium sulphate by dialysis and redissolved in phosphate buffer. It was then chromatographed on a column of Sephadex G75 (65 × 4·5 cm) using a solution of mercaptoethanol (10 mM) in phosphate buffer (50 mM) at pH 7·5 for development and 10 ml fractions were collected. The effect of the proteins in each fraction on the exchange process was then tested by adding them to a suspension of ^{32}P-labelled microsomes and unlabelled mitochondria, incubating them together and then reisolating the mitochondria and measuring

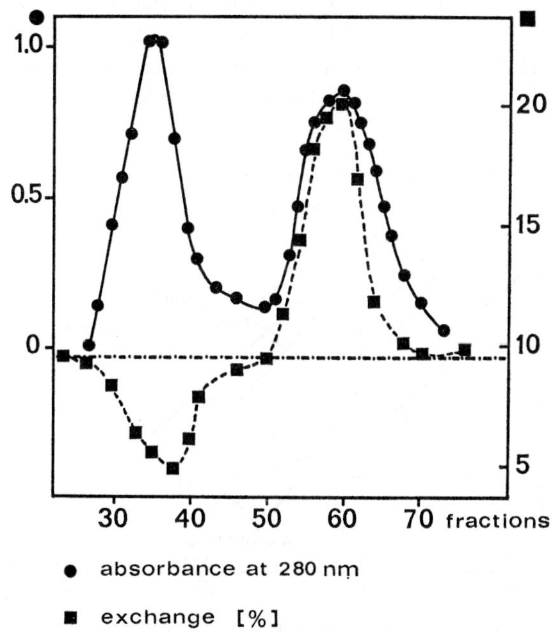

FIG. 11. Separation of a protein fraction from potato tuber cytoplasm which stimulates lipid exchange. Circles: absorbance at 280 nm giving the elution pattern of cytoplasmic proteins (1750 mg) on Sephadex G75 using mercaptoethanol (10 mM) in phosphate buffer (50 mM, pH 7·5) for development (flow rate, 48 ml/h). Squares: percentage of phospholipids exchanged, *in vitro*, between [^{32}P-lipid]microsomes and unlabelled mitochondria in the presence of that particular cytoplasmic protein fraction. The horizontal dashed line indicates the percentage of phospholipids exchanged, *in vitro*, between [^{32}P-lipid]microsomes and unlabelled mitochondria in the presence of phosphate buffer.

their ^{32}P content. In this work a correction was made foı the cross-contamination of the re-isolated mitochondria by the ^{32}P-lipid-labelled microsomes with which they had been incubated; failure to have done this would have led to an erroneously high assessment of the degree of lipid exchange. The correction was made by mixing the ^{32}P-labelled microsomes with ^{3}H-leucine-labelled microsomes before incubation with the unlabelled mitochondria and protein fractions. Making the assumption that no exchange of ^{3}H-leucine-labelled compounds (principally protein) takes place between microsomes and mitochondria, the quantity of radioactivity due to ^{3}H in the re-isolated mitochondria is a measure of their contamination by microsomes; it was found that about 6% of the added ^{3}H was recovered in the re-isolated mitochondria. Accordingly the radioactivity due to ^{32}P in the re-isolated mitochondria was reduced by the same percentage. The residual ^{32}P in the re-isolated mitochondria is that due to genuine exchange of lipid between the two organelles. Figure 11 shows the results of this experiment; some fractions markedly stimulated the exchange process whilst others inhibited it. The stimulatory activity was particularly marked in fractions 60–62 (Fig. 11); the addition of 2 ml of these fractions to the experimental exchange system described earlier resulted in a lipid exchange 80% greater than that observed in the controls. The total stimulatory activity was recovered in a single elution peak (Fig. 11) which approximately followed a peak of absorbancy at 280 nm.

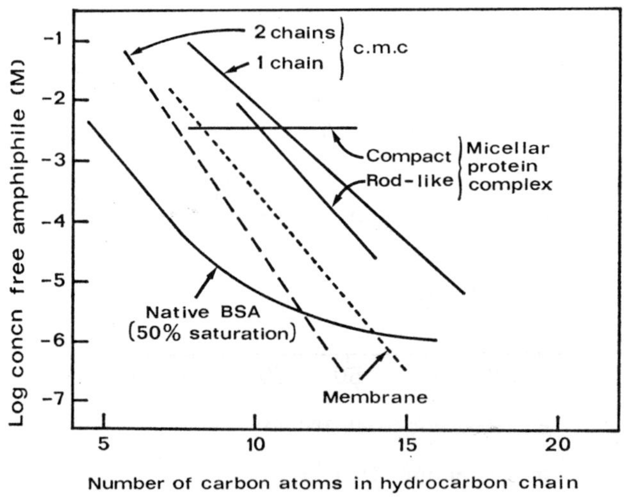

FIG. 12. Thermodynamic stabilities of various lipid–protein associations. All data refer to alkyl sulphates with a simple hydrocarbon chain, except that the dashed line is an estimate of the critical micellar concentration (c.m.c.) of alkyl sulphates with two such chains. The dotted line labelled "membrane" is the critical concentration for incorporation into a hypothetical membrane-like bilayer containing a small amount of alkyl sulphate. It is clear that 50% saturation of serum albumin (BSA) can be attained at free amphiphile concentrations *below* the critical micellar concentration. (Reproduced by permission of C. Tanford (1972).)

When increasing amounts of the active protein were introduced into the experimental exchange mixtures, increasing quantities of labelled phosphatidylcholine and phosphatidylethanolamine were transferred from microsomes to mitochondria. Assuming that the stimulatory protein is globular, its molecular weight is 22 000 ± 1000, a value similar to those obtained by Wirtz *et al.* (1972), Ehnholm and Zilversmit (1973) and Kamp *et al.* (1973) for a functionally similar protein in animal cells.

VIII. Conclusion: Physical Nature of the Exchange Process

Tanford (1972) summarized in diagrammatic form (Fig. 12) all the thermodynamic changes occurring when lipids, either dispersed or in micellar organization, are mixed with proteins in the same aqueous phase. One main conclusion drawn from this diagram is that a protein like serum albumin, for example, or a protein with a much larger hydrophobic binding site, will combine avidly with molecules having long hydrocarbon chains and that this process will *compete* with membrane formation. Such proteins have the effect of removing lipids from membranes under certain conditions. However, under other conditions the attachment of lipids to membranes is favoured. One can, therefore, imagine that the exchange processes involved in lipid distribution in the cell, from sites of biosynthesis towards all cellular membranes and organelles, is a purely physical process, depending only on the existence of exchange proteins such as those already found in liver, brain or plant cells.

REFERENCES

Bahl, J., Guillot-Salomon, T. and Douce, R. (1970). *Physiol. Veg.* **8**, 55–74.

Ben Abdelkader, A. (1969). *C.r. hebd. Séanc. Acad. Sci., Paris* **268**, 2406–2409.

Ben Abdelkader, A. (1972). Doctorate thesis, University of Paris.

Ben Abdelkader, A. (1973). *C.r. hebd. Séanc. Acad. Sci., Paris* **277**, 1455–1458.

Ben Abdelkader, A. and Mazliak, P. (1970). *Eur. J. Biochem.* **15**, 250–262.

Ben Abdelkader, A., Cherif, A., Demandre, D. and Mazliak, P. (1973). *Eur. J. Biochem.* **32**, 155–165.

Bradbeer, C. and Stumpf, P. K. (1960). *J. Lipid Res.* **1**, 214–220.

Cheniae, G. M. (1965). *Pl. Physiol., Lancaster* **40**, 236–243.

Cooper, T. G. and Beevers, H. (1969). *J. biol. Chem.* **244**, 3514–3520.

Devor, K. A. and Mudd, J. B. (1971). *J. Lipid Res.* **12**, 390–419.

Douce, R. (1968). *C.r. hebd. Séanc. Acad. Sci., Paris* **267**, 534–537.

Ehnhohm, C. and Zilversmit, D. B. (1973). *J. biol. Chem.* **248**, 1719–1724.

Jolliot, A. and Mazliak, P. (1973). *Pl. Sci. Lett.* **1**, 21–29.

Kamp, H. H., Wirtz, K. W. A. and van Deenen, L. L. M. (1973). *Biochim. biophys. Acta* **318**, 313–325.

Marshall, M. O. and Kates, M. (1972). *Biochim. biophys. Acta* **260**, 558–570.

McMahon, V. A. and Stumpf, P. K. (1966). *Pl. Physiol., Lancaster* **41**, 148–156.

Mazliak, P. and Ben Abdelkader, A. (1971). *Phytochemistry* **10**, 2879–2890.

Mazliak, P., Stoll, U. and Ben Abdelkader, A. (1968). *Biochim. biophys. Acta* **152**, 414–417.

Mazliak, P., Oursel, A., Ben Abdelkader, A. and Grosbois, M. (1972). *Eur. J. Biochem.* **28**, 399–411.

Meunier, D. and Mazliak, P. (1972). *C.r. hebd. Séanc. Acad. Sci., Paris* **275**, 213–216.

Moore, T. S., Lord, J. M., Kagawa, T. and Beevers, H. (1973). *Pl. Physiol., Lancaster* **52**, 50–53.

Morre, D. J., Nyquist, S. and Rivera, E. (1970). *Pl. Physiol., Lancaster* **45**, 800–804.

Sastry, P. S. and Kates, M. (1966). *Can. J. Biochem.* **44**, 459–467.

Singer, S. J. and Nicholson, G. L. (1972). *Science* **175**, 720–730.

Sumida, A. and Mudd, J. B. (1970). *Pl. Physiol., Lancaster* **45**, 712–718.

Tanford, C. (1972). *J. molec. Biol.* **67**, 59–74.

Vandor, S. L. and Richardson, K. E. (1968). *Can. J. Biochem.* **46**, 1309–1316.

Vijay, I. K. and Stumpf, P. K. (1971). *J. biol. Chem.* **246**, 2910–2917.

Wirtz, K. W. A. and Zilversmit, D. B. (1969). *Biochim, biophys. Acta* **193**, 105–116.

Wirtz, K. W. A., Kamp, H. H. and van Deenen, L. L. M. (1972). *Biochim. biophys. Acta* **274**, 606–617.

CHAPTER 11

Degradation of Plant Lipids by Hydrolytic and Oxidative Enzymes

T. GALLIARD

Agricultural Research Council Food Research Institute, Norwich, England

I. Introduction

The enzymatic breakdown of lipids in plants can be conveniently divided into hydrolytic and oxidative processes and these are manifested under physiological conditions and in disrupted tissues where a net loss of lipid occurs.

Under certain physiological conditions the lipid content of plant tissues is decreased. In germinating seeds the reserve triglycerides are hydrolysed by lipase action to release fatty acids which are subsequently used in gluconeogenesis (see Chapter 9). Also in germinating seeds, membrane lipids are lost from the tissue, for example the phospholipid content of pea cotyledons was shown to fall during germination (Quarles and Dawson, 1969). In senescing leaf tissue loss of chloroplast lipids is accompanied by an increase in free fatty acid levels (Draper, 1969; Ferguson and Simon, 1973). During the ripening of fruits losses of polar lipids have been reported (Galliard, 1968; Kalra and Brooks, 1973).

Disruption by damage or infection of plant tissue may lead to major losses of endogenous lipids resulting in breakdown of lipoprotein membrane structures. Fatty acids released by hydrolytic enzymes may be attacked by oxidative enzymes and, in the case of the lipoxygenases, volatile products with characteristic flavour properties may result. Although desirable flavours in edible plants may be produced by such processes (e.g. the flavour of cucumber), frequently the result of these degradative processes is unsound material and undesirable off-flavours. Hence recent interest in lipid degradation in plants has tended to be somewhat concentrated in laboratories involved with research on edible plant tissues.

This chapter does not claim to be a comprehensive review. Rather, the intention has been to cover recent developments in the subject area. The uneven treatment given to the different aspects is partly a reflection of the volume of research in these areas and is partly due to the author's own research interests. The last five years have seen a large amount of research effort in attempting to understand the biochemistry of the peroxidative breakdown of unsaturated fatty acids catalysed by lipoxygenases and related enzymes and no apology is made for the emphasis on these processes in the latter part of this chapter.

II. Lipolytic Enzymes

A. Lipase

1. *Definition of a lipase*

The early literature on plant lipases is confused by the wide range of substrates which have been used in assays for these enzymes. Desnuelle and Savary (1963) listed many substrates used for plant "lipases" amongst which were various methyl, ethyl, *p*-nitrophenyl and naphthyl esters of fatty acids and the water-soluble triacetyl glycerol as well as short- and long-chain triglycerides and natural oils. The term "lipase" has also been used loosely for any enzyme that liberates free fatty acids from acyl lipids. A true lipase (triacylglycerol acyl hydrolase, E.C. 3.1.1.3) may be defined as an enzyme that hydrolyses emulsified triacylglycerols at an oil–water interface (Desnuelle and Savary, 1963; Jenson, 1971).

2. *Occurrence of lipases in plants*

Practically all the information on true lipases in plants comes from studies on various seeds. Early studies were reviewed by Barron (1964) and more recent studies have been made on seeds of castor bean, peanut and cotton (St. Angelo and Altschul, 1964), conifer (Nyman, 1965; Ching, 1968), wheat

(Stauffer and Glass, 1966; Drapon *et al*, 1969; Tavener and Laidman, 1972), oats, corn, soyabean, safflower and pinion nuts (Berner and Hammond, 1970), rice (Funatsu *et al.*, 1971; Aizono *et al.*, 1973). The most detailed studies have been made on the acid lipase from castor bean and this work has been reviewed by Ory (1969).

Seed lipases catalyse the initial reaction in the utilization of seed oils in germinating seedlings (p. 288). Although present in the dormant seed, lipase activity increases markedly during an early stage of germination (Black and Altschul, 1965; Tavener and Laidman, 1972). Studies with wheat have demonstrated a complex hormonal control of lipase activity (Tavener and Laidman, 1972) and phytochrome control of lipase in *Sinapsis alba* seeds was shown by Hacker and Stöhr (1966).

Lipase activity in seeds is associated with the oil storage bodies (Ory *et al.*, 1960, 1968; Ching, 1968) and is present in the membranes of these bodies (spherosomes) in the castor bean (p. 293).

3. *Properties of lipases*

(*a*) *Castor bean lipase.* Ory and co-workers (Ory, 1969) have established that the acid lipase of castor beans has a low pH optimum (pH 4·1), is moderately heat-stable and requires a novel co-factor together with a low molecular weight, heat-stable glycoprotein for full activity. The co-factor was identified as a cyclic tetramer of ricinoleic acid; it is interesting to note that ricinoleic acid is the major fatty acid component of the castor bean seed oil, representing about 90% of the total fatty acids.

The mode of action and specificity of the enzyme on triglyceride substrates has been a subject of recent interest. The castor bean enzyme catalyses the complete hydrolysis of triglycerides to glycerol and fatty acids without the accumulation of 2-monoglycerides and is thus different from pancreatic lipase. However, Ory *et al.* (1969) showed that the lipase did not attack secondary ester bonds of triglycerides and it was proposed that the first intermediate, 1,2-diglyceride, was isomerized to a 1,3-diglyceride which was then further hydrolysed. Although a 1,3-diglyceride does appear as an intermediate product, Noma and Borgström (1971) have provided evidence that this is not formed by isomerization but is produced by an acyl transfer reaction (4) in the following sequence of reactions:

(1) Triglyceride → 1,2-diglyceride + fatty acid
(2) 1,2-diglyceride → 1- and 2-monoglyceride + fatty acid
(3) 1- and 2-monoglycerides → glycerol + fatty acid
(4) 1-monoglyceride + fatty acid ↔ 1,3-diglyceride

Thus, although the primary ester group of the triglycerides is attacked in preference to the secondary group, the enzyme hydrolyses 1,2- and 1,3-diglycerides at similar rates and the overall reaction is not specific for the position of the fatty acid (Noma and Borgström, 1971).

(*b*) *Other seed lipases.* Information on the properties of true lipases from other plants is limited. Specificity studies on lipases from a range of seeds (Berner and Hammond, 1970; Williams and Bowden, 1973) support a conclusion that seed lipases in general are non-specific for position of acyl ester bonds in triglyceride hydrolysis although oat lipase showed some specificity for individual fatty acids by discriminating against stearic acid in favour of linoleic acid regardless of the position occupied by these acids in triglyceride substrates (Berner and Hammond, 1970). The lipase of *Vernonia anthelmintica* seeds appears to resemble the castor bean enzyme in its mode of action (Olney *et al.*, 1968), contrary to a previous suggestion (Krewson *et al.*, 1962) of a positionally specific lipase in these seeds.

In addition to an acid lipase, douglas fir seeds also contain a neutral lipase activity (Ching, 1968) and wheat germ lipase has a neutral pH optimum (Stauffer and Glass, 1966). Ory and St. Angelo (1971) were unable to confirm a previous report of a neutral lipase in castor bean (Yamada, 1957) and pointed out the possible danger of confusing activities due to contaminating fungi in studies on seeds. Wheat germ (Stauffer and Glass, 1966) and castor bean (p. 292) also contain lipolytic enzymes which attack monoglycerides rather than triglycerides and should be classed as monoglyceride acyl hydrolases rather than "lipases".

B. ENZYMES ATTACKING ACYL GROUPS OF POLAR LIPIDS

1. *Nomenclature of lipolytic acyl hydrolases*

The problem of nomenclature and definition of true lipases (triacylglycerol acyl hydrolases) have already been stated (p. 320). Unfortunately, there is even greater confusion in the area of acyl hydrolases acting on polar lipids; i.e. enzymes deacylating glycerolipids with the following general formula:

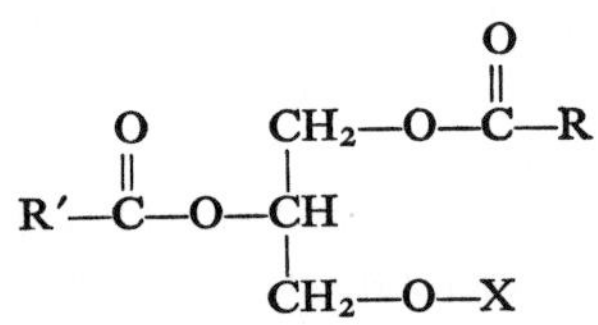

where R and R′ represent fatty acyl residues and X represents a phosphate ester in phospholipids or a glycosidic moiety in glycolipids. Depending on the substrates chosen for assays, various acyl hydrolase activities in plant tissues have been described as phospholipases, galactolipases, sulpholipases, monoglyceride lipases, lipases and esterases. Rarely have the substrate specificities of these activities been fully established with the result that the same enzyme has possibly been given different names by different authors. Furthermore, later sections in this chapter will demonstrate that a given lipolytic enzyme will show marked differences in properties and specificity under different assay conditions (e.g. pH, presence of detergents or metal ions) and with different physiological conditions of the plant. For these reasons, this discussion will avoid specific classification of the acyl hydrolases.

2. *Lipolytic acyl hydrolase activities in plants*

A range of acyl hydrolases with various polar lipid substrates is listed in Table I. In no case has an absolute specificity for a given class of lipid been established. For example, the enzyme termed sulpholipase (Yagi and Benson, 1962) also hydrolysed phospholipids, at least to the *lyso*-phospholipid level (i.e. phospholipase A-type activity). The galactolipid hydrolysing enzyme from *Phaseolus* leaves (Sastry and Kates, 1964; Helmsing, 1967, 1969) and known as galactolipase (E.C. 3.1.1.26) shows a higher specific activity with the *lyso*-phospholipids and monoglycerides as substrates (Galliard and Dennis, 1974a) and is probably identical with the non-specific lipolytic acyl hydrolase described below (P. J. Helmsing, personal communication).

Although specific phospholipid acyl hydrolases are known to occur in animals and micro-organisms (Dawson, 1973), no such specificity has yet been conclusively demonstrated with plant enzymes. An enzyme found in barley seeds and named phospholipase B (Acker and Bücking, 1956) removes both fatty acids from phosphatidylcholine although the monoacyl (*lyso*)-derivative was formed as a transient intermediate (Acker and Geyer, 1969). Confusion in terminology again arises here because, although the term "phospholipase B" was originally applied to enzyme activity releasing both fatty acids from diacyl phospholipids, available evidence (Ansell and Hawthorne, 1964) indicated that this activity could be ascribed to a mixture of two enzymes, phospholipases A_1 [E.C. 3.1.1.32] and A_2[E.C. 3.1.1.4], each specific for one of the two acyl ester bonds. A further complication arises since the term "phospholipase B" is currently quoted as an alternative name for *lyso*-phospholipase [E.C. 3.1.1.5] (Dawson, 1973; Florkin and Stotz, 1973). Complete deacylation of both phospholipids and *lyso*-phospholipids is catalysed by the non-specific lipolytic acyl hydrolase from plants as described below.

Monoglycerides may be grouped with the polar lipids in relation to hydrolysis with acyl hydrolases. The deacylation of monoglycerides by enzymes distinct from true lipases has been described in seeds of wheat (Stauffer and Glass, 1966) and castor bean (Muto and Beevers, 1974), in potato tuber (Galliard, 1971a) and in leaves of potato and *Phaseolus vulgaris* (T. Galliard, unpublished observations). The full substrate specificities of these enzyme activities have not been established, except for the potato enzyme which is a nonspecific lipolytic acyl hydrolase but which shows its highest specific activity with monoglyceride as substrate (see Table II).

3. *A non-specific acyl hydrolase*

Studies in the author's laboratory (Galliard, 1970; 1971a; Galliard and

TABLE I

Polar lipid acyl hydrolase activities in plants

Substrate	Source	Description of enzyme	Products formed in addition to fatty acids	Other lipids hydrolyzed	Reference
mono- and digalactosyl-diglycerides	*Phaseolus* spp. leaves	galactolipase	mono- and digalactosyl-glycerol	phosphatidyl-choline diglyceride	Sastry and Kates (1964)
mono- and digalactosyl-diglycerides	potato tubers and leaves *Phaseolus vulgaris* leaves	lipolytic acyl hydrolase	mono- and digalactosyl-glycerol (*lyso*-derivative as intermediate)	see Table II	Galliard (1971a) Galliard and Dennis (1974a)
sulphoquinovosyldiglyceride (sulpholipid)	*Scenedesmus obliquus* *Chlorella* spp. alfalfa and corn roots alfalfa leaves	sulpholipase	sulphoquinovosylglycerol (*lyso*-derivative as intermediate)	diacyl phospho-lipids	Yagi and Benson (1962)
phosphatidylcholine	germinating barley	phospho-lipase B	glycerylphosphorylcholine (*lyso*-derivative as intermediate)	—	Acker and Bücking (1956) Acker and Geyer (1969) Rebman and Acker (1973)
monoacyl (i.e. *lyso*) and diacyl phospholipids	potato tuber leaves *Phaseolus vulgaris* leaves	lipolytic acyl hydrolase	glycerylphosphoryl esters	see Table II	Galliard (1971a) Galliard and Dennis (1974a)
monoglycerides	wheat germ	lipase	glycerol	(NOT tri-glycerides	Stauffer and Glass (1966)
monoglycerides	castor bean glyoxy-somes	lipase	glycerol	—	Muto and Beevers (1974)
monoglycerides	potato tuber and leaves *Phaseolus vulgaris* leaves	lipolytic acyl hydrolase		see Table II	Galliard (1971a) Galliard and Dennis (1974a)

Dennis, 1974a) have led to the isolation and characterization of a lipolytic acyl hydrolase from plants. Potato tubers are the richest source so far found and the enzyme occurs in leaves and sprouts of potato and leaves of *Phaseolus vulgaris*; similar activity, although not yet fully characterized, is found in a range of plant tissues. The phospholipid acyl hydrolase activity of this enzyme has been reviewed recently (Galliard, 1973).

Table II lists the substrates attacked by the enzyme from potato tubers. Monoglycerides and *lyso*-phospholipids are most readily hydrolysed but diacyl phospholipids and galactolipids, diglycerides and methyl-, ethyl-, *p*-nitrophenyl- and naphthyl-esters of fatty acids are also hydrolysed. Little or no activity is observed with triglycerides (cf. true lipases), wax esters or sterol esters. Surface active agents have a marked effect on the activity and pH characteristics with some substrates. For instance, the purified enzyme is inactive on aqueous dispersions of phosphatidylcholine (liposomes) and on membrane-bound polar lipids; addition of a surface active agent to these systems produces hydrolytic activity. A natural surface active agent present in crude extracts of potato tuber was identified as free fatty acid and its effect was reversed by bovine serum albumin (Galliard, 1971b). Fatty acids (particularly unsaturated long-chain fatty acids) were more effective at low concentrations than other ionic or non-ionic surfactants. The fatty acid was shown to disrupt the bilayer structure of liposomes and presumably exposed the acyl ester bonds to enzyme attack. A similar effect may explain the activation by free fatty acids in the hydrolysis of membrane-bound lipids of mitochondria (Galliard, 1971b) and disruption of chloroplast electron transport

TABLE II

Substrate specificity of purified lipolytic acyl hydrolase from potato tubers (Galliard, 1971a)

Lipid class	Substrate	Enzyme activity (μequiv. fatty acid released/ min/mg of protein)
monoglyceride	monolein	39·0
lyso-phospholipid	*lyso*-phosphatidylcholine	28·0
monogalactolipid	monogalactosyldiglyceride	12·0
methyl ester	methyl oleate	10·8
diglyceride	diolein	8·2
digalactolipid	digalactosyldiglyceride	6·8
phospholipid	phosphatidylcholine	5·1
triglyceride	triolein	<0·1
wax ester	octadecyl palmitate	<0·1
sterol ester	cholesteryl oleate	<0·1
("esterase" substrates)	*p*-nitrophenyl palmitate	15·5
	p-nitrophenyl acetate	0·14

cytochrome system (Davenport, 1975). This type of enzyme action is presumably involved in the production of free fatty acids in homogenates of tissues from many plants and the requirements of high pH and added bovine serum albumin (Galliard, 1974a) protect the activities of subcellular organelles from plant tissues, e.g. mitochondria (Dalgarno and Birt, 1963; Baddeley and Simon, 1969), chloroplasts (McCarty and Jagendorf, 1965; Friedlander and Neumann, 1968; Constantopoulos and Kenyon, 1968) and microsomal fractions (Ben Abdelkader *et al.*, 1973).

In the particular case of the potato tubers, the enzyme is so active (e.g. 40 μmol of monogalactosyldiglyceride hydrolysed per min/g fresh weight) that most of the endogenous membrane-bound polar lipids are destroyed immediately the tissue is disrupted, even at 0°C (Galliard, 1970).

Unlike the barley phospholipid acyl hydrolase activity, which increases to a maximum during germination and then falls to a low level (Rebmann and Acker, 1973), the lipolytic acyl hydrolase activity of potato tubers remains constant throughout the life-cycle (Berkeley and Galliard, 1974). Isoenzymic forms of the potato enzyme have been separated (Galliard and Dennis, 1974b) and these show electrophoretic profiles characteristic for given varieties. Although the activity was very high in over twenty different commercial varieties of potato examined, in one common European variety (Desirée) the enzyme level was 100–500-fold lower than in the other varieties (Galliard and Matthew, 1973) and the electrophoretic profile was markedly different. This low activity variety has been useful in this laboratory for the preparation of subcellular organelles from potato tuber. It has also illustrated the danger of generalizations in defining enzyme contents of cultivated varieties of plants.

4. *Acyl transferase activities of lipolytic enzymes*

The acyl transferase activity of esterases is well established (Hofstee, 1960). Plant acyl hydrolases acting on lipid substrates also show acyl transferase activity. Noma and Bergström (1971) have shown that the castor bean lipase also functions as a hydrophobic acyl transferase.

When methanol was used to solubilize galactolipid substrates (Sastry and Kates, 1964) in the study of the polar lipid acyl hydrolase from potato tubers (Galliard, 1970) and leaves of potato and *Phaseolus vulgaris* (Galliard, unpublished observations), fatty acid methyl esters were obtained in the products. Subsequent studies (Galliard, 1971a; Galliard and Dennis, 1974a) demonstrated that the enzyme will transfer fatty acids from acyl lipid donors to short-chain alcohol acceptors (e.g. methanol and ethanol). Figure 1 shows the effect of methanol concentration in the formation of methyl esters from *lyso*-phosphatidylcholine as the acyl donor; the results indicate that the affinity of the enzyme for methanol as acyl acceptor is a factor of 10 higher than for water as an acceptor.

Enzymic transfer of fatty acids from endogenous donors (phospholipids

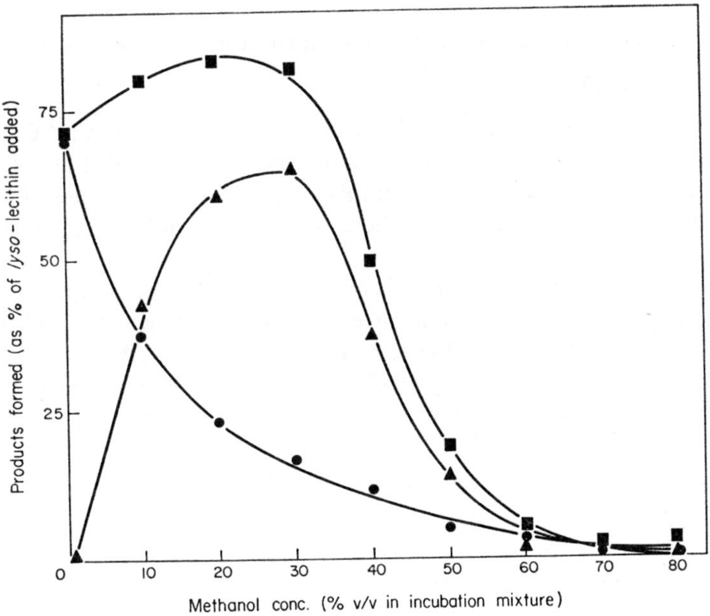

FIG. 1. Effect of methanol concentration on acyl transferase activity of the lipolytic acyl hydrolase enzyme from potato tuber. Amounts of free fatty acid (circles) and fatty acid methyl ester formed (triangles) or *lyso*-phosphatidylcholine deacylated (squares) in 10 min incubations at 25°C are given as percentage of substrate (*lyso*-phosphatidylcholine) added. [From Galliard and Dennis (1974a).]

and galactolipids) to galactolipid acceptors has been demonstrated in aqueous extracts of plant tissues (Heinz, 1967). The 6-position of the terminal galactose residue of mono- and di-galactosyldiglycerides is acylated, resulting in the formation of acylated monogalactosyldiglyceride (Heinz, 1973) and acylated digalactosyldiglyceride (E. Heinz, personal communictaion). Partial deacylation of the galactolipids by acyl hydrolase activity in the extracts, followed by acyl transferase activity, produces acylated mono- and digalactosylmonoglycerides (Critchley and Heinz, 1973). It is not yet clear whether the acyl transferase activity is due to an enzyme different from the acyl hydrolases described above.

5. *Fatty acid thiolester hydrolase*

All the acyl hydrolases described above used acyl glycerolipid (i.e. carboxylic ester substrates). Recently Huang and Stumpf (1971) found a thiolesterase in potato tubers; the partially purified enzyme was specific for coenzyme A thiolesters of C_{16} and C_{18} chain length fatty acids. This activity was responsible for the rapid hydrolysis of fatty acid CoA derivatives in studies on the fatty acid synthetase of this tissue.

C. ENZYMES ATTACKING THE POLAR GROUPS OF POLAR LIPIDS

1. *Phospholipase D*

Phospholipase D (phosphatidylcholine phosphatidohydrolase; E.C. 3.1.4.4) is the only fully characterized enzyme from plant tissues which attacks polar groups of acyl lipids. Information on this enzyme has been reviewed recently (Galliard, 1973) and the discussion here will attempt only to cover the most recent developments in the area.

Although phospholipase D has been considered to be unique to higher plants, amongst which its occurrence is widespread, recent studies have demonstrated phospholipase D-type activity in a red alga (Antia *et al.*, 1970), slime mould (Comes and Kleinig, 1973), yeast (Dharmalingam and Jayaraman 1971), *Euglena gracilis* (Calvayrac and Douce, 1970) and bacteria (Soucek *et al.*, 1971; Astrachan, 1973; Cole *et al.*, 1974).

The reaction catalysed by the enzyme may be generalized thus:

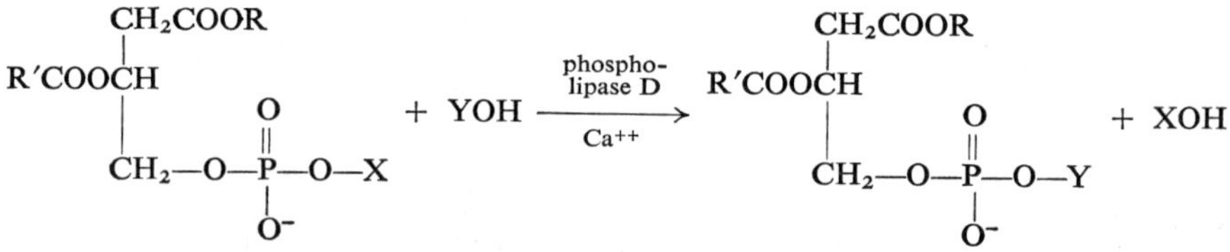

where R and R′ represent fatty acid residues and YOH may be water or an alcohol acceptor. Thus the enzyme catalyses both hydrolysis and transphosphatidylation reactions.

The transphosphatidylation reaction is responsible for the formation of artifacts during the extraction of lipids from incompletely "killed" plant tissues, e.g. phosphatidyl methanol (Douce *et al.*, 1966). It has also been shown that phosphatidyl transfer reactions produce phosphatidylglycerol and diphosphatidylglycerol (cardiolipin) by the following reactions:

$$\text{phosphatidylcholine} + \text{glycerol} \xrightarrow{\text{phospholipase D}} \text{phosphatidylglycerol} + \text{choline}$$

(Yang *et al.*, 1967; Dawson, 1967).

$$2\ \text{phosphatidylglycerol} \xrightarrow{\text{phospholipase D}} \text{diphosphatidylglycerol} + \text{glycerol}$$

(Stanacev *et al.*, 1973).

However, the physiological significance of these transphosphatidylation reactions in phospholipid biosynthesis is in question because, for example, the phosphatidylglycerol formed by the above reaction is phosphatidyl-*rac*-1′-glycerol (Yang *et al.*, 1967), whereas the naturally occurring lipid is phosphatidyl-1′*sn*-glycerol (Haverkate and van Deenen, 1965). Studies on phosphatidylglycerol biosynthesis in *Euglena gracilis* have shown that transphosphatidylation is not involved (Schantz *et al.*, 1972).

Phospholipase D has been shown to catalyse the hydrolysis of most of the common phospholipids with the exception of the phosphoinositides but the reaction conditions for hydrolysis *in vitro* vary markedly for different substrates (see reviews by Dawson, 1973; Galliard, 1973). The physical state of the substrate dispersion is critical for enzyme activity and this is dependent on the nature of the polar group, chain length and unsaturation of the fatty acyl moieties, concentration of substrate, pH and the presence of counter ions or surface agents (Chen and Barton, 1971; Dawson, 1973).

The physiological role of phospholipase D is not known. Although the activity is highest in storage tissues, e.g. the stalk of cabbage and celery, the root of carrot, seeds of pea and marrow and the tuber of Jerusalem artichokes, this generalization is not rigid because the enzyme is absent or low in potato tuber and many fruits (Quarles and Dawson, 1969). The enzyme has recently been demonstrated in the seeds of peanut (Tzur and Shapiro, 1972) and the roots of lupin and *Vicia faba* (Oursel *et al.*, 1973).

The subcellular localization of the enzyme is not unequivocally established. Activity has been found in various particulate fractions and in particle-free supernatant fractions. It is unlikely that the enzyme would be a truly cytoplasmic enzyme and the recent observation of high specific activity of phospholipase D in a lysosome-like particulate fraction from cauliflower (Clermont, 1972) is interesting. The subcellular location of lipid-degrading enzymes in general will be discussed later (p. 348 *et seq*).

2. *Other hydrolytic activities with polar lipids*

Evidence for plant enzymes other than phospholipase D attacking polar groups of phospholipids and other polar lipids is very scarce. Phospholipase C (E.C. 3.1.4.3) which splits phospholipids to diglycerides and phosphate esters has not been characterized in plants although Kates (1955) obtained some evidence for its presence in carrot root and spinach leaves.

Complete degradation of galactosyldiglyceride to galactose, glycerol and fatty acids has been demonstrated in *Phaseolus* leaves but the evidence showed that the lipid was first deacylated by an acyl hydrolase and the resulting mono- or digalactosylglycerol was subsequently attacked by galactosidase (Sastry and Kates, 1964).

III. Oxidative Enzymes in Lipid Metabolism

This section will discuss the enzymic reactions involved in the oxidative and peroxidative breakdown of the fatty acyl moieties of plant lipids. Enzymic oxidation processes, in contrast to autoxidation, generally require the prior release of fatty acids from lipids and certainly most biochemical studies have involved the use of either the free acids or coenzyme A thioesters. The different types of reactions involved in oxidation or oxygenation of fatty acids by plant systems are summarized in Table III. The following discussion will review recent studies in this area but does not claim to be a comprehensive coverage, particularly of those aspects which have been fully reviewed very recently as indicated.

A. ALPHA-OXIDATION

According to the literature to date (for reviews see Stumpf, 1970; Hitchcock and Nichols, 1971; Mazliak, 1973), two apparently different processes of α-oxidation in plants have been described. In both systems a free fatty acid is oxidized at the carboxyl terminal end, with the loss of one carbon unit, to the aldehyde of the corresponding C_{n-1} acid; further oxidation of the aldehyde produces the C_{n-1} fatty acid.

One system, first described in peanut cotyledons by Stumpf and co-workers (see Markovetz *et al.*, 1972), appeared to be a peroxidase process in which a hydrogen peroxide-generating system was required. A second system, studied in leaf tissue by Hitchcock and colleagues (see Hitchcock and Nichols, 1971), did not appear to require hydrogen peroxide but used molecular oxygen. Both peanut and leaf systems were inhibited by imidazole and formed C_n-2-hydroxy acids in addition to C_{n-1}-aldehydes and in both systems NAD was required for oxidation of the C_{n-1}-aldehyde to C_{n-1} fatty acid. The 2-hydroxy acids isolated from these systems were shown to be optically active 2-D-hydroxy acid enantiomers; the 2-L-hydroxy acid was thought to be a transient intermediate in the formation of the C_{n-1}-aldehyde.

However, recent work by Shine and Stumpf (1974) has provided evidence that a common system exists in both seeds and leaves. These authors propose that a flavoprotein-catalysed oxidation produces a 2-D-hydroperoxy intermediate which may be reduced to form the 2-D-hydroxy acid or may decarboxylate to the C_{n-1}-aldehyde (see Fig. 2); in the proposed system the 2-L-hydroxy acids are not implicated in the α-oxidation process.

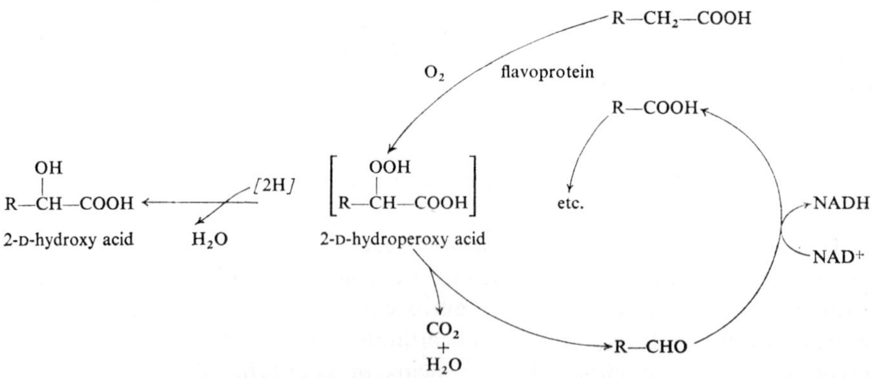

FIG. 2. Proposed pathway for α-oxidation in plants (adapted from Shine and Stumpf, 1974).

TABLE III

Summary of fatty acid oxidation and oxygenation reactions in plants

Reaction system	Substrate	Co-factors	Products
α-oxidation	free fatty acid	(a) O_2, flavoprotein (b) NAD	(a) 2-D-hydroxy-fatty acid (b) C_{n-1}-fatty acid + CO_2
β-oxidation	fatty acyl CoA	NAD^+, flavoproteins	acetyl CoA ↗ gluconeogenesis in glyoxysomes acetyl CoA ↘ CO_2 in mitochondria
ω-oxidation	(a) fatty acyl CoA	O_2, mixed function oxygenase	ω-hydroxy fatty acid
	(b) free ω-hydroxyacid	NADP	α,ω-dicarboxylic acid
β-hydroxylation	fatty acyl CoA (ACP)[a]	O_2, ? mixed function oxygenase	3-D-hydroxy-fatty acid[a]
in-chain hydroxylation	fatty acyl CoA	O_2, mixed function oxygenase	fatty acids with in-chain hydroxyl groups
peroxidation	free fatty acid with ω-*cis*-6, *cis*-9-diene system	O_2, lipoxygenase	ω-6- or ω-10-hydroperoxy derivatives

[a] The β-hydroxylation reaction has only been demonstrated with medium chain-length fatty acids (8–12 carbons). For details see text.

The universal nature of β-oxidation (see below) and its established role in certain plant tissues has possibly overshadowed the importance of α-oxidation in plants. It is likely, but not yet established, that the naturally occurring 2-D-hydroxy acids in plants (the major fatty acid components of cerebrosides) are formed by α-oxidation. Also, the odd chain-length fatty acids can be formed from the more common even chain acids by α-oxidation. However, it is possible that α-oxidation may play a further major role in metabolism of plants. For example, added fatty acids are more rapidly oxidized by α- than β-oxidation in young leaves (Hitchcock and Nichols, 1971) and Laties *et al.* (1972) have provided evidence that the CO_2 formed by respiration induced by cutting potato tuber tissue is formed by α-oxidation of endogenous lipids.

B. BETA-OXIDATION AND BETA-HYDROXYLATION

1. *Beta-oxidation*

The detailed mechanisms of conventional β-oxidation are well established and will not be discussed here. Mazliak (1973), Hitchcock and Nichols (1971) and Stumpf (1970) have reviewed this subject recently.

In plants, two functionally and physically separate roles for β-oxidation of fatty acids exist. In seeds containing storage lipid, β-oxidation is involved in the conversion of seed oil via acetyl CoA to products of the glyoxylate cycle and is associated with the glyoxysomes of these tissues (see Chapter 9). In this role β-oxidation is involved in the net conversion of lipid to carbohydrate. Most studies on β-oxidation in plants have concentrated on seed tissues. Less attention has been paid to the conventional β-oxidation system in non-seed tissue, although available evidence indicates that the energy-producing role of β-oxidation is similar to the mammalian systems and is associated with mitochondria. In animal cells, carnitine is involved in transporting fatty acids across the mitochondrial membrane. Although carnitine is probably not involved in the glyoxysomal β-oxidation process, the occurrence of relatively large amounts of carnitine in mitochondria-rich tissues suggests that the mitochondrial β-oxidation system of plants also uses carnitine as acyl carrier (Panter and Mudd, 1969).

2. *Oxidation of unsaturated and hydroxylated acids*

Unsaturated fatty acids predominate in plant tissues, both in storage and somatic lipids and oxygenated fatty acids occur as major components of some seed oils. The classical β-oxidation enzymes cannot directly metabolize some of the intermediates in the β-oxidation sequence from these fatty acids. For example, oleic acid, 18:1(9*c*), would produce as intermediate the 12:1(3*c*) derivative, whereas the normal β-oxidation sequence involves a *trans*-2-enoyl CoA intermediate; similarly the *cis*-unsaturation at the C-12 position of linoleic and linolenic acids would lead to a 8:1(2*c*) intermediate. Stoffel's group and others have demonstrated, in animal mitochondria, isomerase and epimerase enzymes which convert the above anomalous intermediates into normal β-oxidation intermediates (Wakil, 1970).

These enzymes have not been directly demonstrated in mitochondria from higher plants although they have been shown to occur in a slime mould (Davidoff and Korn, 1964) and in yeast (Okui *et al.*, 1963). Hutton and Stumpf (1971) have recently studied the breakdown of ricinoleic acid, 12-D-hydroxy-18:1(9*c*), in germinating castor bean. Normal β-oxidation would produce a *cis*-3-intermediate at the C_{12} level (due to the *cis*-9 bond in the original acid) and a 2-D-hydroxy-8:0 intermediate (due to the 12-D-hydroxy group in ricinoleate). The authors postulated the mechanism outlined in Fig. 3 to account for the complete breakdown of ricinoleate which occurs without accumulation of intermediates in germinating castor bean seeds. The oxidation and isomerase activities were associated with a non-mitochondrial "cytosome" fraction, presumably the spherosomes (Chapter 9) involved in the conversion of ricinoleate-rich triglycerides to sugar in the germinating castor bean.

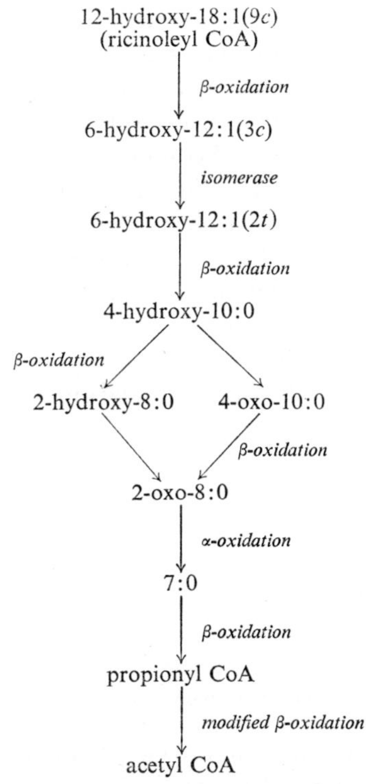

FIG. 3. Proposed pathway for complete degradation of ricinoleic acid in germinating castor beans (adapted from Hutton and Stumpf, 1971). [β-oxidation steps liberate C_2 units as acetyl CoA.]

3. *β-hydroxylation*

The β-hydroxy acid intermediates in the β-oxidation system have the 3-L(+)-hydroxy configuration. Hydroxylation of medium chain length fatty acids at the β-position to form the enantiomeric form (3-D(−)-hydroxy acids), is catalysed by an enzyme system from avocado mesocarp; this β-hydroxylation system and its possible role as an intermediate in the biosynthesis of unsaturated fatty acids is described by Harwood in Chapter 3 (p. 71).

C. OXIDATION AT THE METHYL TERMINAL (OMEGA) GROUP OF FATTY ACIDS

Fatty acids with oxygen functions (alcohol, oxo or carboxyl groups) at the ω-position occur as components of the polymeric structures in cutin and suberin from plants (see Chapter 7).

Kolattukudy (1970) has studied the incorporation of palmitic acid into 16-hydroxypalmitic acid in apple skin tissue. Oxygen was required for the incorporation which was blocked by the iron-chelator, 1,10-phenanthroline. These results indicated that a mixed-function oxygenase was involved, as in ω-oxidation systems in animals and micro-organisms. All such systems require molecular oxygen, a non-haem iron protein and a hydrogen donor (usually NADH or NADPH). In mammalian systems (Lu *et al.*, 1969) and in yeast (Lebault *et al.*, 1971), cytochrome P-450 is involved. Recent results (Kolattukudy *et al.*, 1973) suggest that an ω-oxidation process is involved in the biosynthesis of the C_{18} hydroxy acid components of cutin.

The formation of dicarboxylic acids (characteristic of suberin) from the corresponding ω-hydroxy acid has been demonstrated in cell-free extracts from the epidermal tissue of *Vicia faba* leaves; Kolattukudy and Walton (1972) obtained NADP-specific dehydrogenase activity in a particle-free supernatant fraction from this tissue.

The probable route, according to Kolattukudy and Walton (1972), for the formation of ω-hydroxy and dicarboxylic acids may be summarized thus:

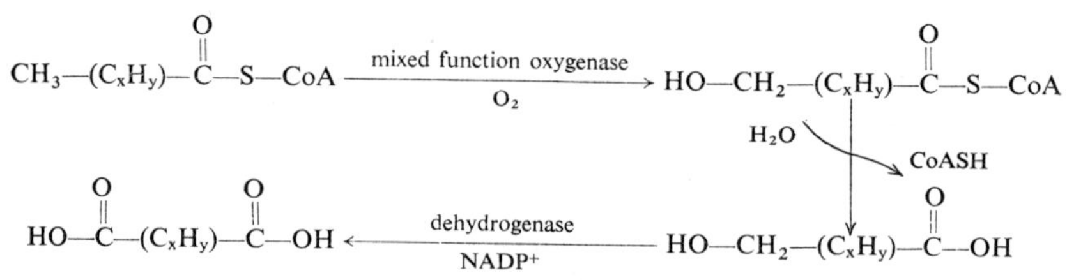

D. IN-CHAIN (NON-TERMINAL) OXIDATION REACTIONS

Fatty acids with non-terminal hydroxyl groups are major components of some seed oils and of cutin (see Chapter 1). Other sections of this volume have reviewed the hydroxylation reactions involved in the formation of hydroxy-acids of seeds (Chapter 3) and of cutin (Chapter 7). All such in-chain hydroxylations appear to be mixed function oxygenase reactions involving the coenzyme A thiolesters of the precursors and a direct involvement of molecular oxygen. Although the hydroxylation of oleyl CoA to ricinoleic acid in developing castor bean seeds is highly specific for both substrate and for the position of oxygenation (Galliard and Stumpf, 1966), the positional specificity of the hydroxylation of cutin components appears to be somewhat less specific (Kolattukudy and Walton, 1972).

The alternative mechanism for the introduction of a hydroxyl group, i.e. hydration of an ethylenic bond, has not been demonstrated in the formation of in-chain hydroxyl groups in plant tissues. However, this hydration reaction is involved in the conversion of linoleic acid to ricinoleic acid in the fungus, *Claviceps purpurea* (Morris *et al.*, 1966).

Introduction of secondary hydroxyl and oxo groups into the long-chain hydrocarbons associated with cuticular structures of plants appears to be by a direct, mixed-function oxygenase type of oxidation system (see Chapter 7):

$$\text{alkane} \xrightarrow{O_2} \text{secondary alcohol} \longrightarrow \text{ketone}$$

E. LIPOXYGENASE-CATALYSED PEROXIDATION OF FATTY ACIDS

1. *Distribution and characterization of lipoxygenase*

The various enzymes grouped under the term lipoxygenase [E.C. 1.13.11.12] have been subjected to extensive studies in many laboratories in recent years. The reasons for renewed interest in these enzymes are due in part to their rôle in the production of both desirable and undesirable flavour components in edible plant products (Grosch, 1972).

Although lipoxygenase enzymes have been known for many years, and the enzyme from soyabeans was one of the first enzymes obtained in a crystalline form (Theorell *et al.*, 1947), its physiological rôle remains enigmatical. First thought to be confined to the seeds of legumes and some cereals (Tappel, 1963), lipoxygenase enzymes are now known to be distributed widely throughout higher plants (Pinsky *et al.*, 1971; Grosch, 1972). Particularly rich sources are potato tubers (Galliard and Phillips, 1971) and egg plant (Grossman *et al.*, 1972b); lipoxygenase also occurs in leaves (Holden, 1970; Grossman *et al.*, 1972a; Anstis and Friend, 1974b) and in green algae (Zimmerman and Vick, 1973).

Lipoxygenases isolated from different sources differ in several respects including substrate specificity, pH optima, effects of inhibitors and, most significantly, in the isomeric form of the products. Even in a single source there exist several isoenzymic forms with different properties, e.g. in soyabean (Christopher and Axelrod, 1971; Christopher *et al.*, 1972; Grosch *et al.*, 1972; Verhue and Francke, 1972) and in peas (Eriksson and Svensson, 1970; Arens *et al.*, 1973b); Anstis and Friend, 1974a, 1974b).

2. *Specificity and mechanism of lipoxygenase action*

All lipoxygenase enzymes specifically attack fatty acids with *cis,cis*-1,4-pentadiene structures. Hamberg and Samuelsson (1965) demonstrated that for the soya bean lipoxygenase, unsaturation at the ω-6 (e.g. Δ-12-position in C_{18} acids) was necessary for activity. With linoleic or α-linolenic acids (the predominant endogenous substrates from plants) optically active 9-D- or 13-L-hydroperoxydiene derivatives are formed although Roza and Francke (1973a) have shown that one isoenzyme from soyabeans preferentially oxygenates α- and γ-linolenic acid at the ends of the unsaturation system (i.e. 9- or 15- and 6- or 12-positions respectively). The proportions of 9- and 13-hydroperoxy derivatives of linoleic acid with different lipoxygenase preparations are listed in Table IV which also illustrates examples of the effect of pH, temperature and O_2 tension on these preparations.

Although the detailed mechanism of the lipoxygenase catalysed reactions is not yet elucidated, Hamberg and Samuelsson (1967) using soyabean enzyme have demonstrated that the initial process involves the stereospecific abstraction of H at the C-11 methylene group of linoleic acid (see Fig. 4). Removal of the L_S-hydrogen at this position results in the stereospecific formation of the 13-L-hydroperoxide. Subsequently Egmond *et al.* (1972) showed that removal of the D_R-hydrogen at C-11 by the lipoxygenase from corn resulted in formation of the 9-D hydroperoxide isomer. Recent results of de Groot *et al.* (1973) indicate that free radical intermediates are involved in the lipoxygenase reaction.

Until very recently, no evidence was obtained for any prosthetic groups nor for the involvement of transition metal ions in the lipoxygenase reaction, even with the crystalline enzyme (Tappel, 1963). However, reasoning on the basis of the conservation of spin of angular momentum, Chan (1971, 1973) argued that the allowed product between molecular oxygen in its ground state (a triplet) and an organic molecule (singlet) is a triplet and a high activation energy is required for reaction. He pointed out that, for an enzymic reaction, the normal way of reducing this activation energy of oxygenation reactions imposed by the spin barrier is the formation of a complex between oxygen and a transition metal ion in the enzyme. Thus Chan re-examined the metal content of highly purified soyabean lipoxygenase using more sensitive techniques. He was able to demonstrate (Chan, 1972, 1973) that the enzyme did in fact contain just one atom of iron per molecule of enzyme (molecular weight approx. 10^5). Subsequently, independent work in two other laboratories (Roza and Francke, 1973b; Pistorius and Axelrod, 1974) has confirmed this observation and moreover the latter group have shown that each of the three main isoenzymes of lipoxygenase in soyabean contains one atom of iron per molecule. The nature of the bonding of iron to the enzyme is not known. It is present as non-haem ferric iron (Pistorius and Axelrod, 1974), but is not easily inhibited by complexing agents, requiring prolonged treatment in the presence of thiols, and the inhibition is not reversible. To date there is no *direct* evidence that the iron is actually at the active site of the enzyme or

N

TABLE IV

Influence of enzyme source and reaction conditions on the ratio of 9- to 13-hydroperoxide derivatives of linoleic acid produced by lipoxygenase enzyme preparations

Enzyme sources	Reaction conditions			Ratio of 9:13 hydroperoxides	Reference
	pH	gas phase	temp. (°C)		
soyabean mixed isoenzymes	9·0	O_2	0	0–30:70–100	Several groups (see text)
soyabean mixed isoenzymes	5·5	air	25	54:46	Galliard and Phillips (1971)
soyabean purified isoenzyme I	9·0	O_2	0	10:90	Christopher *et al.* (1972).
soyabean purified isoenzyme I	9·0	air	25	52:48	
soyabean purified isoenzyme II	9·0	air	25	55:45	
soyabean purified isoenzyme II	7·0	air	25	70:30	
soyabean purified "acid isoenzyme"	7·0	O_2	ambient	*ca.* 50:50	Roza and Francke (1973a)
soyabean purified "acid isoenzyme"	9·0	O_2	0	5:95	
corn germ	6·5	O_2	25	83–93:7–17	Hamberg (1971) Gardner and Weisleder (1970)
corn germ	9·0	O_2	0	15:85	Veldink *et al.* (1972)
potato tuber	5·5	air	25	95:5	Galliard and Phillips (1971)
potato tuber	9·0	air	25	no reaction	
alfalfa seed	6·8	O_2	ambient	50:50	Chang *et al.* (1971)
flax seed	6·9	air	24	80:20	Zimmerman and Vick (1970)
Dimorphotheca sinuata seed	6·9	air	23	0:100	Gardner *et al.* (1973)
pea seed (purified enzyme)	6·75	1% O_2 100% O_2	0	55:45	Leu and Eriksson (1972)
oat seed	7·0	O_2	ambient	*ca.* 90:10	Heimann *et al.* (1973a)
barley seed	6·8	air	ambient	*ca.* 80:20	Graveland *et al.* (1972)

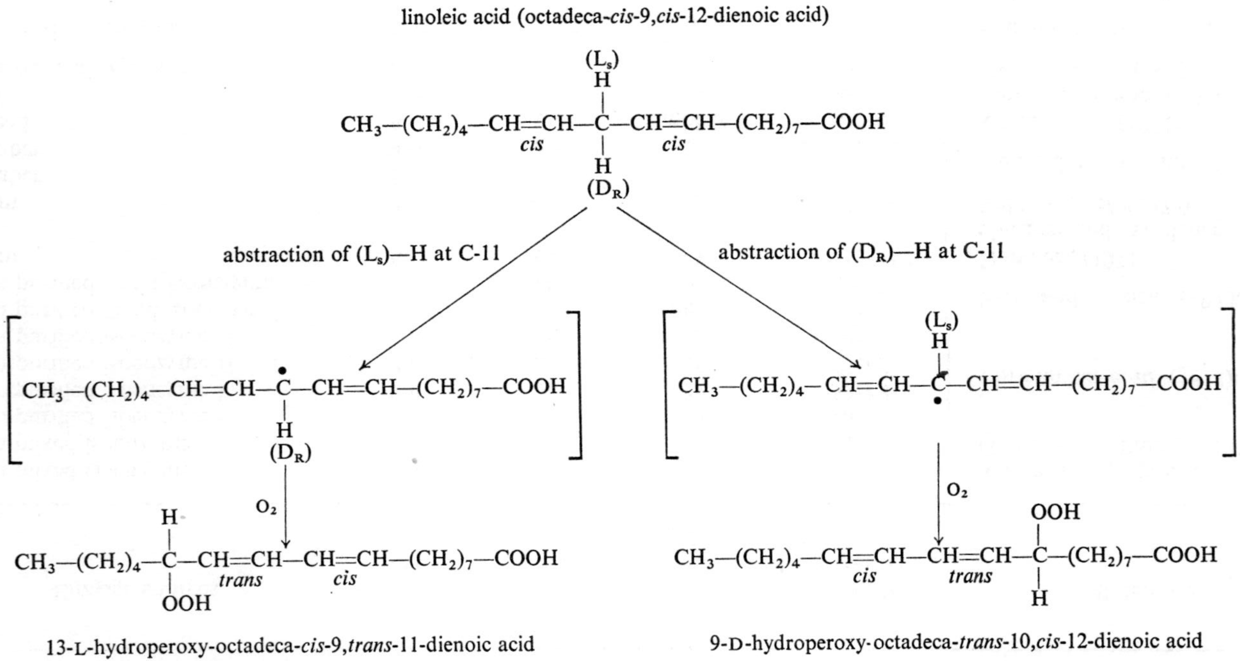

FIG. 4. Formation of 13-L and 9-D-hydroperoxy-derivatives of linoleic acid by the action of lipoxygenases. Stereospecific abstraction of hydrogen atoms at C-11 was shown by Hamberg and Samuelsson (1967) and Egmond *et al.* (1972).

directly involved in oxygen binding. The possible participation of tryptophan residues of the enzyme in oxygen binding was indicated by Finazzi-Agro *et al.* (1973) who demonstrated fluorescence enhancement of the enzyme by oxygen and quenching by substrate. The authors proposed that a novel oxygen–enzyme interaction could be involved. A possibly similar system in which ferric iron and tryptophan residues are involved in the dioxygenase, pyrocatechase, has recently been demonstrated by Nagami (1973).

On the basis of comparisons with model systems, Chan (1971) proposed that, in the active lipoxygenase–oxygen–substrate complex, the oxygen was present in an activated state (analogous to singlet oxygen) and this was subsequently confirmed by Finazzi-Agro *et al.* (1973). Thus, the recent developments tend to suggest that the lipoxygenase reaction is a free-radical process involving a novel type of interaction between enzyme and both substrate and oxygen. The following hypothetical mechanism is compatible with data currently available for lipoxygenase acting on linoleic acid (LH).

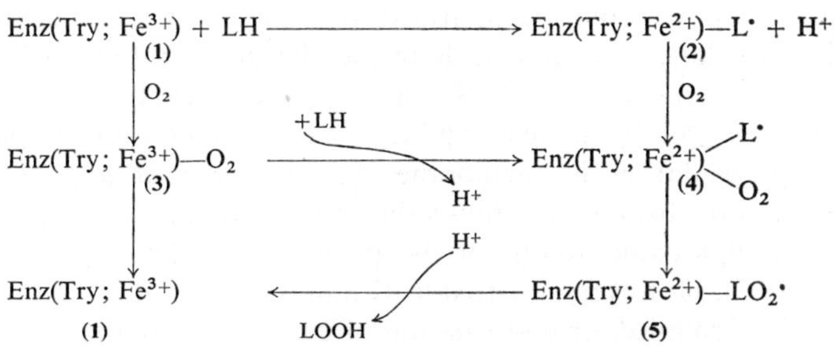

FIG. 5. Hypothetical scheme for lipoxygenase action.

The proposed intermediates (**4**) and (**5**) would be likely to show properties of singlet oxygen by analogy with inorganic model systems for dioxygenases (Chan, 1971).

3. *Lipoxygenase-catalysed co-oxidation of pigments*

The oxidation (bleaching) of natural pigments in systems where peroxidation of fatty acids is occurring is well known. Although pigment bleaching occurs during non-enzymic peroxidation catalysed by haemoproteins, e.g. catalase and peroxidase (Blain, 1970), lipoxygenase-catalysed reactions also give rise to co-oxidation of pigments.

Both lipoxygenase and its fatty acid substrate are required together with oxygen for co-oxidation to take place; neither lipoxygenase, fatty acid alone, hydroperoxide nor inactivated enzyme plus substrate are active. Thus the co-oxidation is due to an active intermediate in the system.

Several recent publications have established the role of lipoxygenases and their isoenzymic forms in the oxidation of carotenoids (Ben Aziz *et al.*, 1971a, b; Firn and Friend, 1972; Arens *et al.*, 1973b; Weber *et al.*, 1973a, b;

Anstis and Friend, 1974b) and of chlorophyll (Orthoefer and Dugan, 1973). However, some isoenzyme forms of lipoxygenase acting at high pH have low pigment bleaching properties (Weber *et al.*, 1973b). There is evidence (Orthoefer and Dugan, 1973) that co-oxidation of pigments requires mixed micelles of pigment and fatty acid substrate to be present with the lipoxygenase; this could explain the low carotene bleaching activity (Kies *et al.*, 1969; Weber *et al.*, 1973b) of the soyabean lipoxygenase isoenzyme acting at pH9—at which pH the fatty acid substrate would be present as a soap solution rather than as a micellar dispersion. These results indicate that care should be exercised in assaying for lipoxygenase using a carotenoid bleaching assay (see Blain, 1970).

The mechanism of the co-oxidation reaction is not known. The co-oxidation of model substrates with 1,3-dienoid structures in lipoxygenase systems was analogous to their oxidation by singlet oxygen (Chan, 1971) and intermediates in the lipoxygenase system such as those postulated (**4** and **5**) in Fig. 5 could be the active principles involved.

Photo-oxidation reactions are catalysed by singlet oxygen (Clements *et al.*, 1973) and peroxidation of fatty acids in membranes is catalysed by singlet oxygen (Pederson and Aust, 1973). Superoxide anions (O_2^-), which dismutate to form singlet oxygen by the reaction $2O_2^- + 2H^+ \rightarrow H_2O_2 + O_2$ (singlet), also cause lipid peroxidation in membranes (Zimmerman *et al.*, 1973). The enzyme, superoxide dismutase, inhibits the above autoxidation systems either by destroying superoxide anions or by quenching singlet oxygen (Finazzi-Agro *et al.*, 1972). However, superoxide dismutase does not inhibit fatty acid hydroperoxide formation or co-oxidation of carotenoid by the lipoxygenase enzyme from potato tubers (Berkeley and Galliard, unpublished) and crude enzyme preparations from this tissue contain significant superoxide dismutase activity. Nevertheless participation of singlet oxygen *at the active site* of the enzyme cannot be excluded.

5. *Physiological role of lipoxygenases*

The actual role of lipoxygenases in plants remains obscure. A fall in lipoxygenase activity in various seeds during the early stages of germination has been demonstrated (Tappel, 1963; Guss *et al.*, 1968) although Anstis and Friend (1974b) have recently shown that an early fall in lipoxygenase activity in germinating pea seedlings is followed by a subsequent increase in the activities of several lipoxygenase isoenzymes at later stages of germination. Phytochrome control of lipoxygenase synthesis has been demonstrated in mustard cotyledons by Oelze-Karow *et al.* (1970). In the potato tuber, after an initial rise in very immature tubers, lipoxygenase activity (on a fresh weight basis) remains relatively constant throughout growth, dormancy and sprouting of the tubers (Berkeley and Galliard, 1974). Garssen *et al.* (1971) have suggested that lipoxygenase may play a role in maintaining a low O_2-tension in seeds at certain physiological stages.

Lipoxygenase may well have a function in the wound response mechanism in plants. Activity of the enzyme increases in damaged tissues (Heinen and Brand, 1963; Bredmeijer and Heinen, 1968; Lipetz, 1970) and is possibly involved in the formation of the "wound hormone", traumatic acid, in damage of tissues (Hitchcock and Nichols, 1971). Increases in lipoxygenase activity in potato tubers in response to wounding and fungal infection were similar to the increases in phenol oxidase and peroxidase under these conditions (Galliard and Oliver, unpublished results) and increases in activities of the latter enzymes are characteristic of plant responses to wounding or infection (Lipetz, 1970).

Lipoxygenase has been implicated in the production of volatile compounds in intact and damaged plant tissues. In ripening apple fruits, lipoxygenase activity increases at the same stage as the evolution of volatile flavour compounds and ethylene (Wooltorton *et al.*, 1965) and the loss of endogenous galactolipids, lipids rich in linoleic and linolenic acids (Galliard, 1968). In both model systems (Lieberman and Mapson, 1964) and a cell-free extract from apple fruits (Galliard *et al.*, 1968), linolenic acid was converted to ethylene via a lipoxygenase-catalysed reaction. However, present evidence (Baur and Yang, 1972) supports the conclusion that the major physiological process of ethylene formation in plants is via the methionine pathway. Nevertheless, the ethylene evolution which takes place on wounding may be via an alternative route involving linolenic acid and lipoxygenase.

Volatile aldehydes with characteristic flavour properties are formed from lipoxygenase substrates in damaged plant tissues. For example, hexanal and hexenal are produced from linoleic acid and linolenic acid respectively in damaged leaves, fruits, and vegetables (for references see Tressl and Drawert, 1973; Gardner, 1975). These aldehydes are also formed during incubations of purified lipoxygenase with the fatty acids (Arens *et al.*, 1973a). Other aldehydes formed in these systems and from macerated plant tissues include saturated and unsaturated aldehydes with chain lengths from five to ten carbons. Non-2-enal and nona-2,4-dienal are characteristic odour compounds in the flavour of cucumber and are formed enzymically from linoleic and linolenic acids in homogenates of this tissue (Grosch and Schwarz, 1971). Although hydroperoxides of these fatty acids decompose to give such aldehydes (Gardner, 1975), some workers believe that the enzymic process involves a direct fission of the fatty acids rather than via a lipoxygenase catalysed hydroperoxide pathway (Grosch and Schwarz, 1971; Major and Thomas, 1972).

Some naturally-occurring fatty acids have conjugated ethylenic and acetylenic functions with or without oxygenated groups in the conjugated system (see Chapter 1). Some of these have a formal resemblance to the known products of lipoxygenase action. For example, *S*-coriolic acid, 13-L-hydroxy-18:2(9*c*, 11*t*), a major component of the seed oil *Monnina emarginata* (Phillips *et al.*, 1970), is analogous to 13-L-hydroperoxy-18:2(9*c*,11*t*), one of the isomeric structures obtained by lipoxygenase action on linoleic acid and dimorphecolic acid, 9-D-hydroxy-18:2(10*t*,12*t*), is similar, except for ethylenic configuration at position 12, to the other isomer formed by lipoxygenase (9-D-hydroperoxy-18:2(10*t*,12*c*). Dimorphecolic acid is the major component of seed oil of *Dimorphotheca sinuata* and Gardner *et al.* (1973) investigated the possibility that lipoxygenase was involved in the biosynthesis of this fatty acid. Although the seeds contained lipoxygenase, the enzyme showed specificity for the 13-hydroperoxy isomer and formed the usual *cis*, *trans*-diene system. However, the authors pointed out that the mature seeds were used in their studies and that a different type of enzyme system could be involved in the developing seed when the seed oil was being produced.

F. METABOLISM OF FATTY ACID HYDROPEROXIDES

1. *Enzymic and non-enzymic breakdown of fatty acid hydroperoxides*

Fatty acid hydroperoxides, formed by the action of lipoxygenase or by autoxidation, are toxic to living tissues and plant and animal cells contain systems which metabolise these hydroperoxides. The hydroperoxides are potentially unstable (although in a purified form can be kept in solution for several weeks) and decompose by non-enzymic processes which have been intensively studied in relation to the oils and fats industry.

Enzymic and non-enzymic decomposition of fatty acid hydroperoxides have been compared in a recent review by Gardner (1975) and the present discussion refers only to recent work on the enzymic breakdown of hydroperoxides in plant tissues. Reference to the rôle of lipoxygenases in the production of volatile aldehydes etc. was made in the previous section. Many laboratories have recently taken renewed interest in the enzymic breakdown of unsaturated fatty acids via hydroperoxide intermediates, but, as yet, there is little known about the complete pathways from fatty acids to the final fragmentation products.

2. *Reduction of fatty acid hydroperoxides*

Conversion of fatty acid hydroperoxides to the corresponding hydroxydiene derivatives has been described in several plant tissues including cereal seeds (Graveland, 1970a, 1973a, b; Graveland *et al.*, 1972; Heimann and Schreier, 1970, 1971; Heimann *et al.*, 1973a, b), soyabeans (Schormüller *et al.*, 1968; Grosch *et al.*, 1971, 1972; Heimann *et al.*, 1973b), *Dimorphotheca sinuata* seeds (Gardner *et al.*, 1973) and potato tubers (Galliard and Phillips, 1972; Galliard *et al.*, 1975).

However, in some cases the reduction appears, at least in part, to be non-enzymic, e.g. in cereals (Graveland, 1973a; Heimann *et al.*, 1973a) and potato tuber extracts (Galliard *et al.*, 1975). It was suggested that non-specific reduction by endogenous protein thiols was responsible for hydroperoxide reduction in flour suspensions (Graveland, 1970a) but thiol blocking agents did not prevent the reduction in cereal suspensions (Graveland, 1973b) or in potato extracts (Galliard *et al.*, 1975).

Fatty acid hydroperoxides are reduced by Fe^{2+} ions or by Fe^{3+} in the presence of thiols although the yield of hydroxydiene is low compared with other products (Gardner *et al.*, 1974) and glutathione is oxidized during lipoxygenase action on fatty acid in extracts of peas (Mapson and Moustafa, 1955). However, there is no direct evidence in plants to date for reduction of hydroperoxides by the glutathione–peroxidase enzyme system present in animals (Christopherson, 1969).

However, "lipoperoxidase" enzyme activity with a variety of electron-donors has been described in plants. Heimann *et al.* (1972, 1973b) have described an enzyme from oats which reduces fatty acid hydroperoxides to the corresponding hydroxy acid in the presence of *p*-phenylenediamine; these workers were unable to separate the peroxidase enzyme from lipoxygenase and proposed that a bifunctional lipoxygenase–lipoperoxidase complex was involved in the metabolism of polyunsaturated fatty acids. A similar bifunctional complex was observed in soyabeans by Grosch *et al.* (1972) who used guaiacol as electron donor and this was confirmed by Heimann *et al.* (1973b) using *p*-phenylenediamine.

3. *Fatty acid hydroperoxide isomerase enzymes*

Enzymic isomerization of fatty acid hydroperoxides to ketols was first described in flax seed extracts by Zimmerman (1966). Subsequent work in several laboratories has established ketol formation from hydroperoxides as a widespread phenomenon. Two types of isomerization reaction exist and different plants may have either one or both of these systems.

These isomerase activities and the products formed from linoleic acid are summarized in Fig. 6. In one system (α), 9- and 13-hydroperoxides are converted to the α-ketols III and IV respectively. In the second system (γ), the γ-ketols IV and V are formed.

The α-ketol system is present in flax seed (Zimmerman and Vick, 1970) and wheat (Zimmerman and Vick, 1970; Graveland, 1973a) whereas alfalfa seedlings (Esselman and Clagett, 1974) form the γ-ketols. Both α- and γ-ketols are formed in extracts of corn germ (Gardner, 1970) and barley (Graveland *et al.*, 1972).

The flax seed system was reported to convert both 9- and 13-hydroperoxides to the corresponding α-ketols (Zimmerman and Vick, 1970) but Veldink *et al.* (1970a) found that only the 13-hydroperoxide of linoleic acid was isomerized by the flax enzyme. The isomerases from corn germ (Gardner, 1970) and potato (Galliard *et al.*, 1975) metabolize both 9- and 13-hydroperoxides.

The mechanism of the α-isomerase reaction was investigated by Veldink *et al.* (1970b) using ^{18}O-labelled 13-hydroperoxide. Their results showed that only the oxygen in the 12-oxo group was derived from the 13-hydroperoxide

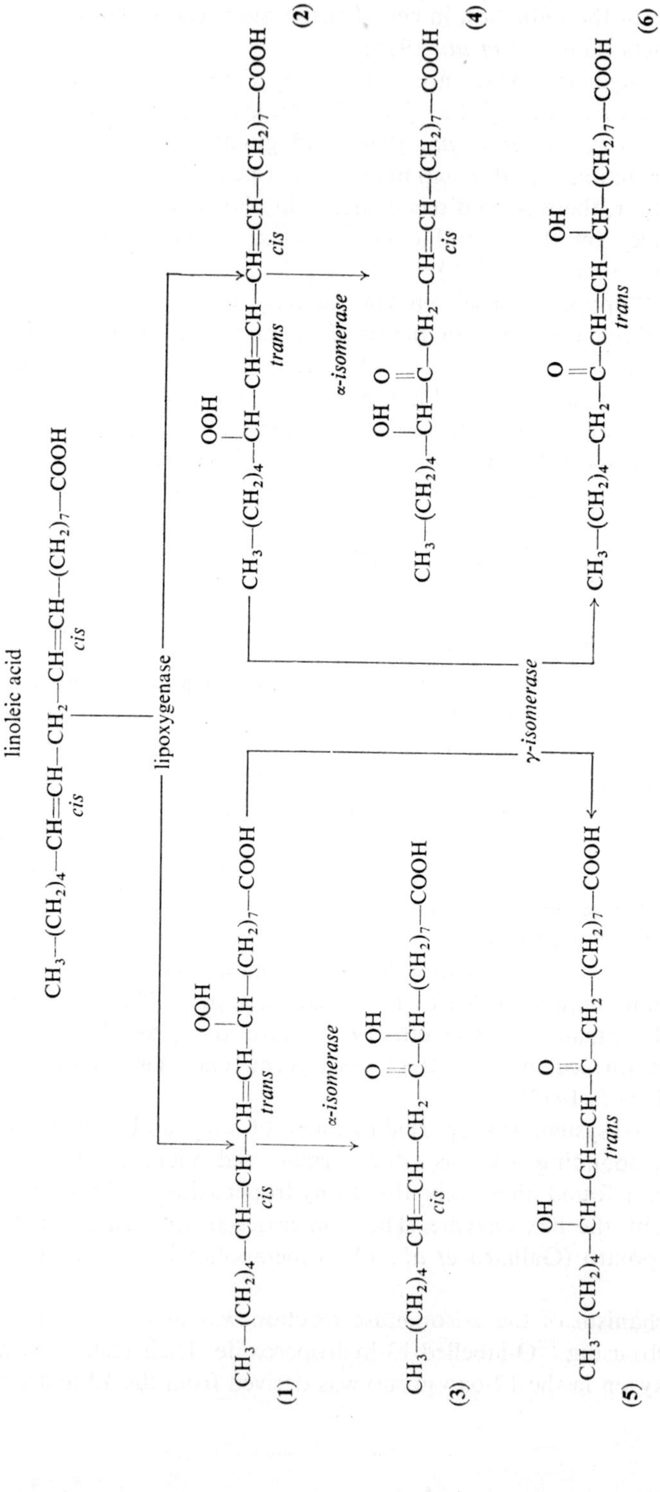

FIG. 6. Fatty acid hydroperoxide isomerase reactions forming α- and γ-ketol derivatives.

group and that the 13-hydroxy group (at the position of the original hydroperoxide function) was not labelled and the hydroxyl group was thus assumed to have been derived from water. The authors proposed that an intermediate with a cyclic oxygen function could explain the oxygen transfer although subsequent work from their laboratory (Veldink *et al.*, 1970a) indicated that epoxide intermediates were not involved; a cyclic peroxide or epiperoxide remained as possible intermediates.

By contrast, the γ-isomerase reaction in alfalfa extracts involves retention of both oxygens from the hydroperoxide group in the resulting γ-ketol and Esselman and Clagett (1974) have proposed that the reaction could involve the intermediate formation of a cyclic epiperoxide and rearrangement to form the γ-ketol.

Hydroperoxide isomerase activity has been reported in other plants including soyabeans and mung beans (Zimmerman and Vick, 1970), peanuts (St. Angelo and Ory, 1972), linseed (Grosch *et al.*, 1971), watermelon and sunflower seedlings and cauliflower buds (Zimmerman *et al.*, 1974). In these studies, assays for isomerase activity were based on the disappearance of fatty acid hydroperoxides as shown by loss of the 234 nm absorption of the conjugated diene system. However, hydroperoxide can disappear with loss of chromophore by mechanisms other than isomerase activity (see below) and can also be reduced to hydroxydiene derivatives with retention of 234 nm absorption. In the author's experience, criteria other than E_{234} are required to establish isomerase activity, e.g. identification and determination of the reaction products.

4. *Enzymic conversion of hydroperoxides to vinyl ether derivatives*

Analysis of the products formed on homogenization of potato tubers at pH 6–8 showed the lipid component to comprise a mixture of two novel fatty acid derivatives containing butadienylvinyl ether structures, colneleic acid (see **3** in Fig. 7) and colnelenic acid, an analogous compound containing an additional *cis*-ethylenic bond at the 15-position (Galliard *et al.* 1973). It was established that these ethers were formed enzymically from linoleic and linolenic acids respectively via the hydroperoxide intermediates (Galliard and Phillips, 1972). Subsequent work with linoleic acid has shown that the conversion is specific for the 9-hydroperoxide; the 13-hydroperoxide isomer is not converted to an ether derivative (Galliard and Matthew, 1975).

A significant aspect of the conversion of hydroperoxide to the ether is that the carbon chain of the fatty acid is broken and an atom of oxygen is inserted into the chain. The origin of this oxygen atom is not yet established although the reaction takes place under anaerobic conditions. It is possible that one of the hydroperoxide oxygens in incorporated, analagous to the isomerase reactions discussed above. Formally the ether formation can be considered as an isomerization/dehydration reaction although the mechanism could be a concerted mechanism analogous to a Baeyer-Villiger rearrangement as pro-

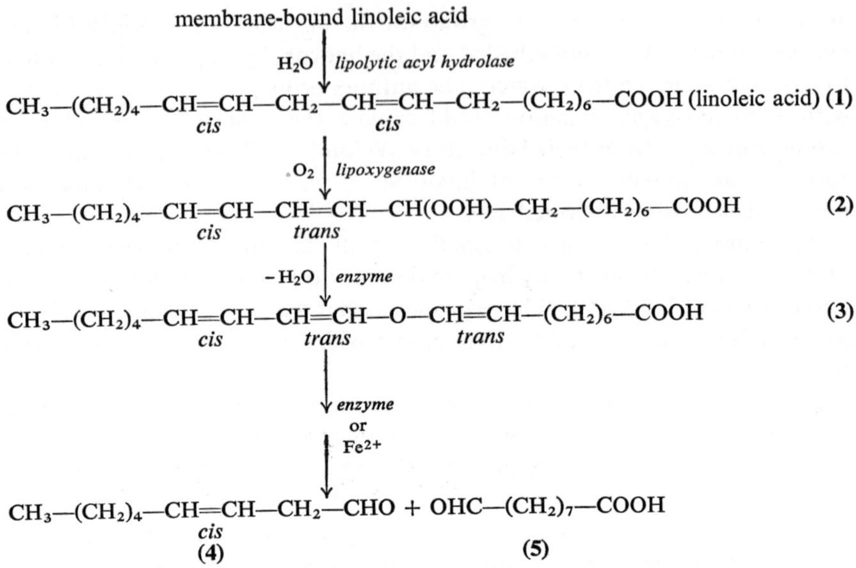

FIG. 7. Enzymic degradation of linoleic acid *via* hydroperoxide and divinyl ether derivatives in extracts of potato tuber (Galliard *et al.* – see text).

posed by Gunstone (1966) as a possible mechanism for the formation of stillingic acid which contains an internal ester function.

The appearance of these ethers in potato tissue extracts is due to a sequence of enzymic reactions (see Fig. 7) which is initiated by cellular disruption: (a) membrane-bound lipids are deacylated by endogenous lipolytic acyl hydrolase activity (see p. 323) releasing free fatty acids, predominantly linoleic acid (60%), linolenic acid (15%) and palmitic acid (20%); (b) the polyunsaturated acids are then attacked by lipoxygenase to form specifically the 9-hydroperoxides (see p. 336); (c) the 9-hydroperoxides are converted to the ether derivatives which can be isolated at high pH levels but, at pH values below 6; (d) the ethers are further degraded as described below.

An enzyme present in potato extracts catalyses the breakdown of colneleic acid via an unidentified unstable intermediate to carbon fragments (see Fig. 7). Although the active factor from potatoes is a heat-labile, non-dialysable, high molecular weight compound, the same reaction proceeds non-enzymically in the presence of catalytic amounts (1–5 μM) of Fe^{2+} ions (Galliard *et al.*, 1974). Other metal ions are either inactive or, in the case of Fe^{3+} and Cu^{2+}, are inhibitory. Haemoproteins are inactive but the non-haem protein, ferredoxin, is as active (on a Fe content basis) as Fe^{2+} ions. Both enzymic and non-enzymic reactions require oxygen with approximately 1:1 stoichiometry of oxygen uptake to ether degraded.

The ether-degrading enzyme activity has been partially purified from potato tubers and, to date, no separation of the enzyme from lipoxygenase has been achieved. Both enzymic and Fe^{2+}-catalysed reactions appear to be free radical reactions which are inhibited by low concentrations of antioxidants (e.g. 5 μM butylated hydroxytoluene, BHT) and some chelating agents (e.g. 5 μM dithizone).

The physiological role of the degradation pathway involving hydroperoxides and vinyl ethers is not known. In the potato tuber extracts this pathway exists together with the hydroperoxide isomerase activity. Although formation of both hydroperoxide and vinyl ether derivatives is specific for the 9-position, the isomerase activity is equally effective on both 9- and 13-hydroperoxide isomers (Galliard *et al.*, 1975).

The carbonyl fragmentation products from the ether compound have characteristic flavour properties. For example, colnelenic acid, the vinyl ether derivative from linolenic acid, is degraded to nona-3,6-dienal, the characteristic component of cucumber flavour; although this aldehyde is known to be formed from linolenic acid in cucumber extracts (Grosch and Schwartz, 1971), no direct evidence has yet been obtained for the degradation pathway involved.

8. *Epoxy- and polyhydroxy-acid formation*

A variety of fatty acid derivatives containing dihydroxy, hydroxy-epoxy, oxo-epoxy and trihydroxy groups are formed from linoleic and linolenic acids in cereal flour-water mixtures (Graveland, 1970a, b; 1973a, b; Graveland *et al.*, 1972; Heimann *et al.*, 1973a). Extracts from potato tubers convert 9- and 13-hydroperoxides of linoleic acid to a mixture of products including hydroxy-epoxy- and trihydroxy-monoenoic acids (Galliard *et al.*, 1975). Although these reactions are heat-labile and appear to be enzymic, similar products are formed non-enzymically from fatty acid hydroperoxides. These enzymic and non-enzymic reactions have been reviewed recently by Gardner (1975).

The mechanisms of the enzymic reactions have not yet been elucidated in detail. It is probable that lipoxygenase catalyses the initial reaction on the fatty acids and either free hydroperoxides or possibly peroxy radical intermediates are then further oxidized. For example, Graveland (1973a) has proposed that the epoxy derivatives are formed by epoxidation at the conjugated diene region of the hydroperoxide. The epoxyhydroxy acids are readily hydrolysed by a water-soluble factor in wheat to form trihydroxy derivatives (Graveland, 1970a).

6. *Anaerobic reactions with fatty acid hydroperoxides*

As discussed earlier, the physiological rôle of lipoxygenase is not yet understood. Most *in vitro* studies with the enzyme are performed in high oxygen concentrations under which conditions free fatty acid hydroperoxides are the major products. However, under physiological conditions, oxygen levels may be low. Boldingh's group in Utrecht have studied the action of lipoxygenase under low oxygen tension and anaerobic conditions and have made the interesting observation that lipoxygenase is catalytically active, under anaerobic conditions, in forming products other than hydroperoxides (Garssen *et al.*, 1971).

When soyabean lipoxygenase was incubated anaerobically with linoleic acid and its hydroperoxide, a mixture of products was formed including 13-oxo-octadecadienoic acid, C_{13} and C_{15} fragmentation products of the C_{18} substrates, together with oxygenated and non-oxygenated dimeric compounds. To explain these results, Garssen *et al.* (1971, 1972) have proposed that the initial reaction involves the removal by lipoxygenase of a hydrogen atom from linoleic acid to form a fatty acid radical which then undergoes various addition and radical dismutation reactions. Evidence for the presence of linoleic acid radicals in this anaerobic system was obtained by de Groot *et al.* (1973).

It should be noted that these anaerobic conditions require the presence of fatty acid hydroperoxide (the product of *aerobic* lipoxygenase action). Smith and Lands (1972) have proposed that the lipoxygenase enzyme contains a product-binding site, separate from the substrate site. The nature of the various interactions between substrate, product, oxygen and the enzyme with its iron and tryptophan residues (see p. 336) is as yet unknown.

Under both aerobic and anaerobic conditions, fatty acid hydroperoxides are also readily attacked by *non-enzymic* reactions. The processes of lipid oxidation involving hydroperoxide intermediates have been widely studied and recently reviewed (Gardner, 1975). In addition to fission products and polymeric compounds, fatty acid derivatives with structures similar to the products of the enzymic reactions are formed. Thus definitive studies of the enzymic breakdown of fatty acids via hydroperoxide intermediates are complicated by the simultaneous occurrence of non-enzymic reactions.

IV. Subcellular Localization of Lipid-Degrading Enzymes

The fragile nature of subcellular structures in plant tissues is now widely recognized and, for example, Beevers in this volume (p. 289) has demonstrated the requirement for the most sensitive disruption techniques to obtain good preparations of glyoxysomes. Thus enzyme activities which have previously been described as "soluble" or cytoplasmic may in fact be localized within discrete but fragile organelles. This would seem likely for degradative enzymes potentially damaging to structural components, particularly since plants often contain very high amounts of these activities, e.g. lipolytic acyl hydrolases, phospholipase D and lipoxygenases.

A. Hydrolytic Enzymes

1. *Lipolytic acyl hydrolases*

The lipolytic acyl hydrolase from potato tubers is obtained almost exclusively in the particle-free supernatant in conventional isotonic centrifugation of homogenates. However, as described earlier (p. 325), the enzyme will attack the membranes of subcellular organelles and the enzyme activity is so high in this tissue that, if the mixing of endogenous substrate and enzyme took place, all the polar lipids of the tissues could be hydrolysed in a few milliseconds! Thus the enzyme must either be inactivated *in vivo* or is localized within an organelle. When tissues with different amounts of lipolytic acyl hydrolase were examined, it was noted that the higher the total enzyme activity in the tissue, the lower the proportions of this which could be obtained in a particulate fraction. This is illustrated for different tissues from the potato plant in Fig. 8 (D. A. Wardale and T. Galliard, in preparation). A possible explanation for these results is that disruption of the tissue causes damage to organelles containing the enzyme and, on release, an autocatalytic destruction of organelles takes place. Thus, to study the subcellular localization of the enzyme we have been forced to take the unusual measure of selecting tissues with relatively low levels of the enzyme. Such studies with cauliflower floret tissue (D. A. Wardale and T. Galliard, in preparation) have indicated that the particulate activity is associated with a low-density fraction in sucrose density gradients and that the activity profile is coincident with

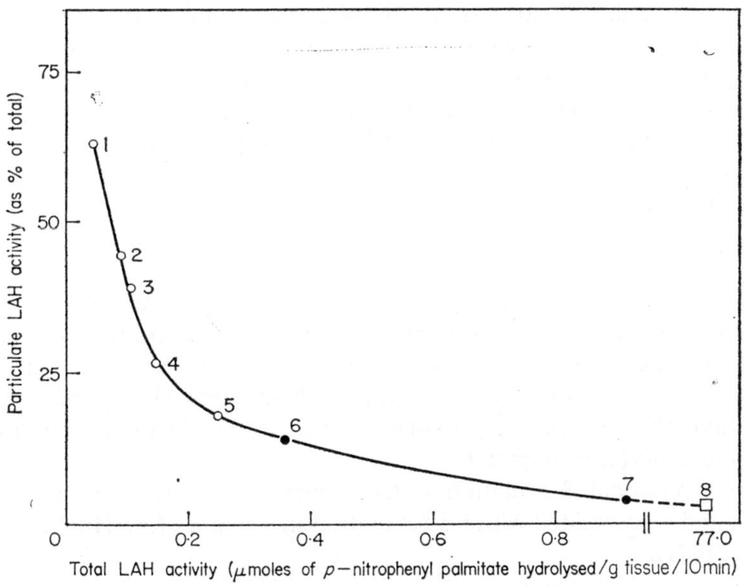

FIG. 8. Relationship between total lipolytic acyl hydrolase (LAH) activity in extracts of tissue from potato and the percentage of the LAH activity obtained in a particulate fraction. Points 1–5 represent extracts from sprouts of tubers from different varieties of potato; points 6 and 7 represent extracts from immature and mature tubers from a variety (Desirée) with low LAH activity; point 8 represents a tuber from a variety (Golden Wonder) with a high LAH activity. (From D. A. Wardale and T. Galliard, in preparation.)

those of other acid hydrolases, e.g. acid phosphatase and RNase. It is likely therefore that the acyl hydrolase enzyme is located in a lyosome-like organelle.

2. *Phospholipase D*

Phospholipase D activity has been found in various particulate fractions of plant extracts as well as in particle-free supernatants (for recent review see Galliard, 1973). In a recent study by Clermont (1972) the particulate phospholipase D of cauliflower floret tissue was distributed in density-gradients predominantly with the microsomal fraction, although the specific activity was higher in a fraction with a density equivalent to lysosome-like organelles.

3. *Lipases*

Triglyceride lipases in oil seeds are associated with the oil-storage bodies (spherosomes) of those tissues as described elsewhere in this volume (pp. 276 and 293).

B. OXIDATIVE ENZYMES

1. *α-oxidation*

The subcellular localization of the α-oxidation system has not been rigorously established. In extracts of germinating peanuts, the activity was distributed amongst mitochondrial, microsomal and particle-free supernatant fractions (Martin and Stumpf, 1959) whereas the α-oxidation system in leaves was reported to be mainly particulate (Hitchcock and Nichols, 1971). In view of the uncertainty about the relative importance of α- or β-oxidation in non-seed tissues (p. 332) further investigations on the localization of this system could be informative.*

2. *β-oxidation*

As described earlier (p. 332), β-oxidation plays two separate roles in plants and the two systems are spacially separated. The β-oxidation system, which is part of the gluconeogenesis pathway in germinating seeds, is present in the membrane of the glyoxysomes, together with the glyoxylate cycle enzymes which take the product of β-oxidation, i.e. acetyl CoA, through to the sugar precursors (see Chapter 9).

The conventional β-oxidation system appears to be identical to that in animals and is associated with mitochondrial membranes (see Hitchcock and Nichols, 1971).

3. *Lipoxygenase*

Despite the large volume of work on lipoxygenases, particularly in recent years, their subcellular distribution has received relatively little attention.

* See Shine and Stumpf (1974) for a recent re-examination of the subcellular localization α-oxidation activities in peanut cotyledons and pea leaves.

Knowledge of their location within plant cells could throw some light on the possible rôle(s) of these enzymes. Published information to date presents a confusing picture of lipoxygenase localization. Grossman *et al.* (1972a) used differential centrifugation methods with alfalfa leaf tissue and found lipoxygenase activity associated with mitochondria and chloroplasts in addition to the particle-free supernatant. Holden (1970) found that much of the lipoxygenase activity in wheat leaves was associated with a plastid fraction. Haydar and Hadziyev (1973), working with pea seedlings, concluded that lipoxygenase was not confined to a specific organelle but was distributed amongst several fractions. After density-gradient purification of the fractions, lipoxygenase-like activity was demonstrated in mitochondrial, peroxisomal and plastid fractions as well as in the particle-free supernatant. However, partial sensitivity to cyanide in these fractions indicated that part of the linoleate-peroxidizing activity may have been due to reactions catalysed by haemoproteins, e.g. catalase and peroxidase.

On the basis of inhibitor studies, Blain (1970) has suggested that haemoproteins are responsible for lipoxygenase-like activities in many non-seed tissues although several groups have recently demonstrated that true (non-haemoprotein) lipoxygenases are responsible for fatty acid peroxidation in tissues other than seeds (e.g. Holden, 1970; Pinsky *et al.*, 1971; Galliard and Phillips, 1971). Nevertheless, studies with isolated organelles, particularly those with relatively high levels of haemoproteins (mitochondria and peroxisomes), should take careful note of their possible complications.

Recent studies with barley seedlings have indicated that lipoxygenase activity is present in subcellular fractions containing peroxisomes (Friend, 1974). As part of a general study of the subcellular distributions of lipid-degrading enzymes in plants (D. A. Wardale and T. Galliard, in preparation), we have found that approximately 30% of the lipoxygenase activity of cauliflower floret tissue can be obtained in a particulate form and that the particulate activity is concentrated in a fraction of high (1·24 g/cm^3) isopycnic density, very similar to, but not identical with, intact plastids. Microbody, mitochondrial and microsomal fractions contained little or no lipoxygenase activity. In pea roots, lipoxygenase was localized in two separate fractions (isopycnic densities 1·17 and 1·19 g/cm^3) and gave profiles in the sucrose gradients coincident with acid phosphatase but separate from plastids (1·23 g/cm^3). These results indicated that lipoxygenase is located in fragile organelles which are separate from mitochondria, microbodies and plastids.

4. *Fatty acid hydroperoxide-degrading systems*

The subcellular localizations of enzyme systems catalysing the further conversion of fatty acid hydroperoxides, formed by lipoxygenase activity, viz. reduction to hydroxydienoic acids (p. 342), isomerization to ketols (p. 343), and conversion to vinyl ethers (p. 345) or epoxy- and polyhydroxy-acid derivatives (p. 347), are not well established. In general, these activities are obtained in particle-free preparations and in no case has the location of observed activity been restricted to a discrete subcellular organelle. In a recent study of the distribution of hydroperoxide isomerase activity in several plants, Zimmerman *et a.* (1974) found most activity associated with particle-free preparations, although some loss of the conjugated diene chromophore of hydroperoxy acids was catalysed by mitochondrial fractions, purified by density-gradient centrifugation.

REFERENCES

Acker, L. and Bücking, H. (1956). *Z. Lebensm. Untersuch. Forsch.* **104**, 423–428.

Acker, L. and Geyer, J. (1969). *Z. Lebensm. Untersuch. Forsch.* **140**, 269–275.

Aizono, Y., Funatsu, M., Sugano, M., Hayashi, K. and Fujiki, Y. (1973). *Agric. biol. Chem.* **37**, 2031–2036.

Ansell, G. B. and Hawthorne, J. N. (1964). "Phospholipids", pp. 152–174. Elsevier, Amsterdam.

Anstis, P. J. P. and Friend, J. (1974a). *Phytochemistry* **13**, 567–573.

Anstis, P. J. P. and Friend, J. (1974b). *Planta* **115**, 329–335.

Antia, N. J., Bilinski, E. and Lau, Y. C. (1970). *Can. J. Biochem.* **48**, 643–648.

Arens, D., Laskawy, G. and Grosch, W. (1973a). *Z. Lebensm. Untersuch. Forsch.* **151**, 162–166.

Arens, D., Seilmeier, W., Weber, F., Kloos, G. and Grosch, W. (1973b). *Biochim. biophys. Acta* **327**, 295–305.

Astrachan, L. (1973). *Biochim. biophys. Acta* **296**, 79–88.

Baddeley, M. S. and Simon, E. W. (1969). *J. exp. Bot.* **20**, 94–99.

Barron, E. J. (1964). *In* "Modern Methods of Plant Analysis" (H. F. Linskens, B. D. Sanwall and M. J. Tracey, eds), Vol. 7, pp. 448–453. Springer-Verlag, Berlin.

Baur, A. H. and Yang, S. F. (1972). *Phytochemistry* **11**, 3207–3214.

Ben Abdelkader, A., Cherif, A., Demandre, C. and Mazliak, P. (1973). *Eur. J. Biochem.* **32**, 155–165.

Ben-Aziz, A., Grossman, S., Ascarelli, I. and Budowski, P. (1971a). *Phytochemistry* **10**, 1445–1452.

Ben-Aziz, A., Grossman, S., Budowski, P. and Ascarelli, I. (1971b). *Phytochemistry* **10**, 1823–1830.

Berkeley, H. D. and Galliard, T. (1974). *J. Sci. Fd Agric.* **25**, 869–873.

Berner, D. L. and Hammond, E. G. (1970). *Lipids* **5**, 572–573.

Black, H. S. and Altshul, A. M. (1965). *Biochem. biophys. Res. Commun.* **19**, 661–664.

Blain, J. A. (1970). *J. Sci. Fd Agric.* **21**, 35–38.

Bredemeijer, G. and Heinen, W. (1968). *Acta bot. neerl.* **17**, 15–25.

Calvayrac, R. and Douce, R. (1970). *FEBS Letters* **7**, 259–262.

Chan, H. W-S. (1971). *J. Am. chem. Soc.* **93**, 2357–2358.

Chan, H. W-S. (1972). Communication to 11th World Congress of the International Society for Fat Research, Göteborg, Sweden.

Chan, H. W-S. (1973). *Biochim. biophys. Acta* **327**, 32–35.

Chang, C. C., Esselman, W. J. and Clagett, C. O. (1971). *Lipids* **6**, 100–106.

Chen, J. S. and Barton, P. G. (1971). *Can. J. Biochem.* **49**, 1362–1375.

Ching, T. M. (1968). *Lipids* **3**, 482–488.
Christopher, J. and Axelrod, B. (1971). *Biochem. biophys. Res. Commun.* **44**, 731–736.
Christopher, J. P., Pistorius, E. K. and Axelrod, B. (1972). *Biochim. biophys. Acta* **284**, 54–62.
Christopherson, B. D. (1969). *Biochim. biophys. Acta* **176**, 463–470.
Clements, A. H., Van Den Engh, R. H., Frost, D. J., Hoogenhout, K. and Nooi, J. R. (1973). *J. Am. Oil Chem. Soc.* **50**, 325–330.
Clermont, H. (1972). *Physiol. Végétale* **10**, 153–158.
Cole, R., Benns, G. and Proulx, P. (1974). *Biochim. biophys. Acta* **337**, 325–332.
Comes, P. and Kleinig, H. (1973). *Biochim. biophys. Acta* **316**, 13–18.
Constantopoulos, G. and Kenyon, C. N. (1968). *Pl. Physiol., Lancaster* **43**, 531–536.
Critchley, C. and Heinz, E. (1973). *Biochim. biophys. Acta* **326**, 184–193.
Dalgarno, L. and Birt, L. M. (1963). *Biochem. J.* **87**, 586–596.
Dawson, R. M. C. (1967). *Biochem. J.* **102**, 205–210.
Dawson, R. M. C. (1973). *In* "Form and Function of Phospholipids" (G. B. Ansell, J. N. Hawthorne and R. M. C. Dawson, eds), pp. 97–116. Elsevier, Amsterdam.
Davenport, H. B. (1975). Personal communication.
Davidoff, F. and Korn, E. D. (1964). *J. biol. Chem.* **239**, 2496–2506.
Desnuelle, P. and Savary, P. (1963). *J. Lipid Res.* **4**, 369–384.
Dharmalingam, K. and Jayaraman, J. (1971). *Biochem. biophys. Res. Commun.* **45**, 1115–1118.
Douce, R., Faure, M. and Marechal, J. (1966). *C.r. hebd. Séanc. Acad. Sci., Paris Sér.* C. **262**, 1549–1552.
Draper, S. R. (1969). *Phytochemistry* **8**, 1641–1647.
Drapon, R., N'Guyen Xuan Anh, N., Launay, B. and Guilbot, A. (1969). *Cereal Chem.* **46**, 647–655.
Egmond, M. R., Vliegenthart, J. F. G. and Boldingh, J. (1972). *Biochem. biophys. Res. Commun.* **48**, 1055–1060.
Eriksson, C. E. and Svensson, S. G. (1970). *Biochim. biophys. Acta* **198**, 449–459.
Esselman, W. J. and Clagett, C. O. (1974). *J. Lipid Res.* **15**, 173–178.
Ferguson, C. H. R. and Simon, E. W. (1973). *J. exp. Bot.* **24**, 307–316.
Finazzi-Agro, A., Giovagnoli, C., DeSole, P., Calabrese, L., Rotilio, G. and Mondovi, B. (1972). *FEBS Letters* **21**, 183–185.
Finazzi-Agro, A., Avigliano, L., Veldink, G. A., Vliegenthart, J. F. G. and Boldingh, J. (1973). *Biochim. biophys. Acta* **326**, 462–470.
Firn, R. D. and Friend, J. (1972). *Planta* **103**, 263–266.
Florkin, M. and Stotz, E. H. (1973). *In* "Comprehensive Biochemistry (Enzyme Nomenclature)" (M. Florkin and E. H. Stotz, eds), Vol. 13 (3rd edition), p. 193. Elsevier, Amsterdam.
Friedlander, M. and Neumann, J. (1968). *Pl. Physiol., Lancaster* **43**, 1249–1254.
Friend, J. (1974). Personal communication.
Funatsu, M., Aizono, Y., Hayashi, K., Watanabe, M. and Eto, M. (1971). *Agric. biol. Chem.* **35**, 734–742.
Galliard, T. (1968). *Phytochemistry* **7**, 1915–1922.
Galliard, T. (1970). *Phytochemistry* **9**, 1725–1734.
Galliard, T. (1971a). *Biochem. J.* **121**, 379–390.
Galliard, T. (1971b). *Eur. J. Biochem.* **21**, 90–98.
Galliard, T. (1973). *In* "Form and Function of Phospholipids" (G. B. Ansell, J. N. Hawthorne and R. M. C. Dawson, eds), pp. 253–288. Elsevier, Amsterdam.
Galliard, T. (1974a). *In* "Methods in Enzymology" (S. P. Colowick and N. O.

Kaplan, eds), Vol. 31, "Biomembranes" (S. Flescher and L. Packer, eds), pp. 520–528. Academic Press, New York and London.

Galliard, T. (1974b). *In* "Recent Advances in Phytochemistry" (V. C. Runeckles and E. E. Conn, eds), Vol. 8, pp. 209–241. Academic Press, New York and London.

Galliard, T. and Dennis, S. (1974a). *Phytochemistry* **13**, 1731–1735.

Galliard, T. and Dennis, S. (1974b). *Phytochemistry* **13**, 2463–2468.

Galliard, T. and Matthew, J. A. (1973). *J. Sci. Fd Agric.* **24**, 623–627.

Galliard, T. and Matthew, J. A. (1975). *Biochim. biophys. Acta* (in press).

Galliard, T. and Phillips, D. R. (1971). *Biochem. J.* **124**, 431–438.

Galliard, T. and Phillips, D. R. (1972). *Biochem. J.* **129**, 743–753.

Galliard, T., Phillips, D. R. and Matthew, J. A. (1975). In preparation.

Galliard, T. and Stumpf, P. K. (1966). *J. biol. Chem.* **241**, 5806–5812.

Galliard, T., Hulme, A. C., Rhodes, M. J. C. and Wooltorton, L. S. C. (1968). *FEBS Letters* **1**, 283–286.

Galliard, T., Phillips, D. R. and Frost, D. J. (1973). *Chem. Phys. Lipids* **11**, 173–180.

Galliard, T., Wardale, D. A. and Matthew, J. A. (1974). *Biochem. J.* **138**, 23–31.

Gardner, H. W. (1970). *J. Lipid Res.* **11**, 311–321.

Gardner, H. W. (1975). *J. Agric. Fd. Chem.* (in press).

Gardner, H. W. and Weisleder, D. (1970). *Lipids* **5**, 678–683.

Gardner, H. W., Christianson, D. D. and Kleiman, R. (1973). *Lipids* **8**, 271–276.

Gardner, H. W., Kleiman, R. and Weisleder, D. (1974). *Lipids* **9**, 696–706.

Garssen, G. J., Vliegenthart, J. F. G. and Boldingh, J. (1971). *Biochem. J.* **122**, 327–332.

Garssen, G. J., Vliegenthart, J. F. G. and Boldingh, J. (1972). *Biochem. J.* **130**, 435–442.

Graveland, A. (1970a). *J. Am. Oil Chem. Soc.* **47**, 352–361.

Graveland, A. (1970b). *Biochem. biophys. Res. Commun.* **41**, 427–434.

Graveland, A. (1973a). *Lipids* **8**, 606–611.

Graveland, A. (1973b). *Lipids* **8**, 599–605.

Graveland, A., Pegman, L. and Erde, P. van (1972). *Tech. Quart. Master Brew. Ass. Amer.* **9**, 98–104.

de Groot, J. J. M. C., Garssen, G. J., Vliegenthart, J. F. G. and Boldingh, J. (1973). *Biochim. biophys. Acta* **326**, 279–284.

Grosch, W. (1972). *Fette, Seifen, Anstrichmittel* **74**, 375–381.

Grosch, W. and Schwarz, J. M. (1971). *Lipids* **6**, 351–352.

Grosch, W., Schwarz, J. and Schormüller, J. (1971). *Fette, Seifen, Anstrichmittel* **73**, 433–437.

Grosch, W., Höxer, B., Stan, H. J. and Schormüller, J. (1972). *Fette, Seifen, Anstrichmittel* **74**, 16–20.

Grossman, S., Ben-Aziz, A., Ascarelli, I. and Budowski, P. (1972a). *Phytochemistry* **11**, 509–514.

Grossman, S., Trop, M., Avtalion, R. and Pinsky, A. (1972b). *Lipids* **7**, 467–473.

Gunstone, F. D. (1966). *Chemy Ind.* 1551–1554.

Guss, P. L., Macko, V., Richardson, T. and Stahmann, M. A. (1968). *Pl. Cell Physiol.* **9**, 415–422.

Häcker, M. and Stöhr, H. (1966). *Planta* **68**, 215–224.

Hamberg, M. (1971). *Analyt. Biochem.* **43**, 515–526.

Hamberg, M. and Samuelsson, B. (1965). *Biochem. biophys. Res. Commun.* **21**, 531–536.

Hamberg, M. and Samuelsson, B. (1967). *J. biol. Chem.* **242**, 5329–5335.

Haverkate, F. and van Deenen, L. L. M. (1965). *Biochim. biophys. Acta* **106**, 78–92.

Haydar, M. and Hadziyev, D. (1973). *J. Sci. Fd Agric.* **24**, 1039–1053.

Heimann, W. and Schreier, P. (1970). *Helv. chim. Acta* **53**, 2296–2297.
Heimann, W. and Schreier, P. (1971). *Helv. chim. Acta* **54**, 2794–2803.
Heimann, W., Reinartz, F. and Schreier, P. (1972). *Helv. chim. Acta* **55**, 2257–2262.
Heimann, W., Dresen, P. and Klaiber, V. (1973a). *Z. Lebensm Unters.-Forsch.* **153**, 1–5.
Heimann, W., Dresden, P. and Schreier, P. (1973b). *Z. Lebensm. Unters.-Forsch.* **152**, 147–151.
Heinen, W. and Brand, I. van der (1963). *Z. Naturforsch.* **18b**, 67–79.
Heinz, E. (1967). *Biochim. biophys. Acta* **144**, 333–343.
Heinz, E. (1973). *Z. Pflanzenphysiol.* **69**, 359–376.
Helmsing, P. J. (1967). *Biochim. biophys. Acta* **144**, 470–472.
Helmsing, P. J. (1969). *Biochim. biophys. Acta* **178**, 519–533.
Hitchcock, C. and Nichols, B. W. (1971). "Plant Lipid Biochemistry". Academic Press, London and New York.
Hofstee, B. H. J. (1960). *In* "The Enzymes" (P. D. Boyer, H. Lardy and K. Myrbäck, eds), Vol. 4, p. 485. Academic Press, New York and London.
Holden, M. (1970). *Phytochemistry* **9**, 507–512.
Huang, K. P. and Stumpf, P. K. (1971). *Archs Biochem. Biophys.* **143**, 412–427.
Hutton, D. and Stumpf, P. K. (1971). *Archs Biochem. Biophys.* **142**, 48–60.
Jensen, R. G. (1971). *Progr. Chem. Fats, Lipids* **11**, 349–394.
Kalra, S. K. and Brooks, J. L. (1973). *Phytochemistry* **12**, 487–492.
Kates, M. (1955). *Can. J. Biochem. Physiol.* **33**, 575–589.
Kies, M. W., Haining, J. L., Pistorius, E., Schroeder, D. H. and Axelrod, B. (1969). *Biochem. Biophys. Res. Commun.* **36**, 312–315.
Kolattukudy, P. E. (1970). *Biochem. Biophys. Res. Commun.* **41**, 299–305.
Kolattukudy, P. E. and Walton, T. J. (1972). *Progr. Chem. Fats, Lipids* **13**, 119–175.
Kolattukudy, P. E., Walton, T. J. and Kushwaha, R. P. S. (1973). *Biochemistry* **12**, 4488–4498.
Krewson, C. F., Ard, J. S. and Riemenschneider, R. W. (1962). *J. Am. Oil Chem. Soc.* **39**, 334–340.
Laties, G. G., Hoelle, C. and Jacobson, B. S. (1972). *Phytochemistry* **11**, 3403–3411.
Lebault, J. M., Lode, E. T. and Coon, M. J. (1971). *Biochem. biophys. Res. Commun.* **42**, 413–419.
Leu, K. and Eriksson, C. (1972). Communication to 11th World Congress of the International Society for Fat Research, Göteborg, Sweden.
Lieberman, M. and Mapson, L. W. (1964). *Nature, Lond.* **204**, 343–345.
Lipetz, J. (1970). *Int. Rev. Cytol.*, **27**, 1–28.
Lu, A. Y. H., Junk, K. W. and Coon, M. J. (1969). *J. biol. Chem.* **244**, 3714–3721.
Major, R. T. and Thomas, M. (1972). *Phytochemistry* **11**, 611–617.
Mapson, L. W. and Moustafa, E. M. (1955). *Biochem. J.* **60**, 71–80.
Markovetz, A. J., Stumpf, P. K. and Hammarström, S. (1972). *Lipids* **7**, 159–164.
Martin, R. O. and Stumpf, P. K. (1959). *J. biol. Chem.* **234**, 2548–2554.
Mazliak, P. (1973). *A. Rev. Pl. Physiol.* **24**, 287–310.
McCarty, R. E. and Jagendorf, A. T. (1965). *Pl. Physiol., Lancaster* **40**, 725–735.
Morris, L. J., Hall, S. W. and James, A. T. (1966). *Biochem. J.* **100**, 29c–30c.
Muto, S. and Beevers, H. (1974). *Pl. Physiol., Lancaster* **54**, 23–28.
Nagami, K. (1973). *Biochem. biophys. Res. Commun.* **51**, 364–369.
Noma, A. and Borgström, B. (1971). *Biochim. biophys. Acta* **227**, 106–115.
Nyman, B. (1965). *Physiologia Pl.* **18**, 1085–1094.
Oelze-Karow, H., Schopfer, P. and Mohr, H. (1970). *Proc. natn. Acad. Sci. U.S.A.* **65**, 51–57.
Okui, S., Uchiyama, M., Mizugaki, M. and Sugawara, A. (1963). *Biochim. biophys. Acta* **70**, 346–348.

Olney, C. E., Jensen, R. G., Sampugna, J. and Quinn, J. G. (1968). *Lipids* **3**, 498–502.

Orthoefer, F. T. and Dugan, L. R. (1973). *J. Sci. Fd Agric.* **24**, 357–365.

Ory, R. L. (1969). *Lipids* **4**, 177–185.

Ory, R. L., St. Angelo, A. J. and Altschul, A. M. (1960). *J. Lipid Res.* **1**, 208–213.

Ory, R. L., Yatsu, L. Y. and Kircher, H. W. (1968). *Archs Biochem. Biophys.* **123**, 255–264.

Ory, R. L., Kiser, J., Pradel, P. A. (1969). *Lipids* **4**, 261–264.

Ory, R. L. and St. Angelo, A. J. (1971). *Lipids* **6**, 54–57.

Oursel, A., Lamant, A., Salsac, L. and Mazliak, P. (1973). *Phytochemistry* **12**, 1865–1874.

Panter, R. A. and Mudd, J. B. (1969). *FEBS Letters* **5**, 169–170.

Pederson, T. C. and Aust, S. D. (1973). *Biochem. biophys. Res. Commun.* **52**, 1071–1078.

Phillips, B. E., Smith, C. R. and Tjarks, L. W. (1970). *Biochim. biophys. Acta* **210** 353–359.

Pinsky, A., Grossman, S. and Trop, M. (1971). *J. Fd Sci.* **36**, 571–572.

Pistorius, E. K. and Axelrod, B. (1974). *J. biol. Chem.* **294**, 3183–3186.

Quarles, R. H. and Dawson, R. M. C. (1969). *Biochem. J.* **112**, 787–794.

Rebman, von H. and Acker, L. (1973). *Fette, Seifen, Anstrichmittel* **75**, 409–411.

Roza, M. and Francke, A. (1973a). *Biochim. biophys. Acta* **316**, 76–82.

Roza, M. and Francke, A. (1973b). *Biochim. biophys. Acta* **327**, 24–31.

St. Angelo, A. J. and Altschul, A. M. (1964). *Pl. Physiol., Lancaster* **39**, 880–883.

St. Angelo, A. J. and Ory, R. L. (1972). *In* "Seed Proteins" (G. E. Inglett, ed.), pp. 284–291. AVI Publishing Co., Westport, Conn., U.S.A.

Sastry, P. S. and Kates, M. (1964). *Biochemistry* **3**, 1280–1287.

Schantz, R., Douce, R. and Duranton, H. M. (1972). *FEBS Letters* **20**, 157–161.

Schormüller, J., Weber, J., Höxer, B. and Grosch, W. (1968). *Z. Lebensm. Unters.-Forsch.* **139**, 357–364.

Shine, W. E. and Stumpf, P. K. (1974). *Archs Biochem. Biophys.* **162**, 147–157.

Smith, W. L. and Lands, W. E. M. (1972). *J. Biol. Chem.* **247**, 1038–1047.

Soucek, A., Michalec, C., Souckova, A. (1971). *Biochem. biophys. Acta* **227**, 116–128.

Stanacev, N. Z., Stuhne-Sekalec, L. and Domazet, Z. (1973). *Can. J. Biochem.* **51**, 747–753.

Stauffer, C. E. and Glass, R. L. (1966). *Cereal Chem.* **43**, 644–657.

Stumpf, P. K. (1970). *In* "Lipid Metabolism" (S. J. Wakil, ed.), pp. 79–106. Academic Press, New York and London.

Tappel, A. L. (1963). *In* "The Enzymes" (P. D. Boyer, H. Lardy and K. Myrbäck, eds), Vol. 8, pp. 275–285. Academic Press, New York and London.

Tavener, R. J. A. and Laidman, D. L. (1972). *Phytochemistry* **11**, 981–987, 989–997.

Theorell, H., Holman, R. T. and Åkeson, Å. (1947). *Acta chem. scand.* **1**, 571–576.

Tressl, R. and Drawert, F. (1973). *J. Agric. Fd Chem.* **21**, 560–565.

Tzur, R. and Shapiro, B. (1972). *Biochim. biophys. Acta* **280**, 290–296.

Veldink, G. A., Vliengenthart, J. F. G. and Boldingh, J. (1970a). *Biochem. J.* **120**, 55–60.

Veldink, G. A., Vliegenthart, J. F. G. and Boldingh, J. (1970b). *FEBS Letters* **7**, 188–190.

Veldink, G. A., Garssen, G. J., Vliengenthart, J. F. G. and Boldingh, J. (1972). *Biochem. biophys. Res. Commun.* **47**, 22–26.

Verhue, W. M. and Francke, A. (1972). *Biochim. biophys. Acta* **285**, 43–53.

Wakil, S. J. (1970). *In* "Lipid Metabolism" (S. J. Wakil, ed.), pp. 1–48. Academic Press, New York and London.

Weber, F., Arens, D. and Grosch, W. (1973a). *Z. Lebensm. Unters.-Forsch.* **152**, 152–154.

Weber, F., Laskawy, G. and Grosch, W. (1973*b*). *Z, Lebensm. Unters.-Forsch.* **152**, 324–331.

Williams, P. M. and Bowden, B. N. (1973). *Phytochemistry* **12**, 2821–2827.

Wooltorton, L. S. C., Jones, J. D. and Hulme, A. C. (1965). *Nature, Lond.* **207**, 999–1000.

Yagi, T. and Benson, A. A. (1962). *Biochim. biophys. Acta* **57**, 601–603.

Yang, S. F., Freer, S. and Benson, A. A. (1967). *J. biol. Chem.* **242**, 477–484.

Yamada, M. (1957). *Sci. Pap. Coll. Educ., Univ. Tokyo* **7**, 97–104.

Zimmerman, D. C. and Vick, B. A. (1973). *Lipids* **8**, 264–266.

Zimmerman, D. C. (1966). *Biochem. biophys. Res. Commun.* **23**, 398–402.

Zimmerman, D. C. and Vick, B. A. (1970). *Pl. Physiol., Lancaster* **46**, 445–453.

Zimmerman, D. C., Vick, B. A. and Borg, T. K. (1974). *Pl. Physiol., Lancaster* **53**, 1–4.

Zimmerman, R., Flohe, L., Wester, U. and Hartmann, H. J. (1973). *FEBS Letters* **29**, 117–120.

CHAPTER 12

Role of Lipids in Water and Ion Transport

P. J. C. KUIPER*

Laboratory of Plant Physiological Research, Agricultural University, Wageningen, The Netherlands

I. INTRODUCTION

A discussion on the role of lipids in water and ion transport in plants starts with the question of whether lipids have an essential function in water and ion transport through biological membranes. Biological membranes are characterized by their semipermeability, so that as a rule compounds with a high solubility in lipid solvents penetrate the cell at a faster rate than compounds with a low solubility. However, as Ling (1973) recently pointed out, a lipid membrane *per se* is not a *bona fide* semipermeable membrane. For instance, ethanol is more soluble in lipid than is water, but it penetrates through biological membranes at a much slower rate than water. For this reason the structural aspects of membranes should be carefully discussed, including the possibility that biological membranes primarily consist of a more or less continuous double lipid layer (Davson and Danielli, 1952;

* Present address: Biological Centre, University of Groningen, Haren (Gr.), The Netherlands.

Singer and Nicolson, 1972). For an accurate description of the role of lipids in ion and water transport, the characteristics and interactions of each group of compounds involved—water, ions, membrane lipids, and membrane proteins—should be considered. The following pages deal with these problems.

II. Membrane Compounds and their Interactions

A. Water

1. *Structure of Water in the Bulk Phase*

The structure of water in the bulk phase consists of a hydrogen-bonded framework of clusters of water molecules with an occasional single water molecule in the cavities (Wall and Hornig, 1965). The formation of a single H-bond between two neighbouring water molecules leads to resonance effects, so that, once a H-bond is formed, H-bond formation spreads until a cluster of H-bonded water molecules is formed. Such clusters range in size from 91 molecules at 0°C to 38 at 40°C (Frank and Wen, 1957; Némethy and Scheraga, 1962), and a continuing decrease in cluster size with temperature occurs. All the H-bonds are broken in water vapour.

Such clusters, however, are thermodynamically very unstable; their lifetime in bulk water is only 10^{-11} s and for this reason Frank and Wen (1957) called them "flickering clusters". A most important aspect of this model is that careful statistical mechanical calculations based on it can explain the maximum density of water at 4°C (Némethy and Scheraga, 1962).

2. *Structure of Water near Interfaces*

There is a great deal of evidence for further temperature-mediated changes in the cooperative structure of water. Unlike ordinary phase changes such as melting and boiling these cooperative transitions generally require little or no free energy. They have mostly been observed in water near interfaces, and they generally occur around 15, 30, 45 and 60°C (Drost-Hansen, 1971, 1972). The surface tension–temperature curve shows a kink around 15°C (Timmermans and Bodson, 1937). The specific heat of water shows a minimum around 30°C (Ginnings and Furukawa, 1953). The disjoining pressure of water between a smooth and a slightly curved quartz plate has been studied by Peschel and Adlfinger (1967). They observed maxima in the disjoining pressure at 15, 32 and 45°C when the plates were 100 Å apart. The same authors measured the surface viscosity of the water between the two quartz plates and observed the same maxima at the above temperatures. Mechanical damping of a water-filled quartz capillary has been studied by Forslind (1966); he observed maxima in damping at 30°C and a second minor optimum at 65°C.

Diffusion of thiourea in water is also affected in such a way that irregularities in the temperature response occur at 17, 28, 43 and 60°C (Dreyer *et al.*, 1969). Again these phenomena were observed in water near interfaces: the experiments were carried out in a capillary of 80 μm diameter. Careful measurements in bulk water generally fail to show such thermal transitions. The latter are obviously limited to water near interfaces: air–water, quartz–water, and others.

Surprisingly enough, no specific surface effects on the behaviour of water have been observed so far and Drost-Hansen (1971) concludes that the main functions at the solid–water interface is that the solid phase because of its presence provides a "sink" for structuring water. However, water structure stabilization near a hydrophobic area should be different from that near a hydrophilic area. Water adjacent to an ionic solid surface is strongly dipole ordered; however, the dipole oriented ordering decreases with increasing distance from the surface until only the characteristic flickering clusters of H-bonded water are present.

The water near a hydrophobic surface will also be structured to a large extent; the number of clusters will be larger than in the bulk phase ("hydrophobic hydration water"), because the water molecules cannot make H-bonds with the surface. Size, quantity, and lifetime of the clusters will decrease as the distance from the surface increases until the average values of bulk water are observed.

Pauling (1961) has described the structure of the hydrate crystals of narcotics like xenon, chloroform, and others. The anaesthetizing (partial) pressure of such apolar anaesthetic agents is positively related to the partial pressure of their hydrate crystals, indicating that the electrical resistance of the brain cell membranes is affected by the clathrate structuring of the water near the cell membranes.

This distinction between the dipole-oriented stabilization of water near a hydrophilic/ionic surface and the cluster-oriented (clathrate-like) structure of water near non-polar surfaces may be very useful for the functioning of biomembranes in transport studies, as will be shown further on.

Lipids and the non-polar parts of the amino acids of a membrane protein may form hydrophobic areas in the membrane, while the hydrophilic areas of the membrane are localized at the polar end groups of the lipids and at the polar groups of the amino acids of the membrane proteins.

In this connection it is of great interest that the presence of hydrocarbon chains tends to increase hydration and the hydrogen bonding of water near the polar end groups. This is the case with surface active chemicals (alkyl trimethylammonium chlorides; Nash, 1965) and one may assume that the same observation holds for the polar plant lipids. Careful calculations on the tertiary structure of proteins (Némethy *et al.*, 1963) also show that hydrophobic interactions between non-polar parts of amino acids in a protein molecule increase the number and strength of the internal H-bonds between the amino acids. Also, such interactions enforce structuring of water by

H-bonding around the hydrophilic/ionic groups in the protein molecules. Perhaps this interaction between hydrophobic and hydrophilic areas in water structuring may explain why no obvious differences have as yet been observed between the role of hydrophobic and hydrophilic surfaces in structuring the vicinal water.

3. *Water–Ion Interaction*

All ions affect the structure of water. Two groups of ions are distinguished, water structure-breaking ions and water structure-ordering ions. Living cells generally discriminate between these two groups by allowing the water structure-breaking ions to enter the cell more or less freely and by excluding water structure-ordering ions (Wiggins, 1971). Wiggins attributes this selectivity of cells to the fact that the cytoplasmic water is highly structured and that it is characterized by a lower solubility of the water structure-ordering ions, which are already largely hydrated and which should compete strongly for water molecules, at the same time disrupting the structured cellular water.

The effect that an ion has on the mobility of the neighbouring water molecules can be measured as the average number of water molecules attached to the ion (hydration number), the affinity between the ion and water (free energy of hydration) and by the ion–solvent interaction, measured as viscosity (viscosity B-coefficient). Table I presents data for a few biologically interesting ions.

The viscosity B-coefficient can be calculated from the formula:

$$\eta/\eta_0 = 1+Ac^{\frac{1}{2}}+Bc$$

in which η = viscosity of the ionic solution,
η_0 = viscosity of bulk water,
c = concentration of salt,
A = constant for the anion–cation interaction,
and B = constant for the ion–solvent (= water) interaction.

Negative B-values are characteristic for structure-breaking ions and they have only been observed (calculated) for ion-(H-bonding) solvent interactions: K^+, Rb^+, Cs^+, Cl^-, Br^-.

Positive B-values are measured for structure-ordering ions and the values increase in the order Na^+, Li^+, Ca^{2+}, Mg^{2+}. Free energy of hydration and hydration number increase in the same order. Na^+ and K^+, which are chemically so closely related, show essential differences in their ion–water interactions, which obviously could be related to the striking differences in the cell membrane transport properties of these ions.

4. *Transport through Cellulose Acetate Membranes as related to Water Structures*

Water transport through cellulose acetate desalination membranes has been extensively studied by Schultz and Asunmaa (1970). The cellulose

TABLE I

Water–ion interactions measured as the number of water molecules per ion (hydration number), the free energy of hydration and the viscosity B-coefficient. Data have been taken from Helfferich (1962; hydration number), from Noyes (1963; free energy of hydration), and from Stokes and Mills (1965; viscosity B-coefficient)

Ion	Hydration number	Free energy of hydration (kcal/mol)	Viscosity B-coefficient in aqueous solutions (25°C)
Mg^{2+}	—	568	0·385
Ca^{2+}	5·2	291	0·290
Li^{+}	3·3	154	0·150
Na^{+}	1·5	76	0·086
K^{+}	0·6	42	−0·007
Rb^{+}	—	34	−0·033
Cs^{+}	0	26	−0·045
Cl^{-}	0	7	−0·007

acetate sheet consisted of an irregular array of closely packed, rounded subunits of 188 Å, giving a calculated pore radius of about 21·3 Å. Using the latter figure they determined that the viscosity of the water passing through the membrane was about 37 times higher than that of bulk water and the corresponding average cluster size, using Némethy and Scheraga's calculations (1962, 1964), was about 162 at room temperature, or twice the cluster size of bulk water at room temperature. Evidently the water in such pores is highly structured.

Ling (1973) studied the transport of water and non-electrolytes (THO, methanol, ethanol, propanol, ethyleneglycol, erythritol, xylitol, sorbitol and several sugars) through a cellulose acetate membrane and through frog skin. Permeability of these chemicals across the frog skin and across the cellulose acetate membrane ranged from 0·75 (THO) to 0·00002 (sugar) μmol/cm^2.h.mM, spanning a 10^5-fold range. A plot of the permeability of frog skin to these compounds (log scale) against that of the cellulose acetate membrane (log scale) revealed a very close linear relationship between the two permeabilities. Ling concluded that the ordered water structure in both membranes is responsible for the observed semipermeability. Lipids evidently play no role, since they are absent in the cellulose acetate membrane. Collander (1959) obviously assumed that the diffusion coefficient of all the studied non-electrolytes in the *Nitella* plasma membrane lipid bilayer was constant, an assumption which has not been tested. The important question now arises as to why nature invented the lipids as building stones for biomembranes if ordered water structure alone would be sufficient to explain semi-permeability. For this reason lipid–water interactions and transport through lipid bilayers will be discussed first.

B. LIPIDS

1. *Water–Lipid Interactions*

Phospholipids mostly melt around 200°C. Interestingly enough they have another transitional melting point, often within the physiological temperature range.

Heat absorption at this transition temperature depends on the amount of water present in the lipid preparation (Chapman and Wallach, 1968). Figure 1 shows partial melting only above 20% water, which corresponds to about 20 molecules of water per molecule of 1,2-distearoyl phosphatidylcholine. This water is most likely to be the hydration water of the glycerophosphoryl-

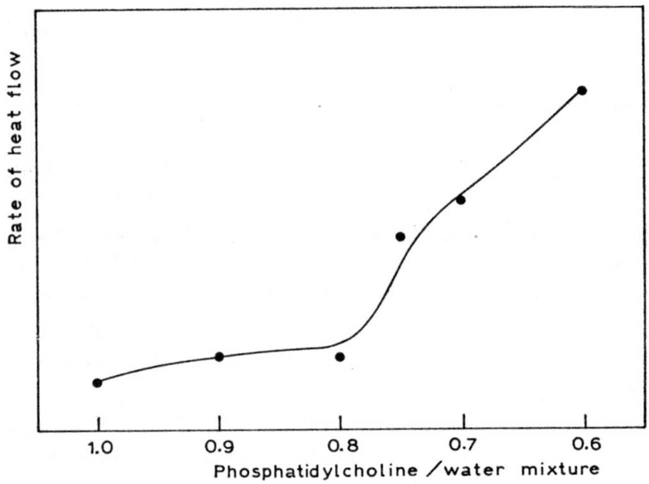

FIG. 1. Rate of heat flow of 1,2-distearoyl-phosphatidylcholine/water mixtures on a weight basis at the transition temperature of about 340°K (after Chapman and Wallach, 1968).

choline group. When more than 20% water is present in the lipid preparation, the transition melting point of the hydrophobic area of the liquid crystal becomes manifest. Evidently, a large quantity of structured water (hydrophobic hydration water in the apolar region) is a requirement for liquefying phosphatidylcholine.

Unsaturated lipids have lower transition temperatures than their saturated analogues, possibly because they break the water structure to a smaller extent.

Heat absorption at the transition temperature disappears when cholesterol is added to 1,2-dipalmitoyl phosphatidylcholine (Ladbrooke *et al.*, 1968). The transition temperature itself is also lowered by cholesterol. No significant changes in the differential scanning calorimetric curve are visible up to a

12·5% addition of cholesterol except that the thickness of the lipid layer increases from 46 to 54 Å and the structural water layer around the polar groups increases from 18·5 to 27·5 Å. Further addition of cholesterol reduces the total layer thickness (water + lipid) with a concomitant sharp decrease in heat absorption at the transition point. The heat absorption curve becomes flattened and broader. Clearly, the addition of cholesterol inhibits the co-operative phenomenon of partial melting, probably because the interaction between cholesterol and phosphatidylcholine has "melted away" the highly structured hydrophobic hydration water, required for the transitional melting. Cholesterol brings the lipid into a semi-fluid state, so that the hydrocarbon chains of the phosphatidylcholine are less mobile than in the liquid–crystal phase, but more flexible than in the rigid crystal state.

2. *Water Transport through Lipid Bilayers*

As stated before, the apolar part of a lipid molecule will induce extensive formation of hydrophobic hydration water, while enforcing H-bonding of water with the polar end-group at the same time.

The electrical resistance of black films (bimolecular lipid films) is high, viz. 10^6–10^8 Ω/cm² because of the hydrophobic area within the film (Hanai *et al.*, 1964; Goldup *et al.*, 1970).

Water permeability of black lipid films has been determined for osmotic water transport and for diffusion, using THO. The isotopic exchange permeability was generally lower than the osmotic permeability coefficient, which could be attributed to unstirred water layers around the film. With improved stirring it can be raised again (Haydon, 1968).

The osmotic permeability coefficient (*Pos*) depends on the lipid species composition of the film and on the fatty acid composition of the lipids present in the bilayers (Goldup *et al.*, 1970; Graziani and Livne, 1972). In Graziani and Livne's experiments (1972) *Pos* increased from 31 μm/s for 1,2-dipalmitoyl phosphatidylcholine to 64 μm/s for a phosphatidylcholine preparation from plants, esterified with linoleic acid (30%) and linolenic acid (36%). According to Finkelstein and Cass (1967) the addition of cholesterol to egg phosphatidylcholine decreases *Pos* from 42 μm/s (no cholesterol added) to about 10 μm/s (phosphatidylcholine:cholesterol = 1:4). It was concluded that addition of cholesterol increases the stability of the membrane and at the same time reduces the osmotic water permeability coefficient.

Table II also shows the temperature dependence of water transport through lipid films, determined as activation energy, *Ea*. Values of *Ea* are obtained as the slope from the linear curve obtained when log *Pos* is plotted against the reciprocal of the absolute temperature. In general the experimental data fall neatly on the linear curve. *Ea* was about equal for the different acyl conjugates of phosphatidylcholine and varied from 10·8 to 13·8 kcal/mol. The higher value of *Ea* observed for monogalactosyl diglyceride

TABLE II

Osmotic water permeability and activation energy of osmotic water permeability of lipid black films
(After Graziani and Livne, 1972)

Lipid species	Number of double bonds per fatty acid	*Pos* (μm/s)	*Ea* (kcal/mol)
PC synthetic	0	31	13·8
PC egg	0·73	55	13·0
PC plant	1·73	64	10·8
PC egg + cholesterol	0·73	25	13·4
MG	2·20	73	17·0
MG + sitosterol	2·20	42	17·6
MG + stigmasterol	2·20	45	17·4
MG + cholesterol	2·20	17	−36·3

PC = phosphatidylcholine; MG = monogalactosyl diglyceride

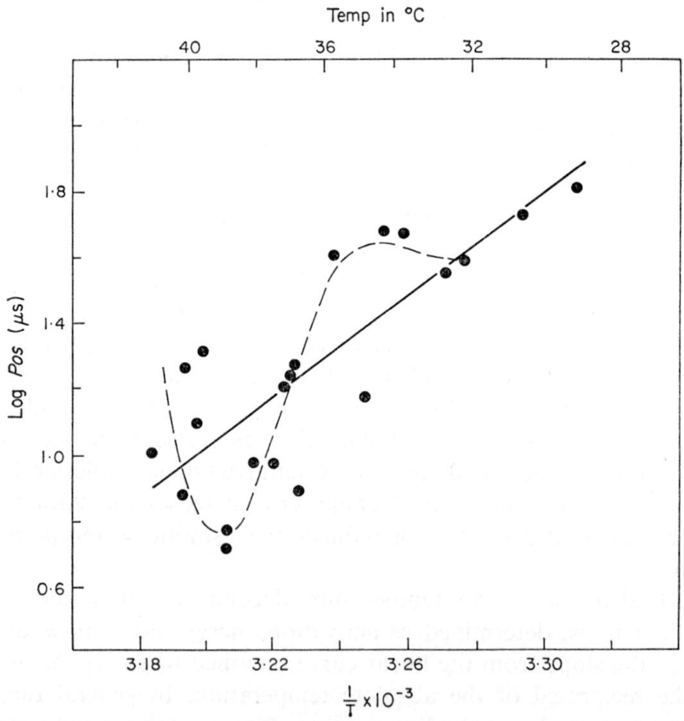

FIG. 2. Effect of temperature on the osmotic permeability of a monogalactosyl diglyceride/cholesterol black film (after Graziani and Livne, 1972). The dotted curve has been drawn to fit the experimental points as closely as possible.

films, 17 kcal/mol, may be attributed to extra water hydrogen bonded to the hydroxyl groups of the sugar moiety.

In general the addition of sterols to phosphatidylcholine or monogalactosyl diglyceride does not affect *Ea* except that the monogalactosyl diglyceride-cholesterol film is characteiized by a more or less negative temperature dependence (Fig. 2). The experimental data no longer fit a linear curve but suggest strongly the presence of the now well known kinks in water structure, which were discussed earlier.

Because the addition of stigmasterol or sitosterol does not affect *Ea* of monogalactosyl diglyceride films, it is the side chain of cholesterol which is responsible for the interesting water structure characteristics presented in Fig. 1. Only flat molecules like cholesterol can penetrate bilayers (Finean, 1953; Van den Heuvel, 1963).

3. *Hormone–Lipid Interactions*

(*a*) *Vasopressin.* Addition of the antidiuretic hormone, vasopressin, to one of the compartments which are separated by a black lipid film, increases the water permeability of the phosphatidylcholine film within 2 min (Graziani and Livne, 1971). It is a hormonal effect and it operates at very low concentrations, 10^{-10}–10^{-8} M. Unlike sterols, vasopressin does affect the activation energy for osmotic water flow through the phosphatidylcholine film lowering it from 14 to 4 kcal/mol. The effects of this hormone on *Pos* and on the temperature dependence of *Pos* are very similar for biomembranes (bladder) and for black films.

(*b*) *Abscisic acid.* The effect of the plant hormone, abscisic acid, on the water permeability of carrot discs has been studied by Glinka and Reinhold (1971). They measured osmotic water flux induced by mannitol and the diffusional flux by THO. In both types of measurement abscisic acid increased water permeability within half an hour of its application. So far no reports on the effect of this hormone on water permeability of black films are available.

Application of abscisic acid to the roots of decapitated tomato plants increased the rate of bleeding (Tal and Imber, 1971). This observation was confirmed for sunflower (Glinka, 1973). When untreated roots or abscisic acid-treated roots were placed in a mannitol solution, the bleeding rate quickly became negative, the highest rate being observed in abscisic acid-treated roots. Cram and Pitman (1973), however, observed inhibition of water flux and potassium flux (uptake) in maize roots and they concluded that abscisic acid reduced water flux because of the reduction in salt transport.

Interestingly enough, *p*-chloromercurbenzoate and carbonylcyanide *m*-chlorophenylhydrazone, supplied together with the required amount of abscisic acid, annihilated the effect of the hormone (Glinka and Reinhold, 1972). If, however, abscisic acid was first allowed to act alone, its effect was stable to subsequent treatment with these metabolic inhibitors, indicating that the increase in water permeability of the carrot cells induced by abscisic acid requires metabolic energy. In this respect the action of abscisic acid is very different from that of vasopressin, already indicated by the liposome studies.

(*c*) *Gibberellic acid.* Application of gibberellic acid induced leaking from glucose-loaded phosphatidylcholine/cholesterol liposomes. Sitosterol could replace cholesterol in the liposomes. In the absence of gibberellic acid addition of sitosterol decreased leakage of glucose from the liposomes, while in the presence of gibberellic acid no differences were observed in liposomes containing 0–30% sitosterol. It should be noted that the experiments were performed with a "crude" phosphatidylcholine preparation and that the effect of gibberellic acid concentration was only substantial with this preparation: liposomes prepared from purified lipids were rather insensitive to treatment with gibberellic acid. Evidently the presence of minor quantities of "other" lipids is important for the increase in permeability of the lipid membrane caused by gibberellic acid. Indeed gibberellic acid may influence the permeability of biomembranes by an effect on the "minor" lipids present in these membranes at physiological concentrations (Wood and Paleg, 1972).

(*d*) *Auxins.* Weigl (1969) showed that auxins like β-indole acetic acid associate strongly with phosphatidylcholine. He attributed this association to ionic interaction of the charged groups of both chemicals. In agreement with this suggestion, β-indole acetic acid has a lower affinity to phosphatidylethanolamine but α-naphthalene acetic acid binds equally well with both phospholipids (Veen, 1974). It is unlikely that the nature of the membrane lipids is important in the primary action of auxins, since both β-indole acetic acid and α-naphthalene acetic acid most probably act on the same (membrane) site and both move in the same polar transport system.

It should be noted that Vendrig (1967) isolated from *Coleus* leaves several interconvertible compounds with a very high auxin activity in the *Avena* curvature test. These compounds most likely are of a triterpenoid or steroid nature.

4. *Ion Transport through Lipid Bilayers*

(*a*) *Bacterial lipids.* Because of their acidic nature most phospholipid bilayers tend to be impermeable to cations and relatively more permeable to anions (Lakshminarayanaiah, 1969; Goldup *et al.*, 1970). The charge on the polar head-group of the lipid molecules of a black film determines the anionic or cationic selectivity of the membrane. The latter was measured by Hopfer *et al.* (1970) as the membrane potential, the latter being a function of the transmembrane salt concentration gradient (KCl, NaCl and HCl). Membranes prepared from bacterial lysyl phosphatidylglycerol have a positive charge and are very selective for anions. Membranes prepared from acidic phospholipids, phosphatidyl glycerol and diphosphatidyl glycerol, are very selective for Na^+, H^+ and K^+. Membranes made from a weakly acidic phospholipid like phosphatidylethanolamine or from a neutral lipid like diglucosyl diglyceride were only slightly cation selective. All these lipids contained very similar fatty acids. Quite clearly there is a very good correlation between the charge of the polar head-groups of the black film lipids and the ion selectivity of the black film, measured as transference numbers.

(*b*) *Animal lipids.* Identical results to those described above for microbial lipids were observed in lipids from animal origin. Similar selectivities between anions and cations had been observed earlier by Andreoli *et al.* (1967), who compared erthrocyte lipids, and by Papahadjapoulos and Bangham (1966), using liposomes. Lev *et al.* (1966) studied the selectivity of membranes made from brain lipids; selectivity followed the Hofmeister series: Li < Na < K < Rb < Cs.

(*c*) *Plant lipids.* Ion transport characteristics of lipids extracted from grape roots (Kuiper, 1968b) were studied as a measure of liquid–membrane permeability, using Schulman's model (Rosano *et al.*, 1962; Agostini and Schulman, 1965). In this model ions are transported by the lipid from one aqueous compartment (A) to another (B) through an *n*-pentanol phase under a concentration gradient (Fig. 3). The *n*-pentanol phase is continuously stirred so that diffusion across it does not limit transport of ions. No detectable amount of Cl^- was transported when *n*-pentanol without lipid was used as the non-aqueous phase. Phosphatidylcholine was then dissolved in the *n*-pentanol phase, the lipid molecules orienting themselves at the solute interfaces. Cl^- transport from compartment A to B could only be measured in the presence of a counter-anion in phase B. Pure water, HCO_3^- and CO_3^{2-} (in

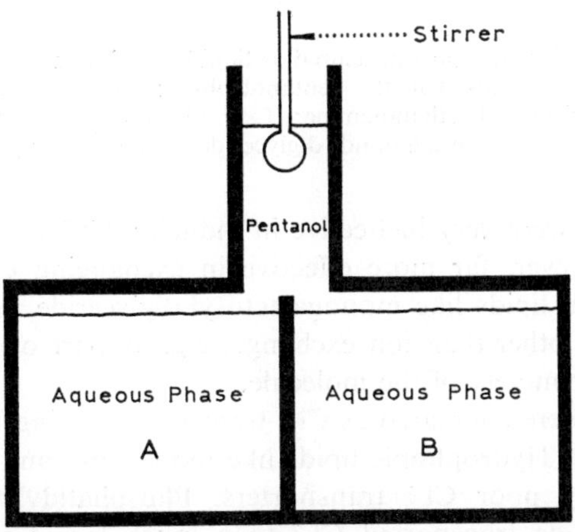

FIG. 3. Apparatus for measuring the liquid–membrane permeability of lipids.

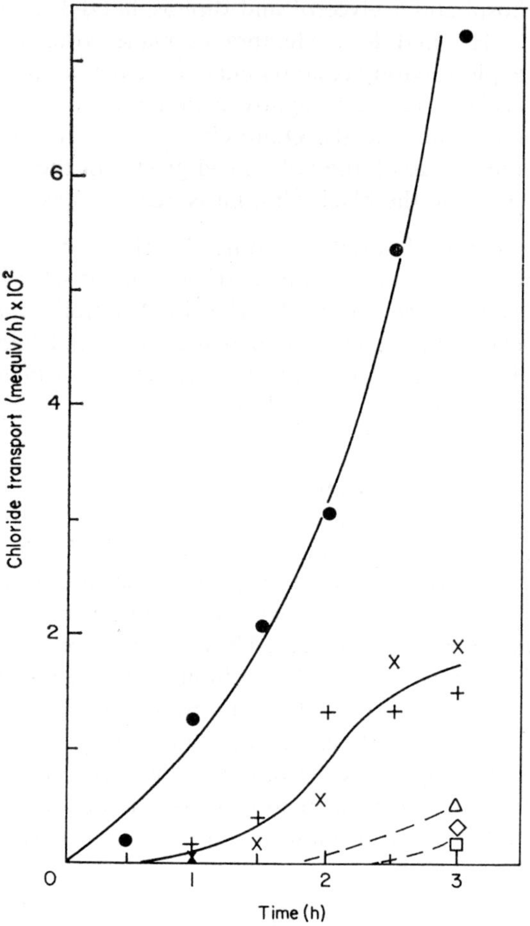

FIG. 4. Rate of chloride transport measured as liquid–membrane permeability with 10 mg of the indicated lipid dissolved in the pentanol phase: (filled circle) monogalactosyl diglyceride; (+) phosphatidylethanolamine; (×) phosphatidylcholine; (open triangle) monoglycerides; (open diamond) diglycerides; (open square) triglycerides.

compartment B) were very ineffective in inducing Cl^- transport. NO_3^- and SO_4^{2-} were, however, far more effective in exchanging Cl^-. It should be noted that neutral lipids like monogalactosyl diglyceride may transport Cl^- by a mechanism other than ion exchange, e.g. as part of the water mantle around the sugar moiety of the molecule.

Several lipids were compared as Cl^- transporters, using NO_3^- for possible exchange (Fig. 4). Hydrophobic lipids like mono-, di-, and triglycerides and sterols were very poor Cl^- transporters. Phosphatidylethanolamine and phosphatidylcholine transported Cl^- to a much larger extent but monogalactosyl diglyceride was by far the most effective Cl^--transporter studied.

The effects of phosphatidylcholine and monogalactosyl diglyceride on Na^+ transport were also compared. Here the picture was reversed: phosphatidylcholine exchanged Na^+ (against K^+ as the counter ion) to a much larger extent than monogalactosyl diglyceride. Phosphatidylcholine thus has more the properties of a cation exchanger, while monogalactosyl diglyceride has those of an anion exchanger, possibly because of the water structure-breaking property of Cl^-. In this connection it is of interest that the sugar moiety of this lipid may lead to regions of low polarity, which may facilitate transport of Cl^- across the non-aqueous phase (Gitler, 1972).

5. *Ion Pair Formation in Phosphatidylcholine Monolayers*

Shah and Schulman (1965) reported that binding of Ca^{2+} to monolayers of phosphatidylcholine depended upon the fatty acid composition of the lipid. Binding was significantly reduced by increased unsaturation of the acyl chains. This could not be attributed to the wider intermolecular spacing of the phosphate groups of the unsaturated lipids alone, since dioleoyl phosphate and dioleoyl phosphatidylcholine, which possess a similar intermolecular spacing, bind Ca^{2+} to a very different degree. Shah and Schulman (1967b) concluded that dioleoyl phosphatidylcholine possessed an internal competitive salt linkage between the phosphate and the choline group. This salt linkage weakened in more saturated phosphatidylcholine analogues, due to an increase in ionic repulsion between neighbouring phosphate groups and between neighbouring choline groups. Addition of cholesterol decreased the binding of Ca^{2+} because the resulting increase in spacing reduces the ionic repulsion of similar groups and thus strengthens the internal salt linkage. In general, cholesterol and Ca^{2+} have opposite effects on phosphatidylcholine monolayers; cholesterol liquefies and Ca^{2+} tends to solidify the monolayer (Shah and Shulman, 1967a).

6. *Protein–Lipid Interactions*

These interactions cover a wide field and have been reviewed extensively. Most work has been done with purified products from animal origin. In this chapter problems of protein–lipid interactions will be restricted to their physiological aspects, and only in so far as they are related to the functioning of membranes at the cellular level. For a general background a few reviews will be mentioned. Dawson (1968) discusses the reactions of proteins, including lipases at the phospholipid–water interface in monolayer studies. Triggle (1970) reviews lipid-dependent membrane enzyme studies, including ion-stimulated ATPases which are involved in active ion transport (see also p. 375). He also reviews experiments on solubilization and reaggregation of membrane particles. Wolman (1970) discusses lamellar and globular structures of biomembranes as influenced by ions. Homogenized myelin membrane particles are hydrophilic in concentrated NaCl and hydrophobic in concentrated $CaCl_2$. Considering relaxation times, biological membranes may exist in both phases so far as transport processes are concerned. Both ionic and hydrophobic lipid–protein interactions are extensively reviewed by Gitler (1972). Kuiper (1972) discusses the mechanism of water transport through plant cell membranes as affected by lipid–membrane protein interactions.

III. Role of Lipids in Membrane Permeability and Transport with Reference to Ecological Adaptations

A. PRINCIPLES

For an understanding of the role of lipids in biomembranes it is essential to find out which criteria a particular membrane has to fulfil. Within the cell the general functions of most organelles are well known, as is their lipid composition. Phospholipids constituted 50–56% of the total lipids in plant mitochondria compared with 90% in beef heart mitochondria (Schwertner and Biale, 1973). Galactolipids and sulpholipid are present in plant mitochondria and are absent from those of animal origin. Chloroplasts are characterized by a large quantity of galactolipid, while typical mitochondrial lipids like phosphatidylethanolamine and phosphatidylcholine are missing or only present in traces (Benson, 1971).

Plasma membranes of plant origin were isolated from oat roots by Keenan *et al.* (1973). Phospholipids as well as glycolipids were detected in this fraction, though these lipids extracted from plasma membranes were generally more saturated and richer in palmitic acid. During purification of this fraction the phospholipid/sterol ratio approached one, as is the case with plasma membranes from animal tissues (Hodges *et al.*, 1972).

One of the great differences between plants and animals is that the former lack the mobility which is so obvious in most animal species. This lack of mobility means that plants have to adapt to changes in the environmental conditions. Adaptation occurs at several levels of organization, including those of the cell and the intracellular organelle. This often means that the biomembranes involved undergo striking changes as a response to environmental stresses. In this connection changes in lipid composition related to environmental changes have often been described as part of the adaptation of the organism. In the following pages such correlations between lipid composition and environmental stresses or changes in the organism itself will be described for low temperature and freezing, for salinity, for the nutritional aspects, and for ageing and senescence. Because several membrane enzymes are lipid-dependent, the adaptation of such lipid-requiring enzymes will also be reviewed.

The role of lipids in ageing and in environmental responses can be studied in several ways. Firstly, a comparison can be made between species or varieties, which are preferably grown under identical environmental conditions. The species are placed in order of adaptation to environmental factors or stresses like salinity, low temperature, etc., and a possible correlation with the lipid composition is made. Secondly, an environmental condition can be varied and one or two species which differ in ecological adaptation to this environmental factor can be compared. A particularly useful experimental approach is to introduce an abrupt environmental change and to follow the ensuing changes in the absolute quantities of the various lipid species present as a function of time. Such observations may give an indication of what role a particular lipid species has in the structure and transport properties of the biomembrane in question, so enabling it to function properly under the changed environmental conditions.

In addition, plant species differ greatly in adaptability to specific environmental changes. Most agricultural crops and weeds can stand relatively large fluctuations in environmental changes. Other plants which can hardly tolerate certain environmental changes, are often very limited in distribution and can only survive under special, stable environmental conditions. Correlation studies on lipid composition may reveal how far this group of compounds is of importance in determining the distribution of plant life on this planet.

B. SALINITY AND LIPIDS

1. *Chloride Transport*

Kuiper (1968a) compared the lipids of the roots of five varieties of grapes, which differed markedly in the extent to which they permitted chloride accumulation in the leaves. When the roots were exposed to 25 mequiv/l Cl^- the most salt-sensitive rootstock accumulated 15 times as much chloride in the leaves as the most resistant variety. The content of monogalactosyl diglyceride of the roots was directly related to chloride transport to the leaves and varied from 9·5 to 39%. It should be noted that this lipid was the most efficient chloride transporter tested in the liquid–membrane model. The content of phosphatidylethanolamine and of phosphatidylcholine were inversely related to chloride accumulation. These contents were 13 and 21% respectively in the most resistant variety and lower values down to 4 and 6% were noted in the more sensitive varieties. It was also found that the most sensitive variety had an unusually low sterol content in the roots, 9% compared with 16–22% in the roots of the other 4 varieties. In conclusion, the charged phospholipids contributed strongly to the observed low chloride transport to the stem and leaves. The importance of long-chain saturated fatty acids in chloride transport was also noted. The amount of lignoceric acid present in the roots decreased from 12% in the most resistant variety to 1% in the most sensitive one. This fatty acid was mainly observed in the phosphatidylcholine and phosphatidylethanolamine fractions. The presence of long-chain fatty acids in phospholipids may well contribute to a high

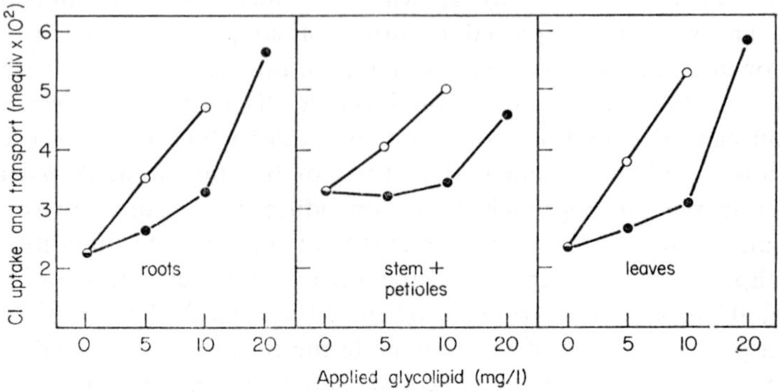

FIG. 5. Effect of monogalactosyl diglyceride (filled circle) and digalactosyl diglyceride (open circle) on Cl⁻-uptake and transport in bean plants.

stability of the membrane in question with a consequent low permeability for chloride.

In further experiments the effect of added lipid to the root environment on chloride transport was studied (Kuiper, 1969). Unfortunately the actual amounts of lipid absorbed by the plant roots were only measured qualitatively and no precise correlation between (charged) lipid composition and chloride transport is available. Galactolipids added to the root environment increased $^{36}Cl^-$-transport into the roots, stem and leaves of bean plants (Fig. 5). Digalactosyl diglyceride was more effective than monogalactosyl diglyceride.

The increased chloride transport by the galactosyl diglyceride into the stem might be explained by the increased chloride absorption of the root tissue. This also applies to chloride transport from stem into leaf tissue. This explanation, however, is improbable. When phosphatidylcholine was added to the root environment, a similar increase in chloride transport into the roots was noted. But this increase in chloride transport was not accompanied by a corresponding increase in transport into the stem and leaves. Thus transport of the glycolipid or a metabolite of it into the plant seems more probable.

The effect of galactosyl glycerol on chloride transport was also studied. No effect of this deacylated glycolipid was noted, possibly because it is not absorbed into or translocated within the plant. Therefore, transport of the intact surface-active glycolipids to the stem and leaves seems to be the most probable explanation for the observed increase in chloride accumulation in the leaves.

Phosphatidylcholine added to the root environment increased $^{36}Cl^-$-transport into the roots, but further transport into the stem and leaves was not affected. When the roots were exposed to [^{32}P]phosphatidylcholine practically no [^{32}P]phospholipid could be detected in the stem and leaves, thus explaining why $^{36}Cl^-$-transport into these tissues remained unaltered. When the deacylated product of [^{32}P]phosphatidylcholine was added to the nutrient solution the [^{32}P]phospholipid content of the roots, stem and leaves were respectively 2, 19 and 13 times higher than when the roots were exposed to [^{32}P]phosphatidylcholine itself. Thus, the addition of glycerophosphorylcholine is a more effective way of incorporating phosphatidylcholine into leaves and stem than addition of phosphatidylcholine itself. Glycerophosphorylcholine added to the root environment reduced chloride accumulation considerably, to 40% of that of the control plants at 18°C. This reduction was temperature dependent (Fig. 6). The experiment again points to the role phosphatidylcholine plays in reducing chloride transport through plasma membranes and to its function in the adaptation of the organism to a chloride-rich environment.

Interestingly, the addition of lipids to the root environment hardly affected $^{22}Na^+$-transport in plants (Kuiper, 1969). The biomembranes of plants are more anion than cation permeable, like black lipid films, indicating the usefulness of black lipid film models in the study of the ion transport properties of plant cell membranes.

2. *Transport of Na^+ and K^+*

ATPases with a specific requirement for $Na^+ + K^+$ have been isolated from beet roots (Hansson and Kylin, 1969) and from *Avicennia* leaves (Kylin and Gee, 1970). Both plants are halophytes, and the $(Na^+ + K^+)$-activated ATPase of such plants has been correlated with transport of these ions as a response to a saline environment. It was of interest to compare the lipid compositions of $(Na^+ + K^+)$-stimulated and non-specific ATPases of plant origin, because plant ATPases are known to be lipid-dependent. Removal of lipid by acetone reduced the activity of bean root ATPase to a negligible

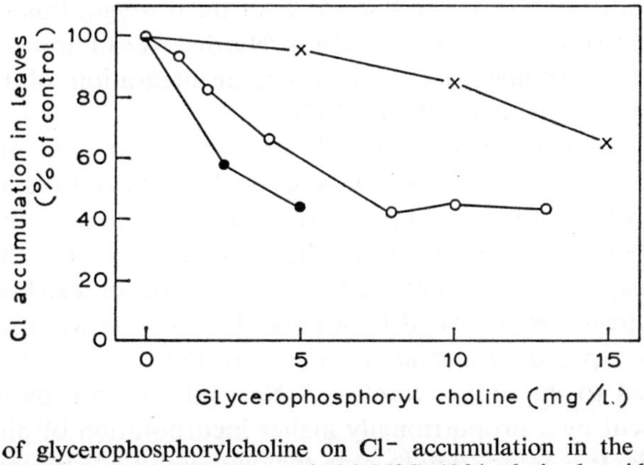

FIG. 6. Effect of glycerophosphorylcholine on Cl^- accumulation in the leaves. (Filled circle) 18°C, 50% relative humidity; (open circle) 25°C, 40% relative humidity; (×) 30°C, 35% relative humidity.

value; its activity could, however, be restored by added lipids. Only phosphatidylcholine (unsaturated analogues) and sulpholipid preparations were effective in this restoration of the enzyme activity (Kuiper, 1972).

The non-specific ATPase activity of 2500–20 000 *g* fraction of beet roots could be made specific for ($Na^+ + K^+$)-stimulation by various treatments: 0·1% deoxycholate for 1 h (Kylin *et al.*, 1972), dithiothreitol (25 mM) for 1 h, and cysteine (10 mM) for 1 h (Hansson *et al.*, 1973). Longer exposure to deoxycholate or combined treatments led to an inactivation of the enzyme preparation.

Such treatments invariably resulted in the loss of lipid from the preparation; this loss was specifically that of phosphatidylcholine, the sulpholipid content remaining unchanged. Overtreatment resulted in additional loss of sulpholipid and in inactivation of the enzyme preparation. Lipid fractionation as induced by the above treatments is important in revealing the ($Na^+ + K^+$)-stimulation, possibly because the negative charge of the remaining sulpholipid, a lipid characteristic of the ($Na^+ + K^+$)-activated preparation, may attract monovalent cations. Indeed, the particles of such induced preparations showed more negative surface charge as determined by particle electrophoresis (Karlsson *et al.*, 1971b). It is of interest that in a ($Na^+ + K^+$)-activated ATPase from an animal origin, the salt gland of the duck, the activation could also be related to the negative charge of a sulphur-containing lipid, viz. sulphatides (Karlsson *et al.*, 1971a).

3. *Lipid Responses to NaCl*

Ferguson (1966) demonstrated that total lipids and phospholipids increased when barley roots were exposed to increasing concentrations of NaCl. The phospholipid content of the roots ranged from 0·0033 to 0·0039% of fresh weight lipid-P when the root solution contained NaCl in the range 0–1·2%. The content of phosphatidylcholine, in particular, responded to the salt stress, viz. from 0·00095 to 0·00161% of fresh weight lipid-P.

In salt sensitive plant species growth is retarded by salt application and the cell membranes become leaky, indicating disintegration of the root cell membranes (Campbell and Pitman, 1971).

Stimulation of plant growth by salt stress has also been observed. The effect of salt stress on leaf lipids was studied by Twersky and Felhendler (1973) in cotton at two developmental stages. Salt stress stimulated both flowers and boll development. During the fruiting stage both the total lipids and the phospholipid content were lower in the brackish-water-treated plants. Salt stress was indicated by a lower K^+/Na^+ ratio in the leaf tissue of the plants exposed to increased salinity. In the fruiting stage the cotton plants adapted to the unfavourable K^+/Na^+ ratio caused by the brackish-water treatment by a proportionally higher incorporation of phosphate into phospholipids. It is interesting to note that successful adaptations to salinity stress as indicated by the greater proportion of flowers and fruits is accompanied by preferential incorporation of phosphate into membrane phospholipids. Clearly, structural adaptations, resulting in more stable membranes, are essential for improved plant growth under salinity stress.

C. MINERAL NUTRITION AND LIPIDS

1. K^+- and PO_4^{3-}-*Deficiency*

Research on the relationships between the nutritional status of plants and their lipid content and lipid composition has just started. Fabian (1969) noted that a temporary deficiency of phosphate caused a decrease in the phospholipids and the saponifiable lipids in the leaves and roots of sunflower. As would be expected, phospholipid content was more sensitive to phosphate deficiency. In a fen sedge, *Carex acutiformis*, a similar effect of phosphate deficiency on the phospholipid content of the chloroplasts was noted (Veerkamp and Kuiper, unpublished results).

Potassium deficiency induced a similar, though smaller decrease in the phospholipid content of roots and shoots of sunflower (Fabian, 1969). Evidently mineral nutrition and especially phosphate nutrition plays an important role in building up the biomembranes functioning in plant cells.

2. *Mg^{2+} and Ca^{2+} and Lipids*

Kylin and Kähr (1973) and Kähr and Kylin (1974) studied the effects of Mg^{2+} and Ca^{2+} on membrane ATPases of oat and wheat roots. Oats prefer acid soils and they have a high demand for Mg^{2+}. Wheat is a crop of more fertile soils and has a demand for Ca^{2+}. In agreement with these ecological requirements, the ATPase fraction from wheat roots was stimulated to a greater extent by Ca^{2+} than by Mg^{2+} (Kylin and Kähr, 1973). The reverse was true for oats roots (Fisher and Hodges, 1969; Kylin and Kähr, 1973). Mineral nutrition of the young seedlings greatly affected the response of the enzyme preparations to Mg^{2+} and Ca^{2+} and opposite effects between the two crops were observed: increasing the strength of the Hoagland solution produced roots with a higher ATPase activity in wheat, while the reverse was true for oat.

Kuiper *et al.* (1974) determined the lipid composition of the above enzyme preparations of plants grown at two nutritional levels. Specific differences between the enzymes present in the roots of the two species were evident, e.g. the lipid content of the oat root preparation was much higher than that of the wheat roots, increasing the concentration of the nutrient solution raised the Ca^{2+}-stimulated ATPase activity as well as the levels of phosphatidylcholine and of sulpholipid. Consequently both lipids could contribute to the Ca^{2+}-activated ATPase activity. The lowered Mg^{2+}-stimulated ATPase activity of oat roots grown at a higher nutritional concentration is followed by a lowered phosphatidylcholine content of the preparation, while the sulpholipid content remained the same. The results indicate that in wheat and in oat phosphatidylcholine is mainly related to the Mg^{2+}, Ca^{2+}-stimulated ATPase activities. Specific differences between wheat and oat as regard the effects of Mg^{2+} and Ca^{2+} may be related to differences in the fatty acid composition of phosphatidylcholine as recent experiments indicate (Kähr, unpublished results).

Inhibition of a Mg^{2+}-moderated, Ca^{2+}-activated microsomal preparation from muscle by phospholipase C can be reversed by addition of phosphatidylcholine or lysophosphatidylcholine with a consequent restoration of the Ca^{2+}-transport (Martonosi, 1968).

Lipids of wheat and oat roots also responded to the level of mineral nutrition (Kuiper *et al.*, 1974), just as was observed for salinity. In wheat phospholipids, phosphatidylcholine and sulpholipid increased with the level of nutrition, while the opposite was true for oat roots. In both species the level of sterols was lowered by increasing concentration of the mineral nutrients. At the same time higher proportions of sterols were incorporated into membranes.

D. ENVIRONMENTAL TEMPERATURE AND LIPIDS

1. *Saturated and Unsaturated Lipids*

Lyons and Raison (1970) observed that succinate oxidation of mitochrondria of tissues which are sensitive to chilling, such as tomato, and cucumber fruits and sweet potato roots, showed broken curves when plotted against temperature. At low temperature a high activation energy can be calculated, which changes abruptly above 9–12°C to a much lower value. The break in these Arrhenius plots can be explained as a phase change of the mitochondrial membrane lipids from a rigid-crystalline phase to a liquid-crystalline phase, as described earlier (p. 364). Such breaks in the temperature curves were absent when mitochondria from tissues which are resistant to chilling, such as cauliflower buds, potato tubers, and beet roots were studied. Mitochondria from these plants contain a larger quantity of polyunsaturated lipids than mitochondria from the plants mentioned earlier which are sensitive to chilling. The absence of a phase shift may be attributed to a lowering of the phase transition temperature due to increased unsaturation of the lipids. Mitochondrial membranes from chilling-resistant plants have a greater flexibility and a higher permeability than the corresponding membranes from chilling-sensitive plants (Lyons *et al.*, 1964), in agreement with the observed greater permeability of black films composed of unsaturated lipids (Table II). Low temperature hardiness is often related to a high membrane permeability (Akabane, 1961).

Similar differences in temperature response to those described above for mitochondria have also been noted for chloroplasts (Nobel, 1974). Osmotic responses of chloroplasts from chilling-resistant species (pea, spinach) and from chilling-sensitive species (bean, tomato) were compared. The reflection coefficient of the osmotic response to glycerol and to erthritol again showed smooth curves for the chilling-resistant plants and broken curves for the chilling-sensitive species, once more indicating a phase change in the chloroplast lipids.

In alfalfa the fatty acid composition of the roots changed during hardening and accumulation of polyunsaturated lipids was noted (Gerloff *et al.*, 1966). At the same time the hardened root cells showed a higher water permeability (Jung and Smith, 1961). Such a change in degree of lipid saturation is accomplished in alfalfa by a change in lipid composition. Lipids, like mono- and digalactosyl diglyceride, phosphatidylcholine and phosphatidylethanolamine, which are predominantly esterified with linoleic and linolenic acids increase and replace more saturated lipids like phosphatidylglycerol, phosphatidylinositol and sulpholipid (Kuiper, 1970). In short-term experiments, differences in the fatty acid composition of corresponding lipids extracted from plants growing at different environmental temperatures, is negligible. In long-term experiments changes in the fatty acid composition of the lipids are evident; in alfalfa the content of linoleic acid decreased during the growing season and the content of palmitic acid increased. This was the case for the phospholipids as well as for the sulpholipid (Klopfenstein and Shigley, 1967).

Wheat seedlings grown at 2°C were capable of developing considerable cold hardiness, which was associated with a high level of unsaturated fatty acids (de la Roche *et al.*, 1972, 1973). The phospholipid composition of seedlings grown at 2°C and at 24°C was similar, indicating that the increased unsaturation at 2°C is the result of an increased desaturation activity and not due to differences in phospholipid synthesis.

2. "*Extra-Long-Chain Fatty Acids*"

Hydrolysis of lipids, followed by methylation of the liberated fatty acids and subsequent analysis by gas–liquid chromatography sometimes reveal the presence of "extra-long-chain fatty acid methyl esters", characterized by extremely long retention times, especially when a column with a (polar) DEGS-coating is used. Such fatty acids tend to be overlooked because of their long retention times. Using DEGS-coated columns chain lengths of 25–28 carbon atoms have been extrapolated for several crude lipid fractions, obtained by DEAE-cellulose column chromatography. Such "extra-long-chain fatty acids" have been observed in the sulpholipid fraction of early-spring plants (Kuiper and Stuiver, 1972), the phosphatidylcholine fraction and sulpholipid fraction of beet roots, ATPase preparations (Kylin *et al.*, 1972), and the mono- and digalactose diglyceride fractions of pine needle chloroplasts (Bervaes *et al.*, 1972). Kuiper and Stuiver (1972) tentatively concluded that cyclopropane rings were present in these fatty acids, because mild hydrogenolysis was ineffective, and bromination resulted in disappearance of the peak in the gas chromatogram. Also, the infrared absorption spectrum showed an absorption peak at 1020 cm^{-1}. Bervaes *et al.* (1972) showed that upon dehardening pine trees such "extra-long-chain fatty acids" in the chloroplasts moved from the digalactosyl diglyceride fraction to the monogalactosyl diglyceride fraction, as collected by DEAE-cellulose column chromatography. Purification of these compounds from the above fractions by thin-layer chromatography, followed by n.m.r. and mass spectrometry indicated that their chemical structure was that of a monomethylester of a cyclic diterpene dicarboxylic acid, pinifolic acid (Bervaes, unpublished experiments). Interconversion of the two forms takes place by internal rearrangement of the molecule by ring closing and opening.

Clearly, more interesting lipids may show up in studies of plants with unusual adaptations. The study of the function of such compounds in the organelles of plants like snowdrop and pine, which are metabolically active at very low environmental temperatures, offers quite a challenge.

3. *Effect of Heat Shock and Senescence*

Tobacco leaf discs exposed for 2 min to a heat shock of 47·5°C showed an increased rate of Cl^--efflux from the discs to the medium (Benzioni and Itai, 1973). This effect was greatly reduced if the discs had been pretreated for 90 min with kinetin before exposure to the heat shock. Such a heat shock greatly reduced ^{32}P incorporation into phospholipids, which could be due to a reduced membrane integrity. The main [^{32}P]phospholipid synthesized was phosphatidylcholine.

Ferguson and Simon (1973) determined the disappearance of lipids from cucumber cotyledons. They found that galactolipids and phosphatidylglycerol have already disappeared by the time the cotyledons have started yellowing, while the major phospholipid, phosphatidylcholine, begins to disappear when the cotyledons have reached their maximal fresh weight. When the tissue starts to lose water rapidly and the cotyledons are demonstrably leaky to electrolytes when submerged in water, 56% of the phosphatidylcholine has already been lost and the levels of the other phospholipids are beginning to decrease. Clearly, the integrity of the plasma membranes strongly depends upon the content of phosphatidylcholine, as was the case in salinity stress. It will be of great interest to study the possible effects of kinetin on the synthesis and breakdown of phosphatidylcholine since this hormone is known to delay the senescence of leaves.

E. STEROLS AND PERMEABILITY

1. *Microorganisms*

Gottlieb *et al.* (1958, 1960) reported that the polyene antibiotic filipin inhibits the growth of many fungi. The plasma membrane is disrupted and ions, sugars, and amino acids leak to the medium. These phenomena can be reversed by addition of cholesterol, ergosterol, sitosterol, or stigmasterol. Sterols with a long side chain were generally most effective in this respect whilst steroids, with the exception of androstane-3β-ol-17-one, were ineffective. Prior growth on a cholesterol-containing medium did not change the susceptibility of yeast to filipin. Evidently this compound complexes with the sterols present in the plasma membrane, and thus makes it leaky.

Olson (1971a,b,c) reported similar effects to those of filipin for the triterpene glycosides, aescin and avenacin. Application of these compounds to the culture medium of *Ophiobolus graminis* and *Neurospora crassa* induced loss of ions and organic compounds. The triterpene glycosides also form complexes with sterols in the plasma membranes. The growth of various fungi whose mycelial sterol content was between 0 and 7 mg/g dry weight, was completely inhibited by concentrations of aescin in the culture medium between >100 and 2 mg/l; the higher the sterol content of the fungi, the more sensitive the fungal cell was for aescin (Olson, 1973). Ca^{2+} reduced the inhibitory effect of the aescin, because it decreased the binding of aescin to the plasma membrane sterols. In this connection it is of interest that an abnormally high cholesterol/phospholipid ratio in a Ca^{2+}-stimulated ATPase preparation from the sarcoplasmic reticulum of a cardiomyopathic hamster is related to a greatly reduced Ca^{2+}-uptake (Owens, *et al.*, 1973); again the interaction between Ca^{2+} and sterol interferes with the functioning of the membrane system.

The addition of plant sterols to the medium affords protection to the cell membrane of the ciliate protozoan, *Tetrahymena pyriformis* against low temperature damage (Reid *et al.*, 1969). Since these sterols were known to protect against the toxicity of triparanol, an inhibitor of cholesterol biosynthesis, it was thought that triparanol might enhance cold-induced injury. This was found to be so; triparanol was more toxic at 0–5°C than at 28°C. Cholesterol, sitosterol and stigmasterol proved to be active as survival factors at low temperatures, possibly because they annulled the toxicity of triparanol.

De Kruyff *et al.*, (1972) studied the effect of incorporated cholesterol upon glycerol and erythritol permeability of the plasma membrane of *Acholeplasma laidlawii*. Cholesterol, incorporated into the growth medium as an ethanolic solution (final concn in medium, 25 mg/l) or in the form of liposomes (final concn in medium, 15 mg/l), was shown to be incorporated into the plasma membrane of *A. laidlawii*. The resulting cholesterol-rich membrane showed reduced permeability to glycerol and erythritol as compared to that of cells grown on a cholesterol-free medium. Identical results were obtained when the permeability of liposomes prepared from the lipids isolated from cells grown on the cholesterol-containing medium was compared with that of liposomes prepared from the lipids of cells grown on a cholesterol-free medium. In intact-membrane studies, incorporated cholesterol reduced the heat absorption at the transition temperature of the lipid phase change, as was observed by Ladbrooke *et al.*, (1968) in phosphatidylcholine liposomes.

2. *Higher Plants*

Grunwald (1968) studied the effect of sterols on the permeability of methanol-treated red beet slices. Methanol (10%) greatly increased the betacyanin efflux from such discs, but very low concentrations of plant sterols (1–10 μM) reduced this effect considerably. Cholesterol was most effective, followed by stigmasterol and sitosterol.

Ergosterol (1 μM), however, stimulated the methanol-induced betacyanin efflux, indicating that these membrane stability effects are very specific with respect to the configuration of the added sterols. In a subsequent paper Grunwald (1971) compared the protecting effects of free sterols, sterol esters, and sterol glycosides on ethanol-induced leakage of electrolytes from barley roots. Sterol esters and sterol glycosides proved to be inactive. Apparently only free sterols are involved in controlling membrane stability and membrane permeability, indicating the importance of the hydroxyl group at C-3 of the nucleus.

F. PERMEABILITY CHANGES INDUCED BY OXIDATION

Ozone-induced changes in membrane permeability are widely recognized as important factors in ozone injury to plants. Application of ozone resulted in an increase in the lysis of *E. coli* cells (Scott and Lesher, 1963), an increase in the ion permeability of *Citrus* leaf cells (Dugger and Palmer, 1969) and in an increase in water and solute leakage from bena leaf cells (Evans and Ting, 1973). Interestingly, in the latter case the water permeability coefficients of both diffusion and osmotic transport, L_d and L_p decreased, and Evans and Ting suggest that this is the result of the increased solute permeability.

The unsaturated lipids of erythrocytes are oxidized upon exposure to ozone (Goldstein and Balchen, 1967). The biosynthesis of chloroplast galactolipids is also strongly suppressed by ozone (Mudd *et al.*, 1971), but the primary effect of ozone is most probably a direct reaction with the double bonds of the polyunsaturated lipids. Peroxidation of oleic and linoleic acids in brain phospholipid membrane films greatly increased the electrical conductivity of these membranes and made them leaky to electrolytes, particularly OH^- (Antonov *et al.*, 1973). The exact mechanism of the action of ozone on cell membranes and specifically on permeability changes, however, remains unclear because ozone also reacts with the sulphydryl groups of membrane proteins.

Similar permeability changes have also been noticed in experiments on photo-oxidation (Dworkin, 1958; Dodge, 1971). Photo-oxidative damage of the chloroplasts of *Cucumis* leaves was observed at low temperatures (1°C). The chloroplast envelope ruptured and vesicles formed in the thylakoids

(Van Hasselt, 1974a). Under the same conditions *Cucumis* leaf discs showed rapid degradation of linolenic acid: it decreased from 73 to 46% in 2 days of light at 1°C (Van Hasselt, 1974b). The linolenic acid content of the dark control was not affected. Blue light was especially effective in the photo-oxidation of unsaturated fatty acids, indicating that carotenoids, in addition to chlorophyll, are involved in the process. The actual mechanism by which absorbed light initiates the oxidation of unsaturated fatty acids, however, is unknown and could be caused by free radical formation or by excited singlet oxygen, or a combination of both (Van Hasselt, 1974b).

REFERENCES

Agostini, A. M. and Schulman, J. H. (1965). *In* "Surface Activity and the Microbial Cell". Soc. Chem. Ind., London, 37–58.

Akabane, N. (1961). Rep. Hokkaido Prefectural Agric. exp. Stn, No. 9, p. 47.

Andreoli, T. E., Bangham, J. A. and Tosteson, D. C. (1967). *J. gen. Physiol.* **50**, 1729–1750.

Antonov, V. F., Vladimirov, Yu. A., Rossels, A. N., Korkina, L. G., Korepanova, E. A. and Trukhmanova, K. J. (1973). *Biofizika* **18**, 668–673.

Benson, A. A. (1971). *In* "Structure and Function of Chloroplasts", pp. 130–148. Springer, New York.

Benzioni, A. and Itai, C. (1973). *Physiologia Pl.* **28**, 493–497.

Bervaes, J. C. A. M., Kuiper, P. J. C., and Kylin, A. (1972). *Physiologia Pl.* **27**, 231–235.

Campbell, L. C. and Pitman, M. G. (1971). *In* "Salinity and Water Use" (T. Talma and J. R. Philip, eds), pp. 207–226. Macmillan, London.

Chapman, D. and Wallach, D. F. H. (1968). *In* "Biological Membranes, Physical Fact and Function" (D. Chapman, ed.), pp. 125–202. Academic Press, New York and London.

Collander, R. (1959). *In* "Plant Physiology" (F. C. Steward, ed.), Vol. II, pp. 3–102. Academic Press, New York and London.

Cram, W. J. and Pitman, M. G. (1972). *Aust. J. biol. Sci.* **25**, 1125–1132.

Davson, H. and Danielli, J. F. (1952). *In* "The Permeability of Natural Membranes" (2nd edition). Cambridge University Press, Cambridge.

Dawson, R. M. C. (1968). *In* "Biological Membranes, Physical Fact and Function" (D. Chapman, ed.), pp. 203–231. Academic Press, New York and London.

Dodge, A. D. (1971). *Endeavour* **30**, 130–135.

Dreyer, G., Kahrig, E., Kirstein, D., Erpenbeck, J. and Lange, F. (1969). *Naturwissenschaften* **56**, 538–559.

Drost-Hansen, W. (1971). *In* "Chemistry of the Cell Interface" (H. D. Brown, ed.), Part B, pp. 1–184. Academic Press, New York and London.

Drost-Hansen, W. (1972). *In* "Proc. XV Solvay Conference on Electrostatic Interactions and the Structure of Water", Institut International de Chemie Solvay, Brussels.

Dugger, W. M. and Palmer, R. L. (1969). *Proc. 1st Int. Citrus Symp.* **2**, 711–715.

Dworkin, M. (1958). *J. gen. Physiol.* **41**, 1099–1112.

Evans, L. S. and Ting, I. P. (1973). *Am. J. Bot.* **60**, 155–162.

Fabian, I. (1969). *Revue roum. biol.* **14**, 293–299.

Ferguson, W. S. (1966). *Can. J. Pl. Sci.* **46**, 639–646.

Ferguson, C. H. R. and Simon, F. W. (1973). *J. exp. Bot.* **24**, 307–316.
Finean, J. B. (1953). *Experientia* **9**, 17–19.
Finkelstein, A. and Cass, A. (1967). *Nature, Lond.* **216**, 717–718.
Fisher, J. and Hodges, T. K. (1969). *Pl. Physiol., Lancaster* **44**, 385–395.
Forslind, E. (1966). *Svenska Naturwetenskap.* **2**, 9–74.
Frank, H. S. and Wen, W. Y. (1957). *Disc. Faraday Soc.* **24**, 133–140.
Gerloff, E. D., Richardson, T. and Stahmann, M. A. (1966). *Pl. Physiol., Lancaster* **41**, 1280–1284.
Ginnings, D. C. and Furukawa, G. T. (1953). *J. Am. chem. Soc.* **75**, 522–527.
Gitler, C. (1972). *A. Rev. Biophys. Bioengng* **1**, 51–91.
Glinka, Z. (1973). *Pl. Physiol., Lancaster* **51**, 217–219.
Glinka, Z. and Reinhold, L. (1971). *Pl. Physiol., Lancaster* **48**, 103–105.
Glinka, Z. and Reinhold, L. (1972). *Pl. Physiol., Lancaster* **49**, 602–606.
Goldstein, B. D. and Balchen, O. J. (1967). *Proc. Soc. exp. Biol. Med.* **126**, 356–358.
Goldup, A., Okhi, J. and Danielli, J. F. (1970). *Recent Progress Surface Sci.* **3**, 193–260.
Gottlieb, D., Carter, H. E., Sloneker, J. H. and Ammann, A. (1958). *Science* **128**, 361.
Gottlieb, D., Carter, H. E., Lung-chi Wu and Sloneker, J. H. (1960). *Phytopathology* **50**, 594–603.
Graziani, Y. and Livne, A. (1971). *Biochem. biophys. Res. Commun.* **45**, 321–326.
Graziani, Y. and Livne, A. (1972). *J. Membrane Biol.* **7**, 275–284.
Grunwald, C. (1968). *Pl. Physiol., Lancaster* **43**, 484–488.
Grunwald, C. (1971). *Pl. Physiol., Lancaster* **48**, 653–655.
Hanai, T., Haydon, D. A. and Taylor, J. (1964). *Proc. R. Soc.* **A281**, 377–391.
Hansson, G., Kuiper, P. J. C. and Kylin, A. (1973). *Physiologia Pl.* **28**, 430–435.
Hansson, G. and Kylin, A. (1969). *Z. Pflanzenphysiol.* **60**, 270–275.
Hasselt, Ph. R. van (1974a). *Kon Ned. Akad. Wetensch. Amsterdam*, **C77**, 37–46.
Hasselt, Ph. R. van (1974b). *Acta bot. néerl.* **23**, 159–169.
Haydon, D. A. (1968). *J. Am. Oil Chem. Soc.* **45**, 230–240.
Helfferich, F. (1962). "Ion Exchange", p. 624. McGraw-Hill, New York.
Heuvel, F. A. van den (1963). *J. Am. Oil Chem. Soc.* **40**, 455–471.
Hodges, T. K., Leonard, R. T., Braiker, C. E. and Keenan, T. W. (1972). *Proc. natn. Acad. Sci. U.S.A.* **69**, 3307–3311.
Hopfer, U., Lehniger, A. L. and Lennart, W. J. (1970). *J. Membrane Biol.* **2**, 41–48.
Jung, G. A. and Smith, D. (1961). *Agron. J.* **53**, 359–366.
Kahr, M. and Kylin, A. (1974). *Physiologia Pl.* (in press).
Karlsson, K. A., Samuelsson, B. E. and Steen, G. O. (1971a). *J. Membrane Biol.* **5**, 169–184.
Karlsson, J., Tribukait, B. and Kylin, A. (1971b). *Proc. 1st Eur. Biophys. Congr.* **111**, 75–79.
Keeran, T. W., Leonard, R. T. and Hodges, T. K. (1973). *Cytobios* **7**, 103–112.
Klopfenstein, W. E. and Shigley, J. W. (1967). *J. Lipid Res.* **8**, 350–351.
Kruyff, B. de, Demel, R. A. and van Deenen, L. L. M. (1972). *Biochim. biophys. Acta* **255**, 331–347.
Kuiper, P. J. C. (1968a). *Pl. Physiol., Lancaster* **43**, 1367–1371.
Kuiper, P. J. C. (1968b). *Pl. Physiol., Lancaster* **43**, 1372–1374.
Kuiper, P. J. C. (1969). *Pl. Physiol., Lancaster* **44**, 968–972.
Kuiper, P. J. C. (1970). *Pl. Physiol., Lancaster* **45**, 684–686.
Kuiper, P. J. C. (1972). *A. Rev. Pl. Physiol.* **23**, 157–172.
Kuiper, P. J. C. (1972). *Physiologia Pl.* **26**, 200–205.
Kuiper, P. J. C. and Stuiver, B. (1972). *Pl. Physiol., Lancaster* **49**, 307–309.

Kylin, A. and Gee, R. (1970). *Pl. Physiol., Lancaster* **45**, 169–172.
Kylin, A. and Kähr, M. (1973). *Physiologia Pl.* **28**, 452–457.
Kuiper, P. J. C., Kähr, M., Stuiver, C. E. E. and Kylin, A. (1974). *Physiologia Pl.* (in press).
Kylin, A., Kuiper, P. J. C. and Hansson, G. (1972). *Physiologia Pl.* **26**, 271–278.
Ladbrooke, D., Williams, R. M. and Chapman, D. (1968). *Biochim. biophys. Acta* **150**, 333–340.
Lakshminarayanaiah, N. (1969). "Transport Phenomena in Membranes". Academic Press, New York and London.
Lev, A. A., Gotlik, V. A. and Buzhinsky, E. P. (1966). *Zh. Evolyutsionnoi. Biokhim. i Fiziol.* **2**, 109–115.
Ling, G. N. (1973). *Biophys. J.* **13**, 807–816.
Lyons, J. M. and Raison, J. K. (1970). *Pl. Physiol., Lancaster* **45**, 386–389.
Lyons, J. M., Wheaton, T. A. and Pratt, H. K. (1964). *Pl. Physiol., Lancaster* **39**, 262–268.
Martonosi, A. (1968). *J. biol. Chem.* **243**, 71–81.
Mudd, J. B., McManus, T. T., Ongun, A. and McCallagh, T. E. (1971). *Pl. Physiol., Lancaster* **48**, 335–339.
Nash, T. (1965). *In* "Surface Activity and the Microbial Cell", Monogr. **19**, 122–135. Soc. Chem. Ind. London.
Némethy, G. and Scheraga, H. A. (1962). *J. Chem. Phys.* **36**, 3382–3400.
Némethy, G. and Scheraga, H. A. (1964). *J. Chem. Phys.* **41**, 680–689.
Némethy, G., Steinberg, I. Z. and Scheraga, H. A. (1963). *Biopolymers* **1**, 43–69.
Nobel, P. S. (1974). *Planta* **115**, 366–372.
Noyes, R. M. (1963). *J. Am. chem. Soc.* **86**, 971–979.
Olson, R. A. (1971a). *Physiologia Pl.* **24**, 534–543.
Olson, R. A. (1971b). *Physiologia Pl.* **25**, 204–212.
Olson, R. A. (1971c). *Physiologia Pl.* **25**, 503–508
Olson, R. A. (1973). *Physiologia Pl.* **28**, 507–515.
Owens, K., Weglicki, W. B., Ruth, R. C., Stam, A. C., Sonnenblick, E. H. (1973). *Biochim. biophys. Acta* **296**, 71–78.
Papahadjopoulos, D. and Bangham, A. D. (1966). *Biochim. biophys. Acta* **126**, 185–188.
Pauling, L. (1961). *Science* **134**, 15–21.
Peschel, G. and Adlfinger, K. H. (1967). *Naturwissenschaften* **54**, 614.
Reid, R., Con, D., Baker, H. and Frank, O. (1969). *J. Prozool.* **16**, 231–235.
Roche, J. A. de la, Andrews, C. J., Pomeroy, M. K., Weinberger, P. and Kates, M. (1972). *Can. J. Bot.* **50**, 2401–2409.
Roche, J. A. de la, Andrews, C. J. and Kates, M. (1973). *Pl. Physiol., Lancaster* **51**, 468–473.
Rosano, H. L., Schiff, H. and Schumann, J. H. (1962). *J. phys. Chem.* **66**, 1928–1932.
Schultz, R. D. and Asunmaa, S. K. (1970). *In* "Recent Progress in Surface Science", Vol. 3, pp. 291–332. Academic Press, New York and London.
Schwertner, H. A. and Biale, J. B. (1973). *J. Lipid Res.* **14**, 235–242.
Scott, D. B. M. and Lesher, F. C. (1963). *J. Bact.* **85**, 567–576.
Shah, D. O. and Shulman, J. H. (1965). *J. Lipid Res.* **6**, 341–349.
Shah, D. O. and Schulman, J. H. (1967a). *J. Lipid Res.* **8**, 215–226.
Shah, D. O. and Schulman, J. H. (1967b). *J. Lipid Res.* **8**, 227–233.
Singer, S. J. and Nicolson, G. L. (1972). *Science* **175**, 720–731.
Stokes, R. H. and Mills, R. (1965). *Int. Encycl. Phys. Chem. and Chem. Phys.* **3**, 22–45.
Tal, M. and Imber, D. (1971). *Pl. Physiol., Lancaster* **47**, 849–850.

Timmermans, J. and Bodson, H. (1937). *C.r. hebd. Séanc. Acad. Sci., Paris Ser. D.* **204**, 1804–1810.
Triggle, D. J. (1970). *In* "Recent Progress in Surface Sciences", **3**, 273–290.
Twersky, M. and Felhendler, R. (1973). *Physiologia Pl.* **29**, 396–401.
Veen, H. (1974). *Z. Naturforsch.* **29C**, 39–41.
Vendrig, J. C. (1967). *Ann. N.Y. Acad. Sci.* **144**, 81–93.
Wall, T. T. and Hornig, D. F. (1965). *J. Chem. Phys.* **43**, 2079–2087.
Weigl, J. (1969). *Z. Naturforsch.* **24b**, 365–366.
Weigl, J. (1969). *Z. Naturforsch.* **24b**, 367–368.
Weigl, J. (1969). *Z. Naturforsch.* **24b**, 1046–1052.
Wiggins, P. M. (1971). *J. theoret. Biol.* **32**, 131–146.
Wolman, M. (1970). *In* "Recent Progress in Surface Science", **3**, 262–272.
Wood, A. and Paleg, L. G. (1972). *Pl. Physiol., Lancaster* **50**, 103–108.

ADDENDUM

After submission of the manuscript of this chapter I received a review article entitled "Phospholipids and plant membrane permeability" by E. W. Simon (1974), which appeared in *New Phytologist* **73**, 377–420. The article extensively reviews the effect of rewetting dry tissues on leakage and its relation to lipids, a subject not covered in my review. In addition, sections on leakage from chilled tissues, on senescence, and on peroxidation of lipids are included.

A review article by A. D. Bangham (1972) on "Lipid bilayers and biomembranes", which appeared in the *Annual Review of Biochemistry* **41**, 753–776, should be mentioned in the context of the discussion on the interactions between lipids and ions and water.

Subject Index

A

D

E

G

H

I

K

L

N

O

R

S

T